SPI Handbook of Technology and Engineering of Reinforced Plastics/Composites

SPI HANDBOOK OF Technology Reinforced

and **Engineering** of **Plastics/Composites**

Second Edition

J. GILBERT MOHR, Editor and Senior Author
Johns-Manville Fiber Glass Reinforcements Division, Waterville, Ohio

SAMUEL S. OLEESKY, D.Sc.
Consultant for RP/C, Los Angeles, California

GERALD D. SHOOK
Consultant for RP/C, Huntington Station, New York

LEONARD S. MEYER, P.E.
Consultant for RP/C, Columbia, South Carolina

VNR **VAN NOSTRAND REINHOLD COMPANY**
New York Cincinnati Toronto London Melbourne

Van Nostrand Reinhold Company Regional Offices:
New York   Cincinnati   Chicago   Millbrae   Dallas

Van Nostrand Reinhold Company International Offices:
London   Toronto   Melbourne

Manufactured in the United States of America

Published by Van Nostrand Reinhold Company
450 West 33rd Street, New York, N.Y. 10001

Published simultaneously in Canada by Van Nostrand Reinhold Ltd.

15   14   13   12   11   10   9   8   7   6   5   4   3   2   1

Library of Congress Cataloging in Publication Data
Mohr, John Gilbert, 1913–

    SPI handbook of technology and engineering of reinforced
plastics/composites.

    Includes bibliographical references.
    1. Reinforced plastics.   I. Society of the Plastics
Industry.   II. Title.
TP1177.M63   668.4'94   73-4797
ISBN 0-442-25448-2

Foreword

The outstanding reception accorded the First Edition of the *Handbook of Reinforced Plastics of the SPI* has prompted the authors to prepare the Second Edition entitled, *SPI Handbook of Technology and Engineering of Reinforced Plastics/Composites*. The authors, together with the publisher, have decided on this new book in view of the dynamic and ever-growing nature of the reinforced plastics/composites industry in the United States, Canada, and overseas.

It is to the authors' credit that in issuing this new edition they have enlisted the support of eminently qualified technicians and management personnel to prepare certain chapters based upon their keen and thorough knowledge of the subject assigned to them.

Without a doubt, this book will become the major technical reference throughout the industry, both to users and potential users as well as in educational circles.

As with the First Edition, the authors have based much of their material on information in the *Proceedings* books issued each year in connection with the Annual Conference of the SPI Reinforced Plastics/Composites Institute. In addition, they used very good judgment in tapping other sources of information in order to make certain that the treatise is as all-encompassing as possible, and also reflective of the industry's wide range of products produced, processes available, and other unique features of these versatile composite materials.

All in all, the authors and all others who contributed to the preparation of this book are to be highly commended for masterminding a most important and technically meaningful volume.

CHARLES L. CONDIT
Senior Staff Administrator
The Society of the Plastics Industry, Inc.

Preface

Continuing the well-established SPI tradition of free exchange and promulgation of technical information within the reinforced plastics/composites industry without trafficking in confidential or proprietary details, this book has been prepared for the purpose of permanently recording the presently known processing, actual molding, and operational methods of the many individual segments which comprise the whole RP/C industry in its present well-advanced state.

Specifically, there are twenty or more methods of processing the various basic materials of RP/C—resins, catalysts, reinforcements, and fillers—each method yielding a tangible, commercially salable item or group of items which together comprise the great family of RP/C products. It has been the intent of your authors to set forth the best known RP/C process information and fabrication methods both as a matter of record and also to spur further growth by pointing up the great strides in process and product development that have taken place since the industry had its inception.

Further, regarding the actual scope of this endeavor, the present volume departs markedly from the format of the original *SPI Handbook of Reinforced Plastics* (Reinhold, 1964). The various methods of molding have been loosely organized into a classification relating to their processing temperature and equipment requirements and then each method has been discussed using essentially the same general outline:

A. Introduction, including history and general survey.
B. Materials and forms of the materials required for processing.
C. Equipment and tooling necessary.
D. Description of the actual processing method.
E. Product properties and performance data.

The use of this outline has put the book preparation on an easier and more logical basis for the authors, and it is believed that it will reduce reading time, facilitate access to the material, and permit easy comparison and cross-referencing for the busy reader.

Discussions of theory are also included where desirable for thorough understanding.

Specialists and knowledgeable persons were invited to prepare discussions representing their specific fields in cases where such contributions were felt necessary to complete the scope of the volume. Their interest and willingness to further advance the RP/C industry are hereby gratefully acknowledged.

The information supplied by the many authors of technical papers in SPI *Proceedings* and other plastics publications from which the authors were permitted to abstract or quote is hereby recognized as a substantial contribution.

J. GILBERT MOHR

SAMUEL S. OLEESKY

GERALD D. SHOOK

LEONARD S. MEYER

Contents

SECTION I

INTRODUCTION

SECTION 1
INTRODUCTION

I-1

What's It All About

The purpose of this book is to set forth in as detailed a fashion as possible the information accrued to date concerning the technology and engineering of all the significant methods of processing to produce reinforced plastics/composites (RP/C). In addition to setting forth a permanent record, this endeavor constitutes a tribute to the persons who have devoted a great deal of time and energy to advance the industry to its present state. It will also hopefully serve to interest future participants.

Reinforced plastics/composites is the generic term for a group of related yet individual enterprises or processes. Each is concerned with some mode of combining a polymerizable but weak resin matrix with a high tensile strength but friable dry filamentary reinforcing media. The combination melds the best properties of each into a superior reinforced plastic or composite, the overall properties of which are unequalled by any other single material.

The RP/C industry did not happen overnight. The cauldron of potentiality bubbled for more than a century, but the two main contributory events came about within the past 35 years: the commercialization of continuous-filament reinforcing fiber glass (1935) and the invention of unsaturated polyester resins polymerizable at low pressures (1939). In the immediately ensuing years and in ways known only to the initiated, an inexorable, step-by-step wresting away of nature's tightly held secrets was begun and has continued to the present. Today, after 25 years of successful invention, development and profitable commercial enterprise, we may safely venture to say that our industry has shown a greater, more

rapid rate of advance than any in recorded mercantile history. Moving steadily ahead by a yearly increase of 15 to 20% in pounds manufactured and sold, the total RP/C volume has reached almost 1 billion pounds in 1972.

Hence, by virtue of impossibilities achieved, each sphere of reinforced plastics activity passed from a humble beginning through minor then major technological accomplishment and to full-blown commercial acceptance and independence. The best is yet to come because new methods of reinforcing plastics and further improvements continue to be fostered and arrive on the scene with amazing regularity.

To help supply a technical understanding for this book, we will consider the following background relating to the emergence of several major reinforced plastics systems.

The present hand-layup method of fabricating reinforced plastics refers to room temperature cure in open molds of a liquid thermoset resin after saturating a suitable fiber glass or other reinforcement. Originally, this method naturally used canvas or other fabrics impregnated with varnish for electrical laminates. An engineering undergraduate with a flair for water-based activities performed as a thesis the construction of model boat hulls using available fabrics plus liquid urea resins. When the prime RP/C materials became available—fiber glass cloth and polyester resin—he performed a natural switch. As a result, this man is now one of the original and largest boat manufacturers in reinforced plastics. His action helped create the industry and paved the way for further improvements in materials and methods, so that hand lay-up is still the most applicable process for outsized structures. Its ultimate growth led to the use of spray-up, or simultaneous ejection of laminating materials through an apparatus and onto the work for reduction of labor and molding time.

In answer to early complaints that single thickness reinforced plastic laminates were too low in flexural and tensile elastic moduli, honeycomb material was adapted and placed into service. By bonding or laminating honeycomb or foam material between two RP/C skins, structures with elastic moduli of 10,000,000 psi or greater were possible. This method is called *bag-molding*.

In later technology, moduli of single skins and laminates were further increased by introduction of stiffer reinforcing materials such as improved glass compositions, fibers from carbon and boron, and metal whiskers.

RP/C molding moved from the long-cycle, room-temperature cure process to the more rapid-interval pressmolding in heated platens via a devious but interesting route. Prior to the time that fiber glass roving was developed, large hanks 4 to 5 in. in diameter were formed of continuous-filament strands, laid parallel, bound, and cut transversely with an abrasive saw into bundles 1 to 1½ in. long weighing 1 to 3 lb each. These were opened by air blast in an intermediate chamber and then fed into a fiber-thrower in which they were randomly deposited upward (anti-gravity) and held by blower suction without binder against the outside of a screen shaped like the item to be molded.

When the blower air was dampered, the deposited glass was carefully taken off the screen in a plastic liner and transferred to the cavity of a mold set. Resin was added, taking care so as not to disturb the glass pattern, and the plug was placed in and clamped until cure. This fumbling process was the precursor of— you guessed it—preform and matched die molding.

With the advent of roving, the roving cutter, and polyester emulsion binders, preforming machines were improved and automated, and handleable preforms became practical. Molding moved from the clamped dies to heated molds in a compression press. All manner of custom-molded items appeared, including inner baskets for washing machines, trays, hard-hats, and welding masks, to be followed by automo-

tive body components and other parts up to 100 sq ft in total area.

Recent improvements comprise precombining resin and glass to form SMC, a moldable composite in sheet form. This makes possible elimination of preforming and handling liquid resin at the press. Resin (low profile) and molding technologies have been further improved to virtually eliminate porosity and provide smooth, paintable surfaces directly out of the mold.

Polyester premix, a related endeavor, was engendered when fiber-washing difficulties were encountered in molding a complicated electrical component after laying loose, precut fiber glass mat parts into a mold and applying fluid resin. Conversely, when the glass was chopped into short lengths and mixed into the resin together with pulverized inorganic filler, this highly thickened blend when molded carried the fiber glass reinforcement uniformly to all remote cavities and channels of the mold. Thus, the now-burgeoning premix segment of the reinforced plastics industry had its inception.

Forward-moving resin technology has provided superficial thickeners and low-shrink resins for further-improved molding properties and finished parts (BMC) which are directly paintable and possess a mirror-like surface quality matching that of the mold. Also now in use is injection molding of premix in addition to the compression and transfer techniques.

Thin, corrugated translucent architectural polyester-glass paneling resulted when attempts were made to enhance the appearance of metal buildings. It was desired to supply attractive, light transmitting, waterproof, strong and functional plastic panels which were free from rust, rot and corrosion. Fluid polyester was used to impregnate glass reinforcement between plastic-film carrier sheets. Curing was accomplished between the films, and many different corrugations were possible merely by preshaping the sandwiched lay-up prior to cure. Oddly, lightning struck twice in different places, because the corrugated panel concept was conceived simultaneously and independently by the personnel of at least two different companies—one in Houston, Texas, and the other in San Diego, California.

The same multi-development occurred in the rod-stock and pultrusion segment of the RP/C industry. The first solid rod stock was made by looping continuous-strand glass fiber between hooks 4 or 5 ft apart, impregnating, and circumferentially wrapping cellophane tape around the entire length. Cure was made first at room temperature and then in ovens. With the commercial availability or roving, continuous throughput rod-making processes concurrently sprang up. In a few short years, metal, bamboo and all other fishing-rod materials except fiber glass were things of the past.

Methods for hollow rod stock were also devised and improvements in resin technology, and curing by radio-frequency ensued. The natural shift for pultrusion was then to larger cross-sectioned structural shapes such as Tees, U- or I-beams and also hollow, heavy-wall and other configurations. Applications penetrated into industrial as well as the recreational markets.

An astute RP/C pioneer, aware of the pultrusion-process capabilities, and casually observing the orderly symmetrical array of glass fiber strands wrapped ribbon-like into a roving package, heeded an inner voice which suggested that the two might be combined. The result was filament winding of RP/C. In this method, impregnated roving is wrapped onto a mandrel using a reciprocating traverse motion programmed with the rotational speed of the mandrel, producing the well-known cylindrical and shell-type structures with the criss-cross reinforcement pattern. Filament winding provides the highest possible glass content, and therefore, highest part strengths in RP/C with the lowest range of variation of these ingredients. With all manner of modifications and wind patterns possible, filament wound components have been one of the basic entities in the great American space program.

Considering thermoplastics, it was quite natural to attempt to draw fiber glass reinforcing roving through a polymeric emulsion or hot melt by die-wiping, and to cure or dice when cold. The consequence of this innovation was a class of injection-moldable pellets which could be handled in standard injection-press equipment with only minor modification. Almost all known thermoplastics have been reinforced with fiber glass, and significant strength and property improvements have resulted in each situation.

So has it gone with the major fields of reinforced plastics and with others too numerous to mention here. To remove the impression, however, that ideas plucked from wherever they occur are an immediate success, in each instance a great deal of theory, systematic investigation and experimentation, blood, sweat, tears and considerable economic bolstering were essential to bring each concept to its present successful status. Also, conversely, countless promising ideas and enterprises generated in the reinforced plastics field were weighed in the balance and found wanting, only to speedily disappear below the horizon. This purports further stature and distinction to the processes and endeavors that actually made the grade.

However, enough of generalities and on to specifics! A total of almost twenty separate processes or methods of molding currently exist in the RP/C industry. This text constitutes a "how-to-do-it" record, describing each process in detail, step-by-step, according to the following outline:

A. Introduction—(purpose, importance, history)
B. Materials
C. Equipment and Tooling
D. Molding Method/Processing Details
E. Product Data (properties, performance, special handling, advantages and disadvantages, markets and uses for the particular RP/C molded parts).

This is probably the last time it will be possible to treat all phases of reinforced plastics under one literary roof, so to speak. The individual processes are becoming large enough in number and of sufficient technical and economic importance to justify their being handled as individual treatises.

Probably the most important single factor in promotion of the learning and idea exchange which has sustained the growth and success of the reinforced plastics field has been the existence of the RP/C Division of the Society of the Plastics Industry. Committees have been formed and meet regularly throughout the country. The Division (now the RP/C Institute) also holds an annual Technical and Management Conference. The technical papers presented at this conference over the years form a substantial portion of the background for this handbook. The annual exhibit of RP/C products which accompanies the conference has shown all the examples of product innovation discussed in this chapter as they occurred, and has served as a springboard to launch these products into complete economic success.

Taking into account the meritorious support contributed by the SPI and its member companies, and bearing in mind the excellent high caliber of those individuals attracted to the industry, we find it extremely encouraging to contemplate the status of Reinforced Plastics/Composites in the year 2000 and beyond.

I-2

Reinforced Plastic Product Types and Markets

INTRODUCTION

Marketing comes into play when a manufacturer has more than one customer for a given product. It forms the basis for the science of modern selling. Many miracles of product generation, growth, evaluation of markets, and creation of a demand for a given product have formed building blocks for the great growth of the American Gross National Product.

Reinforced Plastics have been fortunate to fit into the already established marketing pattern for existing plastic products. This chapter does not intend to present authoritative information on how to conduct market studies, analyze, or create demand, because this has been described elsewhere, especially relating to fiber glass and reinforced plastics.[1] The main purposes are to present factual data concerning the types of RP/C products that comprise the various market categories, and to study in depth the development, growth rate, size, and anticipated future growth of each sphere of selling activity in reinforced plastics.

GROWTH RATE OF REINFORCED PLASTICS

Slightly more than 25 years in age, the reinforced-plastics industry presents an interesting study in industrial growth. Table I-2.1 presents the total annual RP/C production from the inception of the industry to the present.

MARKETS FOR REINFORCED PLASTICS

Several interesting ways of recording reinforced plastics consumption are used in various

Table I-2.1 Reinforced-Plastics Production

Year	Total Est. (Lb)	Average Growth Rate per Yr in 5-Yr Periods (%)
1945	7,500,000	(Entirely for military use)
1946	3,250,000	
1947	1,750,000	
1948	3,750,000	
1949	9,500,000	
1950	17,500,000	26.6
1951	29,000,000	
1952	40,000,000	
1953	60,000,000	
1954	65,000,000	
1955	110,000,000	111.0
1956	160,000,000	
1957	180,000,000	
1958	220,000,000	
1959	264,000,000	
1960	252,000,000	25.8
1961	258,560,000	
1962	282,000,000	
1963	300,000,000	
1964	326,000,000	
1965	340,000,000	7.0
1966	482,000,000	
1967	607,000,000	
1968	785,000,000	
1969	875,000,000	
1970	797,000,000	26.9
1971	978,000,000	
1972	1,200,000,000	

Sources: The Society of the Plastics Industry, N. Y.; W. H. Gottlieb, Morrison-Gottlieb, Inc., Public Relations, N. Y.

phases of the industry. For individual materials such as resins or reinforcements, it is advantageous to record the pounds and percentage of the total channeled into each category or method of molding, because each processing or molding method necessitates a different type of basic product. From such information, a material supplier is able to study and predict the direction in which his business is moving or should be moving in the future.

For a true understanding of commercial and economic factors, however, it is more beneficial to report market activity (sales and purchasing) in degrees or amounts in well-defined channels of product sales or procurement, referred to as markets. In this manner, the percentage consumption for each sphere of activity may be calculated and compared, and the relationship of the volume of sales of RP/C products with that for other materials in the same marketing area may be readily determined and compared. The salient element of market evaluation by trade channels lies in the realization that any of the methods of fabrication or molding which are described in this volume may produce products for any one of the individual market categories.

There are nine markets which singly reflect the action in the reinforced plastics industry. The brief descriptions which follow include examples and listings of special accomplishments that illustrate true progress and the adaptability of the product to the specific market.

Aircraft and Aerospace. The aircraft and aerospace market includes items fabricated directly or indirectly for the construction, equipping and repair of private, commercial, or military aircraft; vehicles for space travel; missiles; radomes in airships and in shore use; and the like.

Of special interest are the stress-designed rotor blades for helicopters; complete RP/C fusilage and wing-construction for small planes; parallel-strand spring assemblies; ablative components; and high-performance, lighter-weight carbon, graphite, and boron-reinforced structures.

Appliances and Equipment. The appliances and equipment category embraces housings, gears, covers, shrouds, and handles, with the market almost equally divided between thermoplastic- and thermoset-reinforced components.

The thermal and dimensional stability imparted by fiber glass reinforcements to the thermoplastic resins has enabled them to become well adapted to gears, bases, and many other parts used in appliance manufacture.

Construction. Included in the construction market are items moving into all branches of the building field such as corrugated and architectural paneling; flat sheet; composite, non-load-bearing modular building-front segments and fascia panels; swimming pools; concrete forming pans; gutters and downspouts; translucent garage doors; integrated bathroom components and fixtures; tub and shower units, shower stalls and receptors; lavatories; marine docks and floats; agricultural structures; cooling tower and sandwich structures (see Figure I-2.1).

Larger and more expansive compages are becoming evident. Some of the latest are 24-ft high fascia panels for a two-story office building, and replacement of turned or lathed wood in complete sectional pillar and rail elements, and fabrication of huge columnar structures (see Chapter II-1). Groups are working concentratedly to establish acceptance of RP/C in various area building codes.

Consumer Goods. Molded utility trays, tote boxes and containers constituted the original consumer goods made with reinforced plastics. Recreational and sporting equipment such as fishing rods, archery bows and arrows, golf clubs, surfboards, skis, ski poles, sleds, snowmobiles, playground equipment, and bowling alley appurtenances have been excellent outlets for the unique properties of the material.

More recently, additional home and houseware items have been developed. Some examples are colorful lampshades, movable and permanent divider screens, and serving and television trays. Following the wide acceptance of shell chairs, items such as home furniture, bedsteads, couches, cabinets, lounges, tables, mass airport seating and stadium-bench and chair seating have been successful.

Corrosion Equipment. The market for corrosion resistant and chemical process equipment embraces all reinforced plastics fabricated and used for above-ground and underground storage and process tanks, reactors, ducts, hoods, piping, housings, fume scrubbers, water cooling towers, fan blades and housings for corrosive atmospheres, and many other items. It is interesting to note that many large cor-

Figure I-2.1 An example of the great growth and acceptance of reinforced-plastic products is demonstrated by their use in the Dana Corporation office building in Toledo, Ohio. All white portions of the building except the window frames were fabricated using RP/C. This includes the top banister and pilings, the decorative trim just below the roof line, the two-story front-porch pillars, and the triangular facing with medallion directly above the porch. (Photo courtesy of Customflex, Inc., Toledo, Ohio and Johns-Manville Fiber Glass Reinforcements Division, Waterville, Ohio.)

porations in industrial manufacturing have specified the use of RP/C corrosion-resistant tanks and equipment instead of metal in all feasible applications. The material is also used as a protective coating for metal corrosion components. The only real limiting end point is the temperature of operation. Rotationally molded RP/C containers, trays, and tanks for the food processing and other chemical industries are fabricated using chopped reinforcing strands of fiber glass plus thermoplastic resins.

Electrical Industry. In the electrical industry, many types of components take advantage of the excellent properties of low dielectric and high arc resistance of reinforced plastics. These components include shaped sheet and rod stock for electronic standoffs and terminal posts, flat sheet for printed circuit boards, motor slot wedges and winding insulation, and also computer parts including exceedingly thin laminates and dimensionally controlled memory disc plates. Larger-scale equipment comprises control switchgear bases and housings, arc chutes, distributor caps, transformer covers, and telephone equipment.

In more recent years, electrical usage of reinforced plastics has expanded to include rod stock of larger diameters (1 to 2 in.) for supports and guy-strain insulators for high-tension lines. Also, the now well-known hot-line maintenance gear, consisting of elevatable and extensible booms plus man-carrying buckets, is used to work on high-voltage-bearing wires,

some of which may be supported by RP/C utility poles, bars or crossarms.

Marine Industry. The marine market represents an almost complete RP/C takeover of the small boat field and a further move to make product serviceability available for construction of watercraft over 70 ft in length. The latter category includes both pleasure boats and the larger fishing and other commercial vessels, and in their construction, reinforced plastics has proved less costly than the closest rivals, steel and wood.

Many industrial and military marine components are being fabricated, such as floating docks, submersible hulls, submarine fairwaters, pontoons and outboard motor shrouds. A new bubble-type, completely enclosed lifeboat is contemplated for widespread usage.

Additional forays by the reinforced plastics industry to maintain its foot in the marine market door have involved mass matched-die molding of one-piece car-top boats—one produced every 4½ minutes and merchandised through one of the largest national retail and mail-order houses. Also fabricated are outsized houseboats as attractive as any automotive housetrailer in existence. The promise of even greater market penetration and marine utilization of reinforced plastics is predicted by the development of an advanced technique by which a thermoformed plastic skin is spray-reinforced with RP/C to form superior components for marine and other uses.

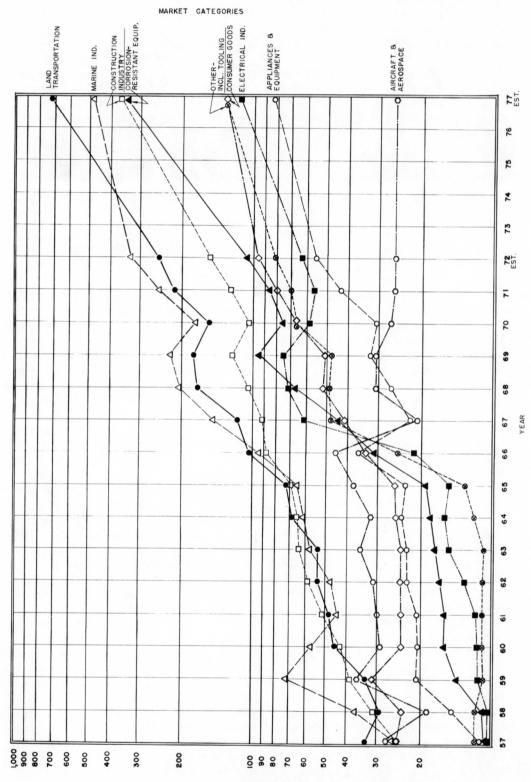

MARKET CATEGORIES

Transportation. One of the most valid methods of evaluating this market is by observing the number of pounds of plastics used in individual cars—a figure which includes RP/C. Starting with the all-fiber-glass-body sports car in 1953, and the use of polyester premix in molded heater housings soon thereafter, there has been a steady increase in the use of reinforced plastics in transportation. Any listing should include parts, equipment and seating for automobiles, trucks, buses, truck-trailers, tank-trucks, railroad and subway cars, and smaller specialty vehicles such as dune-buggies, mail-delivery carts, and motor scooters.

Worthy of special mention are the large-size containerization structures fabricated using a fiber glass-reinforced coating over plywood, and also RP/C skins encapsulating foam-cell insulation made into a monocoque construction. Both types are available in sizes up to 8 ft x 9½ ft x 50 ft in length[2] (see Chapter II-5).

Because of the great adaptability of both reinforced thermosets and thermoplastic materials to design, ease of molding, and resistance to deterioration by rust and corrosion, their largest volume outlets in the years to come will quite possibly be in the transportation and automotive fields.

Other Markets. Other markets and miscellaneous applications of reinforced plastics include industrial shipping, materials handling equipment, containers and trays, machine-tool housings and guards, safety hard hats, welding masks and shields. The largest segment of this outlet for RP/C embodies the tooling industry. To support and assist in all manufacturing in general, tools are fabricated and supplied for metal and plastic forming, for checking fixtures and for dies, hammerforms, master models, foundry patterns, high-temperature tools, and other items (see Chapter II-4).

MARKET PROGRESS SUMMARY

Figure I-2.2 shows the chronological progress in growth and development of each specific field or market of reinforced plastics since the year 1957. Especially in the later years, the total volume presented is based upon actual sales and usage rather than upon materials manufactured and inventoried.

References

1. E. R. Corey, *Development of Markets for New Materials*, p. 70, Cambridge, Mass., Harvard University Graduate School of Business Administration, 1959.

2. W. T. Baker, Litewate Corp., Milwaukee, Wis. 1969. Personal communication.

Figure I-2.2 Chronological plot of reinforced-plastics production in millions of pounds for each major marketing category since 1957. Data extended into 1975 is based on a normal, anticipated growth rate of 10 to 20%, providing an estimated total volume of 1,800,000,000 lb. for that year. (Data from M. Haylin, *Chemical and Engineering News*, 1 February 1971, p. 14; O. A. Hoogkamp, *Kunstoffe* 60 (December 1970): 1091; and E. A. Turner, Johns-Manville Fiber Glass Reinforcements Division, Waterville, Ohio 1970: personal communication.)

SECTION II

PROCESSES INVOLVING LOW-TEMPERATURE CURE

II-1

Hand Lay-Up

INTRODUCTION

Usage. The hand lay-up method of producing reinforced plastics generally comprises impregnating dry fiber glass reinforcement with a liquid, themosettable resin in an open mold and inducing cure at room temperature with little or no external pressure. Fabrication of RP/C parts by hand lay-up procedures is indicated when the size of the piece is obviously too large for matched-metal tooling or for similar types of press molding. Hand lay-up is usually desirable for making prototype parts or a relatively small number of small parts whose size and/or quantity would not justify the expense of production tooling.

Examples: Boats and ships, both motor-driven and sailing craft, currently up to 150 ft long; prefabricated, "unitized" bathroom fixtures and assemblies; tooling structures up to several hundred square feet in area; corrosion-proof and chemical-resistant tanks, ducts, fumehoods, and other items; swimming pools (shell or coating-over-substrate types). These and many similar kinds of unconventional components may be designed and prepared for fabrication by the hand lay-up process.

History. Hand lay-up reinforced plastics were truly born during World War II in answer to the need for materials which could replace scarce metals. The major phases of their development may be itemized as follows:

- Although castable and other thermosetting resins were known, their brittleness and curing difficulties rendered them less than completely satisfactory.
- Glass fibers became commercially available in 1937 as a superior reinforcement for varnishes in electrical and other industrial laminates.
- Liquid polyallomer resins were the precursors of room-temperature cure laminating resins used with glass fiber reinforcement but were costly and still curable only with difficulty.
- The Carleton Ellis patent in 1937, disclosing a mixture of a reactive monomer such as styrene with an alkyd-base resin, made possible an addition-polymerization reaction on curing at room temperature with no byproducts. This laid the basic groundwork for the present-day hand lay-up industry, except that the resins were oxygen-inhibited and the air-exposed surfaces would not cure tack-free.
- Dr. Irving Muscat was the first to incorporate small amounts of paraffin into liquid polyester resins for hand lay-up. The paraffin surfaced during cure and thereby resolved the air-inhibition difficulties.
- Blending in of thixotropic agents to prevent run-down from vertical surfaces plus addition of accessory promoters, new monomers, and more reactive resin bases all served to bring the hand lay-up sphere of activity up to its present level as a full-blown, going enterprise.

Method. Hand lay-up molding typically uses a single mold, either plug or cavity type. The mold may be prepared by use of a gel coat and/or standard release agents. Procedures include both wet lay-up and prepreg molding.

The steps normally required in the hand lay-up procedure include mold preparation, application of release agents, selection of proper resin and reinforcement, insertion of reinforcement and introduction of resin (for wet lay-up), roll-out and distribution of resin, gelation of resin, preliminary finishing, allowance for full cure, removal from mold, and final finishing.

MATERIALS

Mold Releases and Sealers. Because the resins used in reinforced plastics have been designed and selected for their good adhesive and cohesive properties, it is immediately apparent that a method must be provided to prevent the molded product from adhering to and becoming a permanent part of the mold on which it is made. To provide this separability, agents known as mold releases are used. By definition, a *mold-release agent* is any material or substance which, when applied to a substrate, creates a high surface-contact angle, thereby resisting or preventing wetting of the substrate by any material applied over the release agent.

In hand lay-up molding, most releases are applied to the mold and are not of the "internal-release" type. The latter are agents included in the resin mix which are to be expelled during cure to provide separation.

When porous materials, such as wood or plaster, are used for tooling, the surfaces must be completely sealed before mold-release agents are applied. A proper sealing material should penetrate deeply into the wood or plaster surface, filling all voids without causing any degree of material build-up on the surface.

A number of resins, including polyesters and epoxies, provide excellent hard and durable surfaces when applied to wood or plaster molds. Lacquers, shellacs, and varnishes are widely used sealants of this type. Care must be taken in the selection and use of these materials, however, since many are severely degraded when exposed to molding temperatures. Stearates are also satisfactory sealants, although aluminum stearate has been found to inhibit the cure of polyester resins.

In the making of RP/C tooling from plaster master models, care must be exercised in curing the resin while the lay-up is in contact with the plaster. Despite the sealant applied to the plaster mold, calcination and degradation of the gypsum plaster takes place at temperatures in excess of 250°F. Also if exotherm or oven temperatures occur above that level, the laminate structure itself may be adversely affected. Hence, many other elements of the tooling besides the sealant may be deleteriously affected by elevated temperatures.

In general, from 6 to 10 contact-pressure moldings can be made from one master tool before the tool must be retreated and the sealant replaced. Tables II-1.1 and II-1.2 list mold sealers and releases for hand lay-up molding.

Table II-1.1 Mold Sealers for Plaster, Wood, and Other Porous Materials

Cellulose acetate	Varnishes and lacquers
Water-soluble styrene	Urea
Other plastics	Hard plasters
Waxes	Furane resins
Silicones	Stearic acid

Table II-1.2 Mold Releases

Film Formers	Sheet Films	Lubricating Agents	Permanent Types
Cellulose acetate (15% in MEK)	Polyethylene	Teflon coatings (water or solvent dispersion)	Blow-off molded inserts
Polyvinyl alcohol 15% in H₂O-alcohol)	Polyvinyl alcohol (water base)		Plaster additions
Polyvinyl chloride	Polyvinyl chloride	Petroleum oil	Baked-on silicone or Teflon
Sodium alginate (agar-agar)	Cellulose acetate	Lecithin	Bentonite powder
Ammonium alginate	Cellulose triacetate	Waxes	Mica powder
Casein	Mylar	Silicones	Graphite
Carboxy-methyl cellulose	Cellophane	Phosphates	
Methyl cellulose		Lard oil	
		Modeling clay	

Table II-1.3 Types of Gel Coats

Type	Properties	Uses
Boat—neutral, white or colored; isophthalic base	Gel coats are resilient, impact resistant, have good weathering resistance, fair stain and chemical resistance, fair gloss retention, Barcol hardness = 40–45	Boats and general-purpose, 1 or 2 color work
Orthophthalic—general purpose white, or colored	Similar to boat type but lower in cost and in impact strength, Barcol = 40–45	General uses other than boats
Sanitary ware—dense white or color-pigmented	Gel coats are semi-rigid, water and stain resistant, resistant to mild chemicals, high gloss retention, Barcol = 45–55	Bathtubs, unitized or one-piece complete bathroom assemblies
Clear—thixotropic, but non-pigmented	Possess good clarity in thin film, have minimum discoloration, high gloss retention, Barcol = 40–45	Decorative embellishments and imbedments; marble-cast coating; clear, high-gloss layers over pigmented gel coats
Chemical-resistant—chemical-resistant resin with pigment and surfacing mat	Gel coats exhibit durability against acids, alkalis, and solvents equivalent to that for chemical-resistant hand lay-up resins, Barcol = 45–55	Tanks, ducts, fume-hoods piping, non-decorative laboratory fixtures
Isophthalic—General purpose chemical-resistant, pigmented	Gel coats are resilient, abrasion resistant and lower in cost that the chemical-resistant type, Barcol = 40–45	Cement forms; imparts resistance to weak chemicals
Surfacing	Gel coats contain black or grey pigment, sand to visible white, and dust freely, Barcol (sanded) = 40	Used as surface on lay-up part which ultimately must be sanded, primed, and painted
Non-photographing—for gel-coat back-up	Are formulated to contain 40 to 50% inert glass fiber filler to prevent pattern of lay-up or reinforcement from showing in gel-coat surface	Improved decorative appeal of boats, decks, and general purpose lay-ups
Tooling—usually dark or black-pigmented	Hard, surface, non-patterning	Polyester lay-ups for tooling applications
Metal-filled—polyester or epoxy with powered metal filler	Hard surface and are thermally but not electrically conductive	For dissipation of heat due to high-exotherm cure; prevents development of structural strains caused by thermal gradients
Self-leveling—clear or pigmented	Formulated to contain low surface-tension-producing agents which permit glossy, plane surface after cure	Over-spray gel coats and coatings not applied against mold surface, but as a final over-spray or finish on an already-molded item or other substrate

Gel Coats. A *gel coat* is a resinous, mineral-filled, pigmented, non-reinforced layer or coating which is applied first to the mold but which becomes the outer protective layer for hand lay-up laminate when completed.[1] Table II-1.3 summarizes the types of gel coats, their properties and uses.

Laminating Resins. Laminating resins are categorized as unique, low-pressure or room-temperature curing liquids that become solid materials. They permit the formation of high-strength, high-performance articles or structures for many functional purposes. Table II-1.4 provides a summary of hand lay-up resin types with their properties and uses.

Curing Systems. Several types of simple organic liquids added to styrene (or other monomer) modified alkyds (polyesters) generate a process that moves through (1) free-radical initiation, (2) attack of and cross-linking with the unsaturated bonding, and (3) final cure. The ultimate result is a rigid solid in which the matrix resin has joined chemically and mechanically with the reinforcing fibers to provide a synergistic composite structure whose properties are greatly different and sig-

nificantly superior to those of either material alone.

Hand lay-up processes usually employ room-temperature cures based primarily on methyl-ethyl-ketone (MEK) peroxide catalyst plus cobalt-naphthenate promoter. Other catalysts are available for specific purposes. Also, additional or accessory promoters may further speed cure cycles originally established with the MEK-cobalt system. Inhibitors may be added to delay or slow the cure time.

Parenthetically, it may be observed that the term *catalyst* as used in this sense does not necessarily agree with the strict chemical definition of the word. In chemistry, a catalyst is an agent that initiates or accelerates a reaction without entering into the reaction itself. The catalysts discussed here do enter into the curing process.

Many so-called double-promoted or fast-cure resins are ready for MEK addition *as received*, and the promoter systems are frequently proprietary. The individual manufacturer's procedural recommendations should be rigidly observed. Any specific variations or deviations should be thoroughly checked and evaluated in the laboratory and pilot plant prior to final adoption into a production-line procedure.

Table II-1.4 Types of Hand Lay-up Resins

Type	Properties	Uses
Rigid	High modulus, low impact strength, high Barcol hardness	Stationary, non-flexing structures
Resilient orthophthalic or isophthalic type	High impact strength, high Barcol hardness, fast gel-to-cure cycle (double-promoted), good stability and shelf life (aging), freedom from drift of gel and cure times in storage	Boats, bathrooms, general-purpose building material (artificial stone, siding, etc.)
Chemical-resistant, rigid	Usually high viscosity and slow wetting, high cost (consult suppliers)	Tanks, ducts, pipes, fume-hoods, vats, silos, etc.
Chemical-resistant,* flexible	Permits normal viscosity, thix, and cure time	Tank bottoms (vibrated) and corrosion-resistant gel coats
Fire-retardant	Self-extinguishing—will not support combustion, high cost (consult suppliers)	Where required or specified
Epoxy	Slow wet-out, exotherm prior to gelation, good dimensional stability, dark color	Tooling

*E. J. Kerle and J. J. Fisher, SPI 19th RP/C Division Proceedings, 1964, sec. 12-E.

Table II-1.5   Catalyst-Promoter-Inhibitor Systems for Room-Temperature Cure Polyester Resins

Application or End Use	System (%)	Gel Time Starting at Room Temperature (min.)	Approx. Time (hr.) at 70–75°F for Development of Barcol Hardness = 35
Gel coats	MEK peroxide—1.5[a] Cobalt napthenate—0.4[b] (Accessory promoters usually omitted because of tendency to discolor)	30 (High filler content)	6–8 (Can proceed with lay-up over gel coat in 30–45 min.)
For normal lay-up resins	MEK peroxide—1.0 Cobalt naphthenate—0.4	32	6–8
For fast-cure resins	MEK peroxide—1.0 Cobalt naphthenate—0.4 Dimethylaniline—0.1	16	2–2.5
For fast-cure resins	MEK peroxide—1.0 Cobalt naphthenate—0.4 Quaternary ammonium salt—0.1	15	2–2.5
Alternate room-temperature cure	Cyclohexanone peroxide[c]—1.0 Cobalt naphthenate—0.4	30	Approx. 6–8
Alternate room-temperature cure	Bis-l-hydroxy cyclohexyl peroxide[c]—1.0	30	Approx. 6–8
Alternate room-temperature cure	Benzoyl peroxide—1.0 Dimethyl aniline—0.1	20	2
Effect of inhibitor	MEK peroxide—1.0 Cobalt naphthenate—0.4 Hydroquinone—0.1	∞	∞

[a]Percentages based on 100 parts polyester resin.
[b]Concentration cobalt metal 6%.
[c]Peroxides costlier than MEK peroxide.

Using a Barcol hardness of 35 as a reference value, it will be found that curing time of most hand lay-up resins will vary or increase as much as 25% for each month of aging after manufacture.

Typical catalyst-promoter-inhibitor systems for room-temperature cure, hand lay-up resins are shown in Table II-1.5. Table II-1.6 presents typical room-temperature curing systems for epoxy resins. Variations in gel times attributable to shop temperatures and long-term aging for typical fast-cure resins are illustrated in Figure II-1.1.[2]

Fiber Glass Reinforcements. Fibrous glass reinforcing materials have been developed and designed to meet a number of varied and combined requirements of in-process functionality and end-use performance. A discussion of the major types of such reinforcements follows:

Glass Fabrics. Glass fabrics consist of twisted and plied strands of fibrous glass fabricated on textile looms or weaving equipment into usable industrial or commercial fabrics.

Greige (pronounced "gray") *goods* are the fabrics as they come from the loom, with the original starch lubricating sizing still present. In this condition, they are compatible with phenolic and melamine resins without further treatment.

Finished cloth has had the organic material removed by heat or chemical treatment. A polyester-or epoxy-compatible surface treatment is subsequently applied. Typical treatment might consist of a chrome complex or a silane cross-linking agent. Many variables are involved in the design and manufacture of a glass fabric. These include the primary properties of the fabric itself as well as the desired characteristics of the ultimate product into which the fabric will be incorporated as a reinforcement. Typical of such variables are type of yarn used, type of weave, stability of weave,

breaking strength of fibers or yarns, fabric thickness, woven width, fabric weight, porosity, pliability, drape over complex curvatures, resistance to flexure, type of finish applied, modulus, glass content when laminated, and cost. Table II-1.7 indicates the major types of weaves and their general properties.

Because of economic considerations, glass cloth is used in hand lay-up principally as a

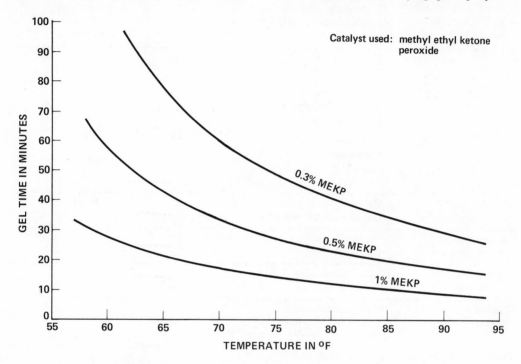

Figure II-1.1 Typical variation in gel times of rapid-cure, double-promoted polyester resins with changes in catalyst and temperature, both catalyzed and uncatalayzed.

Room temperature gel time	
1% MEK peroxide, resin at room temperature	10-20 min.
Catalysed stability	
1% BPO, resin at room temperature	1-14 days
Intermediate temperature stability	
1% BPO, resin held at 150°F	5-30 min.
High-temperature (SPI) gel time	
1% BPO, resin raised to 180°F	1.5-5 min.

Uncatalysed stability
Resin is stored in the dark at 70°F and periodically tested by SPI gel time to observe gel time drift.

Initial reaction
Aging of resin induces a 25 to 50% lengthening or upward drift in gel time, indicating that the inhibitor initially incorporated is chemically converted to a more active, intermediate, inhibiting compound during storage. Time span for upward gel time drift may be from 24 hr. to 4 mo. following manufacture.

Final reaction
Gel time will eventually show shortening or downward drift approaching zero gel time until the entire container of resin gels at room temperature. This indicates that the inhibitor in the resin has ultimately become exhausted. Time span for downward gel time drift may be from 4 to 18 mo. following manufacture. (Data courtesy of Allied Chemical Corporation.)

surfacing or finishing component over glass mat or woven rovings. Where its particular product properties are necessary, as in aerospace applications (radomes for example), glass cloth is used exclusively in bag molding.[3]

Fiber Glass Mats (High Solubility). Fiber glass mats consist of a web or accumulation of randomly oriented glass fiber reinforcing strands chemically or mechanically bonded and designed to provide extremely rapid and thorough wetting and uniform high strength in hand lay-up polyester resin products. Table II-1.8 provides typical information on the two major types of reinforcing mats for hand lay-up.

As in glass fabrics, a number of variables affect the properties of the mat and of the products made from the mat. In the laminate, the variables would include freedom from surface fibers, release of air, rapidity of wet-

out, physical properties, and percentage of wet-strength retention. Table II-1.9 describes the major variables. Table II-1.10 correlates fiber glass reinforcing mat weights per square foot with weights per square yard.

In summary, fiber glass mats constitute the real "workhorse" for hand lay-up reinforced plastics. They provide ease of handling, good performance, and excellent end-use properties at relatively low cost.[4]

Fiber-Glass Woven Roving. In the glass industry, rovings are loosely twisted "ropes" of fibrous glass, containing specified numbers of "ends" or individual strands. Woven rovings, therefore, consist of a type of fabric made by using these loose "ropes" as both warp and fill in a loom. Most of the weaves are of the so-called plain or square-woven type. They are made from rovings which vary from 655 yd

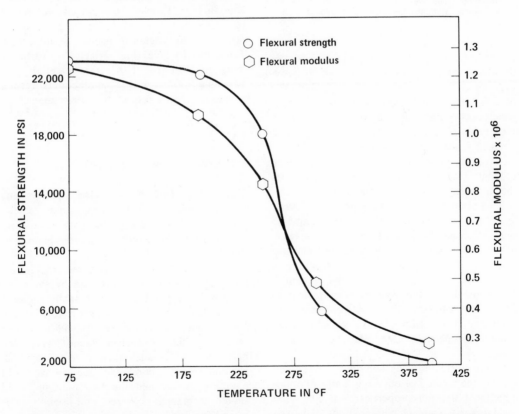

Figure II-1.2 Curves showing degradation of flexural strength and flexural modulus with increasing temperature for typical chemically resistant laminates 5/16 in. thick made according to SPI specification. Due to low thermal conductivity of the material, structures have been exposed to temperatures up to 340°F. for periods up to 12 hr. without weakening or internal condensation. It is recommended, however, that users fabricate sample coupons or prototype structures for exposure to actual chemical and temperature conditions prior to final design decisions. (D. M. Longenecker, Product Development Laboratory, Atlas Chemical Industries, Inc., Wilmington, Delaware, Sept. 9, 1971: personal communication.)

Table II-1.6 Typical Room-Temperature Curing Systems for Epoxy Resins

Hardener System	PHR[a] for Maximum Heat Deflection Temp.	Room-Temperature Pot-Life (min.)
Diethylene triamine	12.5	29
Triethylene tetramine	12.0	30
Amine resin adduct	24.0	22
Amine ethylene oxide adduct	20.0	18
Cyanoethylation product	22.5	42
Amine phenol	16.5	11
Polyamides	64.0	100

[a]Indicates the amount of hardener to be mixed with 100 parts of a general-purpose epoxy resin with an epoxide equivalent of 196.
PHR = parts per hundred of resin for the maximum usable pot life in a container with a resin depth of at least 2 in.
Hardening times will be 1 to 4 times longer on work done at standard laminate thicknesses.

per lb to heavy strands weighing 100 yd per lb. Many variations and combinations of rovings are used in the warp and fill, including a unidirectional, high-modulus woven roving which uses ordinary textile strands in the warp (see Table II-1.11).

As in fabrics and mats, there are a number of variables in the woven rovings and in the products made from them. Table II-1.12 lists the variables in the basic material.

To summarize, woven rovings are an excellent material for application where drape is required in hand lay-up molding. The material has its major strength in the directions of warp and fill, with drop-off at 45°. If optimum strength in all directions is needed, a combination of the woven rovings with mat reinforcements can provide such properties. In laminates made from woven rovings, glass content approaching 65% may be attained.[5]

Mat-Woven-Roving Combinations. As noted earlier, one method of obtaining uniform omni-directional strength properties in hand lay-up moldings is the use of combinations of woven rovings with mat reinforcements. The mat may be either the mechanically (needled) or chemically (binder) joined type. Most combinations are constructed with 1.5-oz per sq ft mat and 24-oz per sq yd woven roving.

The combinations provide major savings of time and labor in large-scale structural lay-ups. Table II-1.13 contains the variables of mat-woven roving combinations.

Veil or Surfacing Mats. Often, is is found that the use of mat reinforcements or woven rovings—and occasionally even woven fabrics—results in an undesirably rough-textured surface on the completed molding. In particular, mat lay-ups have been notorious for individual protruding fibers. To eliminate this problem, surfacing mats are used. These materials are blown or drum-wound attenuated mats in thicknesses from 0.010 to 0.030 in. When applied as surface layers, they provide two major advantages: (1) the fibers are much finer than those of conventional materials; and (2) the construction of the veil is such that it retains substantial amounts of the impregnant, resulting in a resin-rich layer over the laminate surface. Probably the greatest volume of surfacing mat is used in applications where resistance to chemical agents is important. Table II-1.14 gives the variables of surfacing mats. As with any of the basic materials, it is advisable to observe the manufacturers' recommendations with respect to proper usage.[6]

EQUIPMENT AND TOOLING

Molds: Construction and Use. Almost any material including metals, wood, stone, cement, plaster, and even compacted sand and earth may be used for either the master pattern or the basic mold for polyester hand lay-up molding. Usable, well-engineered molds for hand lay-up may be constructed by the exact process described below for actually molding the finished parts. The mold gel coat originally taken from the master pattern will of course be on the side or surface against which the part gel coat is subsequently to be applied. Laminate construction over the gel coat will naturally ensue, and the hand laid-up mold should be braced, bolstered, and liberally reinforced for dimensional stability.

Additional facts about the construction and use of molds for hand lay-up RP/C are summed up in the following discussion.[7] (See also Chapter II-4 for more detailed information about the planning and construction of molds.)

Table II-1.7 Fiber Glass Cloth Types

Weave Type	Description	Performance and Advantages
Plain	Warp and fill strands—1 over, 1 under —available up to 0.022 in. thick	Firm stable weave, porous for ease of air removal in hand lay-up, uniform strength in all fiber directions
Basket	Warp and fill strands—2 over, 2 under	Fairly pliable; flatter, stronger and less stable than plain weave
Crowfoot satin	1 warp strand over 3 and under 1 fill strand	Pliable and suitable for drape over compound curvatures, can be woven into higher warps and fill counts
Long harness satin	1 warp strand over 4 or more and under 1 fill strand	Fairly tight weave; high laminate strength in all directions; highest warp and fill density; most pliable, conforming excellently to compound curvatures
Unidirectional	Strong warp strands plus weak fill strands in any weave configuration	Unidirectional strength, high laminate impact resistance
Leno	Woven with 2 or more interlocking parallel warp ends	Reduced sleaziness or shifting in low-count fabrics
Mock-leno	2 or more warp strands closely spaced with 2 or more fill strands	Good mechanical bonding, lower fabric cost for thicker laminates
3-dimensional	2 or more layers or warp or fill strands blended and combined into thick cloth	Separation of woven surfaces permits construction of high-modulus laminates

The "secret" in the manufacture of a successful RP/C mold lies in the thought and planning that precedes its manufacture. There is no substitute for and no reason to accept less than the best tool suitable for the job.

Steps Involved. Before production molds can be made, it is essential to have: 1. an acceptable design; 2. a tested prototype; and 3. an engineer familiar with reinforced plastics. Where several molds are required, the follow-

Table II-1.8 Fiber Glass Reinforcing Mats

Type	Description and Properties	Performance and Uses
Chemically bonded	Available in weights of 0.75 up to 6 oz per sq ft tight or loose surface, width to 76 in.	Good drapability; rapid wet-out[a] in 3- to 5- poise hand lay-up resins; uniform strength in all directions; often used with other reinforcements
Mechanically bonded	Same as for chemically bonded, usually needled	Better drape than chemically bonded mats, but slightly lower laminate physical properties

[a]A simple wet-out test may be set up for hand lay-up mats by pouring 50 g of uncatalyzed hand lay-up resin on the top surface of a 1-sq ft mat (or other reinforcement) placed on a clear cellophane or mylar film over a black cardboard background. Evaluation consists of noting the time of initial wet-through and also the time of the complete wet-out of the area covered by resin. Comparison should be made with a control standard or photographic standards previously prepared.

Table II-1.9 Major Variables in Fiber Glass Mats

Variable	Description
Glass type	Usually E glass
Weights available, oz per sq ft	0.75–3.0
	(6.0 made for special purposes)
Weight uniformity	±10% on 0.75 and 1.0 oz mat;
	± 5% above 1.5 oz.
Ignition loss,[a] % by weight	4.0–7.0
Styrene solubility,[a] sec (ASTM-D-1529)	Approx. 12–25 or more, depending on mat type and requirements
Roll widths, in.	38 to 80, slit widths available
Roll lengths, ft	To 500 (0.75 oz mat)
Dry-mat tensile strength	Should be sufficient to prevent undue tearing in roll
Drapability (dry)	Should not be stiff and boardy nor have excessive loose surface fibers
Wet-out and conformity	Should wet rapidly and thoroughly and conform to medium compound curvatures; should show high clarity in cured polyester resin

[a]Refers to chemically bonded mats only.

ing sequence is recommended: 1. pattern, 2. master mold, 3. mold master from which the production molds are made.

There are conditions under which variations are desirable. One such set of conditions might involve the tested prototype being used for the pattern and the master mold being used for the manufacture of production parts. Here the steps involved have been reduced to a minimum with a resultant cost savings which could be regarded as substantial. Such a reduction, however, should not be considered without a thorough evaluation of the individual condition.

Requirements for Planning and Engineering. Before any of the basic steps are begun, several questions should be answered to (1) determine whether all of the basic steps are to be involved and to (2) determine the best route in each step. Usually this is done in the estimating stage either for a customer's quotation or as a presentation to management where proprietary items are involved. Naturally, the more detailed the study can be, the closer the estimate will represent actual conditions of manufacture. In determining what steps are to be taken, the following questions should be answered:

1. Will one mold suffice to produce the required units in the given time?
2. Is a full-scale model required for approval and testing? Can it be used as a pattern? Can it be used for more than one mold?

These questions further involve various considerations in making the mold, such as:

1. What is the best approach from the pattern-making point of view?
2. What part requirements will determine how the mold is to be made?
3. Considering production, can molding operations be simplified or can some assembly operations be eliminated?
4. What is best from strength and bracing standpoint?

Most molds will be a result of the best overall compromise, and a clearer picture can be had if all fine points applicable to the condition can be covered, such as:

1. *For Pattern Making*
   a. Should the mold be made in 1 piece?
   b. Male or female?
   c. Will only a half pattern be required due to symmetry or other reasons?
   d. Has full attention been given to items such as: run-off required, sufficient radii, undercuts, part thickness, clearances and tolerances?
   e. What are finish requirements?
2. *Part Requirements*
   a. Must the part be 1 piece?
   b. Where are seams or joints least objectionable or where can they be used to advantage?
   c. Can the mold be made in several parts

Table II-1.10 Fiber Glass Reinforcing Mat Weights per Square Feet and Square Yard

Sq Ft		
Oz	G	Lb
0.25	7.1	0.0156
0.50	14.2	0.0312
0.75	21.3	0.0468
1.00	28.3	0.0624
1.25	35.4	0.0782
1.50	42.5	0.0937
1.75	49.6	0.1093
2.00	56.7	0.1250
2.25	63.8	0.1404
2.50	70.8	0.1562
2.75	77.9	0.1718
3.00	85.0	0.1872
3.50	99.1	0.2184
4.00	113.2	0.2500
4.50	127.3	0.2813
5.00	141.5	0.3125
6.00	170.0	0.3750

Sq Yd		
Oz	G	Lb
2.25	63.4	0.1404
4.50	127.8	0.2808
6.75	191.7	0.4212
9.00	255.9	0.5616
11.25	318.6	0.7038
13.50	382.5	0.8433
15.75	446.4	0.9837
18.00	510.3	1.1250
20.25	574.2	1.2636
22.50	637.2	1.4042
24.75	701.1	1.5462
27.00	765.0	1.6848
31.50	891.9	1.9656
36.00	1018.8	2.2500
40.50	1145.7	2.5317
45.00	1273.5	2.8125
54.00	1530.0	3.3750

to suit molding or assembly convenience?

d. Is the part to be refinished or used as molded?

3. *Production Requirements*

   a. Would it be of advantage to use the mold itself as a jig?

   b. Can a simple jig release a mold to production?

   c. Can some assembly or finishing operations be eliminated by making a 1-piece molding or by splitting the part at the right place without excessive detriment to the molding operations?

   d. Should the molds be stationary or movable?

   e. Can staging or platforms be eliminated by a split mold?

   f. Would it be of advantage to eliminate costly jigs and fixtures at the price of a more expensive molding operation?

   g. Can standard width materials be used to best advantage?

   h. Are all portions of the mold easily accessible to suit the hand lay-up method?

Once these questions have been answered one will have a good picture of what is to be done and the operations required, and a close cost evaluation can be made, since the materials of construction, mold thickness, strength, amount of bracing and so on will have been considered. Consequently, the amount of man hours, materials and special equipment can be specified.

Check Lists and Schedules. Actually, at this point, only the ground work has been laid for the successful large mold. All thinking, calculating and planning will be to no avail if each stage of the operation is not checked. Confer with the persons who are to be responsible for each major step in the operation either singly or together and go over exactly what is to be done, how it is to be done, who will do it. Request suggestions, then develop a check list and schedule of operations, coupled with a realistic time table. Once these are established, stay with them. Provide the supervisor with a list of critical phases which he personally will oversee, direct or approve.

Avoidance of Mold Defects. Pitfalls or mistakes may prevent the use of the mold in production or may require costly rework and refinishing operations on the molds, parts or both. Some of the more prevalent errors and their causes are:

1. Rippled or uneven appearance—Pattern not splined or sanded sufficiently or properly.

2. Rough surface, wood grain, etc.—Pattern not polished sufficiently or surface prepared properly. Parting film or agent not properly applied, dust or dirt on surface before the over-lay is sprayed.

Table II-1.11   Woven-Glass Roving Types

ASTM Type	Weave	Roving Count per in.		Construction (Yd per lb)		Weight (oz per Sq Yd)	No. Plies for 0.125 in. Laminate
		Warp	Fill	Warp	Fill[a]		
1	Plain	10	4	590–655	465–535	13.0	8
2	"	10	4	590–655	300–330	16.0	6
3	"	5	4	210–240	185–215	24.5	4
4	"	5	3	210–240	140–160	24.5	4
5	"	5	2.5	210–240	100–105	27.0	3 (0.110 in. thick)
6	"	7	6	410–470	370–430	18.0	5
7	"	5	4	210–240	230–260	22.0	4
8	"	4	4	210–240	300–330	18.0	5

Source: Reprinted courtesy of ASTM, from ASTM D2150-63T.
[a]Fill yarn is doubled to provide 16 actual types, because of the capability of some machinery to weave 2 picks and count as 1 fill strand.

Table II-1.12   Major Variables in Fiber-Glass Woven Roving[a]

Variables	Description
Glass type	E glass
Weights per sq yd	13 to 27
Ignition loss, %	0.5 to 1.0
Thickness, in.	0.020 (15.0 oz)—0.045 (24.5 oz)
Widths, in.	To 144 in.
Lengths, yd	110 (15.0 oz)—80 (24.5 oz)
Hand	Should be soft and pliable, not stiff and boardy
Reinforcement capability	Provides higher laminate glass content than chopped-strand mat
Drapability	Possesses more flexibility and drape over compound curvatures than chopped-strand mat
Rate of wet-out	Should be rapid and thorough

[a]*Source:* Johns-Manville Fiber Glass Reinforcements Division.

Table II-1.13   Major Variables in Mat-Woven Roving Combinations

Variables	Description
Mat weights (typical), oz per sq yd	Style 18/10—27.5, ± 10%
	Style 24/10—33.5
	Style 22/15—34.9
	Style 24/15—38.0
	Style 48/15—61.5, twill weave
Thickness (24/15), in.	0.075 to 0.080
Widths, in.	38, 50, 60, 80
Roll lengths, yd	50
Ignition loss, %	3.0
Drapability	Possesses soft hand and less drapability than woven roving
Rate of wet-out	Rapid and thorough—comparable with woven roving or high-solubility mat

Table II-1.14   Major Variables of Veil or Surfacing Mats

Variables	Description
Glass type	Usually C (chemical)
Thickness (mil)/weight (lb per square)	10/0.6 (100 sq ft)
	15/0.9
	20/1.2
	25/1.5
	30/1.8
Mat weight, g per sq ft per mil	Approx. 0.272
Roll widths, in.	36–60 as required
Roll lengths, ft	400 or longer
Fiber pattern	Continuous filament, random pattern, usually blown fiber or equivalent
Surface characteristic	Veil-type mat, minimum fuzz; overlay-type mat, very fuzzy (minimum binder)
Ignition loss, %	3–15 as specified
Binder and wet-out rate	Soluble-type binder for hand lay-up veil mat, high wet-out rate

3. Regular lines at periodic intervals—Station marks in the pattern not removed or bracing transfer due to wood bracing lying hard against the outside of the mold.

4. Deformation—Point or line stresses—Excessive weight borne by 1 brace and not transferred to a large area can cause the laminate to yield. Insufficient cure and resulting low strength can also contribute this as well as too thin a mold or improper bracing.

5. Cracks in the overlay—Too rigid an overlay material, too thick an overlay, impact stresses during usage, excessive deformation during usage. Use of wedges to remove part.

6. Pattern transfer—Application of first ply of glass too soon after spraying overlay. Insufficient 'back-up' laminate between overlay and woven roving.

7. Blisters or break through of overlay—Voids during manufacture of the mold. Resin gelled too quickly to prevent proper working out of air. Resin not sufficiently thixotropic allowing too much run off. High exotherm causing "blowing" of resin.

8. Warpage—Too many layers or plies applied at one time or too soon after each other causing strains to be set up due to high exotherm. Improper or insufficient post-cure.

Trouble shooting in large mold making should begin by anticipation before the mold is made. Use a high heat distortion point resin. The back-up laminate should consist of 3 to 5 plies of a fine woven material like 1000 cloth followed by additional plies of 1½-oz mat and 24-oz roving to add strength at moderate cost. The mat is also used as a buffer against pattern transfer and mold thickness is normally ½ in. 1-ply per day is applied with a catalyst system arranged to give lower exotherm and reduce shrinkage. With such a system proper aging on the pattern is a requisite, 2 or 3 days in the sun or an oven post-cure for a similar period at 110 or 120°F is usually sufficient.[7]

Auxiliary Equipment for Hand Lay-up. Additional operational equipment for the successful hand lay-up operation should include the following:

Completely outfitted spray booth or hood with catalyst-injector gun for gel coating

Drum lifts and tilters for handling resin

Storage for catalysts, 150 ft from operation

Resin containers, 1 qt to 5 gal for batch-mixing

Drill press or hand drills for mixing catalyst into resin

Metering graduated plastic containers or calibrated plastic beakers for measuring small quantities

Neoprene and other squeegees

Throwaway brushes

Serrated metal or plastic disc rollers for impregnating

Cut-off equipment: draw knives, drills, and saber saws

Either sheet metal tanks or one or more 55-

gal metal drums split in half lengthwise, with each half set upon legs at working height, open sides facing upward. ½ should be fitted with a hinged cover and filled with acetone solvent for cleaning. The second half should be filled with water, covered with a 3- or 4-in square metal grid and should be used for draining excess catalyzed or gelling resin batches to prevent fires.

MOLDING METHOD/PROCESSING DETAILS

The *hand lay-up process* is designed for low mold and part cost. It provides rapid mold turnover, relative ease of finishing, capability of making very large parts with reasonably close tolerances and good appearance, and many other advantages, most of which apply to economics and time savings.

Mixing and Preparation Equipment Most of the major problems of mixing have usually been handled by the resin manufacturer. In most cases, however, the processor's mixing shop will be required to add catalysts, promoters, fillers, pigments, diluents, and other agents necessary to provide proper curing and desired final properties. Consequently, the processor should supply his shop with the following:

Safe and secure storage and handling facilities for resins, solvents, and other potentially flammable and dangerous materials.

Large scales for weighing batches of resin, styrene, and other major components of the impregnants.

Small scales or balances for weighing catalysts, promoters, pigments, and other additives normally incorporated in relatively small quantities.

Mixers with explosion-proof motors for handling styrene, diluents, colorants, and similar hazardous materials.

Disposable containers (paper tubs or metal cans) in qt, gal, and 5-gallon sizes for batch mixing.

Full and complete protection for plant and personnel against fire, explosion, and injury. This would include items such as the storage facilities mentioned above, fire extinguishers of the proper type, cold storage for catalysts, protective hand creams for employees, safety cans for solvents, and, if the materials warrant such precautions, overhead emergency shower equipment and eye wash facilities.

Glass-Preparation Area For good quality control, it is essential that basic rules of cleanliness be observed. Reinforcing materials should be prepared in a clean area, well separated from the immediate section of the shop where lay-up is performed, even if only by erecting a transparent "tent" of thermoplastic film around the preparation area.

The basic piece of equipment required is a long table, preferably covered with metal or a rigid plastic sheet. The table should be 5 or 6 ft wide, with creeling brackets at one or both ends for handling rolls of cloth, mat and other reinforcements.

Cutting equipment is necessary. This could require simple tools such as scissors, razor blades (in properly safe handles), or large cutters similar to paper cutters. Where a large run of identically shaped pieces is planned, motor-driven tools are available for cutting stacks of reinforcements. These tools are identical with those used in the garment trade for pattern cutting.

Protected storage facilities should be available for patterns, so that they can be kept safe from damage yet be readily obtainable when required.

Rolling carts are useful for transporting the cut reinforcements to the lay-up area. A good industrial vacuum cleaner for both housekeeping and removal of residual contaminants from molds during clean-up is a further aid in maintaining clean working conditions.

Lay-up Equipment. Although the lay-up procedure discussed here has been termed hand lay-up, it is necessary that the workers be provided with a number of tools to perform their job adequately. For preparation of the mold surfaces, waxes or other mold-release agents previously discussed are required. If polished mold surfaces are involved, mechanical buffers are superior to tedious hand-rubbing procedures.

Facilities should be provided for mixing resins and catalysts in small batches (1 qt to perhaps 5 gal. for gel coatings or if a small amount of additional resin is needed to complete a job. Again, the need for observing all safety precautions cannot be overemphasized.

Necessary hand tools include ringed or serrated rollers of all sizes and shapes for flat and curved areas of moldings; paint rollers; brushes; wooden mixing paddles, both chisel-end and pointed; knives for trimming; and other items. Finishing tools and ancillary materials include draw knives; brushes; clean, lint-free rags; and a Barcol gauge to determine surface hardness as an indication of the degree of cure.

Lay-up Procedure.

1. After the mold is built, it must be mounted on a work base. This may consist of a wooden support, metal piping, trunnions, or some similar device whereby all areas of the mold will be completely and readily accessible to the worker. In the case of very deep cavities, it may be necessary to erect a trestle around the mold, from which a "bridge" can be hung to allow the fabricator to descend into the cavity without coming into contact with its walls or base.

2. Assuming that the mold has been properly prepared, as discussed above, by the use of sealants, polishes, or similar surface finishing devices, the next step is the application of the mold-release agent. Care in performance of this step cannot be overemphasized. It has been humorously said that most of the resins used in hand lay-up are not particularly good adhesives, except for holding parts to molds. In the molding of a large-area piece, it does not require much unprotected surface area to make it extremely difficult to remove the molding from the tool without surface damage. Every square inch of the mold surface must be inspected to be sure of complete coverage by the release agent.

3. The gel coat should be applied next. Depending upon the type of formulation, the type of mold, the area, and the practice of the particular fabricator, the gel coat may be applied by brushing, spraying, or any other convenient procedure which will give a smooth and uniform distribution. Proper promotion and catalyzing of the resin will generate room-temperature cure to a leathery, tacky condition. Here it is important that such a full cure be obtained to prevent "styrene strike-through," a condition in which the unpolymerized styrene in the resin formulation remains suspended in

the gel coat to appear later in the molding process as bubbles, surface blemishes, and other undesirable conditions.

4. If the particular procedure and available pot life of the resin permits—and it is most desirable that they should—the impregnating resin should be applied directly over the gel coat, with the reinforcement laid up into the wet resin to assure minimum air entrapment, good wetting, and contact with the gel coat. If this is not feasible, the reinforcement may be laid against the gel coat and the resin may then be applied from the top and worked into the lay-up by hand with spatulas, rollers, and other hand tools. Still another procedure where large sections are involved is to impregnate the reinforcement on a work table and place it, wet, into the mold. This process is one that requires skill in handling, especially where doubly curved surfaces are a part of the mold, because most reinforcements do not submit to stretching and "bridging" can take place. This means that the reinforcement does not contact the surface of the mold at every point, resulting in a subsequent void or delaminated area in the final part. Wet lay-up, however, is used very widely, because it does guarantee that the reinforcement has been completely impregnated prior to lay-up.

5. Curing then takes place. It may be permitted at ambient temperature or may be slightly accelerated with hot-air blowers, heat lamps, or sunlight lamps (if the "sunlight" catalysts have been used). Careful control of temperature is necessary, because an excess will cause rapid boiling or vaporization of solvents, diluents and additive liquids, with resultant bubbles and/or other defects in lay-up. If heat is used, it should only be sufficient to initiate the polymerization process and to speed it along the way. Good ventilation is important for removal of by-product fumes such as styrene vapor, which is inflammable and toxic. The laminate must not be disturbed during this step of the molding until a Barcol hardness tester shows a reading of at least 30. Preliminary trimming may be performed prior to full cure, that is, when the laminate is in a leathery state.

6. After proper hardness has been attained, the part may be removed from the mold. Many techniques have been devised for this important step to avoid damage to the surface of the molding. In many cases, it is possible to slip a flattened nozzle of an air gun between the

edge of the part and the surface of the mold, and then apply compressed air, which will lift the part sufficiently to effect release. Another procedure involves slipping very thin wooden strips between the part and the mold. No metals should be used, as damage to both the mold and the molding will inevitably ensue. A third procedure is the use of water, under pressure, in a manner similar to the compressed-air technique mentioned above. It is impossible to recommend any particular method, because each shop will develop its own, based upon the size of the molding, its shape, its thickness (which will determine the resistance to flexure and possible damage), and the available tools. A nose radome for example, might be removed from its cavity by inverting the mold on trunnions and gently tapping the assembly.

7. The molded part should then be removed to the finishing area, where contamination and damage can be prevented. Trim lines scribed in the mold, precise templates, and other devices should be used to assure the accuracy of trimming and to prevent overcutting.

Adaptations of Hand Lay-up Reinforced-Plastic Molding. General purpose widespread use of hand lay-up molding of reinforced plastics has various specific adaptations. Table II-1.15 summarizes 10 of them and discusses the mold type, construction, finishing, and examples of each. Figure II-1.3 illustrates two of the major outlets for these adaptations of hand lay-up reinforced plastics: marine products and building components. Table II-1.16 outlines trouble-shooting for the hand lay-up process.

Other Adaptations of the Hand Lay-up Process. Specific methods or variations of the hand lay-up technique have evolved to satisfy certain requirements and to fulfill the demands of a ready market. One of the most important of these demands is for RP/C with resistance to corrosive environments. This use will be discussed at length. Also, the well-developed and highly accepted method of applying protective RP/C coatings to metal tanks will be examined.

Reinforced Plastics/Composites for Corrosive Service.[8] One of the major areas of use of hand lay-up, fiber glass reinforced plastics is in

Figure II-1.3 Two of the major outlets for hand-lay-up reinforced plastics composites: marine products and building components. *Top*, 74-ft., reinforced-plastic fishing (shrimp) trawler now in wide, general-purpose commercial usage. The fiber-glass reinforced polyester material makes possible an extremely strong vessel construction that requires little maintenance because of the inherent material strength and the superior gel-coat surface finish. The complete cargo hold of this ship is also of reinforced plastic, making it highly sanitary and easy to clean. (Courtesy of Hatteras Yacht Division, North American Rockwell Corporation and Gibbs and Cox, Inc.) *Bottom*, utilisation of large, fiber glass, reinforced-components in building construction for a two-story office building. The white fascia panels are 24 ft. high, 8 ft. wide and bow out 14 in. at the center. The black scroll-shaped panels bracketing the doorway and building ends are also gel coated, reinforced-plastic material. (Courtesy of R. L. Kuss Co., Findlay, Ohio, and Johns-Manville Fiber Glass Reinforcements Division, Waterville, Ohio.)

corrosion-resistant applications in the chemical industries. The substitution of these materials for stainless steel has reduced initial costs some 40% and has increased the usable lifetime of equipment many times over. It is estimated that this use of reinforced plastics will grow at the

Table II-1.15 Various Adaptations of Hand Lay-up Molding for Other than Corrosion-Resistant Structures

1. Parts with outside surface finished*

Type Mold and Construction: Cavity type, finished inside; limited undercuts, mechanically braced to prevent distortion. Use cut-out portions, flanges, ridges; maintain largest radii possible. Place 3 in. minimum right-angled flange at mold extremities for finishing or joining to exact dimensions.

Recommended Part Construction: Use non-repellent release if part is to be painted. Use high-gloss weather, abrasion, and craze-resistant gel coat. Use non-photographing back-up layer behind the gel coat or in the first reinforcing resin layer. Lay up using good grade resilient resin plus high-solubility fiber glass chopped-strand mat, woven roving or combination—two layers woven roving in normal areas or parts, three layers where extra strength is required. For high part modulus or rigidity, insert bracing or laminated-in stiffeners such as encapsulated foam, end-grain strip balsa, or tesselated balsa, or honeycomb between skins for rigidity. Use fiber glass cloth for fine interior finish if required. Parts can be painted inside to mask fiber patterning if desired. Allow 3 to 4 in. oversize for trim and finishing.

Finishing Required: Trim to line with sharp knife or draw-blade when laminate is gelled and leather-hard (prior to development of observable Barcol hardness). Repair inside and outside defects. Join parts and add hardware.

Remarks and Examples: Simulated logs stone and scenery, boats, car bodies, truck bodies, trailers, marina pontoons, caskets and grave vaults.*

2. Parts with inside surface finished

Type Mold and Construction: Plug-type, braced well on inside to prevent distortion. Allow sufficient sidewall draft or taper for part removal after cure.

Recommended Part Construction: Use chemical-resistant or abrasion-proof gel coat. For heavy structures use two or three layers of woven roving with chopped-strand mat between each for bulk and rigidity. Internal or external service pressure may reach 1200 lb per sq ft. Trim to predetermined lines.

Finishing Required: Remove from mold prior to high Barcol cure. Use air or water pressure to release from mold. Brace unfinished outside if necessary with flanges, laminated-on wood or RP/C gussets. Polish finished side.

Remarks and Examples: Containers, troughs, farm equipment and appliances, cement forms, man-carrying booms and buckets for hot-line repair, swimming pools and pool panels. The man-carrying buckets require electrical insulation and grid protection. Cement forms require a smooth surface and freedom from voids. Swimming pools need colored, non-crazing gel coats and must be installed with non-shifting sand or concrete backup.

3. Sanitary fixtures

Type Mold and Construction: Mold to be made for outside-inside finish on parts, often both surfaces finished in same part. Use multiple-section mold. Brace mold, but allow flexibility for part removal. Minor undercuts are permitted but allow draft in at least one direction for removal.

Recommended Part Construction: Use high-gloss, high Barcol-development, heavily-filled gel coat, usually white. Use filled resinous backup layer to prevent patterning. Use mostly chopped-strand mat lay-up. Use built-up or integral stiffeners behind flat areas subject to high service loads. Extend stiffeners to external base support for load transfer and avoidance of distortion and gel coat defects in high-load areas.

Finished Required: Remove from mold prior to Barcol = 35. Trim to structural or specified dimensions permitted for installation. Allow for anchoring straps to studding.

Remarks and Examples: Sanitary tub and shower units and combinations, modular RP/C bathroom units, sinks and lavatories. Design is almost unlimited except for stool components which are not recommended. Provide for accurate external dimensions of the part to permit installation into space provided at the construction site.

Table II-1.15 Various Adaptations of Hand Lay-up Molding for Other than Corrosion-Resistant Structures (Cont.)

4. *Construction components—large or multiple-section parts†*

Type Mold and Construction: Usually cavity-type and often in several sections. Finished parts are self-supporting, thin shell type, usually large-size. If multiple-section, provide wide mold flanges for proper stress distribution in joining finished parts or panels.

Recommended Part Construction: Use high-gloss, weather and abrasion-resistant gel coats. Use fiber glass mat, woven roving, or combinations. Apply foam, honeycomb, balsa, or plywood sandwich construction where it is required for high modulus where it is not fully provided by part configuration.

Finishing Required: Remove part from multiple-section molds. Finish to proper height or flange-width. Assemble, applying mastic, soft plastic membrane, or laminating resin to join sections. Assemble with fastening hardware preferably 3 to 4 times wider than hole size. Assemble to framing or construction base using slotted holes in dissimilar materials to allow for differential thermal expansion and contraction.

Remarks and Examples: Plastic houses, modular dwelling units, free-form hyperbolic paraboloid and similar thin-shell architectural units or curtain-wall components, also silos and silo covers (finished inside and outside). All structures must be pre-engineered and designed to resist wind-loads, weight of accumulated snow, thermal expansion and contraction, and the forces of internal loads. Parts and panels must be properly moored or supported for permanence during assembly or erection at the construction site.

5. *Parts cast or laid up over inflatable molds*

Type Mold and Construction: Stretched rubber, or inflatable plastic, or bag of plastic-impregnated fabric. Bag is prefabricated to size and kept inflated by low-pressure blower air during the molding process.

Recommended Part Construction: Inflate mold on site (ground, pipe-rim, or prepared cement slab). Gel coats are not usually employed, and lay-up is made directly over mold. Use mostly resin plus chopped strand mat over the inflated bag. Use rapid-cure resin, ultraviolet sensitive catalyst, or heated blower air if erecting in cold or arctic climates. Allow lay-up to cure thoroughly before deflating bag. Scaffolding may be used to reach top part of structure. Apply external gel coat by spray if desired. Finished part thickness should be 1/8 to 3/16 in. for large parts.

Finishing Required: Bolster outside of structure with ridges or arches if necessary. Paint with reflecting beads or metallic paint to resist over-heating from sun's rays. Cut door in one end for removal of bag and for normal access. Apply skin of fiber glass or foam insulation inside for thermal and acoustical protection.

Remarks and Examples: Large and small structures, quonset or igloo-shaped, are erected on site. Units usually possess excellent structural strength provided by the surface of revolution or shell-like shape.

6. *Parts built up using permanent internal ribbing or skeletal frame*

Type Mold and Construction: No formal re-usable mold. Skeleton frame, ribbing or keel plus side members, etc. is erected and becomes integrated into finished part as basic supporting structural elements.

Recommended Part Construction: Build up frame with side supporting ribs to define desired basic shape. Stretch muslin or similar light fabric or holding element like thin plywood tightly over ribs and frame. Apply organic varnish, laquer or "wing dope" to tighten light fabric. Apply medium-heavy boat cloth over treated muslin and draw tight, cutting and lapping to cover the frame. Apply polyester hand lay-up resin to impregnate and remove voids. Apply fiber glass mat and/or woven roving lay-up, impregnate and cure. Apply fiber glass cloth if desired to improve exterior surface finish. Apply exterior gel coat or paint finish.

Finishing Required: Finish by sanding edges and surface to remove rough areas and protruding fibers if desired. Apply paint or colored gel coat. Surface finish resulting from this method of molding will not be as smooth as if part had been fabricated in permanent mold.

Remarks and Examples: One-shot structures only. No permanent mold results and the molded part is not exactly duplicatable. For boats, canoes, housings, where only 1 or 2 of a particular part type are required.

7. Coating substrates such as plywood

Type Mold and Construction: METHOD A—Permit plywood or other porous substrate to act as mold.
METHOD B—Select or fabricate smooth-surfaced mold, preferably gel coated reinforced plastic. Smooth with cement polishing or terrazzo grinding tool to smooth surface. Block mold up for rigidity. Heat mildly and uniformly by embedded elements (80°F) to prevent warpage due to thermal contractions. Mold may be made as wide as possible and up to 200 ft. in length to produce exceptionally long parts using laid-in modular plywood segments.

Recommended Part Construction: METHOD A—Thin resin 10% with styrene monomer to approx. 1.5 to 3.0 poise viscosity. Catalyze and spread saturating coat on the wood surface. Allow prime coat to gel. Follow immediately with lay-up using unthinned resin plus cloth, mat or woven roving reinforcement. Finish by light sanding. For flat work, caul plates may be applied after lay-up while resin is liquid to provide smooth surface. For work involving compound curvatures, sand & gel coat or paint.
METHOD B—Large flat areas. Apply gel coat or pigmented resin layer to mold using travelling drip-trough. Follow with mat and woven roving reinforcement impregnated with room-temperature cure polyester. Laminate in as many plywood components as desired to establish a specified length. Apply woven roving and mat lay-up in reverse order on upper surface of plywood. Finish. Gel coat or smooth resin layer on the mold surface is the exterior finish. Trim and sand edges. The un-gel coated surface constitutes the interior. The finished exterior surface may have rock chips, colored glass cullet, sand, etc., embedded for an appearance factor in exterior construction panels.

Finishing Required: METHOD A—Edge trim.
METHOD B—Edge trim.

Remarks and Examples: METHOD A—Single-sheet plywood for construction and for compound curve surfaces such as covering a wooden boat with fiber glass.
METHOD B—Fabrication of containerization panels or building components. May be made as long or as wide as desired by large mold-size and prefabrication of modular plywood sizes.

8. Prosthetic appliances

Type Mold and Construction: The mold is an internal porous core made to the shape of the extremity to be reproduced, except smaller in diameter at any cross-section by a dimension equal to twice the finished laminate thickness.

Recommended Part Construction: After form is established, cover with dry layer of random, blanket-type saturated polyester fiber mat. Make lap instead of butt joints. Cover with polyvinyl alcohol sock. Tie sock off at bottom, and pour catalyzed resin in top, then tie off open top end of sock. Using hands or soft pliable tool, squeegee on outside of sock to force resin into mat. Rub up and down vertically to work voids out of the lay-up and to distribute the resin. Work the excess resin from the surface to a point beyond the extremity of the part and allow to cure.

Finishing Required: Remove sock and excess resin. Finish surface by light sanding. No paint or external gel coat is required. Apply hardware and connectors as required.

Remarks and Examples: Artificial limbs. The synthetic polyester fiber mat is noted for its smoothness and fine texture when laminated. It provides a smooth, non-irritating molded surface that gives natural appearance and does not injure or aggravate human tissue upon contact.

Table II-1.15 Various Adaptations of Hand Lay-up Molding for Other than Corrosion-Resistant Structures (Cont.)

9. *Lay-up onto vacuum-formed thermoplastic sheet*

Type Mold and Construction: A highly economical method because no elaborate reinforced plastic mold is required. The part which serves as both surface component (replacing gel coat) and mold is thermoformed in a standard vacuum-forming fixture starting with flat acrylic, ABS, vinyl, or other sheeting at least 1/16 in. thick. Form is to shape of inside of finished part. A special holding fixture is needed for the lay-up. Speed of production of vacuum-forming shells is much greater than that of gel coating.

Recommended Part Construction: Set vacuum-formed shell into holding fixture. Lay up using chopped strand fiber glass mat and room-temperature cure impregnating resin. An especially modified resin will be required in some cases to promote maximum adhesion with the thermoformable material used (consult material suppliers). Allow oversize for trim. The vacuum-formed sheet will have become thinned in cross-section especially at corners of deep-draw areas. The RP/C backup, however, provides more than adequate support.

Finishing Required: Finish edges to specified dimension using a wet-grind diamond or abrasive wheel. No finish required for the vacuum-formed surface.

Remarks and Examples: Bath tubs for trailers, sinks, shower stalls, automotive hoods; other large-area external body components; marine and transportation composites.

10. *Miscellaneous*

Type Mold and Construction: Outside finish is required for all parts. Hence, mold requires finished inside surface (cavity). However, some inside finish parts requiring outside finish (plug) molds may be used. Usually applies to smaller custom parts and to some multiple-segment molds to avoid undercuts.

Recommended Part Construction: For gel coats, use flat color, two or more colors, or create 2-color effects by masking. Use glossy gel coats and develop a high-Barcol finish. For laminate construction, use mat, woven roving or combinations. Mold in expanded metal flanges for edge stiffeners and joining, or for required end-use support.

Finishing Required: Finish as necessary, joining parts with servicable fasteners and adding hardware, ribbing, supports, etc.

Remarks and Examples: Housings, machine guards, toys, sporting goods, playground structures, shields, mannequins, etc. Many sizes and types of parts are possible for low mold and finished-part cost. Most are suited to limited quantity production.

*For further information, see Marine Design Manual for Reinforced Plastics, New York, McGraw-Hill, 1965.
†See Figure II-1.3 for illustration.

Table II-1.16 Trouble-Shooting for Hand Lay-up Process

Defect	Probable Cause	Remedy
Gel coat—Sticking to mold (brushed or sprayed)	Improper release agent or application.	Apply release and let cure. If wax, allow to dry thoroughly and buff. If trouble persists, use polyvinyl-alcohol-sprayed film over wax.
Strike-through (fern or leaf-like pattern)	Gel coat swells and separates from mold surface in confined area because of insufficient cure and action of styrene in lay-up resin.	Permit get coat to cure more thoroughly prior to lay-up. Apply gel coat in thicker layer—0.010 to 0.015 in. Use cured intermediate back-up layer between gel coat and lay-up. Add non-coloring accessory promoter to resin.
Hazy or nonglossy surface	Entire part prematurely removed from mold. Contamination of release prior to application of gel coat.	Permit more complete cure of gel coat and lay-up.
Voids under gel coat	Small or large, flat blisters caused by separation of gel coat from lay-up. Gel coat should not cure tack-free in air but should remain sticky for better bond to lay-up.	Allow first lay-up application to cure prior to adding second and third, etc. Inspect closely for blisters after lay-up. Cut out and putty-fill before completing cure. Putty mix: 1 pt resin to 3 pt $CaCO_3$.
Open bubbles, blisters, and pinholes in gel coat surface	Trapped air, free solvent, dirt, or excessively high exotherm in gel coat or lay-up resin.	Avoid mixing air into gel coat when introducing catalyst. Let stand for short period after mixing and before spraying. Keep containers and working area clean.
Cure in thickened rods or strings	Pregelation.	Keep mixing containers clean and free of previously catalyzed gel coat. Use throw-away mixing containers.
Soft areas	Uneven cure.	More thorough mixing of catalyst into gel coat.
Rundown on vertical surface	Too heavy or uneven application or insufficient thixotropic agent.	Revise spray pattern or technique for greater uniformity. Check viscosity of gel coat. Add more thixotropic agent.
Cratering	Use of too high surface-angle release, preventing gel coat from wetting in small spots 1/16 to 1/4 in., so that lay-up shows through gel coat.	More careful selection and application of release agent.
Cracking and fissuring	Larger cracks caused by too thick areas of gel coat or excessive exotherm or thin point in lay-up. Fissuring because of front or reverse impact blow.	More uniform application of gel coat and better mixing with catalyst. Prevent accidental or injuring blows.
Fiber pattern: random fibers from mat or cross-hatch from woven-roving weave	Caused by photographing-through of reinforcement.	Apply thicker gel coat or allow it to cure longer before lay-up. Best solution is application of inter-mediate layer of more rigid, resin-containing Vitro-Strand® fibers.

Table II-1.16 Trouble-Shooting for Hand Lay-up Process (Cont.)

Defect	Probable Cause	Remedy
Lay-up—Draining on vertical surfaces	Resin too low in viscosity; resin with insufficient thixotropic agent; mold or room too warm.	Most probable correction is to increase thix-agent content of resin.
Bubbles	Air entrained in reinforcement after combination with resin.	Add 0.2% green pigment to lay-up resin to see voids. Work lay-up more freely with brushes, squeegees, or serrated rollers. If possible, apply a liberal quantity of resin onto work before applying reinforcement so that resin forces air out from bottom.
Bridging over small-radius curves such as lap-strakes, etc.	Reinforcement too stiff; curves below design-allowables.	Select more highly wettable or soluble mat or woven roving. Use loose-mixed putty to caulk small radii curvatures prior to lay-up. Redesign mold.
Thin areas	Gaps in lapping reinforcement caused by improper placement or short-cutting, etc.	Correct placement and cutting errors. Lay in patches to correct thin spots prior to removal from mold. Try pre-wetting of reinforcement by resin prior to placement in mold.
Fibers protruding from inner lay-up surface	Usually unavoidable if mat is sole reinforcement.	For finish layer, apply woven fabric, woven roving, or veil mat on inside. After cure, sand and apply splatter paint.
Cracked or resin-rich areas usually at bottom or well-point	Drainage of resin in large lay-up to a low point and because of high exotherm, results in cracking, possibly too high a resin-glass ratio.	Introduce more thixotropic agent into resin. Continue to squeegee excess resin out of collection points until gelation occurs. Add additional reinforcement.

rate of 30% per year during the 1970's and will become the number-one market segment in total tonnage by 1980.[9] The total picture, however, is not so bright. The substitution of corrosion-resistant RP/C for other materials, as in the case of new construction materials, has on occasion resulted in disastrous failures which have delayed the full acceptance of RP/C. A basic understanding of the properties and variables which affect the lifetime of RP/C corrosion equipment is therefore necessary for the selection of resin types, fillers and lay-up procedures.

The corrosion resistance of RP/C is largely determined by the resin surface, the chemical structure, and the nature of the corrodant solution. Chemical attack can occur in a number of ways: (1) degradation and disintegration of a physical nature because of permeations and solvent action, (2) hydrolysis, (3) thermal degradation, (4) oxidation, (5) radiation, and (6) a combination of these.

Resin cure plays a very important part in determining the final corrosion resistance of a given polyester resin. To perform satisfactorily, the resin must be fully cured, preferably by oven curing, because room-temperature curing results in a 95% cross-linkage level which increases with time. Resin cure is indicated by surface hardness which should be in the order of 40 to 50 on the Barcol scale.

In general, polar solvents and polar compounds including acids, alkalis, and salt solutions cause a degradation of the polyester linkages by hydrolysis of the polar bond in varying degrees depending upon solution concentration and temperatures. Corrodants such as water and hydrofluoric acid can permeate the resin surface because of low molecular diameter and weaken the resin-to-glass bond, the hydrofluoric acid actively corroding the glass-reinforcing fibers. Non-polar solvents, such as benzene, ethylene dichloride, and other chlorinated solvents attack the non-polar structure of the polyester-resin monomer linkage and styrene cross-linkage by solvent action, causing swelling and permeation by other corrodants.

Radiation damage results largely from the exposure of polyester resins to sunlight-ultraviolet radiations which activate the polyester linkages and unsaturated double bonds in the molecular structure, resulting in an acceleration of chemical attack and rupture of the polyester linkages. Polyester structures exposed to sunlight as well as to ultraviolet radiations must be clad with a pigmented surface to reduce damage from these sources.

Thermal degradation and oxidation occur in varying degrees depending upon temperature and exposure to chemical oxidants such as nitric acid, sodium hypochlorite, chlorine, and concentrated sulfuric acid.

In general, maximum service temperatures for polyester resins seldom exceed 350°F, especially if the structure is under stress, since the flexural and tensile moduli fall off rapidly with temperature. (see Figure II-1.2). RP/C have been successfully used however at cryogenic temperatures because the flexural and tensile moduli increase with decreasing temperature. Metals, in comparison, become brittle at low temperature. Reinforced plastics are also advantageous in that less insulation is required for chemical process vessels because of the low thermal conductivity of the materials.

Polyester resins used for corrosive services vary over a wide composition range and include (1) general-purpose resins, (2) bisphenol and chlorendic-anhydride resins, (3) vinyl-ester resins, (4) isopthalic resins, (5) brominated resins, and (6) epoxy and furfural alcohol resins (see Figure II-1.4). Polyester-resin molecular structure can be varied to build in steric (molecular configuration) protection of ester groups, giving a wide variation in chemical resistance.

General-purpose resins often can be substituted for stainless steel and other metals under mild corrosive conditions. Bisphenol resins are usually specified for corrosive service. Corrosion resistance to acids is improved by partial chlorination of the bisphenol polyester structure which increases bond strength by increasing the strength of the organic acid linkage. Vinyl-ester resins exhibit superior resistance to alkali and acids because of the inert vinyl structure and reduced polyester linkage concentration. Isopthalic resins in general give superior resistance to solvent action because of the higher concentrations of the polar polyester linkage. New furan resins with controllable curing cycles are available for many specific chemical-corrosion applications.[10]

In general, improved resistance to strong acids results in a sacrifice in resistance to strong alkali and vice versa, and improved resistance to solvents is accompanied by reduced resistance to chemical attack. Consequently polyester-resin selection must be tailored to each application.[11] Also, because exposure conditions are varied, it often becomes necessary to clad process vessels with exterior exposure radically varying from the interior exposure. The required fire-resistance level, for example, may impair chemical durability for certain environments. When service conditions are too severe to permit the use of a synergistic additive, composite laminates consisting of an unfilled resin-rich liner and an antimony trioxide-filled shell are feasible. When process changes occur, it is often necessary to surface-coat the interior to meet process-change requirements. When the interior surface is exposed to hydrofluoric acid, the inner cladding should include an inner layer of dynel cloth (PVC-acrylate copolymer). It serves as a vapor barrier and also gives improved corrosion resistance to caustic and strong alkali solutions. The dynel layer is brittle and should not be stressed or otherwise damaged.

For improved chemical resistance it often becomes necessary to clad with other plastics. PVC and saran can easily be bonded to vinyl-ester and chlorinated polyester resins, with vinyl-ester resins serving as a bonding agent for bisphenol and isopthalic resins.[12] Polypropylene and polyethylene can be readily clad to polyester by an intervening glass-cloth lining which is affixed to the polyethylene by heat and bonded to the polyester resin. Elongation upon heating can be a problem, and the thermal expansion of the RP/C should be matched to that of the thermoplastic.

Teflon and Kel F fluorinated liners can also be readily attached by the use of special adhesives in order to render laminate surfaces chemically inert, thus further extending RP/C corrosion-equipment applications.

The abrasion resistance of an RP/C structure can be superior to steel despite the difference in hardness. RP/C ducting, for example, has

Figure II-1.4 An acid sump tank for a vertical pickle line in a Canadian steel mill. The tank is 40 ft. long and 8 ft. wide and weighs approximately 12,000 lb. It replaced rubber-lined steel which was a continuous maintainence problem in the mill. (Courtesy of Precisioneering Ltd., Scarborough, Ontario.)

been substituted for stainless steel in the handling of abrasive coke dust in the steel industry. Equipment lifetime expectancy is on the order of 15 to 20 yr as compared to an average lifetime of 5 yr for the stainless steel. This increased service life stems from the clinging of the coke breeze to the surface of the plastic, which is first softened by organics. The coke breeze becomes embedded in the surface of the resin in order to prevent coke-to-coke abrasion.

When abrasion is encountered in process equipment, the reinforcement of the resin containing flake glass results in improved abrasion resistance over fiber-glass filaments.[13] The flakes are not as easily undercut as are the individual filaments.

Further abrasion resistance can be built into the RP/C surface by including powdered carborundum abrasives and alundum powder. Powdered silica sand is often incorporated into floor lining to act as a filler and for abrasion resistance when acids other than hydrofluoric are involved. Powdered coke is used as a filler for caustic and hydrofluoric-acid service.

Floor linings are usually built up by attach-ment to freshly sand-blasted concrete. The new cleaned floor surface is primed with polyester resin diluted with styrene, followed by a light coating of resin filled heavily with powdered silica sand. A woven-roving reinforcement layer is applied, and then the main floor topping. Concrete surface preparation preferably includes etching with hydrochloric acid to remove carbonates and alkali, rinsing to remove salts, and then drying.

Tank linings and coatings are installed over freshly sand-blasted steel (see below, *Protective Coatings for Metal Tanks*).

RP/C resins will normally support combustion. As a further insurance, however, fire retardancy can be built into the resin structure by chlorination and/or bromination of the polyester resins and by including fire retardant fillers, and additives in the formulation.

Chlorinated resins and vinyl-ester resins give improved fire resistance with self-extinguishing properties when compounded with 5% by weight of antimony trioxide and the newer zinc-borate and hydrated-alumina fire retardants at 5 to 15%.[14] Superior flame resistance has been imparted to other newer resins by bromination of the polyester. In addition to chemical-resistant applications, fire-retardant resins also find use in corrugated sheet, press molding, and so on.

Other methods of improving fire resistance include the cladding of the RP/C surface with glass cloth impregnated with resin containing fire retardants. The function of the fire retardants antimony trioxide and zinc borate is not well understood and lies in the realm of "surface phenomena." The higher cost antimony trioxide, for example, has been extended by vapor coating chemically precipitated silica with antimony trioxide, the fire retardancy of the silica-filled antimony being equivalent to that of antimony trioxide alone on a weight basis. Because the chlorinated polyester resins are nearly endothermic combustion-wise, the gel coating of these resins with fire-retardant resins gives flame-suppressant properties equivalent to those in which fire retardants have been included in the main body of the RP/C structure.

The inclusion of fire retardants can reduce corrosion resistance under certain circumstances. Cabosil and Santocel (precipitated silica: thixotropic agents) and putties containing precipitated silica should not be exposed to caustic and hydrofluoric solutions which actively permeate the resin by dissolving the silica.

Two other parameters that are usually noted in purchase specifications for corrosion gear are void content and mar resistance. Bubbles may effectively be eliminated by mixing a silicone paint additive with the uncured resin.

Mar resistance has been improved via an identical route.[15] Repairs may be easily made and maintenance readily facilitated using hand lay-up techniques on existing corrosion-resistant and industrial gear.[16]

The final selection of an RP/C system must be based either upon previous experience or upon corrosion testing in the plant. Published corrosion-resistance tables are based upon the use of chemically pure reagents under controlled laboratory conditions as outlined in SPI and ASTM test standards. The published exposure test conditions can be used as a rough selection guide.

Data obtained by the use of in-service, corrosion-test coupons must be carefully evaluated when used to predict equipment lifetime. Because these coupons often swell and gain weight, they must be evaluated over a period of time of from 1 to 6 months, with the longer test period preferred for exposure to elevated temperature.

The corrosion rate can then be estimated by means of the equation:

$$\text{mils per yr.} = \frac{534\,W}{DAT}$$

where W = weight loss in mg
A = area of sample in sq in.
D = density of sample
T = time of duration of test in hr

Because end-use requirements are naturally more stringent for corrosion-resistant RP/C structures than for general-purpose hand lay-ups, the lay-up procedures required to maintain the best possible quality are more specific and definitive than those described under *Molding Method*. It is of value, therefore, to list stepwise the lay-up procedures adhered to in the corrosion industry and to point up the logic of their selection.[17]

1. In dispensing resins from tanks or drums, the solids included or suspended such as those which promote fire retardancy have a pronounced tendency to settle. Measures for preventing this settling should be taken. One recommended measure for storage tanks is to recirculate the resin for a definite period of time (30 min., for example) every 24 hr. Another is to periodically stir or rotate the resin received in drums.

2. No prepared or formulated gel coats should be applied as an RP/C structural surface which will eventually contact a corrosive medium or environment. Instead, a resin selected for resistance to the particular medium should be laid up in a heavier-than-normal gel-coat thickness, totaling 0.020 to 0.030 in. In order to gain complete smoothness, this surface coat should preferably be laid up against a prepared and released mold surface. The resin layer should be reinforced using a 0.010- to 0.20-in. C (chemical) glass veil or surfacing mat. The ratio of resin to surfacing mat should be 90:10.

With this procedure, there will be no opportunity for a gel coat to become chemically unbonded and flake or peel off the laminate structure. Synethetic non-wovens such as dynel or orlon may supersede glass for exposure to corrosive environments in which glass is not suitable (hydrofluoric acid, for example).

3. Fiberglass chopped-strand mat in at least a 0.100 in. laminated thickness (minimum of

2 plies, 1½ oz per sq ft) should be used to back up the resin-rich surface layer. Woven roving, cloth, or a filament-wound pass would, in service, tend to trap air and therefore wick the corrosive chemical through to other parts of the laminate. Thus, only glass reinforcing mat should be used to back up the surface layer with no other reinforcement inserted between the plies of mat.

4. If the entire corrosion-resistant laminate is to be less than 3/16 in. thick, only chopped-strand mat should be used as reinforcement. A minimum of 25% reinforcement by weight should be maintained. For laminate structures thicker than 3/16 in., additional reinforcing layers of mat, woven, roving, or fiber glass cloth may be used. Fiber glass reinforcing mat, however, should be used between all woven-roving layers with no 2 of the latter placed together. By this means, air entrapment is held to a minimum, and low laminate shear strength is avoided.

All reinforcing media should be lapped and not butted. Mat should be lapped a minimum of 1 in., and woven roving or cloth, a minimum of 2 in. The overlaps should be staggered throughout the RP/C structure, not superimposed.

5. The outer-surface ply of the structure should consist of reinforcing mat plus surface veil in order to provide a superior, smooth outside-surface finish. As previously explained, the resin type may be varied during or within the lay-up to provide complementing design parameters such as chemical durability (inside) plus fire retardancy (outside). Fire retardancy of the outer surfaces of the tank, pipe, or duct may be further increased or enhanced by increasing the percentage of fiber glass which is laminated in (glass cloth is used instead of veil mat).

6. The structural laminate edge should not be exposed to any corrosive medium or environment. If this is impossible, however, only chopped-strand and veil mats should be used in the edge-exposed lay-up to avoid air entrapment and prevent wicking.

Exposed edges, gaps, or cracks may also be better sealed against wicking by the use of a highly filled, thixotropic resin mix, catalyzed, spread on, and allowed to cure in place. Such a mix may also be impregnated into the mat or veil and be applied to cover exposed edges. All covering operations should preferably be performed as soon as possible to avoid delamination in service, or at least within a maximum of 24 hr after the original lay-up and cure.

7. Fitments such as elbows, tees, and flanges may be fabricated with hand lay-up methods using simple, prepared molds. To establish smoothness, the mold surface should correspond to the inner or exposed part surface. Fitments as small as 2 or 3 in. in diameter may be hand-fabricated. These and smaller ones may also be press-molded using chemically resistant premix formulations.

Specific Design Rules for Corrosion-Resistant, Hand Lay-Up, Laminate Construction[18]

Hand lay-up laminate constructions provide several types of corrosion-resistant reinforced plastics. Table II-1.17 presents the *chemical performance* of these constructions.

From a *mechanical* point of view, several properties of laminates can be discussed.

- The "slack" in woven-roving strands in laminates may be taken up by utilizing more resilient resin or by running the tank or corrosion structure hot to 250°F. Resin ductility results and the expansion takes up the slack. The heated structure has a higher strength than it has at room temperature.
- Resin fragments under localized stress cut the glass filaments and weaken the structure as failure progresses.
- The strain level is one property that must be considered when designing laminates. Table II-1.18 summarizes this aspect.

Testing the critical strain level, and also tensile modulus, and time-temperature dependence should be required for a specific laminate system. Samples from tank cutouts for nozzles, manholes and such should be saved for test samples. Confirmation of specifications during fabrication is desirable.

- The abrasion resistance of reinforced-plastic laminates must be considered. A disadvantage is the removal of the self-built-up protective layer when a high-velocity (or particle-laden) stream is present. An abrasion results in addition to chemical attack. The following design parameters have been found to assist in the elimination of abrasion in chemical-resistant RP/C structures:
  (a) Eliminating right angled turns—use long-radius ells.
  (b) Keep the slurry flow parallel to the wall as much as possible.

Table II-1-17 Typical Flexural Strengths of Laminate Constructions for Corrosion Resistance (Tested in 26% sulfuric acid at 100°C)

Laminate Construction	Original Strength	% Strength after 6 months	% Strength after 12 months
1. 5-ply; chopped-strand mat—26% glass, plus thin glass veil on each surface—15% glass (gel coats not usually used)	100	46 of original	46 of original
2. Woven roving entirely—40% glass—no surface veil	100–125 (Higher original strength than 1.)	—	16 of original
3. Woven roving plus chopped-strand mat interleaved, surface veil both surfaces—35–40 glass	100 plus 15–25	—	75 of original

Source: H. E. Atkinson, "R. P. Chemical Equipment Design and Fabrication for Failure Prevention," *SPE RETEC Technical Papers,* 9 September 1968, p. 30. For further information see United States Department of Commerce, National Bureau of Standards, NBS Voluntary Product Standard PS-15-69.
Notes: Flexural strengths of laminates exposed to 26% H_2SO_4 concentration at 100°C. Major chemical resistance may be provided by crocidolite asbestos liner or surface reinforcement plus structural glass as in (3). Slight solubility of glass surface reduces load-carrying capacity. Surface of laminate B cracked open, and edge penetration and delamination resulted.

(c) Watch out for throttling valves or orifices that may accelerate the velocity tenfold in a localized area and impinge in an erratic manner against the pipe wall.

(d) Where impingement does occur in small localized areas the area may be protected by imbedding in the laminate a 16 gauge metal sheet such as 316 stainless steel or Carpenter 20 or Hastelloy.

(e) All laminated structures need to be cured completely if they are to be used in an abrasive service. The difference between a 27 Barcol and a 38 Barcol may double the life of the RP equipment. This may sound obvious but, unfortunately, RP/C chemical process equipment is still shipped from the fabricator prior to the time that it has developed a complete cure.

(f) For a broad scale increase (450%) in abrasion resistance extending over the entire laminated structure, resin modifications are necessary. The present standard specification for laminates, where an increase in abrasion resistance is required, calls for an exact resin mix using an additive, and fabricated at a prescribed thickness for the wear surface.[19]

Abrasion in RP/C chemical-resistant structures may be systematically evaluated prior to actual service using ASTM D-1044 and/or D-1242.

■ Fire retardancy is an item of major importance, because RP/C ducting systems are

Table II-1.18 Strain-Level Percentage of Static-Loading, Mat and Woven-Roving Laminates

% Ultimate Flexural Strength Applied	% Strain Resulting	Time to Failure
0	0	No failure
10	0.25	No failure
25	0.65	Failed in 40 hr

Notes: Designing to strain levels of 0.12 to 0.15% should provide a life expectancy of up to 15 years. Designing to the strain level of 0.3%, resin fracture occurs and life expectancy is critical.

particularly vulnerable to fires and combustion. The following are recommended as checkpoints to maintain duct systems with flame-spread ratings of 25 or less in addition to the desired chemical durability.

(a) Any resin specified should have a flame spread rating of 25 or less after addition of antimony oxide or the equivalent (ASTM E84-61 Tunnel test).

(b) There should be installed a sprinkler system to provide protection commensurate with the value of the duct system and also the property to be protected.

(c) Periodical cleaning of the duct system both inside and outside should be carried out to prevent fire from combustible residues.

(d) The resin specified should be completely satisfactory in all aspects of the service conditions it will be required to serve. This should include testing of sample coupons prior to fabrication and installation.

(e) As regards smoke damage, the greatest damage from fires occurring in RP/C ductwork and other systems results not from fire damage but from deterioration and impairment caused by the smoke generated. Although held below 25 fire rating, present polymeric materials range in smoke rating from 400 to 2400.

Alternative materials are aluminum, steel, lead linings, polyvinyl chloride, olefins, brick, rubber, coated metal and others. Each presents problems for which RP/C has shown the best possible chances for solution. The industry should work toward systems which provide smoke density ratings of 50 or less which maintain flame ratings of 25 maximum. Resin chemical durability should be improved, and a top resin service temperature of 500°F. Should be strived for.[20]

■ For *construction* using hand lay-up, corrosion-resistant, reinforced-plastic materials, several items need attention:

(a) Woven-roving strength and modulus—max. at 0° to warp; min. at 45° to warp; 90% of maximum at fill direction (90° to warp).

(b) Overlays for joining sections on exterior joints should be as thick as the sections joined and of the same laminate construction.

(c) Interior-joint overlays use liner composition only—C.S. mat or veil mat. Internal overlays are *essential* for nozzles, manholes, and other parts which penetrate through the cross-section of the construction. Failure may occur if they are omitted.

(d) For 3 in. or smaller nozzles and jets, the external portion should be bolstered with 3 or 4 gussets in order to distribute the bending stresses imparted by the connected piping when in use.

(e) Table II-1.19 discusses the methods of joining piping systems.

Protective Coatings for Metal Tanks. A final adaptation of hand lay-up is the method of applying a protective RP/C coating to metal tanks. Protective coatings of RP/C on tanks for underground storage of petroleum, gasoline, and other materials have been found to greatly extend their useful life. Steel tanks buried below ground level are extremely subject to attack and degradation because of "aggressive" soil characteristics, percolating ground waters and acids, voltaic or electrolytic action of stray currents,—particularly in heavily populated areas—and many other causes.

Application of a glass fiber reinforced polyester resin over 100% of the exterior surface of the steel tank provides the required protection. Procedures for preparation of the steel tank and application of the protective coating have been developed by a committee of knowledgeable industry representatives under the guidance of the Steel Tank Institute.[21] Currently a total of 18 approved coaters are operating in the United States. The method of application of RP/C coating to steel tanks involves the following steps: preparation of the tank surface, application of putty, or filler where necessary, application of the coating, electrical testing, and final (flood) coating of the surface.

1. Supporting the Tank. The support of the tank during the RP/C coating operation is of considerable importance. The tank may be set on rollers or rails upon which it can be rotated as the coating operation proceeds. It must be emphasized, however, that the resin must be reasonably well cured before the load of the tank is applied; otherwise, the coating will be damaged. Ideally, the tanks should be suspended at their ends from a central location by a head and tail stock assembly so that free

Table II-1.19 Methods of Joining Piping Systems

Item	Type of Joining System	Advantages	Disadvantages	Comment
1	Butt Joints	a. Tremendously strong. b. Technique can be learned with little training. c. Equal in cost to adhesive joints in up to 6 in. in size and less expensive above that.	a. The joint cannot be taken apart for maintenance purposes. b. Is limited in use to polyester hand laid-up pipe. c. To the "off again-on again" user of this equipment sufficient familiarity with the technique may become lost, producing unreliable joints. d. Should be made in temperatures above 50°F and under dry conditions.	The standard low cost joint with polyester hand laid up pipe.
2	Adhesive Joints	a. Very economical in sizes up to 6 in. b. When properly made makes an extremely strong joint as contact area is considerable. c. Joint can be assembled with low manpower requirements.	a. Above 6 in. in size the cost is more than a butt joint. b. Glue line cracks can be a problem. c. Mixing the adhesive per instructions is of great importance.	Commonly used with filament wound systems as a standard low cost coupling.
3	Flanged Joints (Adhesive-Epoxy)	a. A low cost flanged joint. b. Advances in technology have developed a tremendously strong adhesive. c. It is a tremendously strong flange.	a. Glue line problems in the adhesive can provide a real obstacle in system assembly under hammering and surging.	The use of flanged systems should be limited to that required for maintenance purposes and the replacement of metallic or lined flanged fittings.
4	Polyester Flanges: (A) Press Molded	a. Low cost, comparable in price to epoxy flanges.	a. Must be used with full faced gaskets. b. Cannot be over-torqued or failure at flange neck will result.	Can be used to make less expensive stub ends but must be handled with care on installation.

Table II-1.19 Methods of Joining Piping Systems (Cont.)

Item	Type of Joining System	Advantages	Disadvantages	Comment
	(B) Hand Laid Up	a. Very strong joint. b. Although not recommended, joint can be used without a full faced gasket.	a. A most expensive joint and one which must only be used to provide system assembly and maintenance. b. Inherently a flange joint is weaker than a butt and strap joint. For example, on a 100 lb system flanges may begin to weep at 400 psi but a butt and strap joint may go to 900-1100 psi. Standard specifications are that they will be tight at twice the design pressure.	Unbelievably strong—in thousands of applications have only witnessed one or two failures. Consider also bellows type.
5	O-Ring Type Joints: (A) Bell and Spigot (Flextran* as an example)	a. Commonly used in sewer service. Resists earth movement and tremors. b. Joints will pass a very tight specification on sewer service under 100 gal/in./mile/day at 50 ft. head. c. Joint deflection of 2°-5° permissable depending on size, reducing fitting costs.	a. Normally considered to be a sewer pipe and not recommended for pressure applications above ground.	An integral bell with a spigot end. Rubber ring makes a tight seal.
	(B) Quick connect Couplings (Kwikey†)	a. Satisfactory for moderate chemical service with 225 lb. pressure rating for steady service and 150 lb. for cyclic service. b. Eliminates need for adhesive and provides simple field assembly and disassembly.	a. Temperature limitation of 150°F. b. In severe chemical service where a fitting is required an average of every five feet this joint would lose much of its advantage.	Quick connect coupling, using two O-rings and a double groove and key arrangement with each coupling to provide positive locking. Joint may ultimately be adapted to tough corrosive conditions.

			A type of joint sometimes favored in the epoxy piping system or machine made systems.

| | | a. Commonly disappears in larger size piping. | |
| | | b. Shows weakness in highly corrosive systems and premature failure. | |

	c. Good for temporary application and one where dismantling must be easy.		
	d. Joint is re-usable.		
	e. Good for long, straight piping lines and down-hole piping.		

6	Screwed Joint	a. Quick assembly.	
		b. Easy disassembly.	
		c. An inexpensive joint on small piping.	
		d. Satisfactory for water and mildly corrosive systems.	

Source: J. H. Mallinson, S.P.I. 26th RP/C Proceedings, 1971, Sec. 2-a.
*Trademark of Johns-Manville Corporation.
†Trademark of Fiberglass Resources Corporation.

rotation and access to all parts of the tank is possible. A small block, with a hole to accommodate the head and tail stock, tack-welded to the center of each end should give the required support and can be easily removed after the coating is in place. Weld marks are removed by light grinding, and the support areas are then coated. The complexity of the supporting mechanism will depend upon and/or be justified by the number of tanks to be coated.

2. Surface Preparation. Good surface preparation is essential for all RP/C tank-coating applications. Oil and grease must be removed and the entire tank blast-cleaned to a "commercial-blast" surface. Areas around fittings or manholes, where good adhesion is absolutely necessary, are taken to "white metal." The coating is applied as soon as possible after the tank has been blasted, certainly before rust forms on the cleaned areas.

The RP/C coating will tend to bridge at sharp corners or in irregular areas, and it is necessary to fill or fillet welds, seams, right-angle flanges, connections, and so on in order to create a smooth surface or a relatively large radius which will give uniform support to the coating and prevent bridging, tank-coating separation, and other defects.

3. Filling and Filleting Operation. A putty composed of resin with filler and/or glass fibers may be used in the filling and filleting operation. Putties may be prepared using short fibers or chopped glass in concentrations from 5 to 200 parts per 100 (pph) of resin. Non-fibrous fillers may also be used in concentrations from 10 to 300 (pph). Appreciably higher impact strength is obtained when fibrous reinforcement is included in the putties. One formula indicative of the putty composition is the following: 100 parts resin, 2½ parts fine silica, 190 parts short glass fiber. The materials are mixed thoroughly to give the putty consistency. It must be catalyzed before being applied. In areas of the tank where impact loads or blows are likely to be encountered, such as the knuckle area, fibrous-reinforced putty is definitely recommended. Excessive or irregular areas of putty may be removed by sanding.

4. Processes and Materials for Coating. Two systems may be used for applying a reinforced-plastic coating to a steel tank: (1) the hand lay-up process and (2) the spray-up process. The hand lay-up operation is less demanding and requires less equipment, but it may require

more labor. It will probably be used in most cases until the fabricator has established a need for the faster, more efficient, and more demanding spray-up operation.

Materials required for hand lay-up application are polyester hand lay-up type resin, MEK peroxide catalyst, 1½ oz per sq ft high solubility, chopped-strand fiber glass reinforcing mat, and continuous-filament fiber glass roving.

The method of application for hand lay-up is built around spray application of the resin over the fiber glass mat laid in place. The "bucket-and-brush" method may also be used. (The spray-up technique is discussed in Chapter II-2.)

In both lay-up and spray-up operations, the application of the resin and glass should be made to the upper half of the tank so that the spraying, painting, roller or other technique is done in a downward or, at the worst, a vertical direction.

5. Hand Lay-up Application of the Coating. As we have mentioned, the hand-lay-up operation involves applying by either painting or spraying techniques to that portion of the tank to be coated. A 10- to 15-mil layer of resin is applied and then a ply of 1½ oz mat follows. The glass mat should be placed in position, taking care to insure good contact of the mat with the tank surface at all points.

As soon as the mat is in place, additional resin is applied and worked into it using a roller or brush until the mat has been saturated and the air removed. When all air bubbles are eliminated, the surface of the primed tank is readily visible through the mat. As the glass fibers wet out, they disappear from view and the completely wet out mat is quite transparent. Thus, any bubbles or voids in the coating are quite apparent. Adequate lighting is necessary to make the bubbles readily visible. Excessive quantities of resin are unnecessary, but enough must be used so that the air bubbles can be worked out.

A total of three layers or plies of mat are required, with the second and third plies added prior to gelation of resin in the first ply. The succeeding plies must be staggered and/or overlapped so that the seams occur in different areas. If gelation has occured, the exotherm should be allowed to develop and the heat dissipated prior to applying the succeeding ply; otherwise, the newly applied resin will heat up and may gel before the operator has sufficient time to work the air out of the ply. These plies are again saturated and the air removed in the same manner as the first ply.

When the tank is roller or rail-supported, the resin must be allowed to gel. Sufficient strength must be developed in the coating before the tank can be rotated to its new position where the coating is actually carrying the load of the tank. If the tank is suspended so that the coating will not be required to carry the tank weight during its curing operation, the coating operation may proceed after gelation and cooling without waiting for the curing of the resin.

If painting techniques are used, cleaning the brush is necessary before the resin gels. Acetone, MEK, methyl isobutyl ketone, or ethylene chloride are good cleaning solvents provided that bristles are not synthetic. Brushes cleaned properly and directly after resin application may be used indefinitely. Careless handling, however, makes for extremely short brush life.

The thickness of the coating should be between 0.1 and 0.125 in., with heavier sections applied around the knuckle areas in particular weld seams, fittings, fixtures, and other stress points. A continuous nonporous coating is required to obtain the necessary protection and the lifting lugs, pipe fittings, flanges, and manhole areas must all be covered for optimum protection.

Special efforts and procedures are required to insure adhesion in the fitting areas where there is an "inescapable" hole in the coating. For example, in addition to a white-metal blast in these areas, it may be desirable to wrap the protruding areas of the filling and withdrawal pipe fittings with resin-saturated roving to insure a good, high-strength, permanent bond in these areas.

If the tank contains a manhole, it is necessary to extend the coating up the throat, completely covering the exterior of the flange and the cover contact area to a position inside the sealing gasket. The cover must be coated in the same manner, and sealing gaskets and/or washers should be used at the bolt-head and nut-contact areas.

The manhole should be covered by the lay-up process. One possible method would be to start with discs of mat large enough to cover the top of the manhole; the top, edge, and bottom of the flange; and at least a portion of the throat. The plies of mat should be positioned properly and impregnated in the top flange area. The excess material in the disc should be removed by cutting "darts" from the outer edge of the mat disc to the flange in order to allow the mat to be folded around the flange and throat without encountering excessive overlapping or wrinkling. After impregnation, the mat should be rolled or brushed into position on the bottom of the flange and the throat, where it may be held in place by wrapping the throat with pre-impregnated roving. After gelation, the coating over both the manhole and bolt-hole areas can be easily cut out and removed.

After the coating has been applied, the degree or extent of cure may be determined by the use of a Barcol-hardness meter (Barcol Impressor 934-1). This hand-held indicator will show when the coating has reached its cured state by giving readings of 40 or more.

Air bubbles trapped in the coating will be weak areas and must be eliminated prior to the gel of the resin. If large white areas, indicating air, are encountered, the area should be removed by cutting or sanding and be replaced with air-free coating. Repairs are easily made, and the bond between the original coating and the repair is quite adequate.

6. Preliminary Testing. The effectiveness of the system depends upon the non-porous nature of the coating. Therefore, when hardness tests indicate that the degree of cure is acceptable, the entire tank *must be* tested electrically using a single-point or rake-type Holiday (void) unit. (The wet-sponge Holiday unit is not acceptable.) Special emphasis should be placed on testing the knuckle areas of the tank, because corners are the most difficult area to coat satisfactorily. Any areas found to contain Holidays should be repaired either by addition of resin on the surface or by actually cutting out and replacing extremely bad sections. Let us emphasize again that absolutely no Holidays can be permitted in the tank, and that the repairs have to be made at this point to eliminate any defects in the coating. The Holiday testing must be done only after the resin has cured sufficiently to give satisfactory Barcol readings.

7. Final Coating. After the tank has been thoroughly checked with a Holiday tester and found to be 100% Holiday-free, a final coating of resin is applied. Prior to catalyzation, extra

wax solution is added to the resin to give a smooth, hard, air-cured surface to the tank. If desired, a pigment may also be added to the final coat to give an identifying color. (Pigment pastes are available from the resin suppliers. The pigment paste should be added in concentrations from 1 to 3%, depending upon the color and manufacturer's recommendations.)

The final coating gives an added or even double measure of protection to the already 100% Holiday-free tank and will further insure suitable performance of the tank under extremely demanding conditions.

8. Final Testing. Samples of RP/C coated, ¼ in. tank steel simulating coated tank construction were subjected to the following tests according to AWWA Specification C-203 and others.

- Impact Resistance: Direct (coated side)— 11.5 ft./lb. No separation as determined by high-frequency Holiday tester; 16 sq. in. separation allowed.
- Indirect (metal side)—no cracks or discontinuity.

Table II-1.20 General Design Rules for Hand Lay-up of RP/C

Minimum inside radius	¼ in.
Molded-in holes	Large
Trimmed in mold	No
Built-in cores	Yes
Undercuts	Yes
Minimum recommended draft, deg.	0
Minimum practical thickness, in.	0.06
Maximum practical thickness, in.	0.5
Normal thickness variation, in.	±0.02
Maximum thickness build-up	As desired
Corrugated sections	Yes
Metal inserts	Yes
Surfacing mat	Yes
Limiting size factor	Mold size
Maximum size part (to date) sq ft	Over 3000
Metal edge stiffeners	Yes
Bosses	Yes
Fins	Yes
Molded-in labels	Yes
Raised numbers	Yes
Gel coat surface	Yes
Shape limitations	None
Translucency	Yes
Finished surfaces	One or both
Strength orientation	Random
Typical glass loading, %	20 to 30 by weight

- Mechanical Deflection: An averaged deflection of 0.285 in. was required to produce cracking. Material passed test.
- Tensile Strength: RP/C coating only showed over 6000 psi tensile strength versus 1300 psi for a comparative coal-tar tape wrapping.
- Vibration: RP/C coated, 1¼ in. × 12 in. × ¼ in. thick steel sample was vibrated ¼ in. amplitude over a 10 in. span, 50 cycles per sec., to simulate shipping conditions (5,000,000 cycles). No cracking or disbonding resulted as tested with high-frequency Holiday tester.
- Bond or Shear strength: (ASTM D2339) Interfacial shear bond ranged from 372 to 1146 psi.
- Water Absorption: (RP/C coating material only) Test values were 3.3 to 6.7% water absorbed, higher than the 0.1 to 2.0% permitted. More complete resin cure should be established through aging or mild heating.
- Hydrocarbon Resistance: 7 days' immersion of 1 in. sq samples of RP/C coating on tank steel at 70°F showed the following results: Toluene, acetone, and benzene attacked coating; alcohol, JP-4 jet fuel, aviation gasoline, SAE 10 viscosity lubrication oil, and CCl₄ showed no visible attack.

PRODUCT DATA

The following section encompasses general design rules for the hand lay-up of RP/C (Table II-1.20), representative physical properties for hand lay-up (Table II-1.21), and laminate physical properties (Table II-1.22). Also treated are advantages and disadvantages of hand lay-up procedures, costs and market data, and . . . professionalism! Additional sources for test methods and specifications are presented in a *Further References* section at the end of the chapter. Notes for Table II-1.22 follow:

Orientation vs. Testing force. Reinforcement layer listed first represents outside of structure; layer listed last represents inside of structure, or first reinforcement layer laid-up. Laminates were tested to 5 listed values by applying flexural and impact stresses against the outside (compressive layer), making the inside or bottom the layer in tension.

Variation of flexural strength with thickness. Laminates thinner and thicker than 0.125 in. provide lower measured physical values for same reinforcement content.

Table II-1.21 Representative Physical Properties for Hand Lay-up of RP/C

Property	Values	
	Mat Construction	Glass-Cloth Construction
Specific gravity	1.4–1.8	1.6–2.0
Tensile strength, psi	16,000	40,000
Flexural strength, psi	24,000	60,000
Compressive strength, psi	12,000	45,000
Shear strength (interlaminate), psi	11,000	—
Modulus of elasticity, flexure, psi	1,000,000	3,000,000
Barcol hardness (dry)	40–50 following cure, 50 in 30 days following cure, higher on subsequent aging.	Same

Impact test. Unnotched against "top" laminate side, flatwise on ½ in. specimen.

Resin system. Even though all test panels are oven-cured, 95% of maximum strength will not develop until 30 days after initial gel and cure.

Woven-roving laminates. In the range of 32 to 62% glass, woven-roving laminates yield higher flexural and impact strengths than those for fiber-glass cloth laminates. Cloth laminates, however, show higher flex modulus than woven-roving laminates.

Mat laminates. Because of bulking, it is not possible to attain greater than 33% glass in chopped-strand mat laminates.

Combinations. By combining 2 or all 3 types of reinforcement, physical properties improve over reinforcements used singly (no. 7, 8, 12). Glass fiber is strongest in tension, therefore flexural strength increases with increased glass concentration on "tension" side.

Delamination. Mat placed between skin layers of woven roving decreases delamination and forces the higher density reinforcement to the surface for greater effectiveness.

Economy. Woven roving plus mat provides the most economical laminate construction.

Source: SPI 13th RP/C Proceedings, 1958, Sec. 16-E.

Advantages and Limitations of the Hand Lay-up Method

The hand lay-up method maintains the following three major advantages over other fabricating processes for reinforced plastics:

1. Large, complex parts can be constructed which are impossible to press-mold or complete by spray-up.

2. Part strength is proportional to the glass content, with the highest strengths obtainable up to 60% levels. Fibers may be selectively oriented in local areas to satisfy distribution demands.

3. Mold and part costs are low and frequent style changes are easily affected.

However, some limitations also exist:

1. Production rate is limited, and reduced die or mold cost may be lost in increased labor costs.

2. Skilled or experienced labor is a necessity if uniformity of the glass/resin ratio and laminate quality are to persist.

3. Properties must sometimes be sacrificed for cost. Cloth for example—the most expensive reinforcement—gives the highest glass content and strength properties but may of necessity be superseded by cheaper woven roving and mat.

Generally, hand-lay-up molding methods and resultant parts have found markets because (1) only a short lead time is necessary to tool up for a molding job, and (2) products made with hand lay-up reinforced plastics provide more satisfactory performance for a multitude of uses than previously used metals, wood, or other materials.

Costs and Market Data.

Cured-out costs for hand lay-up laminate construction vary between $0.45 (mat) and $0.60 (mixed reinforcements) per lb. All-cloth laminates cost from $0.90 to $1.00 per lb. Hand lay-up laminates ⅛ in. thick weigh approximately 16 to 17 oz. per sq ft, so that the amount of dollars and the number of sq ft are practically equivalent. Selling prices vary depending upon local

Table II-1.22 Laminate Physical Properties Resulting from Varying Reinforcement Construction

No.	Reinforcement Construction Required for 1/8 in. Laminate		Flexural Strength (psi)	Flexural Modulus (psi) $\times 10-6$	Impact Strength (ft lb/in.)	% Glass Content
	Layers	Material				
1	2	2 oz mat	24,000	1.0	2.3	24
2	10	10 oz cloth	44,000	2.3	5.0	55
3	3	24 oz woven roving	41,000	1.6	6.7	48
4	1	woven roving	40,000	1.9	6.3	37
	1	1 oz mat				
	1	woven roving				
5	1	¾ oz mat	44,000	1.1	10.0	37
	1	woven roving				
	1	woven roving				
6	1	woven roving	22,000	1.1	4.2	37
	1	woven roving				
	1	¾ oz mat				
7	1	¾ oz mat	57,000	1.6	7.3	42
	1	woven roving				
	1	1 oz mat				
	1	woven roving				
8	2	2 oz mat	39,000	0.9	6.2	29
	1	woven roving				
9	1	1 oz mat	22,000	0.8	4.4	27
	1	woven roving				
	1	1 oz mat				
10	1	10 oz cloth	30,000	1.4	3.5	34
	2	2 oz mat				
	1	10 oz cloth				
11	1	1 oz mat	28,000	0.9	2.5	24
	2	2 oz mat				
	1	10 oz cloth				
12	1	¾ oz mat	49,000	1.4	7.3	43
	1	10 oz cloth				
	2	woven roving				
13	2	woven roving	22,000	1.4	4.2	43
	1	10 oz cloth				
	1	¾ oz mat				

Notes: Cure—All laminates are cured for 2 hr at room temperature plus 40 min. at 220°F.

labor conditions, direct and indirect burden, and amount of actual finishing work required.

Hand lay-up activity accounts for approximately 35% of the total annual poundage consumption of reinforced-plastics. Products are fabricated for each major marketing area of RP/C activity: agriculture, aircraft and aero-space, appliance parts plus equipment and housings, construction, consumer product, electrical supplies, marine craft and accessories, piping including ducts and tanks, transportation, and miscellaneous uses.

The industry expects the continued use of parts fabricated by the hand lay-up method and an expansion commensurate with the normal growth rate of the entire reinforced-plastics field—between 15 and 25% per yr.

PROFESSIONALISM

Although not a rabid *aficionado,* one of the things I have always found fascinating about bullfighting is that instant of time in the fight associated with the oft-quoted phrase, the so-called 'moment of truth." If you are like me, you may have used this catchy phrase many times, only vaguely realizing to what it really refers. In any case, as explained to me, the moment of truth is that point in the fight, after all the veronicas and preliminary passes, when the matador must place himself squarely in the path of the onrushing bull and kill him with a single thrust of his sword through a one- or two-square-inch target area on the back of the bull's neck. If it is done expertly and accurately, the sword will pass through the bull's heart and stop him dead; if the sword misses, it is normally deflected by the shoulder bones of the bull, the bull keeps coming, and he most probably gets to gore the matador.

Obviously this process tends to weed out the real bullfighters from those who only pretend to have the skill and is hence called the moment of truth— the moment of testing—that reveals one's skills and abilities, not simply those one claims to have. In addition, failing the test involves penalties, rather severe ones in the cases of the bullfighter who bears the full responsibility for any lack in his competence.

Even though there are certainly other factors involved in professionalism, we think the job of bullfighting proves a dramatic example of at least two of the major elements involved in being a professional. One of these is a *high degree of demonstrated competence* and the other is the willingness to assume *personal accountability* for doing the job right.

Recently we heard of a company that hired a man who claimed to be a plastics engineer, an expert in the field, a professional. He was hired to specify and oversee the construction of a reinforced plastics tank to hold molten sodium chlorate. He went ahead and had the tank built; got the job done; or so he thought. Apparently not knowing enough about chemistry or resins to know that a general-purpose polyester resin would not do the job, he put the tank in service and filled it with the molten salt. In a short time the tank failed catastrophically, spilling the hot chemical around the plant and injuring nearby workers. It is obvious this plastics engineer was an amateur trying to do a professional's job. Yet how was the company who hired him to know; they were forced to simply rely on his word. Unlike the bullfighter, the bungling amateur also escaped any personal responsibility for his incompetence. In addition to the specific injury caused to the company and its personnel, such a failure of a plastic product is also injurious to the plastics industry image as well.

This is a young and vigorously growing industry and mistakes in the past could be excused in part by the almost universal greenness on the part of many of us exploring a whole new technology. However, we are growing up, and with adulthood comes the need for the assumption of the responsibility for our actions. In line with this, it has been suggested that plastics engineers turn their attention to making their discipline a truly professional one. To do this, it has been suggested that plastics engineers working on projects having to do with public safety and welfare be licensed as professional engineers like engineers in mechanical, civil, electrical, and chemical engineering. This would involve the development of testing procedures to evaluate the competence of candidates for the title of professional plastics engineer and procedures for their certification. This sounds like a worthwhile idea. . . . How do you feel about it?[22] Reprinted from the *SPE Journal,* **27,** no. 3, March, 1971.

References

1. *Modern Plastics,* 47, Nov. 1970, p. 82.
2. Additional information on curing systems may be obtained from the following specifications: Resin, Low Pressure, Laminating MIL-R-7575; Resin Quality Standards QPL-7575-1.
3. Additional data and information on the types, properties, and uses of glass cloth may be found in the following references:

Tests for Construction Characteristics, Woven Fabrics	ASTM D578-61
Glass Fabrics, Woven	ASTM D579-64
Glass Tapes, Woven	ASTM D580-49
Glass Fabrics, Griege	
(Unfinished)	MIL-Y-1140C
Glass Fabrics, (Finished)	MIL-F-9084B
Finishes for Glass Fabrics	MIL-F-9118
Fabrics, Woven Glass, (Finished)	MIL-F-25611
Finish for Glass Fabrics for Heat-Resistant Laminates	MIL-F-2552
Handbook of Industrial Fabrics	United Merchants

4. Additional information concerning fiber-glass mats may be obtained from Fiber-Glass Mats (Tentative) ASTM D1529T; Fibrous-Glass

Mats for Reinforcing Plastics MIL-M-15617A; Mats, Reinforcing, Glass Fiber MIL-M-43248 (MR).

5. For further information and data regarding fiber-glass woven roving, reference may be made to: Tentative Specification for Woven-Roving Glass Fabric for Polyester Glass Laminates ASTM D2150-63T; Cloth, Glass, Woven Roving, for Plastic Laminates MIL-C-19663B.

6. J. A. Schlarb, *SPI 19th RP/C Division Proceedings,* 1964, Sec. 4–A.

7. Abstracted and adapted by permission of author E. P. Weaver, *SPE Journal* 187–188, (Feb. 1960). *See also* W. K. Fischer, *Mold Tooling Fabrication.*

8. R. L. Huntington, consultant, Van Buren, O., personal communication. *See also* J. H. Mallinson, *Chemical Plant Design with Reinforced Plastics,* McGraw-Hill, 1969.

9. J. E. Browning, *Chemical Engineering,* March 22, 1971, p. 48.

10. K. B. Bozer, et al., *SPI 26th RP/C Division Proceedings,* 1971, Sec. 2–C.

11. Excellent corrosive media versus temperature data are available from suppliers of the specific corrosion-resistant and fire-retardant resins.

12. Permastran® fiber-glass composite pipe discussed in Johns-Manville Products Corp. Technical Bulletin TR-593-A. *See also* Chapter V-1 in this text.

13. B. I. Zolin, *Chemical Engineering Progress* **66,** no. 8 (August 1970).

14. E. M. Barrantine et al., *SPI 26th RP/ Division Proceedings,* 1971, sec. 2-D. S. M. Draganov, et al., *ibid.,* Sec. 2-E.

15. Dow Corning Corp., Technical Bulletins for nos. 6 and 11 paint additives.

16. P. W. Hill and K. G. Lefevre, *Chemical Engineering Progress* **66,** no. 8 (Aug. 1970).

17. R. McMahon, Ceilcote Co., Berea, Ohio, 2 April 1972: personal communication. *See also* United States Department of Commerce, Bureau of Standards, NBS Voluntary Standard PS-15-69 (June 1970); and the joint symposium of SPI, SPE, and NACE *Problem Solving with Plastics* (Houston, National Association of Corrosion Engineers, 1971).

18. General information in this section is from J. H. Mallinson, SPI 26th RP/C Division Proceedings, 1971, sec. 2-A.

19. *Ibid.*

20. *Ibid.*

21. Steel Tank Institute, Chicago, Ill., DTL Report no. 711159-G (January 29, 1968).

22. G. Smoluk, "Are You a Professional?" *SPE Journal* **27,** No. 3 (Mar. 1971).

Further References

A wealth of information on hand lay-up is available in the literature. Among the sources for data on processing, performance, and properties of molded parts are the following:

Specification Title or Reference	Number or Source
Commercial Standard — Gel Coated Glass Fiber-Reinforced Polyester Resin Bathtubs	CS-221-59
Commercial Standard — Gel Coated Glass Fiber-Reinforced Polyester Resin Shower Receptors	CS-222-59
Fiber Glass and Glass Laminates	OTS-CTR-292 and Supplement
Plastic Laminates, Fibrous-Glass Reinforced, Marine Structural	MIL-P-17549C
Plastic Materials, Glass-Fabric Base, Low-Pressure Laminated, Aircraft Structural	MIL-P-8013
Procedure for Use in Preparing and Installing Reinforced-Plastic-Coated Steel Underground Petroleum Storage Tanks	Steel Tank Institute Specifications (2 available)
Reinforced-Resin Fabrication Ducts, Hoods, and Stacks	E. I. duPont Co. Specification SW-10-S
RP/C for Bathroom Components	19th SPI (1964) Sec. 15-B
Plastic Material, Laminated, Thermosetting, Sheets, Glass-Cloth, Melamine-Resin	MIL-P-15037A
Plastic Sheets, Laminated, Thermosetting, Glass-Mat, Melamine-Resin	MIL-P-17221C
Plastic Materials, Heat-Resistant, Glass-Fiber Base, Polyester Resin	MIL-P-25395
Low-Pressure, Epoxide-Resin Laminates	MIL-P-25421
Glass-Fiber Base, Phenolic Resin, Low-Pressure Laminates	MIL-P-25315
Plastic Materials, Low-Pressure Laminates, Glass-Fiber Base, Silicone Resins	MIL-P-25518

Specification Title or Reference	Number or Source	Specification Title or Reference	Number or Source
Resin, Phenolic, Low-Pressure Laminating	MIL-R-9299	Accelerated Corrosion Testing of RP/C	22nd SPI (1967) Sec. 16-D
Resin, Epoxide, Low-Pressure Laminating	MIL-R-9300A	Vitro-Strand® Fiber in Improving Marine Surfaces	22nd SPI (1967) Sec. 17-D
Resin, Polyester, High-Temperature Resistant, Low-Pressure Laminating	MIL-R-25042	RP/C Coating on Plywood	22nd SPI (1967) Sec. 19-A
Design and Stress Analysis on Road and Rail Transport Trucks	20th SPI (1965) Sec. 18-F	RP/C Silos for External Storage of Specific Products	22nd SPI (1967) Sec. 19-C
Fiber-Glass, Gel-Coated Bathtubs, Shower Units: Design, Analysis, Economics	21st SPI (1966) Sec. 2-C	Analysis and Cause of Variability in Hand Lay-up	22nd SPI (1967) Sec. 19-D
World's Fair Structures	21st SPI (1966) Sec. 2-D	SPI Specifications for Corrosion-Resistance Testing	SPI Committee on Corrosion-Resistant Structures
Suggested Glass-Fiber Fabric System	21st SPI (1966) Sec. 10-A	Chemical Resistance of Thermosetting Resins Used in RP/C	ASTM C581-65-T
Double-Shuttle Weaving Techniques	21st SPI (1966) Sec. 10-C	A Discussion of the Standard, Including Chemical-Resistant Construction, Physical Properties, and Laminate Quality	23rd SPI (1968) Sec. 8-B
Roller Gun for RP/C Lay-up	21st SPI (1966) Sec. 15-C	A Discussion of the Standard Relating to Reinforced-Polyester Round and Rectangular Duct Design	23rd SPI (1968) Sec. 8-C
Corrosion in an RP/C Polyester-Resin Tank	21st SPI (1966) Sec. 16-D	A Discussion of Those Portions of the Standard Relating to Design Considerations for Reinforced Fiber Glass Pipe	23rd SPI (1968) Sec. 8-D
"Custom Contact Molded Reinforced-Polyester NBS Voluntary Product Standard PS 15-69 Chemical-Resistant Process Equipment"		A Discussion of Those Portions of the Standard Relating to Tank Design	23rd SPI (1968) Sec. 8-E
RP/C in Condenser Water Boxes, Heat Exchanger Cover, and Other Heavy Plant Equipment	21st SPI (1966) Sec. 20-A	Future Work of the SPI Reinforced-Plastics Corrosion-Resistant Structures Subcommittee	23rd SPI (1968) Sec. 8-F
Glass-Reinforced Plastic in Freight Container	21st SPI (1966) Sec. 20-D	Chemical Resistance of Polyester Resins Cured under Field Conditions	23rd SPI (1968) Sec. 12-A
Paintable Primer Gel Coat for RP/C	22nd SPI (1967) Sec. 1-D	Advanced Design Considerations in RP/C Process Equipment	23rd SPI (1968) Sec. 12-E
RP Bathroom Modules for Hospital Modernization [sic]	22nd SPI (1967) Sec. 2-A	Properties of Resilient Polyester-Glass Composites and Application of this Data to Design of Corrosion-Resistant Equipment	23rd SPI (1968) Sec. 13-A
RP/C in Mexican Architecture	22nd SPI (1967) Sec. 2-B	New Semi-Rigid, Corrosion-Resistant Resins	23rd SPI (1968) Sec. 13-D
Reinforced Plastics in Multi-Story Buildings	22nd SPI (1967) Sec. 2-C	Fabrication of Large RP/C Trawlers	23rd SPI (1968) Sec. 15-A
Buildings in Barrels	22nd SPI (1967) Sec. 2-E	Training Course for U.S. Navy Reinforced Plastics Boat Construction	23rd SPI (1968) Sec. 15-B
Design and Construction of Boats to 120 ft. in Length	22nd SPI (1967) Sec. 11-A		
Reinforced Plastics in U.S. Navy Minesweeper	22nd SPI (1967) Sec. 11-B		
Reinforced Plastics Extended to Advanced Ship Structures	22nd SPI (1967) Sec. 11-C		
Building Boats for a Market	22nd SPI (1967) Sec. 11-D		
Surfacing Mats vs. Surface Deterioration in RP/C	22nd SPI (1967) Sec. 16-D		

Specification Title or Reference	Number or Source
Evaluation of Fiber-Glass Construction for Commercial Fishing Hulls	23rd SPI (1968) Sec. 15-C
Flammability Characteristics of Polyester Resins Based on Tetrabromophthalic and Tetrachlorophthalic Anhydrides	23rd SPI (1968) Sec. 19-A
Effect of Organic-Peroxide Molecular Structure and Environment on Polyester-Resin Cure	23rd SPI (1968) Sec. 19-C
Accelerators for Organic-Peroxide Curing of Polyesters and Factors Influencing Their Behavior	23rd SPI (1968) Sec. 19-D
Gibbs and Cox, Engineers, Marine Design Manual, New York, McGraw-Hill, 1960.	

II-2

Spray-Up

INTRODUCTION

Usage. The spray-up method is used interchangeably with hand lay-up, although several differences exist. It may supersede hand lay-up in cases of exceptionally large molds where prepared reinforcement would be too heavy or bulky to handle. Spray-up is also much faster and employs unique equipment which makes possible fresh combinations of glass and resin at the work site. Part requirements for spray-up are less stringent than for hand lay-up. Process variables do not permit control equivalent to that possible in hand lay-up. Hence, product uniformity is not as reliable and does not approach the same close tolerances.

Examples: Spray-up is readily adaptable to coating applications and may be automated. It may also be used to repair and fabricate metal-tank coatings, to line interior and exterior walls of buildings, and to line walls of boxcars and trucks and at-the-site for swimming pools and corrosion-resistant structures. Spray-ups can be used further for panels, machine guards, bathroom fixtures and component units, motor homes, trailers, automotive replacement service parts (fenders, engine hoods, truck decks, and other items).

MATERIALS

Mold Releases. Essentially the same as those used for hand lay-up.

Gel Coats. The same as those applied by brush or standard spray equipment for hand lay-up.[1]

Resins and Curing Systems. The resins used in spray-up are polyesters containing a styrene monomer for cross-linking reactivity. They are diluted to achieve sprayable viscosity but contain thixotropic inducing chemicals or particulate fillers to prevent run-off from vertical surfaces. Table II-2.1 provides a summary of typical resin properties. Table II-2.2 lists gel times for resin blends for two-pot systems.

Although polyester resin promoted with cobalt napthanate plus accessory promoters, such as di-ethylaniline or di-methylaniline, and catalyzed with methyl-ethyl-ketone peroxide constitute the largest percentage of room-temperature resin-curing systems, others are available:[2]

Benzoyl peroxide catalyst (1.0% used with dimethyl aniline promoter (0.05 to 0.10%)

Other promoters, such as:
  Colbalt metal dispersions
  Vanadium pentoxide dissolved in phosphates
  Manganese compounds
  Teritary amines
  Quaternary ammonium salts

Used with other catalysts, such as:
  Hydroperoxides
  Peresters
  Diacyl peroxides

Diluents. Various diluents may be used as carriers for the methyl-ethyl-ketone peroxide usually used. Diluents act to prevent excessive loss of and contamination by the catalyst in the air during spray. Being reactive monomers, they copolymerize during gelation and cure.

Table II-2.1 Types of Spray-Up Resins

Type	Appearance	Viscosity, poise at 78 F	Specific Gravity at 78 F	% Polymerizable	Stability	Use
Thixotropic resilient promoted	Cloudy, pink in color	3 to 6	1.1	100	Uncatalyzed[a]—3 mo.	Singly as material for one-pot spray-up systems or with resin no. 2 in equal amounts in two-pot system as "cold" side resin (would require addition of extra promoter by mfgr. for use in two-pot system)
Thixotropic resilient unpromoted	Cloudy, straw-colored	3 to 6	1.1	100	Uncatalyzed—3 mo. Catalyzed at 78°F: 2% MEK—12 hr 1% MEK—22 hr ½% MEK—45 hr Stability at 90°F: reduce 78 F values by 50% Stability at 65°F: increase 78 F values by 50%	With resin no. 2 in equal amounts in two-pot systems as "hot" side resin (containing catalyst only)

[a]For catalyzed stability, see Table II-1.5 in Chapter II-1 (Hand Lay-up) which provides gel time variations with temperature and percentage of catalyst added for typical performance of promoted resin. This applies to use of single-pot system.

Table II-2.2 Gel Times for Equal Quantity Resin Blend[a] for Two-Pot Systems

Resin Ambient Temperature (°F)	% MEKP Added to Resin No. 2[a]	Gel Time (min.)
65	1.0	69
65	1.5	44
65	2.0	30
65	3.0	22
65	4.0	18
77	0.5	60
77	1.0	28
77	1.5	19
77	2.0	14
90	0.5	36
90	1.0	19
90	1.5	12

[a]Equal quantity blend of thixotropic resilient promoted resin (no. 1) and thixotropic resilient unpromoted resin (no. 2).

Notes: Cure time is tentatively reached when Barcol hardness of the laminate produced reaches 30 to 35.

Glass-fiber content has little if any effect upon gel times, provided it is held at shop temperature at the start of the spray-up operation.

Laminates from room-temperature cure systems do not arrive at full cure (when left at room temperature) until 30 to 60 days after gelation.

Table II-2.3 presents the physical properties of diluents.

The optimum ratio or diluent to catalyst is 2:1. Using this ratio, Table II-2.4 records the effect of diluents on the gel and cure times and the peak exotherm. Table II-2.5 measures the effect of a 10:1 ratio of diluent on the gel and cure times.

Glass-Fiber Reinforcement.

Glass fiber in roving form comprises the sole reinforcement for spray-up. Material is supplied in 35-lb multi-end packages, and 1, 2, or 3 are fed into the spray-up gun.

The major requirements of the reinforcement are:

- High ribbonization for traversing boom and guide eyes without excessive fuzzing.
- Complete break-up into individual strands when chopped for rapid and thorough wet-out and dispersion into the resin without trapping air and without allowing excessive resin drainage, especially on vertical surfaces.
- Non-inhibition of cure.
- Development of full physical properties of laminate when cured (requires at least 30 days at room temperature): (See Table II-2.6.)

EQUIPMENT AND TOOLING

Spray-Up Equipment[3].

Requirements for Physical Size and Arrangement of Spray-Up Equipment. All equipment on the market today will have a gun and chopper of approximately equal size because of the limitations of the operator and the weight he can handle during a day's production. There is a wide variation in the size of the pumps and catalyst-dispensing devices, and this is dependent upon the size of the part being made, the production rate, and the limitations of factory space.

Rates of resin/glass deposition are as high as 30 lb per min. in the large systems, and an average of 8 lb per min. for the small shop with the most popular system. The rate is dependent upon the men available to roll out the deposit and the gel time.

A hood or well ventilated area is recommended for spray-up systems that use air to atomize the catalyst and to mix the ingredients. Systems that use any form of airless mixing, either internal or external, are better and can be used without a hood or exhaust fan. The best system for freedom from spray hazard is the system that is airless and internally injects the catalyst and mixes it before the mixture leaves the nozzle tip.

Glass-conveying guides should be as few as possible and yet get the roving to the chopper without snags. Care should be taken to set the equipment so that the roving path is the same for the complete span of the boom. This insures an even cutting speed. If the roving has a different path, it will have a different drag. Guides should be as large as feasible in inner diameter and funneled if necessary to create and maintain proper tension.

The location of the catalyst injector and the drape and length of the catalyst-air hose are vitally important in the systems using this form of catalyst introduction (Fig. II-2.1). Distances of more than 25 ft, and loops (catenary) will cause pools of catalyst to collect in the low points in the static condition and give erroneous ratios when the system is started up. Also, concentrated charges of catalyst come through the nozzle. The solution to this condition is to bleed the system for a few seconds and then introduce the resin.

Location of the glass package should be as close to the chopper as possible to avoid variations in deposition rate.

Table II-2.3   Physical Properties of Diluents

Diluent	Abbreviation	Boiling Point (°C)	Vapor Pressure (mm.Hg/20°C)	Flash Point (°F)	Evaporation Rate[a]	Specific Gravity (d 20)	Viscosity cps/20°C
Ethyl acetate	Et Ac	77	76	56	4.2	0.90	0.5
n-Propyl acetate	Pr Ac	102	25	65	2.2	0.89	0.6
n-Butyl acetate	Bu Ac	126	8	90	1.0	0.88	0.7
Acetone[b]	Acetone	56	186	15	7.7	0.79	0.4
Methyl ethyl ketone	MEK	80	70	38	4.6	0.81	0.4
Methyl isobutyl ketone	MIBK	116	15	74	1.6	0.80	0.6
Ethylene glycol mono-methyl ether	EGMME	124	6	110	0.5	0.97	1.7
Dimethyl phthalate	DMP	282	13/150°C	300		1.19	14/25°C
Dibutyl phthalate	DBP	340	1/150°C	340		1.05	16/25°C
Butyl benzyl phthalate	BBP	370	0.16/150°C	390		1.12	40/25°C
Di-2-methoxyethyl phthalate	DMEP	340	1/150°C	385		1.17	32/25°C
Tricresyl phosphate	TCP	420	0.5/200°C	470		1.16	103/25°C
Tris-B-chloroethyl phthalate	t-CEP	230°/10mm.	0.0	475		1.43	21/25°C
Diallyl phthalate	DAP	300	2.4/150°C	330		1.12	12/25°C
Dibutyl fumarate	DBF	285		285		0.99	4/100°F
Dibutyl maleate	DBM	281		285		0.99	4/100°F

Source: N. S. Estrada and J. D. Malkemus, SPI 20th RP/C Division Proceedings, 1965, Sec. 2-G.
[a]Relative to n-butyl acetate at 1.0.
[b]Included for comparison purposes only; not recommended.

Table II-2.4   Effect of Diluents on the Gel Times and Peak Exotherm Temperature
(SPI Procedure, 180°F)

| | | SPI Gel and Cure Test[a] | | |
| | Gel Time | | Peak Exotherm | |
Diluent	150-190°F		Time	Temp. °F
None	3′ 50″		12′ 00″	275
Ethyl acetate	3′ 55″		12′ 30″	235
n-Propyl acetate	4′ 05″		13′ 00″	244
n-Butyl acetate	4′ 00″		13′ 00″	238
Methyl ethyl ketone	4′ 30″		13′ 00″	232
Dimethyl phthalate	4′ 00″		12′ 30″	247
Dibutyl phthalate	3′ 45″		12′ 40″	247
Di-2-methoxyethyl phthalate	4′ 20″		13′ 20″	237
Tricresyl phosphate	4′ 10″		13′ 30″	234
Tris-B-chloroethyl phthalate	4′ 05″		13′ 30″	240
Diallyl phthalate	4′ 10″		13′ 00″	245
Dibutyl fumarate[b]	4′ 45″		14′ 05″	240
Dibutyl maleate	4′ 30″		13′ 00″	246

Source: N. S. Estrada and J. D. Malkemus, SPI 20th RP/C Division Proceedings, 1965, Sec. 2-G.
[a]One part of MEK Peroxide "60" or three parts of a 2:1 dilution was added to 100 parts of standard polyester spray-up resin.
[b]When this solution was mixed with the polyester, the mixture was slightly hazy and it became very hazy after the peak exotherm.

It is good practice either to locate the spray-up system so that the molds are moved past the spray point, thus limiting any chances for deposition variations, or to arrange the molds peripherally around the unit so that the operator can pass from one to the other with the roller men following. In the later installations, such a set-up easily permits multiple passes; by the time the operator makes the full circle, gelation has occurred on the first part and he can continue for the next layer of spray-up.

Resin-Handling Equipment. Resin-dispensing systems are numerous and represent the preferences of different manufacturers of equipment. They can be described by the manner in which the catalyst is introduced.

Two-Pot System. Twice the percent of catalyst is mixed into one half of the resin, and twice the percent of promoter is mixed into the other half of the resin. The two components can then be:

1. Pumped at low pressure and sprayed through separate airless nozzles to converge and mix on route to the mold surface. (see Figure II-2.1-A).
2. Pumped via air pressure into a tank where they are mixed with air in separate nozzles, converged in mid-air, mixed, and conveyed to the mold surface.
3. Combinations of both of above.

One-Pot, Catalyst Injection System. The resin is promoted with the regular amount of catalyst (less than 1%) and pumped via direct air pressure or a high- or low-pressure piston

Table II-2.5   Effect of 10% of Diluent on the Gel and Cure Times
(100 g, Pot Cure Tests at 25°C[a])

Diluent	Gel Time	Cure Time
None	13′ 15″	29′ 30″
Ethyl acetate	27′ 10″	93′ 00″
Acetone	34′ 00″	120′ 00″
Methyl ethyl ketone	29′ 10″	93′ 00″
Ethylene glycol mono- methyl ether	15′ 20″	48′ 00″
Dimethyl phthalate	15′ 45″	40′ 30″
Dibutyl phthalate	17′ 50″	45′ 30″
Butyl benzyl phthalate	16′ 50″	39′ 00″
Tricresyl phosphate	19′ 50″	45′ 00″
Tris-B-chloroethyl phthalate	20′ 30″	44′ 15″
Diallyl phthalate	18′ 10″	44′ 05″
Dibutyl fumarate	21′ 30″	46′ 00″
Dibutyl maleate	21′ 30″	53′ 00″

Source: N. S. Estrada and J. D. Malkemus, SPI 20th RP/C Division Proceedings, 1965, Sec. 2-G.
[a]In these tests, the components were mixed in the following order:
First 100 g. standard, general-purpose polyester spray-up resin, then 0.5 g. 6% cobalt naphthanate, 10 g. diluent (except in case of blank) 1 g. MEK perioxide "60".

Table II-2.6   Typical Properties of Glass-Fiber Roving for Spray-up

Property	Description
Strand integrity	Should be as high as possible. Individual strands should form a unit thread without individual filaments readily visible or evident.
Chopping properties	Should cut cleanly when passed through the chopping gun. Chopper blades should be frequently replaced to prevent chopping of long lengths.
Dispersion when chopped	Roving should separate completely when chopped into resin stream without clumping, "grass hoppers," or excessive filamentation. No static electricity should be evident.
Distribution on work	Roving is supplied with or without a colored tracer strand. One colored strand in the multiple number of reinforcing strands permits immediate inspection of uniformity and randomness of distribution of the sprayed reinforcement. It also permits the operator to immediately judge whether or not he is maintaining the required laminate thickness.
Wet-out in polyester resin	Rapid and thorough coalescing of chopped glass strands into the resin is desirable. Any "whiteness" of the strands should disappear immediately and not return or persist following gelation and cure. After roll-out, fibers should maintain complete conformation to curvatures and sharp inside and outside radii without "spring back."
Roving package types usually supplied	35 lb, inside delivery, approximately 10¼ in. diameter, 10 in. high, 3 in. inner diameter, 2½ waywind. Shore durometer hardness— 30 to 45. Material should be free from loops, snarls, dirt and any other defects which would interfere with free pull-out or smoothness in chopping through the gun or uniform application to the spray-up work.
Roving yd per lb[a]	With or without tracer—193 to 236. This parameter relates directly to the amount of glass delivered per given unit of time of operation. Hence it should be closely controlled by the glass supplier to maintain desired glass-to-resin ratio on the work.
% ignition loss (of acceptable gun rovings)	Range from 1.0 to 1.4%
% moisture contained[a]	0 to 0.10
Ribbonization[a]	4.0 to 7.0
Stiffness[a]	4.4 to 5.4
Chopping index[a]	0.5 to 1.0

[a]See appropriate roving test procedures available from Johns-Manville Fiber Glass Reinforcements Division, Waterville, O.

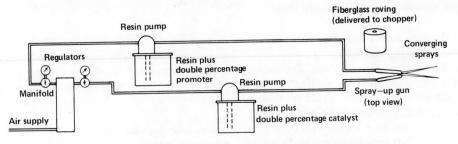

A. AIRLESS TWO—POT SYSTEM, EXTERNAL MIXING

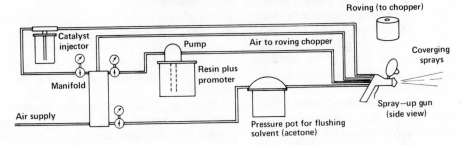

B. AIR ATOMIZATION, INTERNAL MIXING

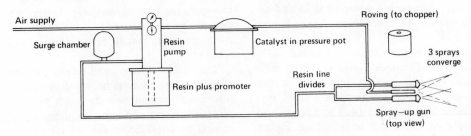

C. AIRLESS ATOMIZATION, EXTERNAL MIXING

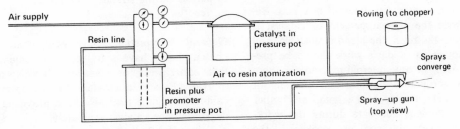

D. AIR ATOMIZATION, EXTERNAL MIXING--SIDEARM CATALYST SUPPLY

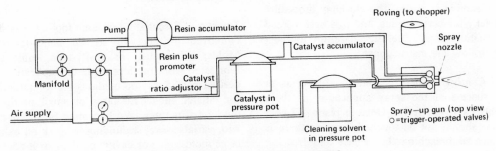

E. AIRLESS ATOMIZATION, INTERNAL MIXING

Figure II-2.1 Schematic representation of five different types of resin-dispensing systems used in the spray-up process.

pump to the gun. Raw catalyst is added using one of the following methods:

1. Air Atomization-Internal Mix. The catalyst is introduced into the air line by way of a venturi. It is then mixed with the resin internally at the nozzle and sprayed onto the mold. (see Figure II-2.1-B).
2. Direct Catalyst Injection-External Mix (3 nozzle). Raw catalyst is pumped under pressure through an airless nozzle between 2 airless resin nozzles. Mixing is accomplished by the converging of the resin sprays onto both sides of the catalyst spray on route to the mold surface. (see Figure II-2.1-C).
3. Direct Catalyst Injection-External Mix (sidearm type). Raw catalyst is introduced into either air-atomized or airless promoted resin spray through a small hollow needle at right angles to the resin nozzle. Mixing is accomplished in mid-air on route to the mold (see Figure II-2.1-D).
4. Direct Catalyst Injection-Internal Mix (airless). Promoted resin is mixed with raw metered catalyst internally at the nozzle and pumped through a small orifice at high pressure (400 to 880 psi) to apply to the mold surface (see Figure II-2.1-E).

Certain precautions are necessary in two-pot systems and in any internal mix gun. Because twice the normal percent of catalyst is mixed into the resin, the blend is unstable, and only enough for a day's work should be prepared. Also, in any internal-mix nozzle there must be scrupulous flushing and cleaning of the complete pump, hoses, gun, and other parts. Flushing is adequate during the workday, but overnight and on weekends, the nozzle must be dismantled and a visual check on cleanliness made. The flushing procedure is tricky for certain guns, and it is best to read the manufacturer's literature carefully and rigidly follow his instructions.

Gel Coating. There is some confusion as to the success of the spray application of these highly catalyzed resins. Once in place, gel coat resins should not be touched, because there is chance of breaching the release and freezing the part to the mold. The main consideration is that the catalyst must be THOROUGHLY and EVENLY DISPENSED WITHIN THE RESIN, otherwise the coat will have areas of over- and under-catalization with resultant hard and soft spots. Internal-mix nozzles are more positive in mixing than the external mix or convergent types.

Glass-chopping Equipment. Glass choppers or cutters are based upon the "Brenner" design, wherein the rovings pass between two rollers. One is rubber or urethane-covered, and the other has a variety of knives protruding from its periphery. This results in the "breaking" of the glass strands as they are folded over the sharp blade edge pressing into the elastomer.

Small air motors are usually employed to drive the chopper, with the motor exhaust vented into the roller area in order to propel the chopped strands toward the resin spray emanating from the gun nozzle. A popular chopper employs throw-away cutter wheels, thus saving man hours and cut fingers resulting from changing small razor blades in most wheels.

The razor blades and elastomer-covered wheel tend to wear rapidly and a routine replacement program will do much to eliminate down time and variations in desposition rate caused by overloading the air motor through worn blades. Most cutter wheels have slots to permit cutting strands into ½, ¾, 1, or 1¼ in. lengths.

General Information. Any equipment purchased should have a trigger arrangement so that either resin or glass can be dispensed individually and alternately for purposes of calibration. Some guns also have provision for dispensing the catalyst separately.

Table II-2.7 lists the major sources for spray-up systems.

Tooling. Essentially the same tooling may be employed for spray-up as for hand lay-up. Molds, either plug- or cavity-type, must be fabricated from masters made to duplicate the desired part. Spray-up methods are a "natural" for inflated molds for large or small unitized outdoor structures. Workers may be lifted up with variable-level scaffolding to work on extra large molds, plug or cavity, outdoor or indoor.[4] The work itself can form the mold or spray-up surface as in coating swimming pools and other items.

Table II-2.7 Sources of Spray-up Systems—1972
(A partial list of systems available throughout the world)

Type of System	Company and Address
Airless, external mix, three axis-automatic, machines	Coudenhove Kunststoffe Maschinen GMBH Vienna, Austria
Many types: airless, two-pot, automatic guns	Binks Manufacturing Co. Chicago, Ill. 60612
Two-pot, external mix	Peterson Products, Inc. Belmont, Calif. 94002
Purchased, Glasmate and Glas-craft Three head, airless, external mix. Air atomize, internal mix, pumps, fittings, etc.	Ransburg Electrocoating Corp. Indianapolis, Ind. 46206 Glas-mate Division Glas-craft Division
Air atomize, internal mix	Robinson Gun-Synthetic Resin, Pty. Sidney and Melbourne, Australia
Dual head, air atomize, high capacity	Spray Bilt, Co. Santa Monica, Calif. 90406
Hydraulic injection, internal mix, three-axis automatic machines	Venus Products, Inc. Kent, Wash. 98031
Hydraulic injection, internal mix, two-pot, full line	Zaco—Apparaten en Instrumentenfabriek Noord-Scarwode, Holland

For inside shop work, molds should be mounted on wheeled carts. They can then be moved in and out of work stations for gel coating, spray-up, cure, finishing, cleaning, and so on. It is easier to move molds than to move most spray-up rigs.

The accepted minimum size for attempting any part in spray-up is 1 cu ft.

Metal molds have been devised and tried experimentally in which facilities were included for rapid heating and rapid cooling. The following times are practical for a 16 in. × 21 in. deep mold;[5]

Gel-coat cure	3.9 min.
Spray-up cure by heat	6.0 min.
Spray-up cooled	3.5 min.

Total cycle time for heated molds was 25.7 min. For the same part with normal cycle time less heating the time was 120 min.

MOLDING METHOD/PROCESSING DETAILS

Spray-up Process. The following steps comprise the procedure for spray-up:

1. *Preparation:* Prepare and clean mold. Apply release agent and then apply gel coat. Allow to cure.

Prepare for spray-up by checking glass, resin, and catalyst delivery systems:

- Check amount of glass chopped per minute or fraction thereof.
- Check amount of resin delivered in a 10- or 30-sec interval and compare. Strive for glass/resin ratio of 25:75 nominal.
- Check catalyst metered in actual quantity, or check gel time of resin (the latter is preferable).

(Refer to temperature versus gel time curves for hand lay-up in Chapter II-1).

2. *Spray-up:* Spray resin and glass onto mold or work. Check distribution of sprayed material with red tracer in roving, colored mold background or fugitive dyes in the resin. (See Figure II-2.2 for sequence of spray-up procedure in fabrication of RP/C tub and shower-stall units.)

If application of a non-patterning, filled resinous layer between the gel coat and lay-up is desired, a time-saving system has been worked out. It is possible, as described previ-

Figure II-2.2 Photographic sequence illustrating a typical production-line spray-up operation to fabricate tub and shower bath modules. *Upper left:* mold-cleaning and preparation for molding; *Upper right:* spray-up over mold and white gel coat which was previously applied; *Middle left:* roll-out operation to coalesce fiber glass and resin and to densify; *Middle right:* application of stiffeners and spray-up coverage; *Lower left:* removal from mold; *Lower right:* final inspection and finishing. (Courtesy of Universal Rundle Corp., New Castle, Penna.)

ously, to apply over the cured gel coat either a layer of resin (0.012-0.020 in. thick) containing 20 to 40% of a free-flowing, non-balling type of short-length reinforcement (Vitrostrand® for example) or a skin-coat of short-length chopped fibers. When cured, this layer possesses higher dimensional stability than unfilled resin and lessens resin shrinkage around the major fibers. It is the shrinkage which contributes to patterning, that is, showing through of the major fiber pattern or weave into the outer gel coat surface after final cure and removal from the mold. Although this separate protective layer is highly desirable for work in which the smoothest possible exterior finish is required, both the extra labor required for application and the curing time are economically undesirable. However, the same non-patterning protection without the extra labor step may be accomplished as follows:

- Arrange for extra on resin pot or pots to contain regularly used spray-up resin to which approximately 20% of the non-balling fiber has been added.

- Start the first spray-up pass over the cured gel coat using this filled resin sprayed together with the chopped fiber-glass reinforcement.

- For the second spray-up pass, switch to the tank containing the normal, unfilled spray-up resin. If desired, or for insignificant extra gel coat, the filled resin may be employed for the entire spray-up operation.

Make one spray-up pass approximately $\frac{1}{16}$ in. thick and roll out. Make a second pass at the same thickness and roll out. Avoid shifting or disruption of the wet glass-resin mass while rolling out (this is fundamentally important for the retention of uniform physical strengths in the laminate), but strive to remove air voids as thoroughly as possible. Check spray-up thicknesses with needles or wire stuck through a cork for an inexpensive depth gage. Keep the spray-gun handy and use it to add a burst of resin and catalyst to dry or starved resin areas. Keep grounded, covered, metal, solvent tanks handy for cleaning rollers and brushes. These containers are commercially available and have two or three dividers providing three or four compartments. Put clogged rollers or brushes in the dirtiest solvent section, progressing to cleaner sections. Usually the cleanest material is kept for hand cleaning since slightly dirty solvent will leave hands sticky.

3. *Roll-Out Procedure.* Roll-out is a very important step because it consolidates, coalesces, densifies, and orients the reinforcement. Poor roll-out can induce structural weakness through dislocation of fibers or starving of resin. Good roll-out must be practiced in three stages:

1. Distribution of Materials. Roll-out contributes some mixing of resin and catalyst, especially when external mix nozzles are used. Brushes and rollers should be used together. Use the brush with a stipple, not a wiping, action. Rollers may be made from mohair, felt, similar to household paint rollers, or from plastic or metal discs. Rollers will gently push extra resin onto resin-starved areas. Brush stipple action is also good for pressing the mix into corners.

Roller handling is critical. Skidding and slipping can displace reinforcement and should therefore be avoided. Remember that the major strength of the laminate depends on the *fiber* orientation and density, not on the resin.

2. Consolidation and densification. This means that the fibers are laid down into the resin and onto the mold, and also that the entrapped air is worked out through the fibers and resin escapes from the surface of the mix. In order to force the chopped fibers and resin to completely coalesce, use commercially available grooved or serrated rollers and brushes of a diameter, width, and shape consistent with the mold surface. Contoured plastic disc rollers are available to match curvatures.

Small-diameter rollers work best with fabrics, and larger rollers should be used with chopped fibers and mats to avoid skidding (mats and woven roving are frequently laid-up over spray-up).

Use the roller in a criss-cross pattern when densifying. Rapid movement with light pressure seems to do a better job than slow, heavy pressure. When rolling, use lighter pressure pushing away and heavier pressure on the return stroke. Better shops train their men to use rollers in both hands, thus getting faster wet-out with less fatigue.

3. Touch-up. This is usually done with a small roller and large brush (after steps 1 and 2). The brush will push down fibers where the roller may pick up and tangle them.

Small rollers, especially corner types, can be used much faster than a brush if carefully placed to densify and touch up extensive sharp-corner areas.

Rollers should be well cared for in order to maintain long life. Do not throw rollers because the fins may become bent or marred, causing them to grab and tangle the fibers, thus reducing their effectiveness as a reinforcement.

Clean the roller directly after use with a brush and acetone. This is easier than trying to clean it after the resin has cured. If resin has set up, however, use a strong solvent, such as paint remover or methylene chloride, or burn resin off with a blow torch (not welding torch), being careful not to burn the thin fins (in metal rollers only). Keep a wire brush handy and remove the resin as soon as it becomes soft.

Saturate the felt and mohair rollers in at least two cans of acetone—the first, dirty acetone for the major resin removal, and the second, rather clean acetone to remove the remaining resin. After a thorough rinse, "milk down" the roller between the thumb and first finger, and spin the roller to remove the last traces of acetone. It is sometimes cheaper on big jobs to throw away the felt or mohair rollers.

4. *Additional Reinforcement.* Reinforce the *inside* corners on molded parts because these will be the *outside* corners of the product. Add provisions for removal of part from molds. This is especially important on large, complicated parts, because they are flexible, hard to grasp, and easy to fracture. Loops of rope or cable should be affixed by laminating in areas shown by experience to logically withstand stress. Remove part from mold using a hoist or crane. The loops may be ground off later.

5. *Finishing and Trimming.* The finishing and removal from mold are conducted in the same way for hand lay-up. The primary knife trim should be planned while parts are in the mold at the leather-hard cure condition. Actual removal takes place before the Barcol hardness reaches 30 to 35.

The final trim of the hardened spray-up is accomplished using carbide-tipped drills, tungsten-carbide or diamond-granule faced discs, saw, and routers—all now commercially available.

Tips on Quality Control. In addition to glass-resin and catalyst-metering and thickness-gaging, the following are helpful in the surveillance of the spray-up operation that yields a consistent quality product.

Glass/Resin Ratio. Resin viscosity and resin and shop temperatures should be kept as uniform as possible to hold this ratio within a few percent. Glass/resin ratios for spray-up are not as critical or tightly held as those for prepreg or filament winding but should be kept at the narrowest tolerance permitted. This is usually limited by how closely the operator can read the flow meter or move the walking beam.

Quantity of Catalyst. The gun operator may insure uniform part cure and quality by continually checking to make sure that catalyst introduction is being maintained. Watching the black-ball level on the flow meter on some types and watching the sideways surge of catalyst bubble in the loop supply tube on others are about the only ways to know while actually spraying. Some advanced operations have photocells attached to the catalyst-injector bubble tube, and interruption of the beam by a falling black ball will signal the operator that flow has been slowed or stopped.

Factors Affecting Cure. The factors which affect the cure are the catalyst ratio, thickness of laminate, resin-rich pools, shop temperature, and mold temperature. The latter is the most critical.

Thickness. An experienced operator should be able to control the weight from part to part within $1\frac{1}{2}\%$. Thickness will vary despite normal efforts, however, unless the management is willing to gamble a considerable number of parts so that the operator can experiment with his technique until he has mastered the motions and speed of his traverse. Some operations have added timers that are activated by the gun's trigger, causing a horn to blow at intervals with the gun turned off at a certain count. Operators find that better uniformity is attained by planning one pass in a certain direction, and the second pass at right angles to it. Some operators may even plan a third pass, with all passes at 120° to each other.

A good operator controls thickness by keeping the gun at a constant distance from the surface, pointed nearly perpendicular to the surface, and moving at an almost constant sideways rate, regardless of the angle. One of the best ways to approach uniform thickness is to arrange to immediately weigh every part sprayed-up. This instant feedback will help train the operator because his mode of action is still fresh in his mind. The weighing also keeps him on his toes. A good thickness tolerance for a ⅛ in. part would be $+\frac{3}{64}$ in. $-\frac{1}{64}$ in.

Another way to avoid thickness variation is to re-design and re-machine the spud in the resin-delivery nozzle or nozzles in order to correct and eliminate resin overspray.

Resin handling. By all means, any accidental reverse flow of the catalyst or catalyzed resin into the uncatalyzed promoted resin piping or tank should be avoided. This type of accident is possible in catalyst-injection, internal-mix systems, but not at all in dual-or triple-head external-mix systems. Carefully read the manufacture's directions regarding any flushing operations and then follow them to the letter.

Applications of the Spray-up Method. The spray-up method is generally used for large, complex-shaped parts of 50 lb or more, with gel coated surfaces. Maximum production rarely exceeds 1000 per month.

The sprayed-up fiber glass method is a desirable way for a man to enter into the RP/C business without too large a space requirement or too great an economic commitment. He can then proceed to other methods while also maintaining the spray-up business. Some examples of products using sprayed-up fiber glass are: *in situ* coating or structuring, as for corrosion-proofing, boats, and smaller shapes; RP/C forms for concrete products;[6] wall-coating for building renovation;[7] outdoor and indoor swimming pools and buildings;[8] and flat panels using automated spray-up operation.[9]

Spray-up portrays the ingenuity of engineers in adapting a new material, reinforced plastics, for a unique method of fabrication. Further process improvements are underway which promise new levels of uniformity (where the need is great). One example is the automated spray-up of non-uniform parts.

Advantages:[10]
- *Cost.* Spray-up represents the lowest cost for labor and raw materials of any comparable RP/C process. In 1967 dollars, spray-up as coating (0.040 in. thick) equaled $0.45 per sq ft. Hand lay-up equaled $0.46 (possibly more); wallboard was $.20; and plastering was $0.60 per sq ft.
- *Efficiency.* Spray-up is efficient because it lends itself to the "team concept" of manufacturing. One operator can spray while another does the roll-out, manipulates the molds, trims, removes parts, and so on. Different mold shapes or mix can be handled in line sequentially without disturbing the process.
- *Drape.* Sprayed-up fiber glass is unlike the dry, two-dimensional fiber glass prepared mats and fabrics and will follow the contour of the mold without the formation of laps or folds.
- *Part Size.* Spray-up permits fabrication of parts too large for matched-die work and for parts that may have return shapes requiring inserts or other additions. There is really no limitation to the size of a spray-up part (sprayed-up roads is one for example).
- *Waste.* In the reinforcement procedure, there is little waste in parts that are at least 1 ft sq. There are savings on material for parts which have holes in them and for those whose outer contour is other than square. The worst waste occurs when an expensive catalyst is used in the air-atomized, internal-mix systems. The airless, internal-mix systems are most economical in the use of the catalyst, showing a large savings in high production.
- *Local Reinforcement.* Highly stressed areas may be readily built up by the addition of more chopped mat or fabric wet-out with only resin. Metal tapping plates, stiffeners, nailing inserts, and other parts can be easily sprayed in as desired.
- *Speed of Production.* Because there is no need to mix the catalyst separately, this operation is eliminated. Spray-up is therefore faster than hand lay-up. Furthermore, higher concentrations of catalyst are feasible, thus speeding the cure and getting a better cure as a bonus.
- *Wetting.* Better wetting of the fibers is possible because of the driving action of the spray—especially with the airless systems—and because the chopped fibers are surrounded by resin on the way to the mold surface.
- *Automation.* Spray-up systems automated for flat-panel work have been used successfully, especially in combination with auto-

mated spray-up of contoured surfaces. (See Process Variation 2.)

- *Fillers.* Fillers and/or aggregates can be added between spray-up skins to form cores for insulation, chemical resistance, and other types of reinforcement. Some frequently used fillers are sand, cork, vermiculite, balsa, marble chips, and foam.
- *Surface.* Part surfaces rarely show the pattern or print of any fabric reinforcement effects because random mats can be applied behind the gel coat.

Disadvantages
- *Thickness Control.* One of the most difficult things to control in spray-up is thickness. Control can only be achieved by trained and skilled operators working in conjunction with well-trained roll-out personnel. The thickness can be checked with probe gages and, secondarily, by color or weight after trim.
- *Strengths.* Ultimate strengths in spray-up are lower and have wider variations than those in hand lay-up because spray-up reinforcement is controlled by the operators, and hand lay-up utilizes pre-made materials such as mats and fabrics.
- *Safety and Cleanliness.* The overspray and free catalyst fumes in the air-atomizing systems make it necessary to do the work either in or near a fan-exhausted spray booth. Paper face masks are recommended and should be frequently changed. In airless systems, especially the internal-mix types, free solvent, catalyst, and styrene fumes are at a minimum. Common practice shows that workers are willing to spray and roll down with only the masks, rarely requesting or needing exhaust-type spray booths. The fume level of airless spray is very low. Human tolerance levels of styrene, etc., should be continually checked using appropriate air-sampling equipment.
- *Planning.* Spray-up can be uneconomical if utilization of the machine is low. The work mix and the area should be planned so that the operator and his roll-out crew can move from mold to mold, keeping the equipment working most of the time. It is advisable to keep spare sets of critical parts, gun, resin pump, seals, hoses) ready for replacement with minimum down time, should a malfunction occur.

Generally speaking there are many improvements and additions which can be realized in the existing spray-up machines and guns. Some examples[11] are:
- A foolproof test for and indication at the gun of the presence of the correct amount of catalyst.
- Oscillating, roving eyelets at the cutter which will even out the wear on the blades and greatly increase their use between replacement.
- A visual timer on the gun so that the operator knows how much time he has left to spray. (Use nixie tubes and a digital readout).
- Scales on the glass supply set to cut off the gun when a predetermined amount of reinforcement has been applied.
- Audio instructions piped to the operator's ear which will tell him what to do as he sprays—an excellent idea for the operator doing several different jobs during the day. The instructions can remind him of the operation, as each mold arrives at the spray-up station.
- An automatic brake on the rovings so that loops at the gun do not form when the cutter is stopped.
- A back-pack roving supply which would allow the operator to move about with no tangling of the roving.
- A plug-in, replaceable chopper for standby and quick change in the event of fouling of the roving strand or dull cutter blades.

There are many other devices, but the purpose of this list is to point out that it is desirable to eliminate as many possibilities for variance of the product as can be found. Spray-up is still an art (1973) and as such is still dependent upon the skill of the operator.

Spray-up Process Variations. The following two variations of the spray-up process help to simplify it and to promote economy: (1) spray-up into a thermoformed thermoplastic shape, a procedure generally termed "rigidizing," and (2) complete automation of the spray-up operation. The following discussion treats these processes and gives several examples, advantages, and disadvantages.

Spray-up into Thermoformed Thermoplastic Shell—Rigidizing (Variation 1) This process was presaged by developers in the United Kingdom[12] who originated substantial improvements in applications for corrosion resistance in chemical plants, refrigerator cars, and other items. They applied reinforcing, wet fiber-glass-polyester lay-up onto rigid polyvinyl chloride which had been previously shaped by welding, bending, or other fabrication means. The PVC became the external or "exposed" surface. The liquid polyester resin contained a modifer which promoted adhesion between the PVC and the cured laminate. Also in fabricating, a resin-rich layer was first applied to the PVC substrate to avoid dry, unbonded areas.

In the current, surviving process variation, licensable as a process for "rigidizing" thermoplastic shapes, acrylic or other thermoplastic material is thermoformed to provide a shell or receptacle in which lay-up or spray-up may be done.[13] Unmodified acrylic is the preferred thermoplastic, providing the best post-cure bond and surface resistance to abrasion. Modified (plasticized) acrylic is also used, however. The shell serves both as mold and as gel coat, since it amalgamates with the polyester spray-up material and becomes part of the finished item. Cure, of course, proceeds via room-temperature mechanisms. The steps in this process are sequentially illustrated in Figure II-2.3.

Additional applications of this variation are found in products such as: houseboat hulls, boat bodies and hulls, light-weight bathtubs, sinks and lavatories, dune buggies, locomotive windows (fabricated components), railroad-car door liners, refrigerator-car and truck-trailer

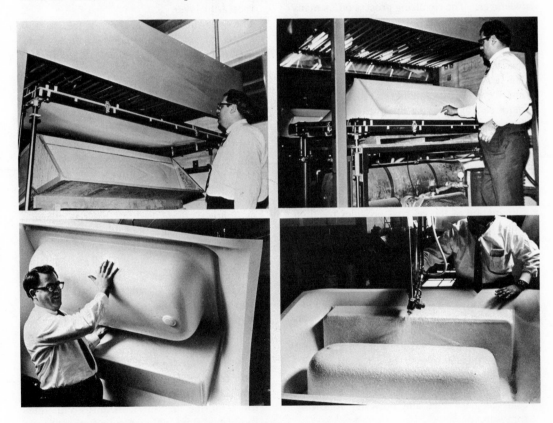

Figure II-2.3 Photo sequence illustrating spray-up into or over a vacuum formed thermoplastic shell (rigidizing). *Upper left:* An acrylic sheet is clamped in a frame, heated under electric-resistance heating rods to the proper softening point, and then lowered over the mold. *Upper right:* A full vacuum is drawn and the softened thermoplastic sheet assumes the shape of the outside of the mold. Exact dimensional details are picked up and surface texture may be embossed into the mold for pick-up onto the part if desired. *Lower left:* The formed sheet—now a bathtub—is quickly cooled and easily lifted from the mold. *Lower right:* This view shows the acrylic sheet serving as both mold and permanent surface coating for the spray-up operation. Applied using standard spray-up equipment are chopped fiber-glass spray-up roving, plus polyester resin containing glass spheres and modified with a special monomer. (Courtesy of Federal Huber Corp., Plano, Ill.)

facing for insulated linings, trailertops and extensions, trailer bodies, marine tool cabinets and chests, agricultural feeding stations and equipment, sports equipment, automotive replacement parts (fenders and other components), furniture, freight pallets, tote boxes, and concrete casting forms.

Although finished part size is somewhat restricted by the current limitation of the largest available thermoforming equipment (8 ft × 30 ft), acrylic sheet stock is purchasable in widths up to 112 in. Many fabricators are therefore, taking advantage of the availability of this process and are looking for future expansions in possible products and technology improvements.

Advantages. The rigidizing process offers many advantages:

- As stated above, the acrylic skin serves both as an inexpensive mold and a gel coat for the finished part. Hence, it is unnecessary for the fabricator to construct molds from normal RP/C or other mold material. All that is required for spraying-up into the vacuum-formed shape is a simple back-up frame. The mold cost is therefore much lower than that in a normal spray-up operation.
- The polyester resin systems available include a general-purpose type, a fire-retardant type, and a water-extendable type for lower cost.
- Elimination of the gel coat provides freedom from the usual gel coat difficulties and defects and also the associated mold release problems present when gel coats are applied directly to the mold.
- Acrylic thermoplastic is innately weather-resistant and should provide performance superior to that of a filled polyester gel coat.
- Waxing for release and stripping for cleanliness of regular molds are not required, and the touch-up and repair of gel coats following molding and cure are eliminated.
- A lesser amount of roll-out is required to coalesce glass fiber and polyester following spray-up.
- Difficult color matching required for gel coats from part to part and batch to batch is eliminated.
- The presence of voids in the RP/C layer under the gel coat is not as critical when using the thermoplastic shell. Voids are more easily removed in roll-out because of the glass spheres in the lay-up resin.

- Many varied and heretofore unachievable decorative effects may be created by embossing the thermoplastic sheet or mold surface. Color streaks and mottled effects may be incorporated by use of the thermoformed thermoplastic skin.
- Acrylic plastic, especially when unmodified with plasticizers, is more highly resistant to abrasion than is a polyester gel coat.
- Properties of the thermoplastic film or skin such as gloss, impact strength, stress cracking, and heat-deflection temperature are improved by the presence of the spray-up backing. Therefore, properties of the spray-up RP/C layer such as weatherability, modulus of elasticity, surface cracking and crazing, and shear strength are enhanced by the thermoplastic layer.
- The following thermoforming rates are possible: 60 per hr for 0.030 in. thick sheet, 40 per hr for 0.060 in. sheet, 24 per hr for 0.090 in. sheet, 30 per hr for 0.120 in. sheet, and 12 per hr for 0.250 in. sheet. The thickness selected is governed by the required final thickness. A 0.060-in. sheet draws down to 0.020 in. in a mold 3 ft deep. A shallower 1 ft draw-down will permit the use of a 0.030 in. sheet, ending up with a 0.020 in. minimum thickness. In some requirements, a minimum of 0.030 in. is preferred.
- In actual practice, large forms were produced as follows: three bathtub cavities thermoformed at one shot and sprayed up at the rate of 39 per hr. A large boat hull 30 ft long × 8 ft wide × 4 ft deep, vacuum-formed and sprayed up at the rate of 6 per hr. Process time is substantially shortened. The main problems associated with the use of polyester gel coating and molds are waiting for the gel coat to cure and waiting for the mold to be freed for re-use. Gel coat cure may take from 30 to 90 min. and only 1 or 2 parts (at the most for all but the smallest parts) may be turned over in RP/C molds during an 8-hr workshift. With thermoformed skins, only a simple support frame is required.
- Process costs are estimated to be 15 to 30% under those for the gel coat method.
- If both surfaces must be finished, two skins may be vacuum-formed with allowance for lay-up thickness—the RP/C material sprayed-up onto one skin and the second gloved together into the lay-up to bond firmly.

- A lower capital investment is required than for gel coating using polyester molds. A large thermoformer to produce shapes 8 ft × 30 ft costs $11,000, and a spray-up system costs approximately $2,000.
- Other economies realized are the improvement in labor productivity resulting from the "stationing" of work cycles, the reduction of unit tooling costs, and better quality control with probable reduction or elimination of "seconds" and reject parts.
- Lead time between order securement, engineering, mold construction, and part production is shorter than for processes involving gel coating. This factor gets the product on the market fast and also facilitates mid-season model changes and design improvements.
- For further economy, vacuum-formed shapes may be rush-produced and inventoried for later use and time saving. Also, thermoformed shapes may be nested and economically shipped for spray-up backing and final production to sites near the markets, thereby saving weight in shipping.
- Success of this operation of process is indicated by the fact that there were over 60 licensees during the first year or two of operation (1973).

Disadvantages. Conversely, there are several disadvantages of the rigidizing process.
- There is some difficulty encountered in holding exact skin dimensions, because thermoforming is not a precise forming method. Areas of deepest draw are reduced to one fourth of the original sheet thickness, or even less.
- It is difficult to eliminate all voids and resin-rich areas.
- Polyester gel coats are generally glossier than thermoformed acrylic unless the latter is highly polished after processing—a procedure which adds to the cost.
- Problems with adhesion are critical, and no dry areas may be tolerated in the interface between the skin and lay-up.
- A perfect match of thermal expansion between the thermoplastic skin and RP/C spray-up is difficult to establish. A poor thermal expansion may result in part separation or delamination during the period of usage. The expansion of the fiber glass-reinforced polyester resin layer is greater than that of the thermoplastic sheet. This is coun-

teracted by incorporation of 10 to 30% of 30 μ (325 mesh) glass spheres into the spray-up resin, a costly procedure because the beads are expensive and must be silane-treated for highest efficiency in resin bonding. The beads are difficult to work with and require constant stirring and agitation in the resin bath or tank to prevent settling. However, they are reputed to help lock in the reinforcing fibers, improve stress distribution, break up voids and other gas or air entrapments, and lessen the flow of moisture along the reinforcing fibers during service of the molded part.
- Because the polyester resin must be modified with specialized monomeric ingredients for the most satisfactory bonding to the acrylic substrate or shell, general-purpose polyesters cannot be used. The source of resins is therefore limited.
- Spray-up resin exotherm must not be so great that distortion of the thermoplastic shell results. Exotherm is lowered by controlling resin composition, limiting catalyst concentration, and incorporating glass beads and other fillers. The shell is particularly subject to distortion by exothermic heat in the areas of thinnest dimension because of furthest draw-down. These thinnest film areas are also the most vulnerable to the transfer or telegraphing through of the fiber pattern from the polyester spray-up.
- The process of spray-up over thermoformed shells is no panacea and has size, shape, and depth limitations. The technology of the thermoformed spray-up rigidizing process has been further advanced in an effort to reduce process time and produce parts with still greater uniformity and consistency of properties.[14] One result is SMC-type composition (see Chapter IV-2) containing resin, filler, and reinforcement placed between two of the thermoformable acrylic skins. These skins are processed in the thermoforming equipment in the normal manner. The leather-hard, pseudo-B-staged SMC material is softened and flows together with the skins during the heating cycle and draw-down when the vacuum is applied. The part is permitted to dwell in the thermoforming mold long enough to chill the thermoplastic and substantiate cure of the SMC material.

Table II-2.8 presents typical formula for use

Table II-2.8   Formula for SMC

Ingredient	% by weight
Polyester resin*	22.0
Syntactic foam beads‡	12.0
Asbestos floats	0.6
Perlite diatomaceous earth	5.4
Glass spheres—30 μ‡	10.0
Calcium carbide filler‡	28.0
Fiber glass reinforcement	21.0
(short length for high flow)	

*Resin and catalyst are proprietary and controlled by by the licensing activity
‡The total filler content of 56% may be adjusted to produce compound densities between the extremes of 26 pcf (high syntactic foam, low glass, and $CaCO_3$ filler), and 98 pcf density (low syntactic foam, high glass, and mineral filler). The two different densities will be desirable for different types of finished product application.

as SMC between 2 0.080 in. thermoformable acrylic skins.

The major problem with this vacuum-formed SMC version of rigidizing is the avoidance of any out-gassing of the SMC material between the skins once they have been compacted together and the combined vacuum forming and curing process commences. Out-gassing would result from too rapid heating and concomitant boiling of the monomer, excessive heating to heat-shock resin or catalyst prior to final gel and cure, and other causes.

Automated Spray-up Process (Variation 2). This variation involves a system in which the spray-up head is held by a mechanical device programmed to guide it over the mold in a controlled manner so that the material laid down will be of uniform quality.

The beginnings of this process variation were pioneered by such companies as Binks and De-Vilbiss and had definite effect on later work. Another influence was the well-known industrial robot. Mechanical, cam-operated types of robots were used by mass-producers long before the spray-up heads were available to industry.

In 1963, the author designed and built a machine to produce large panels for the Bell System building at the 1964-65 World's Fair in New York City.[15] These panels were as large as 14 ft by 40 ft. They were made on an automatic machine that sprayed-up the reinforce-ment and rolled it down onto flat molds on wheels that passed under a reciprocating head on a trolley bridging the molds. The rollers were mounted to the bridge and oscillated on the laminate. They did a good job of densification. There were problems with the machine, and the greater part of the production was consumated with mat as a reinforcement because the early chopper units were incapable of withstanding the constant use of 8 hr, 6 days per week. The machine was quite sophisticated for the time inasmuch as it had an electric eye connected to an alarm and pointed at the head of the gun. The photocell would sense when glass was not being cut, thus signaling the operator to stop the machine. Also, a tachometer was mounted on the glass cutter which, as the speed gradually lowered to a set minimum, told the operator when to change blades. There was a console containing all of the dials and gages that recorded the required settings for easy location of any malfunctions. Pump pressures, table speeds, reciprocating head speed, and roll-out speed were also displayed. Soon after this machine was publicized, many others generally with improvements,[16] were announced and incorporated into a system which included gel coating, spray-up, roll-down, curing, and trim.

As the containerization market grew, more companies on both coasts adopted such machines for making large panels. In 1970, the Owens Corning Fiberglass Corp.[17] wedded a standard industrial robot[18] with an airless spray-up head that had an automatic trigger and flush valve.[19] The engineers proceeded to teach this robot to spray-up into a small dinghy mold. They then made several dinghies using this robot-spray-up gun combination, with only the roll-down done with human assistance. This effort was reported in a recent discussion of open-mold technology[20] (see Figure II-2.4).

The process of training a robot to spray-up a mold has several steps. It is advisable to use a robot with a tape-deck memory and continuous-path control which is essentially a servo-controlled linear-actuator with several degrees of freedom, depending upon the job to be done. Each degree of freedom is guided by one channel on the magnetic tape. The machine has different modes: learning, replay, run, repeat, retrace, and others. In learning mode, the machine is guided through the process via a "Joy

Figure II-2.4 Automatic spray-up of dinghy hull. A view of the beginning of a spray-up pass on a dinghy hull. The gun is doing the transom area. A knob on the end of the horizontal arm is the "Joy-stick," used by the operator for the initial training of the machine. Resin, catalyst, air, and flush lines are suspended from a common overhead boom. (Courtesy of Owens Corning Technical Center, Granville, Ohio.)

Stick," which stimulates the actuators to follow slight pressure from the training operator and which simultaneously imprints the tape with the motion parameters. On playback or run, the machine follows the direction from the tape, and the operator is free for other functions. With reruns, it is obvious that the machine can then repeat the function without deviation, ad infinitum. The machine can be triggered through appropriate signals on the tape to start, stop, delay, repeat, and for functions such as operating the cutter, flush, and resin dispenser. What is now possible is a machine that will work beside the human helper, doing uniform spray-up every time and relieving the human operator of the boredom and tedium of repeated operations. The machine will make a spray-up and clean itself after each job, needing help only with the roll-down and the changing of molds.[21,22,23]

From the success of the above program, it is easy to predict that there will be a rash of applications of automation to spray-up and that these operations will be more sophisticated than what has been reported to date (1973). Some advantages of this system are:

- Absolutely uniform spray-up from mold to mold.

- Guaranteed production rate.

- Uniform weight of parts.

- No problems with operators, no breaks, no money difficulties.

- With several tapes, the machine can spray-up different sizes and complexities of part.

There are some disadvantages, of course. The high potential production rates are based upon high output from the roll-down men. It is still a challenge to the designers to make a machine that will automatically roll down the spray-up on compound curves.[24,25,26]

The scarcity of skilled operators needed for effective, second-generation production has shown the industry that something must be done. Automation, therefore, has to be seriously considered. In the construction of large products such as the 150 to 200 ft ship, the automatic spray-up head can be used to advantage. A further benefit of automation is that with the uniformity comes a lowering of the safety factor which had to be high enough to still be adequate when the spray-up was poor. A lowered safety factor means lower weight, more effectiveness, and attractive pricing.

PRODUCT DATA

To summarize the spray-up process, information is presented here concerning design rules, finished physical properties, cost estimating, and new developments in spray-up.

Design Rules for Spray-Up of RP/C. In order to insure desired characteristics after molding with the spray-up RP/C method, the following design criteria should be followed.[27]

Draft. Negative draft can be used where the part is large and could be flexed off the mold, or where the mold is sufficiently flexible. Otherwise, 1 degree of draft especially on male mold forms can be incorporated. A 1-ft cube is the smallest feasible shape for spray-up.

Stress Distribution in Part. Tapered thickness of wall sections and built-up radii may be easily built in so that stresses are distributed evenly. Different glass/resin ratios may also be incorporated within the same part where they are needed in high-stress areas.

Fittings, Stiffners, Attachments. Embedded attachment fittings or stiffeners of wood, paper tubing, plastic, or perforated or expanded metal for local stiffening or coupling to other structures may be easily added or laminated on with additional sprayed-on reinforced resin. Sandwich construction (foam, balsa wood), wood members, or other materials are easily added for increasing modulus, vibration damping and thermal insulation.

Curvatures. Unusual or multi-compound curves and contours are possible in spray-up that otherwise would be prohibitive to machine in metal dies for molding, or into which it would be difficult to drape reinforcement in the hand lay-up process. Reentrant corners may be produced although it may be necessary to shorten fiber length.

Joining. It is possible to join by spray-up RP/C parts too large or complex to make in a single assembly.

Coatings. It is possible to apply coatings on metal or cement parts of a structure to prevent corrosion or leakage.

Mold and Part Component Savings. Multipart sheet or other metal fabrications can be superseded by the spray-up of RP/C parts. By combining and designing properly, 1 part may be made to cover the total area of several metal parts. Painting to prevent rust is unnecessary.

Holes. Holes may be formed either by building a void in the mold or by outlining with line in mold for cutting out later when the laminate is leather-hard (knife trimming).

Delamination. Spray-up methods avoid delamination, but provide slightly lower strengths than those of products made by hand lay-up. It is also easier to remove voids in spray-up.

Feasibility. Spray-up should not be used where characteristic RP/C strength is not called for or where cheaper materials will do the job.[23]

Table II-2.9 Representative Physical Properties for Spray-up of RP/C

Property	Values for Spray-up
Specific gravity	1.4–1.6
Tensile strength, psi	9000–18000
Tensile modulus, psi	0.8×10^6–1.8×10^6
Compressive strength, psi	15000–25000
Flexural strength, psi	16000–28000
Flexural modulus, psi	1×10^6–1.2×10^6
Impact strength, ft-lb per in.	5–15
Hardness, Barcol	40–80
Moisture pickup, %	.05–1.0
Heat resistance, continuous °F	150–350
Resistance to acids and alkalies	Fair to excellent
Resistance to solvents	Good to excellent
Machining qualities	Fair, use carbide or diamond tools
Resin content, %	65–75

Finished Physical Properties. Finished product properties of spray-up RP/C, including physical strength values, are presented in Table II-2.9.

Cost Estimating. It is of interest to illustrate the manner in which costs for a part to be produced by the spray-up method may be developed.

For any job or part the following information must be known: part area in square feet, individual material costs, labor rate, fabricated material weight per square feet. In addition the following items must be estimated or determined: total weight of each unit, total material cost for each unit, overhead or indirect costs, total ultimate cost for each unit, tooling cost.

In the example below, it is assumed that a part is to be fabricated which is 61 sq ft in area (including trim waste), and of uncomplicated shape that complies with hand lay-up and spray-up design rules. For purposes of bracing and edge reinforcement, it is assumed that a portion (5% of part weight) is included as fiber-glass, chopped-strand reinforcing mat.

Laminate weight is 1.42 lb per sq ft (including gel coat,) making the total part weight 86.6 lb. Part thickness is 0.125 in. Glass/resin ratio is 25:75. It is necessary to determine the

Table II-2.10   Determination of Total Material Cost[a] (1973 data)

Process Step	% of Material Required	Weight in Part lb	Material Price per lb (U.S.)	Unit Material Cost
Gel coat	9.0	7.7	0.650	$ 5.00
Spray-up and lay-up resin	68.0	58.4	0.285	16.65
MEK peroxide (based on resin)	(1.0)	0.6	0.800	0.48
Glass-spray-up	18.0	15.4	0.340	5.24
Glass-mat	5.0	4.3	0.460	1.98
Sub totals	100.0	86.4		$29.35
Plus stiffeners, ribs, and splatter paint				10.00
Total material cost				$39.35

[a]Based on 61 sq ft body surface; laminate weight = 1.42 lb per sq ft (0.125 in. thick + gel).

total unit cost and the tooling cost for annual part volumes of 2,000 and 5,000 per yr. Table II-2.10 presents all information needed to calculate the total material cost. Man hours required for each unit are determined in Table II-2.11. Table II-2.12 lists the items which figure in the total unit cost. Lastly, the tooling-cost is calculated in Table II-2.13.

Table II-2.11   Manhours Labor Required for Each Unit (In U.S.A.)[a]

Cycle	Manhours Required
1. Mix resin, fill, and calibrate gun	2.0
2. Wax 6 molds and make ready	1.5
3. Gel coat 6 molds	1.5
4. Spray-up glass and resin and roll-out	6.0
5. Cut mat, lay-up mat in frame, ribs, etc.	3.0
6. Trim	1.5
7. Remove parts from mold	1.5
8. Clean-up, paint parts inside	7.0
Total	24.0

[a]Based on: 1. 3 laborers turning out a total of 6 units per 8-hr day. 3 men at 8 hr = 24 hr per 6 units = 4 hr per unit.
2. $3.00 hourly labor rate.

Table II-2.12   Determination of Total Unit Cost (In U.S.A.)

Material cost per unit	$39.35
Labor cost per unit	12.00
Indirect costs (est.)	40.00
Subtotal	$91.35
Contingency—10% (for equipment maintenance, rework damaged and defective parts, resurfacing molds after each 25 parts, etc.)	9.14
Total cost per unit	$100.49

Table II-2.13   Tooling Costs

Parts per Year	Molds Necessary	Price per Mold	Cost
2000	8	$400	$3,200
5000 (1-shift operation)	19 or 20	400	8,000
5000 (2-shift operation)	10	400	4,000

References

1. H. Ailes, *SPI 21st RP/C Division Proceedings,* 1966, Sec. 15-F; E. B. Euchner, et al., *SPI 20th RP/C Proceedings,* 1965, Sec. 13-B; J. D. Henry, *SPE RP/C RETEC Technical Papers,* (Sept. 1968), 43.

2. *SPI 23rd RP/C Division Proceedings,* 1968, Sec. 19-D.

3. Information this section is from: Sprayed Reinforced Plastics, E. S. Mylis, Glass Industry, Oct. 1961—PP581.

Guide to Sprayup Equipment for RP.-R.C. Hosford, *Plastics Technology* April 1964, PP36.

Sprayup Process Improvement—H. B. Ailes, 21st Proceedings, 1966, Sec. 15-F.

"GRP Technology," W. S. Penn, McLaren & Stone, Ltd., 1966 PP184-194.

4. Western Plastics, July, 1968, p. 16.

5. SPI 21st RP/C Division Proceedings, 1966, Sec. 15-F.

6. SPI 22nd RP/C Division Proceedings, 1966, Sec. 2-F.

7. SPI 22nd RP/C Div. Proceedings, 1967, Sec. 2-D.

8. NSPI Minimum Standards for Fiber Glass Swimming Pools, National Swimming Pool Institute, Washington, D.C., 1960. Prepared by J. G. Mohr et al.

9. *Plastics Technology,* Apr. 1964, page 37. SPI 19th RP/C Div. Proceedings, 1964, Sec. 15-A, Shook, G. D.

10. *SPE Journal,* June, 1961, page 561.

11. *Ibid.*

12. SPI 16th RP/C Proceedings, 1961, sec. 6-B, Ader, G., "Handbook of Reinforced Plastics," Oleesky-Mohr, Van Nostrand Reinhold, 1964, p. 107–108.

13. SPI 25th RP/C Div. Proceedings, 1970, sec. 15-A, Stayner, V., *See also* patent application USSN 700,212. *See also* "Comoform Process," SPI 27th RP/C Proceedings, 1972, Sec. 1-B, Zion, E. M., and Williams, G. L.

14. Personal Communication, V. Stayner, Federal Huber Corp., Plano, Ill.

15. SPI RP/C Div. Proceedings, 19th-1964, Lunn Laminates, Inc., Wyandanch, N.Y., Shook, G. D.

16. Oleesky and Mohr, "Handbook of Reinforced Plastics of the SPI," PP49, New York, Van Nostrand Reinhold, 1964.

17. Zion, E. M., RP Market Dev. Eng'r. Owens-Corning Fiberglas, Technical Center, Granville, Ohio.

18. Sutherland, J. M., Prod. Mgr. Versatran Div., AMF Corporation, Stamford, Conn.

19. Ives, F. Pres, Venus Products, Inc., Kent, Wash.

Chironis, N. P., Prod. Eng'r. Magazine-PP83 "Industrial Robots," Feb. 16, 1970.

Plastics Design & Processing-PP26, "Robots in Plastic Processing," May 1969.

20. Versatran Div. AMF Corporation, 695 Hope St., Stamford, Conn. 06907.

Venus Products, Inc., 1862 Ives Ave., Kent, Wash. 98031.

Owens-Corning Fiberglas Corp. Tech. Center. Granville, Ohio 43023.

Amos, H., *et al,* U.S. Patent No. 2,870,054-Jan. 20, 1959.

Wiltshire, A. J., U.S. Patent No. 3,012,922, Dec. 12, 1961.

SPI 26th RP/C Div. Proceedings, 1971, Sec. 11, Panel Discussion.

21. Personal Communication, Zion, E. OCF Technical Center, Granville, Ohio 43023 (1972).

22. Technical Data Bulletin-"Automatic resin-fiber spraying unit," Coudenhove Kunstoffe Maschinen, Vienna, Austria, 1972.

23. Venus Products, Inc., Kent, Wash. 98031, 1972.

24. Technical Bulletin: "T-10 Tape-Controlled Spray-up Machine," McClean Anderson Corp., Milwaukee, Wis. 53212.

25. Shook, G. D., 19th SPI-RP/C Div. Proceedings, 1964, Sec. 15-A.

26. Ailes, H., 21st SPI-RP/C Div. Proceedings, 1966, Sec. 15-F. *See also* Shook, G. D., SPI 27th RP/C Div. Proceedings, 1972, Sec. 18-C.

27. *Product Engineering,* Jan. 18, 1958, p. 17.

II-3

Vacuum-Bag Molding

INTRODUCTION

Vacuum-bag molding is the procedure by which RP/C parts are made with a single mold, either male or female. The device which makes the part conform to the mold shape is a flexible membrane. Withdrawal of air from the space between the membrane and the mold causes atmospheric pressure to be applied uniformly to the outer surface of the membrane, forcing it and the enclosed plastic part against the rigid mold. Maintenance of the vacuum holds the part to the desired shape while the impregnating resin is cured, either at ambient temperature or by means of heat. Pressures induced by the application of a vacuum, usually between 10 and 15 lb per sq in., are sufficient to remove voids, accomplish good densification, and provide better dimensional control than simple hand lay-up techniques. The process is substantially more rapid than contact molding but somewhat slower than plunger-assisted or matched-die procedures.

There are a number of advantages accruing in the use of vacuum-bag molding. Not the least of course is the economy of using only one mold. Parts as large as 30 or 40 ft in size have been made in quantities. Compound curves offer no real problems. Sandwich construction is readily handled, despite slight mismatches in core thicknesses which could not be handled in matched tooling.

There are a number of variations in techniques and procedures in the vacuum-bag process. All of those in current use will be analyzed and compared.

MATERIALS

Glass and Reinforcements. The selection of reinforcing materials for bag-mold products will depend upon the properties desired in the product. Surface finish, strength, cost, size, and shape all have a bearing on the choice. For example, if an extremely smooth finish is required for subsequent painting or surface treatment, it might be preferable to use a thin woven fabric, rather than a fiber glass chopped mat. Where strength combined with double curvatures (as in a nose radome) are requirements, a crow's-foot or satin-weave cloth would probably be used. However, if a prototype automobile fender were to be made on plaster tooling, a simple mat lay-up would probably suffice to satisfy the aesthetics or design potential of the molding.

Woven fabrics are, of course, more costly than chopped rovings, so this factor must be taken into consideration. Generally, fabrics are selected when dimensional tolerances are important, where good mechanical properties are required, and low-cost is not a dictating consideration.

Both fabrics and mats may be obtained in the "dry" and the pre-impregnated forms. The pre-impregnated material, known as "prepreg," is loaded with a pre-determined amount of resin which has been partially advanced in cure, or "B-staged."

Use of the prepreg material enables the fabricator to maintain close tolerances over resin concentration, resin flow, volatile content, and a number of other factors which would be subject to a great variation if he were to use wet lay-up. A number of companies provide pre-preg materials using epoxy, phenolic, silicone, polyester, polyimide, and other resin types in standard widths and roll lengths. The fabricator or molder can specifically ask for a desired resin content and be reasonably certain that the material he receives will not vary from that level by more than a couple of percent. When standard materials are not suitable for a particular application, special runs may be made by the coater.

Most of these materials are molded into shapes by lay-up in molds, with subsequent application of pressure and heat, which advance the B-staged resin impregnant to a condition of final cure. However, there are also materials available containing catalysts and/or promoters which are sensitive to ultraviolet radiation and which may be cured without the use of heat. Generally speaking, the final ultimate properties of the cured material are not attained when only the ultraviolet cure is used, and a subsequent postcure at elevated temperature is required for best performance. In many products where precise thickness tolerance is not a problem and rapid build-up is indicated, the use of woven roving is common. Such products as boats, containers, and structural panels fall into this category. In many cases, surfacing mat is applied to the mold or platen, followed by one or two layers of chopped mat or woven fabric. This procedure provides a relatively smooth exposed surface, suitable for fine finishing. Over this preliminary lay-up, the fabricator adds as many plies of woven roving as may be necessary to attain his desired wall thickness. If necessary, this is followed by a final finishing lay-up of the same materials used at the start, such as mat, cloth, and/or surfacing mat.

Resins. Impregnants used for vacuum-bag molding depend, just as the reinforcement does, on the properties desired in the finished part and on cost considerations. The curing procedure also has a strong bearing on the choice of resin. At present time, and probably for a number of years to come, the preponderance of volume in the bag-molding market will lie with polyester resins of various types. The primary considerations are their relatively low price in comparison with the more exotic types and their ease of handling and cure at ambient or elevated temperature. Catalysts are available for room-temperature cure, ultraviolet cure, sunlight cure, and heat cure. The resins are relatively low in viscosity, and wet the reinforcement readily. They are resistant to most chemicals for reasonable periods of time, and they provide good mechanical properties when properly impregnated and cured with the correct reinforcement.

When better mechanical properties or temperature resistance may be needed, epoxy resins are often selected for use. Epoxies provide good wetting, but require more care in handling than polyester resins because of the possibility of undesirable skin reactions induced by the resins and the amine hardeners. These side effects can be avoided by using protective clothing, rubber gloves, skin creams, and other protective media. Like polyesters, epoxies will cure at room temperature if the proper hardener is used. Also, like polyesters, their final molded properties are greatly enhanced by subsequent heat postcure. They are more expensive than polyesters, but because resin cost is not normally a major consideration in the final price of the molded part, this should not be a governing factor in the selection of the impregnant.

There are several other resin types used in bag-molding, but they are not normally acceptable for the wet lay-up procedure. Phenolic, diallyl phthalate, silicone, and other more exotic types are available only in prepreg form. These materials are perfectly satisfactory for bag-molding but do not attain their best strengths without subsequent elevated temperature postcure.

Mold Release Agents. It has often been facetiously remarked that most of the resins used in vacuum-bag molding are not really good adhesives, except in holding molded parts to the molds. Consequently, a major investment in the development of procedures is the selection of the proper release agents. These are covered elsewhere in this book, but it is of paramount importance that they are not overlooked in vacuum-bag molding.

Honeycomb Materials. One of the procedures for producing light-weight, structurally stiff composites is the fabrication of sandwich shapes, both flat and curved. Because most of the rigidity of a panel or beam is dependent on the properties of the outer fibers or structural elements (e.g., an "I" beam), it is not necessary that the panel be of uniform density. Thus, in a sandwich, dense, rigid outer skins are bonded to lightweight core materials to develop desired properties without undue addition of dead weight. The skins, of course, are of resin-impregnated glass or other high-modulus fibers, and the core may be a lightweight foam or a honeycomb structure.

Honeycomb materials are supplied from several sources with glass, paper, aluminum, or other material in the cell walls, in sizes ranging from about a ⅛ in. hexagonal cell up to ½ in. or more. Aluminum and nylon-phenolic are the two materials most widely used.

EQUIPMENT AND TOOLING

Molds: Construction and Use. In the vacuum-bag method of molding, tooling is a relatively simple problem when compared, for instance to matched-die molding. The latter process is designed for relatively long runs of close-tolerance parts and requires that the molder procure sets of machines or cast tools which conform to the inner and outer contours of the molding, usually including integral heating devices, hardened cutoff edges, blowout holes or ejection pins, and other sophisticated (and expensive) adjuncts to production processing.

All of these procedures are not essential for the fabrication of RP/C parts by vacuum-bag techniques. The type of mold is selected on the basis of which surface of the part is critical. For example, if a nose radome for an aircraft is to be molded, it is obvious that the outer surface is controlling. The molder will therefore select a hollow mold (called a *female*) in which to lay up the part. Thus, the radome, when removed, will have an outer surface which conforms to the contour of the inner surface of the mold. Conversely, if a chair shell is the product, it is necessary that the inner surface be the smooth one. In this example, a male mold would be the choice, with the laminate laid over the mold and the vacuum bag fitted against the outer, non-critical surface of the laminate.

Because the materials being molded start in the fluid phase as flowing resins, and pressure is exerted against the lay-up during the curing process, the final molded part will conform very accurately to the surface of the mold against which it has cured. Finish requirements for the mold surface, therefore, must be predicated on the desired surface smoothness of the product.

The material from which the mold is made depends upon several variables, included among which are the number of pieces to be molded, the type of resin to be used, the cure temperature and hence thermal resistance necessary in the mold material, the desired surface finish, and the time available to produce the part. A wide selection of mold materials is available. In general order of preference, they are: machined steel, cast or laminated epoxy resin, well-seasoned hardwood, cast plaster or gypsum, cast phenolic resin, cast or machined aluminum, melt-out or wash-out casting salts, other alloys or material combinations (wood, metal, etc.).

In addition to the actual molding surface against which the part is formed, the mold must have extensions, such as flanges or flat areas, to which the vacuum bag may be sealed. Vacuum outlets must be provided. Reinforcements and braces are necessary to avoid any bending or distortion when the mold is heated. Sometimes integral heating devices must be included, such as hollow tubing for circulation of steam, hot water, or heating oil.

Critical dimensions may be inscribed in the mold surface. This may include trim lines, location of holes or attachment points, and so on. The surface of the part will conform precisely to that of the mold, and all such marks will be transposed to the part when it is removed from the mold, thereby making finishing less of a problem.

Sealants. As earlier described, vacuum-bag molding entails the use of a membrane over the molding from which air is removed. The ambient pressure of the atmosphere is then used to hold the part against the mold long enough for resin polymerization to take place. If this procedure is to be effective, it is of utmost importance that the vacuum be maintained throughout the cure cycle. To accomplish that aim, several methods are used to seal the membrane securely against the entrance of air into the evacuated portion of the assembly. Probably the most common device is the use of viscous materials as interfaces between the bag and the mold. One example is a zinc chromate paste, which may be extruded or formed around the periphery of the mold. When the vacuum bag is pressed into the paste, an effective seal is formed which is relatively insensitive to the normal molding temperatures encountered in cure. Rubber hoses used as vacuum lines may be wrapped with the zinc chromate where they enter the bag to attain a seal at that point. After completion of the cure cycle, the bag is stripped from the mold and clean-up is accomplished without too much difficulty.

A second procedure involves a mechanical seal, whereby the outer edge of the mold is routed or in some manner provided with a channel, normally about ½ in. wide and ½ in. deep. After the vacuum bag has been carefully fitted to avoid "bridging," i.e., stretching across filleted corners or depressions in the mold, a piece of ½ in. outside diameter flexible hose is laid over the routed channel and driven with a mallet into the channel. The bag material is of course between the tubing and the mold so that a perfectly viable seal is formed. In some cases, the vacuum lines are brought up to built-in exhaust ports inside the seal area. In other designs, short sections are left without the channel and zinc chromate paste is used to complete the seal.

Several other systems are in use, including double-faced, pressure-sensitive tapes (handy especially for flat-sheet laminating), formed and extruded mastics, and even wooden strips laid over the bag and held in place by C-clamps. The choice of procedure is wide and is dictated primarily by the size of the job, the type of mold, the number of parts involved, and any number of economic considerations. The object is to keep air out of the evacuated bag, and the method is unimportant as long as it is effective.

Bleeder Materials. The bags or membranes used to seal the molded part from the atmosphere during evacuation of air from the lay-up are usually polyvinyl acetate, polyvinyl alcohol, nylon, or similar flexible films. With the removal of air and the concomitant external pressure exerted by the atmosphere, there is a tendency for the bag to seal itself against the lay-up in a manner which prevents the ready flow of internal air toward the exhaust ports. The result of this can be bridging or sectioning of the mold where the bag is not firmly pressed against the part.

To prevent this fatal condition, a course interlayer is placed between the lay-up and the vacuum bag. Materials such as burlap, loosely woven glass cloth, glass mat and porous fabrics fill the requirement admirably. These materials are referred to as 'bleeders" because they permit the air to "bleed" through them without allowing the bag membrane to seal itself against the part.

To prevent adhesion of the bleeder layer to the molded part, it is customary to interpose an additional layer, often a perforated film, between the part and the bleeder. This technique permits escape of the air and excess resin, without introducing problems of clean-up after the cure cycle is completed. The film permits peeling the ancillary materials from the molding with a minimum of labor. It is necessary of course to place the bleeder material in the lay-up so that it does not add to the molding problems by creating bridges itself.

Another useful material is known as "release cloth." This is a woven fabric, usually between 5 and 10 mils thick, which has been coated with silicone resin or other non-sticking material. The release cloth is normally colored or tinted (usually pink) to enable the shop personnel to recognize it readily during the post-molding cleanup operations. The release cloth strips readily from the molded part, leaving a clean surface for subsequent fiinishing operations.

Vacuum-Bag Materials. As in almost every other area of vacuum-bag molding, the operator has a relatively wide choice of materials for use as a membrane to exclude air from the lay-up. Usually the selection is predicated on economic factors. When a large number of parts are planned, a permanent or semi-permanent bag may be made for re-use over a run. Where only a few pieces are involved, less expensive materials are indicated.

Molding methods must also be considered in choosing vacuum-bag materials. For example, if a part is to be made using sunlight or ultraviolet cure, high temperatures will not be expected, and a simple material may be selected for the bag. If high temperatures or autoclave procedures will be involved, however, a more reliable and stable membrane is obviously required.

Another consideration is inspection procedures. When void-free laminates are being made, for example, it is necessary to impregnate the reinforcement with an excess of resin, which is subsequently worked out by hand, usually with squeegees. This process requires that the operator be able to see the lay-up during the resin removal, both to insure adequacy of the removal and to inspect the part for the absence of voids prior to advancing the resin to a cured state. A transparent membrane is an absolute necessity in this situation. Some of the transparent materials in current use are cellophane, polyvinyl alcohol film, polyvinyl chloride film, and nylon.

Cellophane is an acceptable material for relatively flat or slightly curved parts. However, it has a tendency to be weak, and to tear if not carefully handled, and it will not stand elevated temperatures without charring. It is relatively inexpensive, but is not widely used in the industry primarily because it does not lend itself to stretching over areas of double curvature.

Polyvinyl alcohol film (PVA), usually about 0.003 in. thick, is commonly used for a number of resin systems. It is a water-soluble material, however, and should not be used with condensation type resins such as phenolics which give off water as a by-product during cure. It is flexible, will normally stretch to conform with mold shape, and is not expensive.

Polyvinyl chloride film (PVC), is used in thicknesses of from about 0.004 to 0.008 in. thick. It is less expensive than PVA, but has a tendency to inhibit the cure of some resins. This tendency may be used to advantage on occasion, however, when full cure is not necessarily or immediately required. Such an occasion might be the construction of a honeycomb sandwich with void-free skins, where the skins may be molded separately using PVC film, resulting in a slightly uncured inner surface. This will lend itself to better adhesion when the honeycomb core is added in the next molding step. When other bag materials are used and a complete cure is attained, it is necessary to include a peel ply in the lay-up, or to sand or otherwise roughen the surface prior to bonding the honeycomb.

When autoclave or hydroclave molding is to be used, heavier membrane materials are required to prevent leakage. Various types of natural and synthetic rubbers are available for these purposes.

Because of its good heat stability, neoprene is a common bag material. Precautions must be taken to avoid contact of the neoprene with the lay-up, because, like PVC, it inhibits the resin cure. To avoid this, it is customary to include a layer of cellophane or nylon between the laminate and the bag. Neoprene has a tendency, too, to lose flexibility and "age-crack" after repeated high-temperature cycling. As a consequence, it has limited life.

Natural rubber, cheaper than neoprene, has a longer life and more elasticity than neoprene. Again, because sulfur is used in vulcanizing the rubber, and also for other reasons, rubber also inhibits resin cure and precautions must be taken to avoid contact with the uncured part.

Preformed silicone rubber blankets are used in long runs where their price is justifiable. They have long life and are transparent to ultraviolet radiation, hence may be used with u-v cure cycles.

Nylon film is still another membrane material that is often used. Generally, the same comments that were made about cellophane apply to nylon, except that nylon is considerably stronger. Nylon too is subject to attack by water, so that care must be used in selecting it for certain condensation resin systems.

In summary, the selection of a bag material must be based upon the number of parts to be molded, the type of resin, the temperature of cure, and the shape of the mold.

Clamps and Clamping Mechanisms. Several devices are used for clamping the film to the mold where mechanical means are preferred to the use of viscous pastes or tapes.

When long production runs are in progress or being planned, it is often more economical costwise and laborwise to prepare a sturdy holding fixture which may be repeatedly used with a minimum of effort and expense. Such a device might be a steel or aluminum ring shaped to the mold edge contour and fitted with clamping for ready application of the sealing pressure. The periphery of this ring (or ring segment for larger molds) may be permanently rigged with C-clamps, toggle clamps, spring clamps or similar fasteners. They are permanently or semi-permanently fastened to the mold or the ring in a manner such that only a few seconds are required for application of the clamping force. The bag or membrane is, in these cases, stretched over the mold edge, the ring (sometimes lined with a compression gasket) is dropped into place, and the clamps or toggles are rotated into position.

Pressure is applied quickly to bring the ring down and the seal is complete.

Vacuum Equipment. As explained, the whole technique of vacuum-bag molding is based upon removal of air from the space between the membrane and the lay-up, so that the atmospheric pressure of about 15 psi can be applied to all areas of the bag, forcing the impregnated reinforcement firmly against the mold surface while resin cure is being accomplished. If insufficient air is removed or if a leak exists, the atmospheric pressure will not be brought to bear. The lay-up will not therefore conform rigidly to the mold surface, and voids, bridges, or other imperfections will result. In many cases, the presence of air will also inhibit the cure of the resin. Broadly speaking, insufficient pressure caused by leaks will cause high rejection rates.

Most production shops are provided with master vacuum systems in the same manner that high-pressure air and electrical power systems are distributed. Vacuum outlets are spaced conveniently on walls, partitions, and columns so that relatively short runs of flexible hoses may be attached to the mold exhaust ports and to the outlets. The design of such a vacuum distribution system is a simple matter. The major consideration is the size and capacity of the vacuum pump. It must be large enough to provide a minimum of 25 in. of mercury, which corresponds to about 12 psi of external pressure at the individual mold position. The pump must also be large enough to accommodate the number of molds expected to be in operation. Unfortunately many molders start with small pumps and then find that with increased production schedules their capacity is insufficient for the job. They either cannot run the number of mold positions necessary or their vacuum is too low for the larger number of molds. In either case their profits are unfavorably affected and production schedules suffer. Part of the design procedure involves taking into taking account the length of pipeline from the pump to the outlet because pressure drops are just as valid in vacuum lines as they are in water or airlines. All connections such as elbows, tees and reducers must be fully sealed to prevent leakage and consequent overloading of the pump.

At the point where the vacuum line is attached to the mold, a resin trap should be in-

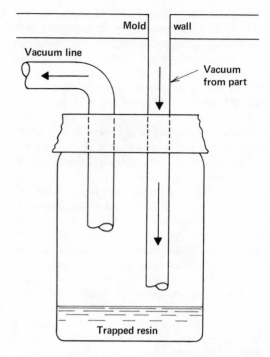

Figure II-3.1   Resin trap.

serted into the system. Obviously, as air is removed from the lay-up, external pressure is applied and excess liquid resin will be drawn into the vacuum line. If this flow of liquid into the line is not prevented, two deleterious conditions can ensue: The first is a clogging of the line, resulting in loss of vacuum to the part. The second, much more serious and costly, results when the viscous resin actually reaches the pump and enters the vaccum cylinders. While viscous, the resin causes pumping trouble. After polymerization, it really generates catastrophe, necessitating complete disassembly of the system. If the resin has reached the pump, it must also be assumed that it is well-distributed through the pipes, junctions, and fittings leading from the pump to the mold. In extreme cases, this means replacement of the entire distribution system and overhaul of the pump. Again, the effect on production schedules can be disastrous.

Hence, the resin trap can insure against all these dire consequences, and it is sufficient to say that a resin trap should be installed at the point where each vacuum line is attached to each mold. As shown in Figure II-3.1, a resin trap is a combination of tubing and a receptacle which prevents the air exhaust issuing from the mold from entering the vacuum line directly. Instead, the bleeder tube by which air leaves the mold cavity is fed into a heavy glass or metal jar. The jar, in turn, is evacuated by the line from the pump. Thus, as the air is drawn from the mold, it goes first into the jar or trap, taking with it any resin or any other solid or liquid contaminant. The undesired materials go to the bottom of the trap, while the air is drawn to the vacuum line.

The accumulated contaminants and resin may be removed by unscrewing or unclamping the trap, cleaning it and replacing it. Obviously, the trap must be large enough to accommodate all resin and residues expected during a single molding cycle, since clean-out during a cycle cannot be accomplished without loss of vacuum. Most trap fittings are securely welded or otherwise sealed to the mold to insure leak-proof connections. Jars or resin receptacles must be fitted with seal rings for the same reasons.

Ovens. Vacuum-bag molding developed early in the history of the "low-pressure" plastics industry as an off-shoot of the cruder contact-molding procedures, wherein the laminates or other lay-ups were merely placed in contact with plaster, wooden, or other inexpensive tooling and allowed to set-up and cure at ambient temperature. Where climatic conditions warranted, resin polymerization was advanced by taking the parts outdoors for exposure to the heat of the sun. This is a major reason why the southern California area became one of the major development regions for the industry.

Unfortunately, two factors mitigate against the continued dependence upon ambient conditions for resin cure. The first is the element of time. Polyesters and other room-temperature resins require time spans on the order of several hours to reach a point where they can be safely handled and removed from the mold. When production runs are involved, such schedules are economically prohibitive. The second major factor is the fact that none of these resins attain their maximum physical properties under conditions of room-temperature cure. It is necessary that they be exposed to the temperatures of at least 200 or 250°F to develop the properties that are inherent in their design potential. The logical solution to this requirement was the use of oven-curing techniques. This method maintained quality control of the product, because repeatable cure cycles could be assured by time—temperature programs developed for the particular resin system used.

Practically every plant that can reasonably be considered a reinforced plastics/composites processor depends upon at least one, and usually a number, of heated chambers which are fitted with instrumentation adequate for modern quality control. The designing of an oven is not difficult. It requires reasonably well-insulated walls and top, with a reliable source of heat. Most ovens today are gas-fired, although electric power is used in areas where cost factors are not prohibitive. It is advisable to have blowers or other means of air circulation to maintain a uniform temperature throughout the chamber. In this way, all parts of the product under cure receive the same treatment and differential thermal stresses are avoided.

Provision should be made for bringing vacuum lines into the oven, either through the walls or doors, or by fitting the oven with internal vacuum-line outlets to which the mold may be connected. Thermocouples, recording thermometers, or other reliable heat-indicating devices must be included in the design to assure adequate control procedures during molding. It is wise, too, to provide vacuum gages in convenient locations outside the oven. Should a major leak develop in the vacuum during early stages of cure, loss of the part may result. Surveillance of temperature and vacuum gages during cure cycles is the best insurance against failure.

In some cases, it may be advisable to provide an additional source of heat besides the circulating hot air. An example might be a deep-draw molding, such as a tank or a long nose radome for an aircraft. In constructing these objects, designers often use integral mold-heating devices. These could be resistance wires or tapes, circulating hot liquids (either water or oils), or infrared radiators in inaccessible areas.

Ovens must be large enough to accomodate the molds without blocking the full circulation of hot air. Adequate space must be left around the part for personnel movement, vacuum and power lines, and good heat distribution.

Miscellaneous Equipment. In addition to all the basic materials and equipment itemized above, there are a number of small items required in any shop to make operations less time-consuming and more profitable. These come under the category of hand tools and ancillary equipment.

The procedure for making a void-free part, as discussed, involves the use of excess resin during impregnation, followed by removal of the excess after bagging and before cure. For such removal, it is customary to provide "squeegees," or semi-flexible strips of rubber, neoprene, Teflon, or a similar material. The tool is usually made in the shop, and is about 3 in. × 5 in. in size, cut from a sheet about ¼ in. thick. It is hand held, with the long edge used to squeeze the resin from the center of the mold toward the drain lines and resin traps.

When pre-impregnated reinforcements are used for lay-ups, it is usually found that forming them to a curved (especially double-curved) surface is not easy. Under these conditions, a heat gun is indispensible. This is a hand-held heat source, normally an electric resistance coil over which air is passed by blower from a small blower motor. Heat guns are available commercially in several different temperature ranges up to 1000°F. When used, they are directed for short periods of time at the prepreg. The heat causes the fabric to become slightly tacky and allows the operator to drape it more readily in areas of curvature. The added tack also enables the workmen to induce adhesion of the layers so that vertical surfaces will not collapse during lay-up. Care must be used to avoid excess heat or over-exposure of the prepreg, because full cure of the resin is certainly not desirable during the lay-up procedure.

Where relatively large flat areas are being bagged, small rollers are handy devices for inducing good interlaminar contact. Depending upon size and shape of the area, the roller (usually rubber or neoprene) may be from 4 to 10 or 12 in. long. As with a squeegee, the roller is used after the bag has been placed over the lay-up.

Wooden wedges for separation of the molded parts from mold surfaces are frequent additions to the specialized equipment in a molding plant. Knives, razor blades, routers, power sanders, saws, mallets, and numerous other hand tools and power tools are required for finishing operations.

Safety cans containing methyl ethyl ketone, acetone, methylene chloride, styrene, and similar organic solvents should be available for the general cleanup of tools and hands. Paper towels, cloth wipers, and driers are advisable. Masking tapes, double-faced tapes, convenient adhesive dispensers are in daily use. No shop can be considered completely equipped without a large assortment of C-clamps.

A stock of coiled wire spring should also be available. This is a necessary item for use around the periphery of a mold to prevent sealing of the bag against the part, and also to permit the flow of air and excess resin into the traps. Coiled spring is also inserted into the end of the vacuum line inside the bag to prevent air pressure from collapsing the hose.

MOLDING METHOD/PROCESSING DETAILS

Vacuum-Bag Molding Process. The following discussion provides a step-by-step procedure for making parts by the vacuum-bag process. While some shortcuts may be developed in a few cases, strict adherence to the process, both with respect to the order of procedure and to the actual need for each step, will assure the molder of a controllable method of making sound, reproducible parts.

1. Check out the mold. Be sure that all dimensions are accurate, that it is leakproof, that the surface has been finished to at least the level of smoothness required, that vacuum outlets are clear with resin traps in place, and that the mold is rigidly mounted.

2. Carefully apply the mold release. Follow the supplier's instructions implicitly. Improper use of release agents has probably caused more loss of molded parts than any single factor in the manufacturing process.

3. Lay the reinforcement and resin in the mold. The customary procedure is to impregnate the reinforcement on a table adjacent to the mold and move the wet material into the mold after cutting to shape. Some materials such as glass mat, however, are not amenable to this procedure and must be placed into the mold dry, with the liquid resin added as a subsequent step. In either case cut the reinforcement to shape as nearly as possible immediately prior to placement in the mold. This has two beneficial effects: handling is easier, and the mold surface is protected against possible damage by cutting tools.

Prepreg materials, normally dry and sometimes "boardy," may be warmed slightly with a heat gun to induce tack and enable the operator to form the reinforcement to the mold contour and hold successive plies in place. Keep laps, overlays and joints to a minimum, both to simplify lay-up and to improve ultimate mechanical properties. Do not superimpose one lap or joint over another, except where they cross at an angle. Where laps are required, they should be between 1 and 2 in. wide. No laps are permissible in outside corners, and those in inside corners should be between ½ and 1 in. wide. Butt joints must be avoided because the action of the vacuum bag in forcing the reinforcement against the mold will normally cause them to separate.

In the case of prepreg molding, it is useful to have the entire mold surface raised to a temperature not exceeding 120°F. This will supply slight tack and some movement to the reinforcement and permit the operator to place it in close conformance to the mold surface. Hold the first layer tightly to that surface so that succeeding plies will not develop wrinkles, bulges, voids and similar defects.

Where recesses, fillets, or other changes in mold surface occur, exercise care to assure contact of the reinforcement and to avoid bridges in the final molding. Use heat guns judiciously so that advancement of the resin does not adversely affect interlaminar adhesion.

4. Lay the specially-treated release fabric, perforated film, or other separating medium over the complete laminate or lay-up.

5. At this point, put the bleeder material into place. The configuration of this step depends upon the size and shape of the part, but its purpose is to guarantee adequate and complete removal of the air from the space between the bag and the part. The bleeder may cover the entire part or may only go around the periphery, with tributary branches leading into all areas of the bag. Use strips, pie-shaped sections, or other convenient shapes. The selection becomes obvious as the configuration of the mold is studied. Where peripheral bleeders are used, they should be in contact with the molding at every possible point.

6. Carefully remove the mold-release agent from the surface of the mold outside the periphery of the bleeder material. Place the bag sealer (zinc chromate, tape, gasket) completely around the mold on the surface just cleaned.

Be sure that all vacuum exhaust ports are inside the seal-off area.

7. Lay the vacuum bag over the entire assembly. By hand, distribute the film so that all wrinkles and bubbles are minimized and so that radii and corrugations are properly covered with sufficient material to avoid subsequent bridging.

Go completely around the mold, working the bag into the sealant or clamping it securely against the gasket to assure a vacuum-tight closure. In handling the bag, be extremely careful to avoid pinholes or punctures.

Apply a partial vacuum to the system to check that all is in proper working order. This will also cause the bag to pull against the lay-up and show areas of potential difficulty, such as wrinkles or bridges. Move the bag by hand where necessary to preclude any such problems and to distribute the membrane uniformly over the surface of the mold. This will prevent differential stretching of the bag, which can cause rupture or leaks.

8. After the bag has been properly placed, check the seal at all points for possible leakage. Tighten clamps, squeeze sealant into folds, and in every way possible insure the vacuum integrity of the system. Cover the outer surface of the bag with lubricating oil or a thin coat of silicone grease. This will protect the surface during subsequent operations.

9. Complete the vacuum system by attaching all lines to the pumps or to vacuum outlets in the plant. Be certain that sufficient lines are provided to handle the volume of air within the lay-up and to maintain the vacuum at a proper level. If the mold is to be placed in an oven or autoclave, bring the lines through the wall of the chamber before making connection to the mold.

10. If a void-free lay-up is desired, use squeegees at this time to move excess resin from the center of the part toward the edges, where it may be drawn by the vacuum into the traps. Use caution not to use too much pressure in squeezing the resin out, or a dry laminate may result.

11. Apply full vacuum—at least 25 in. of mercury. This will assure a minimum of 10 to 13 psi of pressure over the surface of the part during cure. Again check for contact of the bag, bridging, leaks, and any other problems which could cause failure. If a leak develops, clean the bag at that point and repair it im-

mediately. Repair may consist of a simple procedure, such as merely covering with a pressure-sensitive tape or a piece of zinc chromate paste. If a major leak develops, a patch may be required. In really severe cases, it may be best to replace the entire bag.

12. Cure the part. This may mean allowing it to stand at amibent temperature until the resin polymerizes, or placing it into a chamber at elevated temperature and/or pressure. In either case, follow the instructions of the resin supplier or the prepreg coater with respect to time and temperature requirements for optimum molded properties. When the required cycle has been completed, remove the mold from the oven or autoclave and allow it to cool. If the part has been cured at room temperature, prepare to remove it from the mold.

13. After cure has been completed, remove the bag and take the part out of the mold for subsequent finishing operations, such as trimming, clean-up, and painting. Remove from the mold surface all contaminants, such as mold-release film, sealants, bleeder material which may have adhered, and so on. Clean the entire mold surface carefully and prepare to make the next part, starting with the first step described above.

Other Methods of Bag Molding. By following step by step the basic procedures of the bag-molding process, the molder can produce excellent laminates with the highest attainable physical and mechanical properties. There will obviously be circumstances, however, under which the molder needs to produce items other than simple laminates, or use other resins, reinforcements, or materials which do not necessarily lend themselves to optimum handling when molded as described above. Several alternate processes are available for production under these conditions.

Honeycomb Sandwich Construction A common example of RP/C which provides desirable properties without weight penalties is the so-called sandwich configuration. This consists of two relatively dense layers, such as glass fabric laminates, separated by a lightweight medium, usually a honeycomb structure or a low-density foam. Just as the flanges of an I-beam are stiffened by the vertical web between them, so are the skins of the sandwich enhanced by the core separating them. Because

most of the tensile and compressive forces of a beam in bending are exerted on the upper and lower layers, the center layers of the solid laminate may be replaced by a less dense medium without penalizing performance. Consequently, sandwich construction is common in areospace and building.

Fabricating a sandwich is generally analogous to making a laminate. However, it is not usually a one-step cure process. For best results, two or three cure cycles are involved. The procedure is as follows:

1. Lay up the outer skin of the sandwich consisting of the required number of plies of impregnated fabric. Perform all of the operations necessary to obtain a void-free laminate. Apply full vacuum and cure as required by the particular resin system. It is sometimes advantageous to add one additional layer of thin "tear-ply" to the laminate under the bleeder. This is peeled from the lay-up after cure, leaving a rough, toothy surface to which subsequent layers will bond more readily.

2. After the honeycomb core has been cut to the proper thickness and tolerance, prepare it for bonding by any one of several methods. One procedure involves roller coating the surfaces with resin to assure distribution of the impregnant to all contacting areas. Another method involves dipping the sliced honeycomb into a tray containing resin which has been thinned by addition of a volatile solvent such as acetone. The slices are removed from the dip bath and permitted to drain on screens. With both procedures, avoid excessive resin deposits on the honeycomb to prevent filling of the cells during cure.

3. After the honeycomb has been prepared for bonding, it is fitted carefully against the cured outer skin. Interlock the edges by "pegging" or "keyholing." This consists of driving small sections into the butting edges with a mallet, in the manner of the mortar and tenon joint used in cabinet making. Where mold curvature is present, compress the honeycomb cells so that they may resume their normal shape when forced outward against the mold by the vacuum bag. When the entire core has been set in place, lay a thin "tie-ply" of impregnated glass fabric over it to cover the cell structure. This layer is essential to close the cells against drainage of resin when the inner skin is applied at a later stage of the process. Hold the resin content of the tie-ply to a minimum,

consistent with the need for a good bond to the core. Again, bag and cure the assembly.

4. Except for such steps as adding edge build-up, inserting plugs for attachments, and other finishing requirements, the construction of a honeycomb sandwich is basically complete as described. They do not, however, apply to the foam-in-place sandwich. This construction normally uses precured inner and outer skins with matched male-female tooling.

One-step Honeycomb Sandwich Fabrication.
All of the vacuum-bag molding methods and variations described above are generally involved with the laborious task of impregnating reinforcements with viscous resins, laying these material into molds, and removing excess resin prior to cure. Numerous variables are introduced into the product, regardless of care in fabrication and quality control. Among the variables are such items as resin content, flow, volatiles, and others attributable to the resin condition, ambient temperature and humidity, and other conditions not under the direct control of the molder.

One solution to these problems is the use of prereg materials for skins in the sandwich. While the basic material may be more costly than the components of the wet lay-up, it is usually found that reduced labor man-hours compensates for the difference. Uniformity of product is an additional benefit, because prepregs are closely controlled with respect to all of the variables mentioned.

The process for making a sandwich with prepregs is identical to that described above, except that it can be done in one cure cycle.

1. Cut the prepreg fabric to the required pattern and lay it in the mold, using heat guns or other procedures to obtain the necessary drape and tack.
2. Lay in the core, fitting it carefully to conform with mold curvatures.
3. Build up closures, attachment edges, and other parts with the prepreg cut to size and exact shape.
4. Place the inner skin, also carefully precut, against the core, and drape it to fit by gentle and judicious use of heat guns.
5. Bag the assembly and cure it in accordance with the recommended cycle. Heat and pressure will cause sufficient resin flow at the honeycomb cell edges to provide good fillets for structural integrity.

While the use of a vacuum bag, (providing 10 to 14 psi) will usually give a good part, it is recommended that an autoclave cure be used if possible (see p. 88). The additional pressure insures better conformance of the part to the mold surface and higher interlayer adhesion between the skins and core.

Bag Molding—Ultraviolet Cure. Under some circumstances, it is not desirable nor possible to develop cure in resin systems by use of high-temperature. For example, if a thermoplastic material had been included in the lay-up, and application of heat would distort the insert. Other conditions may also preclude application of the 200 to 250° heat necessary for attainment of full cure.

To meet these problems, catalysts and promoters have been developed which are reactive to the presence of ultraviolet light. The following procedure permits fabrication of satisfactory parts without normal heat curing of the resin:

1. Using the ultraviolet-sensitive prepreg, lay up the laminate as with any other prepreg, prewarming the mold to 120°F and using a heat gun to induce drape and tack. As usual, be sure that the first ply is in close contact with the mold surface at all points in order to avoid wrinkles, blisters and voids.
2. After the laminate has been completed, apply a vacuum of 25 in. or more of mercury and hold for a minimum of 30 min. Ultraviolet prepregs become translucent, so that voids, bridges and other defects become visible and may be corrected prior to cure.
3. Move the mold into the ultraviolet cure facility. Broad-spectrum ultraviolet lamps should be stationed about 20 in. from the surface of the lay-up and should be distributed so that all areas are uniformly exposed to the radiation. Cover the entire lay-up with kraft paper. Turn on the lamps and preheat for about 20 min. Remove the kraft paper and continue the cure for the time necessary to complete polymerization. This cycle should have been pre-determined by earlier laboratory test procedures.
4. Turn off the lamps, still maintaining the vacuum and allow the part to cool.
5. Release the vacuum, remove the part

from the mold and perform the necessary clean-up.

6. If the part, because of excessive thickness or complex shape, has not reached full ultimate cure, it may be subjected to a temperature of 275 to 300°F for cure completion. Full gelation and rigidity should be reached at this stage, however.

Pressure-Bag Molding. As previously stated in the description of the vacuum-bag molding process, the procedure of withdrawing air from the space between the membrane and the molding will cause atmospheric pressure to be applied to the exposed area of the part. This pressure is normally sufficient to provide good interlaminar contact and part density. There are times, however, when additional pressure may be necessary, either because additional density is required or because the particular resin system in use demands pressures greater than 10 to 14 psi. One method of obtaining higher pressures is by the use of matched tooling in presses. This is costly, both in time and in money. Also, where only a limited number of parts may be required, matched tools cannot be justified on any basis.

To obviate the need for expensive tooling and procedures, the pressure-bag method of molding meets the need for providing uniformly distributed loading of from 15 to 50 psi over the area of the part. The usual steps follow.

1. Prepare the lay-up exactly as you would for vacuum-bag molding. If the part is small enough, however, the mold may be placed in a compression press. If size does not permit this, the mold must have a circumferential or peripheral flange which will accomodate a heavy flat plate.

2. At some convenient spot on the heavy plate, drill and tap a hole and fit an air valve so that compressed air may be passed through the plate. On the face of the plate closest to the mold, attach a flexible elastic bag of membrane so that it will be leakproof when the assembly is completed.

3. Follow the normal vacuum-bag procedure including evacuation, void-free operation and inspection.

4. Lower the plate and pressure bag into place. Securely fasten the plate using clamps, toggles, or (in the case of the

small part) by the platen of the compression press.

5. Force air into the pressure bag causing it to inflate and to exert a force against the vacuum bag. The pressure bag fills the space between the heavy plate and the part, adding its pressure to that already applied to the vacuum bag.

6. Resin cure may be promoted by the use of hot air or steam as the pressure medium.

7. After completion of the recommended cure cycle, relieve the pressure, remove the plate, and take the part from the mold.

Autoclave and Hydroclave Molding. There are times when pressures of only 50 psi are insufficient for adequate compression and cure of a composite part. Specifically, higher pressures may be required for some resins, such as phenolics, or where high densities are desirable. With parts of reasonable size in production quantities, this situation may be handled by matched-die press molding. There are circumstances, however, under which cost, part size and/or shape, or some similar consideration makes matched tooling unfeasible. One example is an exhaust nozzle for a rocket engine, which can be relatively immense and have complex inner and outer contours. For these conditions, a method has been devised by which pressures of 1000 psi or more may be applied to the molding with a minimum of tooling cost. This method uses the autoclave or hydroclave.

An autoclave is a large chamber, usually cylindrical, sturdily constructed to withstand high internal loading, and fitted with means for raising the temperature to levels required for resin curing and post-curing. Internal pressure is attained and maintained by the use of a gas or vapor, such as hot air, nitrogen, or steam.

A hydroclave is a similar chamber, wherein pressure is developed by the use of a liquid, such as hot water or a suitable oil.

Because of the inherent dangers attendant to air compressibility, autoclave pressures are seldom maintained above 100 to 200 psi. Hydroclave pressures, on the contrary, are often as great as 1000 psi or more. From this, it may be appreciated that a hydroclave is an expensive piece of equipment, calling for sturdy steel walls, usually 6 or more in. thick, with hydraulic closure systems, complex seals, and all the appurtenances necessary to assure the safety of plant and personnel. Methods of obtaining high temperatures range from simple immersion heaters to batteries of gas-fired burners.

With this type of high-pressure molding, the part is prepared exactly as for vacuum-bag molding. The bag, of course, must be sturdy and made of a material resistant to high temperature and to moisture. Typical materials are neoprene and silicone rubber. Extreme caution must be taken in sealing the bag, since the pressures used are such that the most minute leak will permit penetration of the pressurizing medium. Such a leak will almost invariably result in a reject part.

After the steps for vacuum-bag molding have been completed, the mold is moved into the autoclave or hydroclave. The vacuum is maintained by connecting the exhaust points to lines which have been brought through the walls of the chamber. Although it is normally not strictly required, it is advisable to maintain full vacuum on the part during cure. This will hold the bag in place and prevent its being torn or otherwise damaged during the cycle.

When condensation resins, such as phenolics, are used, the molding must be permitted to "breathe" prior to completion of cure. That is, the vapors evolved as a condensation by-product must be removed from the bag. If this step is not taken, these vapors may be trapped between the layers of the laminate. At some future exposure of the part to high temperature, the trapped gases will expand violently, resulting in delamination or catastrophic failure of the part.

Parts made by this type of molding will have fewer voids, resulting in lower moisture penetration and better electrical insulating properties.

Cable-Clave Molding. There are times when a part may be so immense in size that its dimensions exceed even the capability of a large pressure chamber. Yet high pressure and/or temperature may be required to develop the required final properties of the piece. One molder has developed a proprietary process for handling such a problem without prohibitive costs. For reasons that will become apparent, he has chosen to call the process "cable-clave" molding.

The most desirable shape for cable-clave molding is cylindrical. To do the job, a strong steel mandrel is machined to shape and properly treated with mold-release agents. The laminate or assembly is laid up on the mandrel, and a two-ply rubber diaphragm is next wrapped around the lay-up. This is followed by a relatively thin, flexible metal plate or caul. Finally, over the caul, a heavy strand of strong metal (usually steel) cable is tightly wrapped and secured.

The mandrel is heated to the required temperature, usually about 350°F., and hot water is forced at about 1000 psi between the layers of the two-ply rubber diaphragm described earlier. Because the entire assembly is restrained by the tightly wrapped cable, the hydraulic force is thereby constrained to being applied to the molding, which is cured in the process.

Rubber-Plunger Molding. Rubber-plunger molding is a process which simulates matched-die molding but uses only one mold, usually a cavity. The male, or plunger, consists of a solid rubber or suitable elastomer which has been cast or otherwise shaped approximately to the inner contour of the part.

In this method, the wet or prepreg lay-up is made in the usual manner in the cavity mold, which is then mounted under the ram of a hydraulic press or similar device. Instead of the vacuum bag, however, the lay-up need only be protected by a separating film, such as cellophane or a similar thin, flexible medium.

The ram of the press is then actuated, and the flexible plug is lowered into the cavity. Pressure is exerted, heat (if necessary) is applied, and the cure cycle proceeds. Upon completion, the ram is lifted and the part is removed from the cavity.

The molded part will have a low void content and a good inner surface when normal pressures of from 50 to 150 psi are used in the molding. When a material with reasonably good elasticity is used for the plunger, undercuts or areas of reverse curvature may be included on the inner surface of the part. This design feature is not usually attainable with matched-metal tooling.

Inflatable Rubber-Bag Molding. When thermosetting resins cure or polymerize, there is always a small amount of shrinkage. Polyesters, for example, often shrink as much as 10% in volume when cast unfilled. Even with the addition of fillers or reinforcement, there is a residual shrinkage of a few percent. This inherent property of the resin and its composites make the problem of constructing a cylinder by the wrapping process a difficult one. If a perfectly cylindrical mandrel is used, the shrinkage of the wrap during cure makes removal of the tube from the mandrel difficult, if not impossible. Tapered mandrels, of course, will not produce cylindrical pipe. Soluble or melt-out mandrels are relatively expensive in labor hours and make the process non-competitive economically.

A proprietary technique has been developed for making pipes, tubes, and tanks with parallel walls. The procedure is based on the use of an inflatable rubber bag which conforms closely to the shape of a cylindrical metal mold with end closures. The cylinder is wrapped with a reinforcement which is dry and contains no resin. After wrapping has been completed, one end of the assembly is dipped into a resin tank. A vacuum is applied to the other end, causing the resin to rise through the reinforcement, removing the entrapped air as it wets the fibers. If a closed tank is planned, end caps may be preformed and set in place, provided that at least one opening is left for removal of the bag at a later stage.

The mold is then clamped shut tightly and pressure is introduced into the inflatable bag. Customarily, the pressure may be as great as 200 psi. The entire assembly is placed in a sling or on hangers and attached to a conveyor system which carries the part through a water bath maintained at about 180°F. The hot water initiates resin cure and also removes the exothermic heat developed by the resin, preventing hot spots and overcure.

This process is used for making small tanks, such as fire extinguishers and water softeners, which can safely operate at internal pressures as high as 100 psi when properly designed.

Mating Molds for Wet Lay-up. In describing the molding procedures where a vacuum is used as a means of applying pressure to the part, it was mentioned that the surface of the piece will reflect the finish that is inherent in that of the mold. Consequently, when using the bag, one side of the molded part will show a smooth finish and the other will exhibit the texture imparted by bleeder fabrics, bag wrinkles, resin accumulations, and similar sources of irregularity.

Press molding with matched tools will, of course, result in parts with finished surfaces on both sides. In many cases, however, the desired results will not justify either the expense involved in making complex matched dies nor the pressure involved in hydraulic press molding. One procedure for obtaining the two-sided finish imparted by matched-die molding without the concomitant cost involves the use of mating molds. These tools are less complex than standard matched dies in that they may be made of either metal or plastic; they do not incorporate ejection mechanisms, cut-off edges, or heating cores; and they are far less rugged in construction. They consist merely of a pair of smooth dies, made to conform with the shape desired. If large parts are necessary, the individual halves of the mold may be sectioned and assembled during the molding process.

There are two methods of making parts with mating molds. One involves lay-up of the assembly outside the tools; the other uses the procedure of lay-up in the tools.

If the first method is selected, the wet lay-up is accomplished on a flat surface between cellophane or similar film sheets. After impregnation or the reinforcement, the entire assembly is moved to the lower half of the mating pair of molds. After all air entrapments and wrinkles have been removed with rollers or by lightly rubbing with squeegees, the second half of the tool is lowered onto the lay-up. Cure is at ambient or slightly elevated temperature, depending on the resin system.

The second procedure requires that the wet lay-up be made in the lower half of the mold. After removal of air and wrinkles, the upper half of the tool is set in place and the process allowed to proceed to its conclusion.

Several problems are inherent in this molding method. Primarily, an excess of resin is needed in the lay-up, so that run-off will carry with it the entrapped air and assure complete wet-out of the reinforcement. Because of this, close tolerance on resin content is difficult, if not impossible. It is also necessary that air "run-back" be avoided. This is a problem that results when insufficient contact is maintained between all areas of the mold surface during gelation and cure. Shifting of the mold halves during this period may permit air to leak back into the lay-up, resulting in bubbles and delamination. Both matching halves must also be completely prepared with mold-release agents. This, of course, is necessary in all molding. Press molding, however, has the advantage of hydraulic rams for tool separation as well as ejection mechanisms, but the procedure described here has no such advantages. Molds must be lifted by hand, by hoist, or by fork-lift, and separation can be difficult if release agents are not meticulously applied. Since no release film is used in the second process, both mold halves may be gel coated, enabling the fabricator to impart color and/or high gloss to either or both surfaces of the part.

Inspection, Testing and Quality Control.
Careful adherence to the processes described above will result in successful production of molded parts only if the processor is first aware of the need for similar adherence to a number of sensible rules for controlling the reproducibility of high quality in those parts.

Confusion often arises as to the precise delineation of the terms: inspection, testing and quality control. In fact, there is no real similarity of meaning among the three.

The term *inspection* is usually accepted to mean those steps in various stages of the molding operation during which the part is subjected to visual examination, dimensional measurement, weighing and determination of specific properties which may, by comparison with previously established norms, indicate the degree to which the part conforms to requirements and specifications.

Testing implies that the molded part is subjected to procedures which will indicate its physical, mechanical, chemical, optical and other critical properties. This includes flexural, tensile, and compressive stressing; thermal conductivity and/or expansion measurement; resistance to degradation by chemical attack, determination of the index of refraction; dielectric properties, and so on.

Quality control is an overall procedure by which adherence to all requirements is assured. It starts even before the raw materials are received by the molder and continues through every step of manufacture until the final product has been delivered to the ultimate consumer. In a manufacturing operation, it is wise to separate these functions from the production department, so that decisions regarding acceptance or rejection of parts because of deviation from specifications may not be controlled by personnel who have responsibility to meet quotas, hold down costs, or otherwise produce parts as a basic part of their jobs. Supervisors of inspection, test, and quality control should report to a responsible executive above the departmental level.

If a really effective quality-control procedure is established at the inception of each production run, all other inspection and test functions can become routine operations which guarantee the proper operation of the quality-control organization.

As previously stated, the procedure begins when the product requirements are established. At that time, material specifications are set up, the purchasing people are alerted to these specifications, and vendors are selected who will meet the indicated levels. Normally, vendors are required to provide specification sheets with each shipment, indicating any tests made on their materials and how the results of those tests meet the specification requirements. Where minor, correctible deficiencies or variations exist, deviations may be requested. If Quality Control feels that no deleterious effect will result from acceptance, the deviation is usually granted, subject to proper performance.

After delivery of the material by the vendor, it is customary for the purchaser to perform tests on selected samples to determine the accuracy of the vendor's specification sheets. Frequently, differences in test conditions will produce different results, so that some adjudication may be necessary. However, Quality Control should be the final arbiter of acceptability of a shipment.

Typical properties which may be subject to inspection are listed below. This is not a complete list. Special considerations may require other tests or may permit omission of any or all of those listed. The tabulation is provided as a simple check for coverage of general requirements.

Resins are normally checked for:
  Viscosity
  Color
  Acid number
  Gel time
  Catalyst effect (quantity, mixing, etc.)
Glass and other reinforcements are tested for:
  Proper style of weave
  Absence of weave defects

Type of distribution (mats)
Prepreg properties:
  Resin content
  Viscosity
  Flow
  Softening temperature
  Gel time
Lay-up procedures:
  Thickness (number of layers, thickness per layer, build-up areas, etc.)
  Wet-out (resin distribution and completeness of reinforcement wetting)
  Voids (complete elimination by all possible means)
Cure cycle:
  Temperature
  Pressure
  Time
Final inspection:
  Release from mold
  Molded thickness
  Mechanical testing
    Non-destructive on part
    Destructive if necessary, on samples or prototypes.
  Voids (inspect by back lighting, transparency, or similar method)
  Cracks, porosity, inclusions, or surface defects.

Most of the items listed are self-explanatory. Many of them are included as part of production part specifications. The molding method may eliminate the requirement for some tests, after uniformity has been established by early measurements, Among these, for instance, may be the dimensions of a part made in matched dies, closed to stops. Here, it would be necessary only to spot-check a few moldings from time to time to assure proper operation of the press.

In this connection, it should be obvious that no quality control procedure can be effective unless all equipment is kept in top operating condition. Mold surfaces must be kept up to standard; mechanical equipment—pumps, motors, presses—must have established maintenance schedules; plant cleanliness must be established and kept up; and the most important piece of equipment, the human operator, must be kept at a reasonably high level of efficiency. Carelessness, slovenliness, and inattention to detail is probably the cause of more rejections than all of the other factors combined. It must always be remembered that every reject, every cent spent on repair of rejected parts, and every man-hour lost by equipment down time is costly in terms of production and profits.

PRODUCT DATA

Design Rules for Vacuum-Bag Molding of RP/C. All of the rules applying to hand lay-up technology are basic to the design of bag-molded parts. This can be noted in Table II-3.1. Table II-3.2 lists the range of physical properties characteristic of bag-molded RP/C.

Economics. As regards resins, the largest percentage of vacuum-bag moled parts are fabricated using non-exotic impregnants. Costs, therefore, are not considered high. At this time, for example, most commercial polyesters may be obtained in the price range of $0.20 to $0.30 per lb.

Special application resins will be higher price, of course, and the addition of flame-resistant materials, ultraviolet inhibitors, and

Table II-3.1 General Design Rules for Vacuum-Bag Molding of RP/C

Minimum inside radius, in.	¼
Molded-in holes	Large
Trimmed in mold	No
Built-in cores	Yes
Undercuts	Yes
Minimum recommended draft, degrees	0
Minimum practical thickness, in.	0.060
Maximum practical thickness, in.	0.500
Normal thickness variation	±10%
Maximum thickness buildup	As desired
Corrugated sections	Yes
Surfacing mats	Yes
Limiting size factor	Mold size
Metal edge stiffeners	Yes
Molded bosses	Yes
Molded fins	Yes
Molded-in labels	Yes
Raised numbers	Yes
Gel coated surface	Yes
Shape limitations	None
Translucency	Yes
Finished surfaces	One (two possible)
Strength orientation	Random
Typical glass loading, %	20 to 50
Metal inserts	Yes

Table II-3.2   Representative Physical Properties for Vacuum-Bag Molding of RP/C

Property	Values	
	Mat Construction	Cloth Construction
Specific gravity	1.2–1.7	1.6–2.0
Tensile strength, psi	10,000–20,000	30,000–70,000
Tensile modulus, psi $\times 10^{-6}$	0.9–2.0	1.5–6.0
Compressive strength, psi	15,000–30,000	30,000–60,000
Flexural strength, psi	20,000–40,000	45,000–75,000
Flexural modulus, psi $\times 10^{-6}$	1.2–1.8	2.0–5.0
Impact strength, ft lb per in.	5.0–25.0	10.0–30.0
Barcol hardness	40–80	45–85
Moisture absorption, %	0.05–1.0	0.05–1.0
Burning rate	Slow to self-extinguishing	Slow to self-extinguishing
Heat resistance, continuous, °F	150–350	150–350
Resistance to acids and alkalis	Fair to excellent	Fair to excellent
Resistance to solvents	Good to excellent	Good to excellent
Machining properties (using carbide or diamond tools)	Fair to good	Fair to good
Resin content, %	50–70	40–55

other ancillary ingredients cannot be accomplished without a concomitant cost differential. However, only a small percentage of bag-molded items use specialty resins, such as the polymides or polybenzimidazoles.

With respect to reinforcements, the preponderance of fibers still lies in the glass field. Chopped mat (as preforms or spray-up), continuous-strand mat, and woven glass fibers do the major job. At present, more emphasis is beginning to appear in the area of high-modulus filaments, including boron, carbon, graphite, and metals. Most of these are extremely costly when compared to glass, however, and their application is currently limited almost exclusively to research, development, and a small amount of experimental production of specialized items for aerospace. Few commercial applications call for such high-strength, high-modulus reinforcements, especially with material costs ranging up to hundreds of dollars per lb.

Because the time factor cannot be ignored in any analysis of economics, the use of simple tooling cuts the lead time tremendously, especially when compared with matched-die molding. In the latter process, each half of the mold must be machined in a contour lathe or milling machine. The two parts must then be juxtaposed in a manner that will enable the machinist to provide precise spacing between them. Methods must be included to maintain the established juxtaposition at all times during use. Guide pins, cut-off edges, ejection mechanisms, and similar adjuncts must be included in the matched-die mold.

Lead times for matched-die tools customarily run from a few months to a year or more, depending upon the complexity of the surface contours, the dimensional tolerances and the tool material. Aluminum tools, for example, can be machined more rapidly than those of steel. When considering dimensional tolerances, we must also take into account the part thickness, which controls mold spacing. Accuracy requires many man-hours of machine time. In contrast, a prototype tool for bag molding may be made in a few days. Certainly, a large tool may require several weeks to fabricate. Generally, however, mold lead-time requirements are only a fraction of those needed for matched dies.

Another important economic consideration is the normal process of change during the planning of the tool and part. While an engineer or designer may have a clear and valid picture in his mind of what the part should look like or how it will appeal to the customer, the normal sequence of events results in a decision to make minor changes in shape, thickness, or some other major characteristic. This decision would be a major catastrophe with matched dies. It may, in fact, mean starting from scratch and retooling. With a single mold, however, it is only necessary to add or remove material from the mold surface at the places where contour changes are required. Thickness changes are handled by adding or removing layers of reinforcement. No adjustment of spacers, guide pins, stops, or cut-off edges is involved. The differentials are handled by variations in the vacuum bag, which adjusts itself automatically to the shape and dimensions of the enclosed lay-up.

In summary, we may say that, with the exception of contact laminating, vacuum-bag molding techniques probably offer the best economics in the entire RP/C field. Tool costs are reasonable, materials are not normally expensive, and highly skilled operators are not indispensable. Moreover, presses are not required and plant facilities are minimal.

Advantages of Vacuum-Bag Molding of RP/C. Many advantages of vacuum-bag molding procedures are inevitably woven into their economics. There are several others, however. Because construction of a mold for bag molding normally involves fabrication of a model or "mock-up" of the proposed part from which "splashes" or casts are taken to develop the actual tool, it is obviously feasible to take a number of casts from the same mock-up and thus produce a number of identical tools for multiple part production. Again, the cost is low because the greatest expenditure of time and effort is in making the mock-up to precise dimensions.

In addition to cost, size also not usually a problem. In making a set of matched dies, the machinist must have available lathes, mills, drill presses, and other precision equipment that can handle the parts involved. No such limitation is imposed upon the maker of a mold for vacuum-bagging. He can build his tool on the floor in any empty space available. Tools 20 to 30 ft long have been made without difficulty. Machines for handling such ungainly items are available in few shops anywhere.

The fact that variations in wall configuration may be handled with ease is another advantage. A single mold shape may be used for a dense solid laminate, a thin honeycomb sandwich, or a thick urethane structure without changing the surface of the mold in any manner. Again, the bag takes care of the variations.

Finally, the process results in finished parts whose density, absence of voids, physical and mechanical properties and surface finish are all far better than those properties obtained by simple contact molding procedures.

References

Further more detailed information on vacuum-bag molding, particularly design and performance of structures, may be found in the following sources:

Marshall, A., and Minaldi, R., "Adhesive Bonding and Design Handbook," Hexcel Division, Rohr Aircraft Corp., Dublin, Calif. 1971.

MIL Handbook 23, ASD, Structural Sandwich Composites Part #1, Nov. 1, 1962, Part #3, Nov. 9, 1961, U.S. Dept. Defense, Washington, D.C.

Technical Service Bulletin #120, Hexcel Division, Rohr Aircraft Corp., Dublin, Calif., 1970.

II-4

Major Applications and Procedures for Plastics in Tooling

INTRODUCTION

The term *plastic tooling* refers to the use of RP/C materials for duplicating models or tools with the required accuracy and durability.[1] Duplication is accomplished by casting, laminating, splining, and other methods. In this discussion, we will cover master models and model duplications, fixtures, metal-forming tools, vacuum tools, injection molds, foaming molds, foundry patterns, and core boxes.

MASTER MODELS AND MODEL DUPLICATION

For fabricating master models, several materials and methods are employed depending upon requirements. Large aircraft and automotive tools are now constructed using a syntactic (paste-like) foam, spread or troweled 1 to 3 in. thick over a core constructed to the approximate finished shape by fabricating honeycomb sheets (similar to templates), and then bonding these sheets together. This form is then cut to the exact contour on a tape-controlled mill and hand-finished. (See Figure II 4.1, upper left.) Large automotive models may be constructed by applying a carvable paste or syntactic foam over a rough-contoured, impregnated wooden core and then computer milling to shape. The final shape is then hand-finished to the bottom extremities of the milled grooves. Plastic materials for splining are used to provide fillets or changes in

This chapter prepared by L. Frank Bogart, President, Tool Chemical Co., Inc., Hazel Park, Mich.

models. These materials are easy-to-spread pastes, which may be easily worked both before and after they harden. After a master model is finished and certified for accuracy, a duplicate may be obtained by computer cutting another part to either the same or opposite hand. So far this method has not proven as accurate as fabricating a laminate female and then constructing a laminate duplicate.

Such a laminate duplicate is lighter in weight than a cast or splined part and is less affected by thermal changes and severe handling. Its construction involves several steps. First, a run-out or fence is built and a sealer coat (lacquer) is applied to the surface to be reproduced. A wax release coat is then applied over this and polished. A film of PVA is brushed or sprayed on. After thorough drying, a surface coat of epoxy resin is brushed over the entire area and left until firm (tack-free). A layer of fiber-glass cloth is applied and butted tightly over the entire surface and into corners. Epoxy resin is then brushed on to wet the cloth. This laminating procedure is repeated by applying glass cloth, butting it down tightly into corners, and again wetting with resin. Care is taken to prevent any excess resin from accumulating in puddles, which upon curing will create excess heat and non-uniform shrinkage and result in poor accuracy in the finished tool.

Often special glass cloth is used, such as three-dimensional fabric. This fabric is self-wicking, thus eliminating resin-rich areas, and is also quite thick, reducing the number of layers required, and consequently labor and time. After applying approximately 14 layers

95

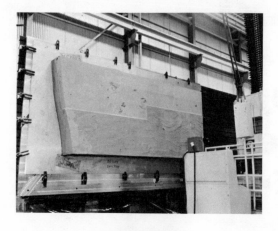

Figure II-4.1 Four examples of plastics in tooling. *Upper left:* Elevation view of machine for numerically controlled (computer-taped) fabrication of aircraft master models from a syntactic foam blank. Present two-axis, numerically controlled machines are limited to a 16 in. rise and fall cut geometry. Development of numerical-cutting use is continuing, however, and it is estimated that 20% of the masters for future aircraft and other components will be created using this technique. (Courtesy of North American Rockwell Space Division, Downey, Calif.) *Upper right:* Large model of an aircraft landing gear fabricated from carvable clay. (Courtesy of Astro-Netics, Inc., Warren, Mich.) *Lower left:* Light-diffuser of cast epoxy from an Ad-Rub® rubber face backed by an epoxy laminate mold, illustrative of model duplication in plastic tooling. (Courtesy of Tool Chemical Co., Hazel Park Mich.) *Lower right:* The Bourdeaux Process. Automotive window-checking fixture of metal-filled epoxy facing over a cast-metal base. The plastic face is designed to withstand extreme abrasion. (Courtesy of Bourdeaux Plastic Industries, Inc., Detroit, Mich.)

of 10 oz cloth or two layers of 10 oz cloth plus two layers of three-dimensional fabric, a thickness of approximately ¼ in. has been layed up. At this point, the laminate should be allowed to set overnight to cure. A back-up structure of epoxy tubing or epoxy-faced honeycomb is then bonded to the laminate exterior for support, using either a non-shrinkable glass-filled epoxy paste, or fiber-glass tape with epoxy resin.

Small models may be carved out of new claylike materials which are either used by themselves or are cast over a core of preimpregnated wood or metal. These materials are strong and easily worked because they have no grain. They are also durable enough to act as prototype parts (see Figure II-4.1, upper right). Cast plastic molds may be taken from small models by casting rigid epoxy resin as a face over impregnated materials or over a mass-cast plastic core. Tooling rubbers, such as silicone or Ad-Rub®, cast over cores of impregnated wood or metal, provide excellent reproduction of surface detail. This is particularly true when using silicone rubber, because silicone requires no release agents either from the part or model or when casting the epoxy duplicate. (See Figure II-4.1, lower left).

Ad-Rub is very good for deep grooves and severe undercuts, because it has excellent tensile strength and no appreciable shrinkage. It requires only a light wax coat separator for best release. Where thin sections are encountered, Ad-Rub has been able to replace the plaster female for duplication. Either silicone or Ad-Rub may be bonded to a base or core by priming the latter with a silicone primer.

For solid casts or large casts, other special tooling resins are employed. Casts up to 3 or 4 in. thick may be made with special epoxy systems specifically designed for that purpose. Large casts usually employ an initial surface coat of tough resin-filled material to provide detailed reproduction without voids. This is followed by a special mass-cast resin system. Granular fillers such as aluminum chips are generally used to lower cost, reduce or eliminate shrinkage, and reduce the danger of distortion caused by resin exotherm. These chips are available in 50 lb. bags and are approximamately $3/32$ in. in size. They are usually mixed in the proportion of two parts to one of resin mix, thus materially reducing the cost of mass casting.

FIXTURES

Plastic fixtures are tools used to hold parts for assembly, for work operations, or for checking contour or size (see Figure II-4.1, lower right). These tools generally are reproductions of the peripheral dimensions of the part and thus often utilize narrow laminates or small castings supported by a rigid frame. These laminates may use narrow glass cloth called "tape," which comes in widths of ½ to 12 in. and in thicknesses of 0.010 to 0.079 in. Great care must be exercised to prevent resin-rich areas which could cause warping and shrinkage.

The back-up structure which employs fiberglass tubing is carefully fitted together and bonded. The special tube lengths have a cast on the bottom side to allow precision machining for correct height and side locations.

Often honeycomb fabrications are used in a box or egg-crate structure to supply rigidity. Machineable bases which provide for accurate location to reference lines on the master are purchasable.

For long, narrow structures, steel is usually used because of its higher flexural modulus of elasticity. Care must be taken with steel weldments to prevent dropping or severe shock. The steel is more likely to take a permanent "set" or warp than are the more resilient plastic laminates.

When utilized for fixtures, cast blocks are usually molded using tough, abrasion-resistant materials, such as epoxy or urethane. These resins are generally cast in small blocks which are bonded to a rigid frame much the same as a laminate skin would be.

When utilized as spotting blocks or spotting racks, the size of the fixture usually dictates the process; small fixtures are usually cast, and medium or large ones almost always laminated by the same process used for a duplicate die model.

A new procedure devised for rapid fabrication of low-cost, accurate fixtures comprises the utilization of a two-component, filled epoxy resin, claylike material[2] that is mixed and packed onto the contour to be reproduced until a skin of approximately 1 in. has been built-up. This clay hardens overnight to a tough, non-shrinking, stable form. Once stable, the form is framed by a special plastic board which can be drilled, tapped, or fitted with self-tapping screws prior to separation from the original contour. This process can reduce fixture fabrication cost by more than 50% and time by two-thirds or more. The clay comes in several forms. One is extremely tough and hard. Another is somewhat lighter, with low resilience, and is used to make spotting racks. This process has been very successful in several tooling programs in the automotive industry.

In many instances a simple template shape will serve adequately as a checking fixture. One material that performs satisfactorily under these circumstances is a single sheet of plastic which has been machined to the desired contour. Many different plastic sheet materials are available: phenolic paper sheet, which is easily machined; epoxy-glass cloth laminates, which are extremely strong and stable; canvas phenolics, which are tough and durable; urethane, which is extremely tough and wear-resistant.

METAL-FORMING TOOLS

The use of plastic materials as a base surface for forming short-run and prototype sheet-metal parts has been very successful. Aircraft stretch tools; hydroform tools; draw, hammer, and other forming dies for the automative industry have been made of plastic for 20 years. Construction varies with the requirement, although three main processes are presently used: (1) cast faces on kirksite cores, (2) cast laminate faces on cast plastic-composite cores, and (3) laminate faces backed by high-density foam cores.

The cost of casting plastic faces on kirksite is often the lowest because of the reclaim value of the kirksite core. The cost of equipment, handling, and set-up however, can offset these savings unless the equipment is already available in the tooling shop. The casting usually shrinks sufficiently to allow the necessary ¼ to ¾ in. clearance on the male plug. Metal clearance is provided by waxing over the plug and ring development; suspending the oversize female die in place; and then applying a ¼ to ¾ in. face cast of a tough, hard-cast epoxy. Today, most automotive panels for prototype and reproduction are formed this way.

Shops without kirksite-casting equipment cast a core of rock and epoxy or aluminum chips (aproximately $\frac{3}{16}$ in. size) also with epoxy. They then either cast a face as before or cast the cores with a mass-cast epoxy mix over heavy, tough surface coats.

Where shear loads are high or where impact loads are likely to exceed the physical properties of face-casting materials, laminates of epoxy and glass may be used. The glass cloth must be impressed directly onto the surface of the tool to prevent occurrence of stress failures in any unsupported sharp edges of plastic which might appear. Hand-hammered parts may be made on tools constructed like the draw and form dies, but the design of metal-working tools must be carried out by people who know the forming and drawing characteristics of metal.

Aircraft stretch dies are getting larger and larger. The core mix of granite chips and mass-cast epoxy over laminate skins has provided

excellent tools for a core cost of aproximately $12.00 per cu ft (6 parts chips to 1 part resin). These tools are heavy and require special handling equipment. Some large tools are now constructed using a heavy laminate surface backed with a puttylike material up to 1 or 2 in. thick and foamed behind with a 10 to 15 lb per cu ft foam. The base is plywood covered with a cast or laminated epoxy to provide strength.

VACUUM TOOLS

Plastic materials have proven very successful in the production of vacuum-forming molds. For use in small- and medium-size tools, the form can be cast solid. For medium or large tools, the surface is usually heavily gel coated with a heat-resistant surface coat backed by a laminate to provide sufficient strength.

This surface laminate system is often vacuum-bag molded to assure a void-free laminate that will not develop the surface and laminate failures common in severe and prolonged use. Behind this face, cooling coils are positioned and cast in place with a mix of aluminum diamonds mixed in an epoxy paste (6 to 8 parts aluminum to 1 part of resin mix), applied in a thickness of approximately 1 in. This cavity is then filled with a core mix of 6 to 8 parts of lightweight Rocklite® pellets to 1 part of resin mix, thus providing a lightweight porous backup. The walls and base of large tools are often enclosed in honeycomb sheet to provide strength plus light weight.

Following fabrication and cure of the tools for vacuum forming, holes must of course be drilled into the surface. In order to avoid drilling into the metal cooling coils encapsulated behind the laminate face, a low-voltage lamp circuit is connected across both the coils and drill. Any proximity to or contact of drill and coils causes the lamp to become incandescent, thereby warning the operator to select a new site for drilling the particular hole.

INJECTION MOLDS

Temperature-resistant (to 250°F) epoxy resins have made possible rapidly fabricated, low-cost, prototype injection molds. Special high-heat and abrasion-resistant surface coats are used, backed by a cast of the same thermally resistant resin on an epoxy and aluminum-chip mix core.

Modifications of this construction include cast plastic faces on metal cores. The use of solid casts for small parts has also been successful, although the molding cycles are necessarily slow because of low heat dissipation of the curing plastic mix. This limits the use of plastic to short-run and engineering prototype parts.

An old but growing process is the use of metal-cast, plated, or electro-deposited shells approximately 0.08 in. thick, backed by epoxy for injection molds. Metal and plastic over concrete have also been proposed and evaluated.[3]

FOAMING MOLDS

The furniture and decorations industries have very recently begun to utilize cast and foamed plastics to replace wood parts. This change has required the use of duplicate tooling molds, often by the hundreds, and mostly for wood grain or undercuts. The molds must be flexible and provide good life and ease of fabricating. Among the most widely used materials are silicones, Ad-Rub®, and urethanes. A mold life of 80 to several hundred parts per mold is typical with mold rubbers. The rubber skins are usually backed up by a cavity or frame to provide proper alignment and leveling of edges. The stiffer material needed for the more dense foams tends to break down in the deeper molds unless thickness is kept to the minimum required for strength.

The use of release agents or self-release barrier coats greatly increases mold life. However, they must not require a cleaning operation for every cycle or the savings can be lost.

Prior application of a silicone emulsion onto all surfaces likely to come into contact with spilled plastic or seepage greatly reduces clean-up time. This emulsion may be used on the cardboard bleed to release it, rather than have a paper back which would cause warping. Finally, room-temperature cure flexible rubber is used extensively for production. A rapid improvement in materials continues.

FOUNDRY PATTERNS AND CORE BOXES

Tooling RP/C used for patterns in the foundry provides fast, accurate duplication for multiple patterns or core boxes. They are usually cast solid using an abrasion-resistant, rough epoxy or urethane. Larger tools may be cast with a metal back-up (often with hinges) plus an abrasion-resistant face. Proper formulation will prevent sticking of sand cavities or cores even when the sand is hot.

Laminates with tough, abrasion-resistant surface-casts provide large, lightweight tools with excellent accuracy. Storing and handling are simplified because of toughness, light weight, and non-deterioration under the worst conditions.

The use of rubber molds for plaster and ceramic cores is widespread. Also available is a type of tooling rubber which is capable of reproducing the finest detail with no shrinkage away from the dimensions of the model.

Plastic casts are used to make wax cores for precision casting. They are tough, strong, and accurate and can be stored for long periods without deterioration.

SUMMARY

The use of RP/C for tooling has steadily increased in both variety of application and sophistication. No shop can overlook the cost savings afforded by plastic tooling. Anyone interested in starting a program or updating his present operation need only contact any of the reliable formulators to obtain the latest information regarding materials and processes.

References

1. Proceedings Twelfth National Plastics for Tooling Conference, Wayne State Univ., Detroit, June 16, 1970.
   Plastics Technology, May, 1971 (entire issue), p. 11 et seq.
   SPE Journal, Vol. 27, March, 1971, p. 24 et seq., Savla, M.
2. Maraclay® (Marblette Corporation) Available from Tool Chemical Co., Inc., 29 East Eight Mile Road, Hazel Park, Mich. 48030.
3. *SPE Journal*, Vol. 27, May, 1971, p. 18, Garner, P. J.
4. *See also* "Glossary of Terms," SPI RP/C Tooling Division, O. D. Lascoe, Purdue University, W. Lafayette, Ind.

II-5

Miscellaneous Processes Related to Low-Temperature Cure

INTRODUCTION

Many interesting and productive variations of the hand lay-up process and low-temperature cure are available. They involve the simplest method of combining resin and reinforcement, varing from open mold work to cold-press molding. All utilize hand-lay-up type resins with room-temperature cure catalyst systems.

HISTORY

These processes were not the first used in RPC work. In most cases, they had to wait for the availability of thixotropic additives to control drainage; accelerators to lower catalyst breakdown to room temperature, such as cobalt napthanate; waxes in the resin to mask air-inhibition so that resins would cure dry and non-tacky to the touch.

When these materials appeared, open molds, molding onto vertical surfaces, and non-tacky cured parts became possible. They eliminated the necessity for heated molds, vacuum or pressure bags, cellophane over the lay-up to keep out air or application of dry powders to prevent tackiness. Large parts, such as boat hulls and tanks, could therefore be manufactured by the open-mold, room-temperature cure method.

GUN-WETTING PROCESS

In the gun-wetting process, woven reinforcement or mat is cut and placed into the mold and the resin is sprayed on. This eliminates the need for the old bucket and brush technique.

Materials:

Glass Reinforcements. Glass mats, woven roving, or glass cloth are used as reinforcements. The maximum reinforcement thickness for single-spray application should be limited to 5 oz per sq ft for fine detailed work, and about 6 oz per sq ft for large-scale work. The finish or binder on the glass should be of intermediate styrene solubility for large-scale work in order to avoid quick disentegration of mat integrity. The highest styrene solubility is satisfactory for work in molds. Double layers of reinforcement can be laid up. Chopped strand shaken onto the work and gun-wet is also used effectively in tooling.

Catalysts and Resins. Standard double-promoted, room-temperature cure resins are satisfactory. If the double-pot system is used, unpromoted resin is necessary. MEK peroxide catalyst, or the equivalent for room-temperature cure, is the most widely used curing system. Catalyst injection systems may be used to permit faster cures.

Equipment and Tooling:

Guns. Some specialized gun-wetting is performed by using a perforated roller with discs through which two-pot system resins are pumped. Final mixing is accomplished within the roller and on the mold. The best procedures, however, use two-pot systems which are airless. The nozzle of the gun is sized to release the resin in a spray like that of a garden sprinkler, thus minimizing overspray and styrene fumes.

Spray systems may be automated with reciprocating resin spray guns, as in the machines used for manufacturing flat RP/C panels.

Rollers and Finishing Equipment. Same as for hand lay-up.

Tooling. Same as for hand lay-up.

Molding Method—Processing Details:

1. Apply release agent and gel coat in the standard manner.
2. Apply one coat of resin onto the cured gel coat.
3. Place the reinforcement into the mold and affix to the mold edges if necessary. Allow at least 2 to 3 in. of overlap at the edges of the pieces of reinforcement. Stagger all breaks and roll this material into the resin previously applied.
4. Commence and complete the wet-out using the gun or applicator, and roll the resin into the reinforcement to remove any trapped air. Install bracing or ribs if required.
5. Allow to cure. Trim at leather-hard condition if feasible, and remove from the mold.

Product Data—Examples:

Tooling. All types of tooling used for hand lay-up work may be applied to the gun-wetting process.

Flat Panels. Flat panels intended for use in building construction are molded by automated spray. Roll-out of the prepared reinforcement is accomplished on molds which are transported on a continously moving platform or belt, passing beneath the point where the sprayheads are located.[1]

Instant Roads. The 4 to 6 oz per sq ft fiberglass mat reinforcement is unrolled from a truck or special vehicle onto earth or ground graded as the vehicle passes. Catalyzed resin is sprayed directly onto the web on the ground and is then rolled into the web.[2] (see Figure II-5.1).

Advantages of Gun-Wetting:
- Reduces waste and spillage of resin.
- Carries the resin to the point of application.
- Permits fast-curing systems because pot life only commences at the point of application, not on the mixing floor, as in the normal hand-lay-up system.
- Permits division of labor and therefore less mess; the gun wet-out man works next to the roll-out man.
- Results in improved uniformity of resin application.
- Easy to monitor the resin usage.
- No catalyst waste. No loose catalyst on the laminating floor.

Disadvantages:
- Equipment is comparatively expensive.
- Equipment requires trained personnel to operate and maintain it.
- Hoses trailing the wet-out man may contact or be dragged across an uncured lay-up and dislocate the reinforcement.
- Hoses get messy and require frequent attention.
- Process is neither efficient nor economical unless sufficient work is available to keep the unit working all day.

PREWETTING PROCESS

Pre-wetting is similar to hand lay-up except that the resin is combined with the reinforcement outside of the mold. Saturation is allowed to take place and the composite mass is then transported to the mold and worked into the mold shape, rolled out to remove voids, and allowed to cure.

Materials:

Glass Reinforcement. Woven roving, Plymat,[3] and glass mats are the primary reinforcements. Cloth is useful as a carrier. Glass mats should not be left unsupported. The binder for mats should have medium to low solubility in styrene so that the fibers will not pull apart during handling.

Catalysts and Resins. Standard types such as MEK peroxide and normal hand lay-up resins are used. A lower catalyst concentration permits longer handling time.

Equipment and Tooling Guns. Applicator guns may be used, preferably the airless spray type, to minimize contamination of the pre-wetting area.

Figure II-5.1 Continuous laying of "Instant Roads." Arriving at a swampy or sandy terrain where a road is to be laid, an applicator device carrying an 11½ ft. x 250 ft. roll of fiber-glass mat plus liquid resin and a catalyst, is towed by a bulldozer which drags its blade to prepare the road surface ahead. As the bulldozer progresses, smoothing the surface, a continuous ribbon of 11½ ft. wide, resin-saturated fiber-glass roadway is trailed out behind. In less than an hour, the entire system hardens to provide the road surface to support the weight of heavily loaded military vehicles. Experiments have been conducted in a wide range of weather conditions. (Courtesy of the United States Marine Corps.)

Applicators. Much of the success of pre-wetting is in the choice of a resin applicator. Squeeze-type roll applicators and doctor-blade types work well for fabrics and woven rovings although the latter drag on mats. Curtain coaters or drip-through applicators work best with reinforcing mats.

Rollers and Finishing Equipment. Same as for hand lay-up.

Tooling. Same as for hand lay-up.

Molding Method:
1. Prepare mold and apply release agent and gel coat, if required.
2. Apply the reinforcement to the pre-wetting table or area. The reinforcement should be cut to shape previously, if feasible. Non-porous cellophane over cardboard may be used for support.
3. Wet out the reinforcement using one of the following methods: Apply the resin from a pail, spray gun, or traversing manifold. After application, wet out the reinforcement using brushes and rollers or other mechanical means. Invert the reinforcement and wet out from reverse side for more thorough treatment. Reinforcement, especially mat, should not be left unsupported or stacked up in contact.
    Transfer the lay-up to the mold either by lifting the cardboard holding the membrane or back-up sheet or by rolling the wetted reinforcement onto a cardboard or wooden tube and then unrolling it onto the work. Cellophane, metal screen, or other non-porous materials may be used to assist in the transferring. Finish the lay-up with rollers and brushes.
4. Repeat to build up the required number of layers according to the specified lay-up

schedule, adding stiffeners. Roll out each application.

5. Allow to gel and trim at leather-hard condition, if feasible. Remove from the mold in the standard manner.

Variations of the Wet-Out Method in the Pre-Wetting Process. When rovings are used to fabricate RP/C parts, such as in the weaving of grills and walking flats for chemical and marine environments, it is common practice to prepare a trough filled with the pre-catalysed resin and to soak the rovings in this trough. Rolls of tapes and other fabrics that will resist unraveling can also be soaked in the resin and applied to the mold by wrapping, as in making tubes, or drawing through an orifice and hanging vertically to cure, as in strain rods, hot sticks, and low-cost rod stock. By this dip method an exact amount of resin may be weighed and prepared for take-up by the required amount of reinforcement.

Product Data[4]

Advantages:
- Pre-wetting provides excellent control of resin content.
- It is particularly good for large-size lay-ups, especially if the wetting is accomplished by machine.
- The automatic nature of the wetting operation saves time.
- Control of the lay-up is easier to maintain because the wet-out crew can keep a regular work schedule, and the lay-up crew merely accepts the material in sequence.
- The process is extremely efficient when there is a great deal of vertical work, for example, in large tanks and boat hulls.
- Pre-wetting may be used in combination with or for fortifying other types of lay-up, such as spray-up.
- An excellent finished molded surface is attained in pre-wetting. Gel coating may be eliminated in some cases.
- The pre-wet lay-up method may be used in compression molding to great advantage.

Disadvantages:
- Equipment costs can be high for large installations.
- Maintainance costs can be high.

- A potential waste exists if carrier film is used only once or twice.

CONFINED-FLOW OR CONFINED CAVITY MOLDING

Paired-plug and cavity molds prepared with a gap between them, appropriately sized for the finished thickness of the RP/C part, form the working structures for confined-flow moldings. The reinforcement is laid in place, with oi without resin, and the mold halves are closed and clamped. Resin is forced throughout the reinforcement by means of pressure, vacuum, or by the force of squeezing as the molds are closed. Large and small parts can be made using this process.

Materials:

Reinforcement. For glass-type reinforcement, high-solubility mat material is used almost exclusively. Several types are available, and attention must be paid to the direction of the loads. Fiber-glass mats are available that contain some parallel-strand roving.[5]

Other materials, sisal, burlap, abestos, and many synthetics can also be used in this process. Some experimenting may be required to select the proper binder or finish for the material to permit good wetting by the resin.

Catalysts and Resins. Standard catalysts and resins for room-temperature cure systems are used. Gel times must be prolonged for full impregnation. For large items, cumene hydroperoxide plus a manganese naphthenate system (ratio 1.0 or 1.5%: 0.4%) is used.

Equipment and Tooling. High-capacity vacuum or pressure pumps are a requirement. Other equipment is roughly the same as for other hand lay-up systems.

Molds for confined-flow molding are generally made in the same manner as those used in the hand lay-up technique—with one important difference. Because vacuum or pressure is used to force in the resin, the molds must be designed like pressure vessels with the back-up structure strong enough to prevent deflection. In vacuum impregnation (the Marco method), a mold which was under designed would permit compression of the reinforcement between the mold halves, preventing the resin from entering and completely saturating. As a result, dry spots would appear in the final part. Conversely, in pressure impregnation, an under designed mold would permit the halves to be forced apart and the part within would tend to be resin-rich, heavy, and prone to easy fracture in its end use.

In the vacuum-impregnation method, a female mold is taken off the master plug. Moldmaker's wax is then applied to the female to the correct part thickness, and a male or inner mold is laid up over the smoothed wax surface. The wax is removed, and the mold halves are made ready for the process with correct reinforcement, addition of vacuum fittings, clamping means, and a reservoir around the juncture of the two molds.

For pressure impregnation, the molds are made in the same manner, clamping means are provided, and pressure pots are required to force the resin into the space with the reinforcement. Pressure sealing is required at the periphery between the two mold halves. Sometimes limited bleed is provided so that all the air can be let out.

In the squeeze-impregnation method, the molds are made as described above, but with the considerable extra strength required to squeeze out the extra resin previously impregnated into the reinforcement. It is common to utilize a press for the squeezing, also providing a simple rim-clamping to hold the pressure until catalyzation has occurred. One press then can be used for several molds.[6]

Molding Method—Processing Details

1. Prepare both molds with the best available grade of release agent. Gel coat one or both mold surfaces.
2. Tailor the mat reinforcement to fit over the male portion of the mold staggering any laps of material in brick fashion to avoid thickness build-up which could cause dry spots in the part.
3. Close and clamp the mold.
4. Apply pressure or vacuum to the resin or mold and allow the resin to permeate the mold cavity, driving all air before it.
5. If squeeze-impregnation is used, pour catalyzed resin over the reinforcement after tailoring it onto the mold surface. Then place the mating half of the mold in position and clamp either half with rim clamps or with an auxiliary press to force the resin through the reinforcement to the periphery of the mold.
6. Allow the clamped assembly to remain until gelation and Barcol development have been achieved.

In all of these processes, remove the excess resin before it progresses past the leather-hard state. Otherwise, considerable time and effort will be necessary to clean up the part and the molds for the next operation.

Product Data

Advantages:
- Molds are usually inexpensive and can handle large shapes.
- The process lends itself to prototype, experimental, or tooling pieces as well as to full-scale production. Monocoque structures are possible, thereby avoiding multi-component parts which require elaborate post-molding assembly.
- Resin is confined and, as a result little or no styrene fumes escape and fire hazards are minimized.
- Both surfaces can be mold surfaces and can be gel coated.
- They can also have different textures and colors.
- Cures can be accelerated either by having a well-insulated mold, or by applying mild heat from an oven or from electric strip heaters with metal molds only.
- In pressure impregnation, metal or RP/C molds may be used.
- In vacuum impregnation, the molds must be very rigid because the vacuum pushes the molds together, thereby squeezing the lamination, restricting the flow of resin, and possibly generating dry spots.
- Small molds for squeeze impregnation can be lightweight as long as they can be put into a vacuum bag. This occurs in a process variation very similar to ordinary vacuum-bag operations, but with molds on both sides of the lamination.

Disadvantages:
- Resin rising through the reinforcement in vacuum impregnation follows the path of least resistance. Therefore, weak molds, overlaps in the reinforcement, or anything that causes strictures may produce starved areas in the finished part.

- There is no way of knowing the quality of the laminate until the resin is cured and the molds are opened.
- Resin content is usually higher than in other room-temperature molding methods for RP/C.
- Excess resin must be used and wasted in order to provide a quantity sufficient to wash out trapped air and to ensure adequate mold fill.
- Neither undercuts nor sharp corners should be built into molds for these processes.
- The method is slow and cannot be successfully forced.
- The method demands the most careful mold-release preparation and the best mold-release materials.
- In large parts, because of the sheer size, the vacuum must be at least 26 in. Hg. Extra pumps should be provided so that failure of one during vacuum or pressure impregnation does not mean that the part will have to be scrapped.
- Vacuum impregnation requires careful placement of the draw cocks for the resin. As impregnation proceeds, it usually becomes necessary to stop off the cocks where the resin arrives first and switch to other cocks for finishing the impregnation. Selecting the sites for these cocks may therefore require several attempts and possibly the loss of several parts.
- Close control of resin temperature, viscosity, and rate of resin flow is a rigid requirement.
- In some parts, there is a tendency for the reinforcement to "wash" (move around). Prior to impregnation it may be necessary to stitch or baste the plies together. This means extra labor.
- In vacuum impregnation, it is critical to remove the moat of resin around the periphery between the male and female mold at the point during cure in which the resin is leather-hard. Removal too early breaks the vacuum, and removal too late requires a great deal of extra labor.

ULTRAVIOLET CURING SYSTEMS

The ultraviolet curing method is based upon resin systems which contain ultraviolet or sunlight-sensitive curing agents which are blended with suitable fiber glass reinforcements. These systems cure with a minimum exotherm when subjected to normal ambient temperatures of 32°F. and upward in the presence of ultraviolet radiation.

Materials:

Reinforcement. Almost any material which is compatable with polyester resin can be used as a reinforcement.

Prepregs. One company manufactures material in the form of preimpregnated fabrics and rovings.

Catalysts. Usually benzil or benzoin in tricresyl phosphate,[8] is added to the resin in 3 to 5% quantity.

Resins. High-reactivity styrene monomer polyesters are recommended, although other types may also be used.

Equipment and Tooling. Guns, tools, and other equipment are the same as those used in the hand lay-up method.

Molding Method—Processing Details. (*Vacuum Bag, Prepreg*)
1. Prepare the mold with the proper release agent.
2. Unroll the preimpregnated, ultraviolet-sensitive reinforcement, and cut to size.
3. Apply the reinforcement to the mold in the pre-determined pattern.
4. Apply the vacuum bag carefully and apply the vacuum, working out the trapped air by rolling. The resin is still in "B" stage.
5. Turn on the ultraviolet light, or move the entire assembly outside into sunlight.
6. Re-work the laminate under the light as the resin commences to flow and work out the remaining voids.
7. Allow the laminate to cure, turning off the lights or moving it inside.
8. Remove the part and trim. Clean the molds.

Molding Method—Prosthetic Body Laminates.
The special technique of prosthetic body laminates using ultraviolet curing technology was put into practice as early as June 1965. It was suggested as a way of making leg inserts for amputees while still under anesthetic. A re-

cently issued patent[9] describes the process in detail. It is titled "Method and Application of Orthopedic Appliances With UV Curable Plastic Bandage."

The sequential steps in the process may be outlined as follows:

1. Prepare the patient's limb with a careful washing and drying. Pull cotton or a synthetic stockinette over the affected area.
2. Tape thin layer of soft, flexible, synthetic foam material over the stockinette.
3. Apply and secure wrap of 2 mil polyethylene tape, 2 in. wide over the foam.
4. Wrap the pre-wet reinforcement (with ultraviolet catalyst added) over the polyethylene to an average thickness of $\frac{1}{16}$ to $\frac{1}{8}$ in. or more in areas where more stress is likely to be applied.
5. Apply the ultraviolet light and rotate it slowly around the limb or rotate the limb under the light, if possible.
6. Cure is achieved in 20 to 30 min. Much faster cures are possible if the correct frequency ultraviolet light (3000Å) is available.[10]

Molding Method—Conventional. The conventional ultraviolet-cure technique was used as early as 1947 in the manufacture of complicated shapes such as ducting for lighter-than-air-craft (blimps). The following steps comprise this method:

1. Wooden or plaster male mandrels are fabricated or selected, then waxed, polished, or sprayed with PVA release agent.
2. Impregnate thin fiber-glass fabric (style 181) with polyester resin plus about 5% Benzil catalyst in TCP. Wrap this around the mandrel in the prescribed fashion.
3. Place the wrapped mandrel under ultraviolet lights, and turn over every 10 min. for a total time of about 20 min. on all portions of the part.
4. Remove the assembly from the ultraviolet area and carefully cut the laminate off the mandrel in a predetermined area or along a premarked line.
5. Reassemble the thin-cured skin with masking tape on the inside mating edges, as far as feasible into the tube.
6. Wrap the part with remaining laminating tape, saturate it with resin catalyzed with 1.0 to 1.5% MEK peroxide, and cure in the normal hand lay-up manner.

7. Trim the cured tube to dimension and remove the masking tape.

Product Data

Advantages:
- No promoter is needed.
- The system is stable if kept in the dark. Unlimited pot life is possible.
- The system cures with very little exotherm. Parts can be cured next to the body without degrading human tissue.
- The process permits manufacture of complex parts with all primary bonds (those in which the mating laminates have uncured resin.)
- Selective cures are possible, such as panels made with an uncured RP/C strip between two membranes or skins. Any time after production, the panels are erected and the uncured strip is exposed to ultraviolet radiation and rigidizes.

Disadvantages:
- Common ultraviolet light sources are very inefficient. Soon after startup they usually emit more visible than ultraviolet light.
- Efficient ultraviolet lights are available[12] but are costly and of limited size and variety. This situation could be improved as the technique becomes more widely used.
- The process is currently limited to special applications. Workers require goggles and protective clothing. Persons with sensitive skin should be excluded from working with this process.
- Prepreg systems require the use of a vacuum bag to get proper laminate density and void-free properties.

FILLED RESIN TECHNIQUES

The manufacture of filled resin systems includes synthetic marble products, wood products, terrazzo, floor and wall coverings, and glasslike translucent window materials. One may argue that *filled* resins are not reinforced resins. Resin and filler composites, however, show a synergistic effect resulting from the mixture of materials.

In the building industry, many applications exist for polyester resins mixed with sand, water, stones, ground granite and limestone, ground shell flour, alpha-cellulose papers, and the like. The result is a bewildering variety of synthetic products that in most cases have substantial advantages over the natural materials. The products are usually lower in cost and weight, more resistant to chemicals, and easy to duplicate. All these factors make mass production an easy and economical matter.

Materials:

Resins. General-purpose polyester resins do a remarkable job in most applications. (Other resins, such as the epoxies, are not generally used because of cost.) Several resin manufacturers make resins specifically for marble, woodlike products, body putties, and the recently developed water-extended polyester resins for low-cost castings.[13]

Catalysts and Accelerators. For room-temperature cures, the common MEK peroxide and cobalt-naphthenate system is adequate. Exotherm is usually quite high with marble mixes, a favorable quality which tends to hasten thorough cures without the need for ovens. Even so, a low-temperature oven post-cure (140° for 2 hr) does a great deal to promote maximum physical properties and allow greater usage of the molds for each day's production.

Additives. Additives vary markedly depending upon the product being made.

Simulated marble products require the following fillers: quartz no. 1, quartz. no. 00, silica 100 mesh, silica 200 mesh, crushed granite no. 00. For veining materials—those that make the striations or grain in the marble—collodial silica, China clay, and pastel color pigment can be used.

Furniture products require calcium carbonate, pecan flour, pearlite, Saran[14] beads, glass beads, pigments, and decorative printed overlays such as those on alpha-cellulose papers.

Tooling and Equipment. The tooling is generally simple for marble products. In large-scale production, much use is made of polished (5 mμ) stainless steel for counter tops, vanity tops, and the like. Cast aluminum is used for the bowls and other shapes not formable with pressbrakes.

In making furniture, mostly silicone rubber, cast urethane, vinyl plastisols, rubber latex, and some other materials are cast from wooden or plaster masters. The flexible skin is usually backed up with RP/C, plaster, or combinations of both. For regular or symmetrically shaped parts, steel backing can be used. Needless to say, there are numerous commercial formulations available that can be used for mold material. We are still (1973) looking for that low-cost, long-life, flexible, and tear-proof mold material.

For mass production of any of the above products it is desirable to use conveyorized molds passing under a prepared compound dispenser and then through a room-temperature (or insulated) adiabatic section to permit the exotherm to build up and hasten the cure. Often, a low-temperature oven is added to shorten the line and the corresponding time cycle.

The mixing dispensers are usually combinations of comercially available equipment[15] and home made adaptations specifically designed for the product at hand.

Resin is usually put into a mix tank, the filler and color are added and the blend is pumped to a nozzle where the catalyst is metered into the resin and mixed before discharging from the nozzle to the mold. For water-extended resins, the catalyst is usually mixed with the water and then blended with the resin and filler at the nozzle. Caution is necessary because such systems exhibit a reaction time of about 2 to 3 min. Using very cold water can extend this time somewhat. (See Table II-5.1.)

Much synthetic marble production is accomplished with the simplest equipment. Basic resin and stone is mixed in a small cement mixer. The veining compound is blended in a small 5 gal. mixer. Both compounds are then catalyzed, blended with each other, and poured into the mold as soon as possible. Rotary or piston vibrators are either clamped to the mold or held against it during the pouring operation to hasten the flow of the viscous mix and to aid in releasing any trapped air. Table II-5.2 presents typical basic marble and vein mixes used in the preparation of finished cast marble. The water behaves like a filler and no inorganic materials are necessary. It may be varied from 40 to 60 parts. As the water increased within this range the cost drops rapidly; the strength, toughness, and durability diminish; the emulsion viscosity increases; the overall curing shrinkage and peak exotherm tempera-

Table II-5.1 Formulation of Resin Mix for Water-Extendable Polyester Casting Mix

Ingredient	Parts
Water-extendable polyester resin	50
Water	50
Promoter (consult resin mfg.)	2
Catalyst (DSW, water-soluble)	0.5–1.0

Notes: Mixing instructions:
1. Dissolve promoters in the resin. These are usually cobalt octotate and dimethyl aniline (1.0% and 0.5% respectively).
2. Emulsify resin by stirring water into the mix slowly and with controlled agitation.
3. Add catalyst and pour. Gel times will occur in 2 to 4 min. Cure time is 15 min. in mold; 1 hr total, with the exotherm at 200°F.

ture are reduced; the water is less permanently held in the ultimate casting.

Two veining mixes are usually used for the desired effect. Approximately 1 cu ft of material will result from 26 lb of resin in the marble formulation and ¼ lb in the veining.

Molding Method—Processing Details: Flat Cast Panels

1. Wash and polish the mold. Some molds have a bake-on silicone varnish, or sintered Teflon® release coating.
2. Apply 10 to 30 mils of clear, chemically resistant gel coat, and allow to cure thoroughly. Be sure to take precautions to prevent any dust or lint fallout on the gel.
3. Skillfully stir the mixed and catalyzed

veining compound into the measured and catalyzed basic compound.
4. Pour this mix onto the mold, applying some vibration. This step also requires a great deal of skill.
5. Allow the mix to gel.
6. Strip the casting from the mold and place on a holding rack for full exotherm and cure. It is advisable to insulate this rack so that full benefit is gained from the exotherm. The precise time for the stripping operation depends upon the size, shape, weight, and production desired.
7. Trim and inspect.
8. Final polish, inspection, and packaging.

Molding Method: Sink and Vanity Top

1. Complete the first four steps as you would for molding flat cast panels. After this, but before complete gelation of the first pour, gently place the female bowl mold in place with its lip tightly against the partially gelled first pour. Be sure bowl overflow tubes are in place.
2. Pour the remaining material in the space between the male and female bowl-shaped molds, using discrete vibration to assist in producing a dense casting with no voids.
3. Allow the material to gel.
4. After a Barcol hardness 10 has been reached, remove the part from the mold and place it in a curing frame.
5. After full cure has been attained (2 to 3

Table II-5.2 Typical Basic Marble and Veining Mixes

Marble Mix		Veining Mix	
Ingredient	%	Ingredient	%
Resin	19.500	Resin	20.000
Catalyst	0.196	Catalyst	0.200
Pigment	1.500	Pigment	3.780
Marble chips 00–000	48.800	Limestone 40–200	37.800
Silica 100 mesh	15.000	Clay	37.800
Limestone 40–200	11.180		
Barytes #1 bleached	3.750		
Total	99.920		99.580

Notes: Instructions:
1. Place all liquids into the mixer and mix thoroughly.
2. Add the dry materials. The veining mix is very viscous, a desirable property which prevents smearing.
3. Add the veining mixture while turning the main marble mix about 4 to 8 times.
4. Pour this veined mix into the mold with a swirl or crisscross pattern to duplicate the marble configuration desired. Success results from experience, practice, and ingenuity.

days at room temperature) trim and inspect, and perform the final touch-up and polishing of the visible surfaces.

6. Dispose for final inspection and packaging.

Molding Method: Water-Extended Resins

1. Prepare the mold and then apply release agent.
2. Measure the resin and water and add the proper amount of catalyst to the water.
3. Install a high-shear mixing blade in the resin pot. The pot should have a capacity large enough to hold three times the required resin volume. Start the mixer and add the catalyzed water *slowly* into the vortex. The mix at this stage should properly exhibit resin around drops of water. The viscosity will increase properly when the water is added. If you have an "inversion," where the water encapsulates the resin the viscosity will not increase, and the batch will be poor.
4. Assuming that you have made a good batch, cease mixing a few seconds after all the water has been added, and pour the mix into the mold.
5. Exotherm is not a problem with these water-extended mixes. In one case, 4 ft × 5 ft × 1½ ft casting weighing over 1800 lb did not exceed 180°F during exotherm![16]

Product Data.[17] The manufacture of synthetic marble products is literally unlimited. They can be used to duplicate many products presently made with the natural material. There are reports that even synthetic grave stones are being made and sold. The role of polyester resins filled with shell flour, Perlite, Saran® beads, glass beads, and the like for furniture applications underwent a major expansion in 1972 and an enormous growth is expected.

Some of the more common applications are terrazzo; marbelized, large and small stoned floor tile; Spanish tile; brick facing; travertine; terra cotta; baseboards; stair treads; thresholds; seamless flooring; roofing materials; wall paneling; mosaics; murals; windows and sills; bathroom fixtures; vanity tops; commodes; flower pots; planter boxes; birdbaths; fountains; statuary; plaques; lamp bases; trays; boutique items, such as cases for lipstick, stamps, and jewelry.

Examples of water-extended polyester resin products are statuary, lamp bases, furniture fronts, legs and plaques. These products have the most uses where slip-cast plaster has been popular.

Advantages of Synthetic Cast Marble:
- More chemically resistant than natural marble.
- Will not break as easily.
- Weighs less.
- Can be duplicated.
- Less costly to fabricate.
- Lower freight costs.
- Less damage in shipment and installation.
- Easier to handle.
- With care, can be repaired.

Disadvantages of Synthetic Cast Marble:
- Requires skilled operators.
- How to market the product is a problem. Good outlets for the material are needed—a big problem for the small manufacturer.
- Mold investment, maintenance, and finishing costs are high.

Advantages of Water-extended Polyester:
- Cost as low as $0.10 per lb.
- Trained operators can easily mix them.
- Has reasonable strength and can be reinforced.
- Can be painted, nailed, drilled or otherwise machined or worked upon.
- At 50% water, the material will withstand freezing.
- There are several suppliers of resins.

Disadvantages of Water-extended Polyesters:
- Low strength.
- An expensive mixing machine is needed for production.
- Indefinite and unclear patent and licensing situation exists.
- High shrinkage in castings in loose-tolerance parts (1973). This could change.
- Poor surface texture.

LARGE, ONE-PIECE SANDWICH CONSTRUCTION

The capability has been developed for actually molding one-piece, monocoque-construction, box-type, encapsulated-foam, fiber glass-reinforced polyester elements as large as 8 ft × 9½ ft × 50 ft (The most popular commercial size is 8 ft × 8 ft × 20 ft) Finished shell-type structural units assume the form of insulating, refrigerator-type boxes which may be truck-mounted as trailer vans or used as cargo containers for shipping.

The forerunner of these ingeniously fabricated RP/C structures was a vacuum-bag molding process in which trying to hold vacuums and prevent leakage during curing caused extreme difficulty. The inventor turned to gravity resin impregnation plus mild pressure impregnation with success. We will discuss this process in detail.[19]

Materials:

Resins. The resinous materials used are 1.5 to 3 poise viscosity semi-rigid polyesters with white pigment paste at 0.5% in the resin, cyclohexanone peroxide catalyst, and cobalt naphthenate.

Reinforcement. Reinforcing materials include 10 to 30 mil fiber glass veil mat, ¾ oz per sq ft chopped-strand mat, 14 and 17 oz per sq yd woven roving—all in combination as wall, bottom, and roof insulation. Plymat, woven roving and chopped-strand mat combined is laid in as corner reinforcing.

The molded-in insulating material is comprised of 2 lb per cu ft density rigid poly-urethane foam precut to required sizes. Foam-in-place techniques are not employed.

Equipment and Tooling. Mold equipment consists of three main elements and miscellaneous gear:

An outside or exterior mold 9½ ft deep, 8 ft wide, and 50 ft long. The sides are I-beam supported steel for easy release and freedom from rust. Shorter sections of the mold may be blocked off and used to produce a box of any desired length—10, 20, or 40 ft.

An inside mandrel fabricated and assembled from components suitable for each job. These components consist of flat sides or panels which contact the reinforcement or lay-up and which are supported and jacked by inner frames. A system is provided for expanding the flat mold sides outward against the impregnated sandwich construction in order to define laminate thickness, apply pressure and enhance both the curing cycle and laminate density.

A cover plate which is placed over the top

of the assembly after initial impregnation and drawn down hydraulically over the roof elements to assist in forcing the resin into the fiber glass reinforcement, thereby encapsulating the foam.

Miscellaneous gear for manipulating the huge pieces of mold metal. This includes a large overhead crane for lowering the inner mandrel into position and for removing completed boxes from the mold.

Molding Method/Processing Details. The following steps comprise the fabricating of a box from start to finish.

1. Self-extinguishing, 2 lb per cu ft poly-urethane foam with a K factor of 0.12 is sized by sawing from large 'logs' into slabs 3 in. thick, 8 ft long, and into various widths (from 3 to 24 in. for positioning in either floor, roof, or walls.

2. All foam slabs are wrapped with either chopped-strand mats or woven fiber glass roving. The number of layers of glass reinforcement over the foam is dictated by the required resistance to stress. This depends upon actual position in the box.

3. Veil mat, woven roving, chopped-strand-mat woven roving, and foam slabs, followed by the same layers of reinforcement in reverse are laid up or draped on floor, and sidewalls. One end of the container is always left open and unmolded to eventually accommodate a refrigerator door. (See Fig II-5.2)

4. Steel framing is placed around each end section only. Removable corner blocks are also inserted in the mold to provide areas in which permanent steel mounting fixtures may ultimately be permanently mounted in the finished box.

5. The inner mandrel is carefully placed into the mold over the fiber glass foam lay-up. The roof lay-up placed on top of the mandrel. The side elements of the mandrel are expanded to place pressure against the sides of the lay-up and to define the wall thickness.

6. The low-viscosity, catalyzed polyester resin is injected from approximately 6 or 7 ports around the mold periphery. Impregnation is not complete at this stage and a pool of resin approximately 3 in. deep builds up on top of the lay-up.

7. The top mold section is applied, sealed, and hydraulically forced down, causing the resin to infiltrate completely through the lay-

Figure II-5.2   Large, one-piece sandwich-construction molding method and the resultant product in its ultimate usage. *Top:* The process of laying-up fiber-glass reinforcing plus foam cores preparatory to molding the one-piece sandwich-construction units. *Bottom:* The completed, fully insulated and fitted RP/C unitized construction element serving, in this case, as a refrigeration transport truck. (Courtesy of Litewate Transport Equipment Corp., Milwaukee, Wis.)

up, impregnating the fiber glass and encapsulating the foam. The entire lay-up becomes virtually void-free.

8. The gel time of the resin is approximately 1 hr. 15 min., permitting adequate working time in the mold. Exotherm is low in order to prevent overheating and introduction of non-uniform stresses. A dwell-time of 12 hr in the mold is allowed for complete cure and relief of any stress build-up caused by uniform and complete cooling to room temperature.

9. The top mold mandrel is removed, the side pressure of the inner mandrel is relieved, and the cured box assembly including the inner mandrel is raised upwards out of the mold using the overhead crane.

10. The inner mandrel is removed, and finishing and addition of the refrigerator hardware ensue. The units are usually placed in service without painting.

Product Data

Insulating Performance:

Urethane Foam: K factor = .12. Heat loss is therefore less than 5000 Btu per hr for 100°F temperature differential for a 40-ft insulated box. Heat loss is 2700 Btu per hr for 100°F temperature differential for a 20-ft box tested after 1 yr of service.

Through-Metal Heat Conduction. None, because only two metal frames exist—one at each end—and do not extend lengthwise or from inside to outside faces.

Interior Air Circulation. Ribs are molded into sides and ceiling. Vanes in the bottom floor permit thorough interior air circulation and may also be molded in.

Physical Properties:

Weight: A 20-ft trailer weighs approximately 4800 lb when it is molded and equipped (without cooling unit).

A 40-ft trailer weighs approximately 10,250 lb.

Thickness: Fiber glass sandwich-construction sidewalls are approximately 3 in. thick overall with I-beam construction (reinforced strips between foam slabs due to fiber glass wrap).

Floor thickness is 5⅜ in.

Front-nose molded end is 4¼ in. thick.

Separate fiber glass laminate thickness is ⅛ in. for sidewalls increasing to ⅜ in. in corners and highly stressed areas.

Interior Capacity. For a 40-ft box the interior capacity is 2050 cu. ft.

Structural Properties:

Laminate Properties. Greater than 2000 lb tensile stress in vertical direction, and greater than 7000 lb in the longitudinal direction.

Tensile strength of fiber glass laminate is 31,000 nominal.

Glass/resin ratio is 40:60 nominal.

Sandwich impact strength is 4.5 ft. lb.

Shear strength equivalent to laminate-to-foam bond.

G Forces:

0.4 g on front bulkhead.

0.6 g on sidewalls.

(Both exceed USASF standards for van cargo containers).

End Load. 160,000 lb. Two 20-ft trailers joined and supported by end restraint.

Compression. International standard requires that boxes withstand seven-high stacking.

References

1. World's Fair Panels, G. D. Shook, SPI Annual Preprint 1964, sec. 15-A.
2. General Motors Exhibit, World's Fair, N. Y. 1964-65, *Materials Engineering* **69**, No. 1 (Jan. 1969), p. 35. (Development by Boeing Aircraft Corp.); Future of RP Processing in Year 2000, G. D. Shook, SPI preprint 1964, sec. 16-F; Allied Chemical Corp. Bulletin 851-46.
3. Registered trademark and product designation, Johns-Manville Fiber Glass Reinforcements Division.
4. Bulletin 851-46, p. 5, "Polyester Resins" Allied Chemical Corp; Technical Data Sheet, "Super Roving Impregnator," Venus Products Inc., Kent, Wash. 1970.
5. Technical Data Sheet, "Rovmat" Form 1-028 Fiberglass Industries, N.Y. 1970.
6. SPI 23rd Preprint, 1968, sec. 13-E.
7. Brochure: "Cordopreg UVFR (Feb. 1, 1966), Ferro/Cordo Corp., 34 Smith St., Norwalk, Conn. 06852.
8. Brochure: "UV-50 Ultra Violet Catalyst," U.S. Peroxygen Corp., 850 Morton Ave., Richmond, Calif.
9. Beightol, L. E., U.S. Patent No. 3,421,501 Jan. 14, 1969.
10. Brochure: "UV Energy Source Module" (June 1, 1966) Ferro/Cordo Corp., 34 Smith St., Norwalk, Conn. 06852.
11. For further reference *see* Horne, V. I. and Novkov, R. L., "How to Process UV Prepregs," *Plastics Technology*, April, 1966. SPI Preprint 21st Annual 1966, sec. 7-C Editor, Fiberglas Canada, RP Report p. 3, Mar. 1970.
12. Brochure: "UV Energy Source Module" (same as for 10.)
13. Examples of water-extended resins are mentioned in the following:
    Technical Data Sheet N1143 "Aropole WEP" Ashland Chem. Co., Mar. 1968.
    Technical Data Sheet LMT 67 "Laminac EPX-289-4" American Cyanamid Co., Mar. 1970.
    Technical Data Sheet "Polylite 32-180", Reichold Chemical, Ind. Sep. 1969.
14. Registered trademark Dow Chemical Co.
15. Kenics, Inc., U.S. Patent No. 3,286,992 (Aug. 1966).
16. Technical Data Sheet N1143 "Aropol WEP," Ashland Chemical Co., Mar. 1968.
17. For further reference *see:*
    Doyle, E. N., "The Development and Use of Polyester Products," New York, McGraw-Hill Co., 1969.
    Wood, Stuart, *Modern Plastics* V46 N8 p50, 1960.
    Plastics Technology V14 N3 p15, 1968.
    Signorini, P. P., *Plastics World* V27 N6 p26, 1968.
    Technical Data Sheet, "Foamable Plastic Spheres" No. 190-101-69 (Brochure) Dow Chemical Co., Oct. 1969.
18. Ashland Chem. Co., American Cyanamid Co., Reichold Chemical, Inc.
19. The process is patented and proprietary but has been licensed for the manufacture of smaller containers of similar construction. Permission has been granted to discuss here the large container manufacturing equipment and sequential process stages (1).
    Personal Communication, Baker, W. T. and Reeves, J. F., Litewate Transport Equipment Corporation, Milwaukee, Wis.

SECTION III

PROCESSES FOR INTERMEDIATE CURE:

ARCHITECTURAL AND INDUSTRIAL PANELING

III-1

Translucent Corrugated and Flat Panels

INTRODUCTION

RP/C panels represent a significant segment of the reinforced-plastics industrial output. Their development started shortly after World War II when knowledge and materials originally destined for military purposes were directed toward commercial and peaceful uses. They found application as light-transmitting building elements made in corrugated shapes to fit with corresponding metal sheets. Other uses were promoted, notably for residential building purposes in the form of patio roofs, fences, room dividers, screen and other decorative applications. Technological advances in raw materials and processing techniques also stimulated new panel uses in commercial and industrial-type buildings.

Several grades of panel are available, ranging in quality from commodity items to highly durable products backed by guarantees of 15 to 20 yr. maintenance-free performance. Flame-retardant panels are made to meet building code requirements. All grades of panel are made to match a variety of sheet-metal panel profiles and to produce decorative structures with pleasing aesthetic qualities.

Three basic processes are used to produce translucent panels: hand lay-up, press, and continuous. Historically, the industry started with the hand lay-up process, and in the face of demand for large-volume production, subsequently evolved press operations and continu-

The authors are particularly indebted to the late George R. Huisman for his invaluable contributions to this chapter.

ous in-line processes. The latter is used throughout the industry, with limited hand lay-up and press methods still in existence.

MATERIALS

Resins. All mass-produced translucent panels are made of a resin in the polyester category. In its simplest form, the resin is a styrene-modified phthalic anhydride/maleic anhydride and propylene glycol ester containing approximately 40% styrene monomer. Other forms might incorporate isophthalic acid; neopentyl glycol; chlorinated, phosphated, and/or brominated compounds; or methyl methacrylate monomer. The more sophisticated formulations are conceived to provide improved weather resistance, flame retardancy, and/or chemical resistance. Although originally problematical, corrugated-panel resins with both high translucence and flame-spread retardancy (25 or less) are available. All panel resins are characterized by the addition of ultraviolet-light stabilizers at about 0.20 to 0.25% concentration for standard resins, and 0.75 to 1.50% concentration for flame-retardant resins.

Technical data have demonstrated an improvement in the weathering resistance of reinforced-polyester panels by partial substitution of methyl methacrylate monomer (MMA) for the commonly used and less expensive styrene monomer. Assuming a 40% monomeric content (60% alkyd), laboratory studies indicate optimum properties with a 20:20 blend of MMA to styrene. Practical economic and process considerations have led to the common

Table III-1.1 Properties of a Typical Light-Stabilized Polyester Resin for Translucent Panel Manufacture

Parameter	Value
Monomer[a]	30% styrene
Viscosity at 77°F, poise	21–25
Color, APHA	75 maximum
Specific gravity	1.14 to 1.15
Acid number	22–25

Source: Koppers Co.
[a]Further reduction in viscosity to 2–3 poise with methyl methacrylate monomer (ratio 45% MMA to 55% styrene) assists in wet-out and significantly improves weathering resistance of the cured resin.

use of an 8:32 to 12:28 blend. Higher quantities of MMA often lead to the formation of voids in the panel caused by volatilization under the heat applied for cure. Table III-1.1 lists the properties of a typical uncured polyester resin for translucent panel manufacture.

Catalysts normally used include benzoyl peroxide (BZP), cumene hydroperoxide (CHP), and methyl-ethyl-ketone peroxide (MEKP). Benzoyl peroxide is used in either granular or paste form. CHP and/or MEKP initiate the cure cycle and are adjusted to provide a balance between gel time under particular cure cycles and pot life of the catalyzed resin mix. BZP, reactive at higher temperatures, assures complete cure. Most panel resins are furnished with integrated inhibitors which help stabilize the uncatalyzed resin and also with promoters which induce accelerated catalyst activity. Promoters are characterized by any of a class of organic chemicals which impart immediate reactivity to the system without inducing discoloration. Typical materials for panel manufacture are benzene-phosphinic acid, mercaptans and vanadium naphthenate dispersed in neutral vehicles such as methyl cellosolve or dioctyl phthalate.

Typical variation in gel time with concentrations of widely used catalyst-promoter systems for translucent panel resins are presented in Table III-1.2.

MEKP and MEC catalysts are useful in hand lay-up operations when the time interval from preparation to cure initiation is long in comparison to that of continuous-process operations, and when the temperature at which gelation occurs is lower.

For a continuous process, handling characteristics of the resin determined from a 50 g sample in a 100-ml beaker in a 180°F water bath are usually set in the following range:

Gel time from 100°F	3-4 min.
Gel temperature	145-155°F
Time to peak from 100°F	6-8 min.
Peak exotherm	390-420°F

Color control is very important to the production of RP/C panels. Pigments must be light-stable, consistent in color and tinting power, and neutral with respect to the cure characteristics of the resin. If paste pigments are used, the vehicle must be compatible with the resin and must not induce yellowing. Several tests are performed to evaluate the color. Pigment color is usually checked by a smear test against a known sample. Tinting power is checked in a letdown solution or a reinforced-laminate form. Effects on the cure are determined from gel time tests. The color stability is established by a fadeometer, a weatherometer or, when time permits, outdoor weathering.

Efficient and successful panel operations are marked by well-defined, precise, and carefully controlled resin-mixing procedures. Colors must match and cure characteristics must be duplicated within very narrow tolerances.

Reinforcements. Glass fibers cut from roving and made into an unbonded mat at the sheet-forming line provide the most common form of reinforcement currently used. Bonded, chopped-strand mat made with a minimum of highly soluble binder is also used, but primarily in hand lay-up operations. In either form, the preferred glass finish is of the silane type compatible with polyester resins.

The usual fiber length is 1½ to 2 in. Fast wetting, good filamentation, soft hand, very limited strand integrity, and low static electrical build-up characterize the roving most useful for RP/C panel mat. Non-tinted, color-clear glass is preferred over the standard "E" glass, especially for white and light pastel colors. Glass/resin ratios are generally 25% glass fiber to 75% resin, with 2.0 oz chopped glass per sq ft, yielding a cured panel $\frac{1}{16}$ in. thick. Table III-1.3 presents typical weight/thickness values for several types of panel configurations. All panels are basically categorized and sold according to nominal weight per sq. ft. Therefore while weight remains constant, average

Table III-1.2 Representative Cure Characteristics for Typical Polyester Resin*

% Cumene Hydro-peroxide	% Benzoyl Peroxide	% Promoter Benzene-phosphinic Acid Type	Gel Time (in min.) and Temperature When Placed in 150°F. Bath				Time to Peak Exotherm (in min.) and Peak Exotherm, Deg. F.				Pot Life at 78°F. (in min.)
			5 Min. After Catalyst Added		60 Min. After Catalyst Added		5 Min. After Catalyst Added		60 Min. After Catalyst Added		
			Time	Gel Temp. F	Time	Gel Temp. F	Time	Temp. F	Time	Temp. F	
1.0	0.25	0.15	19'50"	149	15'55"	150	30'30"	399	26'00"	409	138
1.0	0.25	0.20	15'20"	144	11'25"	140	25'40"	400	21'30"	404	100
1.0	0.25	0.25	11'30"	144	6'25"	130	18'55"	410	16'30"	399	78
1.0	0.25	0.30	9'45"	142	5'00"	124	16'25"	409	14'35"	396	56
0.5	0.25	0.15	22'40"	148	18'50"	150	32'35"	399	27'35"	405	188
0.5	0.25	0.20	18'55"	144	16'45"	142	27'20"	400	25'40"	398	142
0.5	0.25	0.25	14'55"	145	11'30"	140	21'55"	403	19'15"	399	134
0.5	0.25	0.30	13'05"	143	10'05"	138	19'40"	403	17'20"	397	113

*Consisting of: resin, 84 parts. Styrene, 8 parts. Methyl methacrylate, 8 parts.
Gelation test: Standard SPI procedure for gel and exotherm conducted with test-tube in 150°F. oil bath.

Table III-1.3 Typical Weight and Thickness Values for Translucent Panels

| | TYPE I—General-Purpose | | | TYPE II—Fire-Retardant | |
| | Nominal Weight | | | Nominal Weight | |
Configuration	8 oz	6 oz	5 oz	8 oz	6 oz.
Flat	.060″	.045″	.037″	.057″	.043″
2½″ × ½″	.056″	.042″	.035″	.053″	.040″
2.67″ × ⅞″	.050″	.038″	.031″	.048″	.036″
4.2″ × 1-1/16″	.053″	.040″	.033″	.050″	.038″
1¼″ × ¼″	.057″	.043″	.036″	.055″	.041″
5-V crimp	.056″	.042″	.035″	.053″	.040″
2.67 rib	.054″	.041″	.034″	.051″	0.39″

Source: Corrulux Corp., Houston, Tex.
Notes: Thickness tolerance ±0.010 in.

thickness will increase or decrease in proportion to the surface area of a specific configuration. The average thickness of a given weight of flat material will be greater than the thickness of various configurations.

Typical Formulations. In a standard panel, the ratio of glass to resin is typically 25 to 28 parts glass to 72 to 75 parts resin. In flame-retardant panels, the glass content is frequently about 10% higher in the range of 30 to 33 parts glass to 67 to 70 parts resin. Glass, being incombustible, contributes to improved flame retardancy, as measured by flame-spread ratings when tested in accordance with the UL tunnel test.

Pigments are added to produce the desired color and degree of light transmission. A clear panel with little or no pigmentation will show light-transmission values of 85 to 90%. These values vary somewhat inversely with the amount of pigment used. Heat-transmission values generally follow light-transmission values. Pigment concentrations typically amount to less than 1% of the resin-mix weight.

Fillers, such as calcium carbonate, magnesium carbonate, aluminum hydrate, and so on, are used primarily for economic and certain technical benefits. When used for economic purposes, fillers may be in a range of 10 to 20% of the resin-mix weight. At this level, process means must be provided to prevent flocculation and/or settling of the filler while the resin mix exists in liquid form. Generally speaking, most fillers used in these higher quantities also have a deleterious effect on a panel's weather resistance. This condition stems from higher water absorption with subsequent deterioration of the resin-to-glass bond.

Judicious use of fillers in a range of 4 to 10%, depending upon the material used, will impart better weathering resistance, a smoother surface, a measured reduction of light and heat-transmission values, and a slightly improved flame resistance. Furthermore, the presence of such quantities of inert material creates a sensible heat sink, thus permitting slightly higher catalyst levels and corresponding cure rates.

Slight changes in gelation rates will of course be monitored by testing or sampling from complete formulations and then adjusting accordingly.

Typical resin-mix formulations for hand-laid sheets, press molding, and continuous-process production are shown in Table III-1.4.

Special Formulations. Weather resistance of a panel is directly related to the nature of a panel's surface. Commodity panels made by simple combination of glass and resin present a reasonably good surface, but is subject, however, to a degree of degradation upon outdoor exposure. Such panels normally require periodic cleaning and eventual re-surfacing with acrylic modified coatings at 3 to 5 yr intervals to maintain their original appearance for long-term use.

Textured surfaces in which a pattern of high resin content is created demonstrate greatly improved weathering resistance. Smooth, resin-rich surfaces are made by transferring a pre-gelled resin coating from the film package to the panel. Color stripes can be introduced also by transfer of patterned, pre-gelled, pigmented resins from the film to the panel surface.

Blown-glass fiber veil mat (5 to 10 mils) is also used to impart a highly durable resin-rich

Table III-1.4 Typical Resin-Mix Formulations for a White, Medium-Light-Transmission Panel Made by Hand-Lay-Up, Press Molding, or Continuous Process

| Ingredient | Parts Required in Formulation | | |
	Hand Lay-up	Press Molding	Continuous
Resin—light-stabilized, styrene-modified	80.00	83.00	85.00
Monomer—methyl methacrylate	20.00	17.00	15.00
Catalyst—benzoyl peroxide	0.25	0.60	0.50
Catalyst—cumene hydroperoxide	0	0	0.50
Promoter—cobalt octoate (6%)	0.25	0	0
Promoter—vanadium naphthenate (6%)	0	0	0.25
Filler—calcium carbonate	3.00	3.00	3.00
Pigment—titanium dioxide (white paste)	1.00	1.00	1.00

surface. By nature, glass-fiber veil mat has the capacity to hold high percentages of resin (80 to 90%), thereby creating conditions conducive to long-term resistance against surface degradation.

All panels made are processed in a synthetic film envelope. Tedlar®[1] film, when used, becomes an integral part of the panel, thus forming the most durable panel surface known to date. Because of its relatively high cost, Tedlar is generally used only on one side. Panels so constructed are marked in some fashion to assure installation with the correct surface facing the elements.

Printed inserts are used to impart decorative features, such as marble effects and wood-grain representations. The inserts are usually lightweight, synthetic papers printed in appropriate patterns with special inks which are insoluble in the uncured resins and which are formulated to neither accelerate nor delay gelation and ultimate cure.

EQUIPMENT AND TOOLING

Equipment costs for the continuous process are approximately equal to those for the press process. At current prices, a process line, continuous or press, costs from $75,000 to $150,000, excluding engineering costs and depending upon the sophistication of design. Bulk-storage tanks and attendant resin-handling and mixing equipment may cost an additional $10,000 to $20,000. The cost of a hand-lay-up facility is much lower, at $25,000 to $40,000.

Taking practical considerations into account, the typical annual production capacity for each type process on a 3-shift, 5-day week basis is estimated as follows:

Hand lay-up	3.0-5.0 million sq ft
Press	2.0-3.0 million sq ft
Continuous	20-30 million sq ft

The above capacity of a continuous line is typical of some existing machines. It is a practical range under present-day conditions but could be increased considerably simply by lengthening the line. A line producing 20 to 30 million sq ft is 150 to 175 ft in length. Generally, oven length is the determining factor bearing on maximum production potential.

Factory space to produce and warehouse a reasonable stock at the production levels noted above is of the following magnitude:

Hand lay-up	20-25,000 sq ft
Press	15-20,000 sq ft
Continuous	50-60,000 sq ft

Tooling costs vary considerably. A set of fixtures for the continuous process are least expensive, and press models are most expensive. Sheet-metal mold costs for the hand-lay-up process depend upon the number of duplicate sets needed to support production for any given shape.

In this process, some operations have been established on the basis of using commercially available roll-formed sheet-metal molds. For such operations, the hand-lay-up tooling is least expensive. However, when panel shrinkage is considered and extra width and length are provided for trim allowance, the costs are much higher. Availability of this type mold is limited at best, because many sheet-metal shops are unable to make the molds within necessary tolerances.

Tools for the continuous process, often referred to as forming fixtures, consist of wood, masonite, or aluminum strips cut to an appropriate profile and mounted at intervals in the upstream end of the curing oven. The positions extend into the exotherm area. Downstream, after the panel has hardened, simple supports are sufficient. Costs vary across the country depending upon the availability of adequate woodworking or metalworking facilities. A set of fixtures may cost from $500 to $1500 for most configurations but could be from $1000 to $2500 for a sophisticated shape with intricate detail.

MOLDING METHOD/PROCESSING DETAILS

Hand Lay-up. The oldest process still used for making translucent RP/C panels is a hand-lay-up technique commonly called the "batch" or "stacked-panel" process. The following steps are involved in this process:

- Spread properly catalyzed resin with a doctor blade as evenly as possible over a cellophane sheet on a table.
- Cut fiber-glass chopped-strand mat to size, and lay a piece of this mat into the resin.
- Place another sheet of cellophane on top of the resin-mat lay-up. Remove trapped air with a squeegee, or roll out by hand.
- Fold and tape the protruding edges of cellophane film. In some operations, the film is wide enough and long enough to eliminate the folding and taping step. A thin string approximately the same as the panel thickness run peripherally around the panel lay-up will prevent run-back of air.
- Interleave the prepared packages between aluminum sheet molds of the desired shape and then apply 0.3 to 0.5 lb per sq ft weight.
- Batch-cure an appropriate stack of panels. Characteristic oven cycles are 8 to 16 hr at 150 to 180°F in a large oven. In some very small volume operations, it is possible to allow the panels to cure overnight at room temperature.
- Upon removal from the curing facility, trim the sheets to size and strip the cellophane film.

This process is limited to small production requirements. The relatively high labor factor ranges from $0.04 to $0.10 per sq ft, depending upon the size of operation and the degree of sophistication employed in the individual steps of the process.

Although the quality of product can be very good, most hand lay-up panels are not as uniform in thickness as those made by other methods. The cost of production in large volume is higher than that associated with the continuous process, not only because of more expensive labor, but also because of a larger waste of raw materials.

The most efficient use of the batch process is found in a semi-continuous system where the materials preparation and impregnation steps are mechanized, in a manner somewhat similar to the steps described under the press and continuous processes.

A major disadvantage of the hand lay-up process is the procurement and maintenance of the sheet-metal molds. They are large, ungainly pieces subject to handling abuse. Furthermore, they are originally manufactured by press-brake with tolerances which preclude perfect matching of cross-sectional configuration. Cured resin frequently accumulates on the molds and the labor effort needed to keep them in usable condition constitutes a significant factor in the cost of panel manufacture by the hand lay-up method.

Press Process. The press process has certain advantages for making translucent RP/C panels. A heated platen press is the key element of the concept. Surface texture can be imparted by the mold and precise thickness throughout the panel is possible.

Practical considerations, however, seriously limit the usefulness of the process. Cost factors are high and the process is slow. The very advantage of thickness control becomes a hazard in practice. The reinforcement used is not perfectly uniform and irregularities in the mat cause irregularities in the appearance of a panel. For example, areas of high glass content frequently show up with greater fiber pattern in relation to areas of lesser glass content where the fibers blend better with the resin. Another disadvantage is the fact that the press must be fitted with stops or spacers to prevent the excessive loss of resin. In essence, the press system calls for a high degree of precision of an order not required by the more popular and widely used continuous process. The following manufacturing steps comprise the press process:

- Deposit resin on a moving film, usually cellophane.

- Lay bonded or unbonded mat into the bed of resin and compact using serrated roller-disc-type impregnators.
- Introduce top film as the material is drawn through a set of metering rolls.
- Move the package or sandwich into the platen press. Means should be provided to guide the package into a cross-sectional configuration closely approximating that of the finished panel.
- Close the press and allow the panel to cure. Carefully control the cure rate to prevent excessive exotherm which has a tendency, if uncontrolled, to introduce an undesirable degree of opacity in the panel.
- Subsequent to cure, cool, trim, inspect and stack the panels.

A typical press-time cycle includes 15 sec. for loading, 60 sec., for cure, and 10 sec for unloading. Either post-curing ovens or the latent heat of stacked panels may be employed to complete the cure.

Continuous Process. Continuous processing of RP/C panels is widely used in the industry. It is based on the concept of making the best product at the lowest possible cost. In practice, this means a minimum of labor and the highest possible ratio of raw-material input to finished-product output. The system incorporates the following elements into an in-line operation: mat formation from roving, impregnation of mat with resin, consolidation of the material into an encased package, configuration of the package to a predetermined and desirable contour or shape, solidification of the resin, and separation of the issuing continuous ribbon of product into individual sized panels. The steps may be described in greater detail:

- In general, mix resin on a batch basis, following a schedule to maintain a continuous supply at the line. Meter the resin together with doctor blading for thickness control onto a moving film, at a suitable width.
- Deposit an unbonded chopped-strand fiberglass layer or mat, cut from roving, in the bed of resin. Force the resin into the mat with compactors and/or impregnators to thoroughly wet the fibers.
- Introduce an upper film to encase the material, usually at the squeeze rolls. Adjust the rolls to allow passage of a prescribed amount of resin and glass.
- Maintain a small reservoir of resin at the squeeze rolls to act as a seal against the intrusion of air.
- Seal the film envelope along the edges if the machine permits. Many panel machines, however, operate without this feature.
- Move the package over forming fixtures which progressively establish the final corrugated configuration. (Figure III-1.1, also shows the corrugators along the sheet-forming section of a continuous-panel operation.)
- The material moves through a tunnel oven where it is exposed to a programmed time/temperature cycle. Gelation of the resin must be made to take place after the point at which the shape has been created. Typical oven-tunnel sections and temperatures are: resin-film preheat—140°F, gel—180°F, exotherm range—no heat, final cure—250°F.

Heat sources for curing the sheet can be electricity, gas, or oil. Air circulation in the oven is preferred and indeed necessary on high-speed operations, especially in the mid-section of the oven where the resin's exotherm takes place. During exotherm, the resin is not yet fully cured. Therefore, the rate of temperature rise must be controlled to prevent volatilization of the unlocked monomer. Downstream sections of the oven complete the cure.

- Trim the cured panel to width and cut to length. Move flying cross cut saw with the panel to cut off finite length pieces. A traction unit, acting on the cured panel, powers the entire operation.
- Removable film is usually stripped automatically. When it is not, remove it by hand. Cellophane film is most commonly used for the package envelope. Successful panel production is dependent upon proper film treatment. The cellophane is plasticized and contains some moisture which also acts as a plasticizer. The plasticizer and moisture are driven off at cure-cycle temperatures, and the loss of plasticizer causes shrinkage of the film. Excessive shrinkage, especially in the early stages of cure (through the hard-gel state), results in deformation of the panel configuration and introduces objectional surface irregularities. This phenomenon should be prevented by humidity-controlled pre-shrinkage of the film after it is unwound but before it makes contact with the panel ingredients.

The major advantages of the continuous process are uniformity of product, low labor

Figure III-1.1 Illustration of the process for continuous production of translucent, architectural, corrugated RP/C paneling.

Top: Corrugators along the sheet forming section of a continuous panel operation.

Bottom: Examination of finished translucent corrugated paneling over lighted inspection table.

costs, efficient use of raw materials and high production capacity. The major disadvantage is a degree of inflexibility to produce small quantities efficiently. Also, high-production machines, capable of running at 30 to 50 ft per min. are approximately 150 to 200 ft long. Besides length, start-up and shut-down costs on runs of 1000 to 1500 lineal ft. can be prohibitive, unless great care is taken and experienced personnel are available. The continuous process, like the hand lay-up and press processes, is not yet fully automated and there is a definitive need for operator skill to maintain desirable high efficiencies in operating the process.

Process Quality Control. Process quality control for all three processes is especially important in the following areas:
- Control is required for parameters of incoming liquid resin and for integrity plus speed and thoroughness of wet-out of glass fiber products.
- Uniform glass distribution is required to prevent both weak resin-rich areas and unsightly opaque high-glass areas.
- Thorough wet-out must be accomplished to prevent dry areas and areas of mass bubbles or voids.
- Control of resin gelation in the lay-up is vital. Undercured resin results in soft, weak areas, while overcure produces excessive exotherm and consequent opaque or white-fiber areas.
- General cleanliness around the operation is required to prevent rejects caused by visible dirt and entrained either internally or on the surface of the panel.

Special Effects. In all three processes, special effects can be obtained by the use of surfacing mats, inserts, and special films. In the continuous process, let-off units are provided to introduce surfacing mat or printed inserts into the panel at the head end or impregnation section of the line. Dual units are used to assure a continuing source of supply. Film is mounted in a similar manner so that splices between rolls can be made without interrupting production. In the case of Tedlar®-faced panels, the Tedlar simply replaces the commonly used cellophane. Textured-surface panels are made with embossed cellophane.

The press process is capable of producing embossed panels by using embossed or patterned press platens or mold plates. Plain cellophane is still required as cover film for encapsulating the sandwich or lay-up.

A random-patterned surface can be created in panels in any of the three processes by the use of cellulose acetate film which has the capacity for expanding upon absorption of the styrene monomer. This action causes the cellulose acetate film to wrinkle and pucker, thereby forming a pattern of raised ridges. The latter principle is generally not used today because of its non-reproducible nature. Tedlar film coatings, embossed surfaces, and transfer gel coats greatly extend panel service life, and make possible elaborate decorative effects. Some manufacturers' guarantees offer up to 20 years protection.

PRODUCT

Translucent RP/C panels are made in a variety of configurations, types, sizes, and colors. Figure III-1.2 summarizes the produce line offered by one manufacturer. The products represented could have been manufactured by any one of the three processes discussed. Note that the panel lengths available stop at 12 to 14 ft. because of handling limitations.

Translucent RP/C panels are generally made in two grades—general-purpose and fire-retardant—as well as to proprietary and/or customer specification. United States Standard CS214-57 (revised 1964) is the only standard specification used in the industry. It is currently being revised again under the guidance of the SPI Panel Council and is to be reissued by the Department of Commerce as CS-103. The present standard is very limited in scope and does not cover a large number of commonly used panels. For example, only panels weighing 8 oz per sq ft with 4 configurations are mentioned, whereas the industry is producing

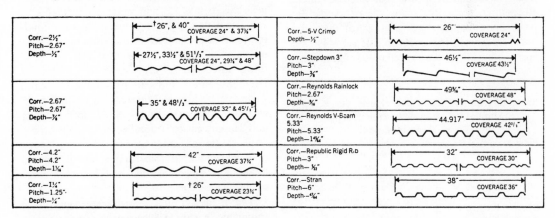

Figure III-1.2 Representative Configurations, Types, Sizes, and Colors of Corrugated and Flat Translucent Panels. *Types:* Markets supplied include general-purpose building—e.g. patios; commercial and farm use—e.g. skylights, matching metal buildings; decorative items; opaque materials—e.g. cooling towers; fire-retardant, translucent materials with UL ICS ratings as low as 20 to 25; and materials with glazing quality—e.g. translucent windows. Finishes may be either smooth or embossed (granitized). All panels except the fire-retardant type contain acrylic monomer modification and/or Tedlar coating on one side for improved weather resistance. *Sizes:* Panel widths are available from 2 to 5 ft. and panel lengths, from strips ½ in. long to panels from 2 to 50 ft. long. *Colors:* Almost any color is available, but the bulk of panels sold are colored white, green, yellow, tan, brown, gray and very slightly shaded for non-glare. Also produced are translucent and near-opaque or opaque panels for service applications. These are colored brown, gray, and other shades and are of course non-light-transmitting. (Courtesy of Corrulux Corp., Houston, Tex.)

Table III-1.5 Load-Bearing Requirements for Panels[a]

Corrugation (in.)	General-Purpose Panels (lb per ft of width)	Fire-Retardant Panels (lb per ft of width)
2½ × ½	300	250
2.67 × ⅞	450	350
4.2	400	300
1¼ × ¼ (on 18 in. span)	200	150

[a]8 oz per sq ft.

panels ranging in weight from 4 to 12 oz per sq ft of some 40 to 50 cross-sectional configurations. Table III-1.5 lists the load-bearing requirements for 8 oz. per sq. ft. panels covered by CS-214.

A load-bearing test is conducted according to ASTM Method D1502-57T using quarter-point loading on a 24 in. simply supported span. Tests on the 1¼ in. × ¼ in. corrugation are run on an 18 in. span because of its flexibility. Another test method, which treats the panel as a continuous beam over two spans and subjects it to uniform loading, is under development by the SPI Panel Council. Recommendations for spans will of course depend upon the type of corrugation and the panel thickness. Table III-1.6 lists suitably safe spans for the more popular configurations.

Table III-1.6 Recommended Spans[a] for Several Standard Corrugated, Translucent-Panel Configurations

Configuration	Nominal Weight per Sq Ft	
	6 oz	8 oz
2½″ × ½″ Corr.	3′6″	4′0″
2.67″ × ⅞″ Corr.	4′0″	4′6″
4.2″ × 1-1/16″ Corr.		4′6″
1¼″ × ¼″ Corr.	2′6″	3′0″
5 V crimp × ½″	2′0″	2′6″
Rigid rib 3″ × 15/32″	3′6″	4′0″
5.33″ V-beam	4′6″	5′0″

Source: Corrulux Corp., Houston, Tex.
[a]The recommended maximum roofing spans are based on a minimum of two continuous spans (3 purlins) for uniform loading of 40 psf with a 2.5 safety factor. (100 lb per sq ft when fastened in accordance with specifications.)

Table III-1.7 Representative Physical and Structural Properties of RP/C Panels

Property	Value
Tensile strength, psi	12,000–15,000
Compressive strength, psi	25,000–30,000
Flexural strength, psi	30,000–35,000
Modulus of elasticity, psi	$0.8-1.2 \times 10^6$
Bearing strength, psi	40,000–50,000
Izod impact, lb per in.	8–12
Flammability (D635) in. per min.	Self-extinguishing to 2.0
Barcol hardness	50–55
Linear thermal expansion, in./in./°F	$1.5-1.6 \times 10^{-5}$

For vertical applications, spans may be increased up to 20%.

Physical and structural properties of RP/C panels vary somewhat depending upon glass content, panel formulation, and thickness. Table III-1.7 presents the typical values.

Fire-retardant panels are rated according to ASTM Method E84, "Surface Burning Characteristics of Building Materials." This is the so-called *tunnel test*. Commercially available fire-retardant panels have flame-spread ratings generally below 75, and some are available at 25. These figures are significant because building-code requirements restrict the use of materials with higher flame-spread ratings. Anything rated at 25 or below is treated as non-combustible and can be used in all categories of buildings. Materials rated above 25 and below 75 are classified as slow burning and are subject to a number of specific limitations. Materials rated above 75 are allowed only in residential construction and in limited quantities for commercial and industrial construction use.

The highest initial light-transmission value found in clear, 6 to 8 oz per sq ft panels is in the range of 85 to 90%. Highly translucent colored panels have light-transmission values in the 40 to 80% range, and more opaque pastel-colored panels are found in the 10 to 40% range. By definition, translucent panels are not opaque, although products with very low light-transmission values, approaching 0%, can be produced. Light transmission of translucent corrugated panels is tested according to ASTM D1494. Figure III-1.3 illustrates an

Figure III-1.3 Illustration showing use of SPI transmissometer for translucent paneling. (Courtesy of Fiberglass Division, Ferro Corp.)

assembly of this equipment. Typical transmission values for a range of panel colors are tabulated in Table III-1.8.

Weather resistance is an important characteristic of RP/C panels because of the many outdoor applications. Panel producers have developed a battery of tests related to this function but ultimately rely on actual outdoor exposure for verification of projected peformance. The tests used include immersion in boiling water and in various chemicals; exposure to heat lamps (fadeometer), to oven heat, to water vapor, and in weatherometers.

Of all these tests, the weatherometer gives results which most closely correspond to actual outdoor exposure experience. The other tests have little or no meaning when considered

separately. The SPI Panel Council has developed a tentative artificial-weathering test method which is expected to be adopted as a standard by the industry. In general, it requires 3000 hr exposure in a carbon arc weatherometer (sunshine or twin-arc). After exposure, specimens are examined and graded visually in comparison with control specimens for (1) change in surface-fiber appearance, (2) apparent color change, and (3) loss of gloss. The changes are assigned numbers which in turn are weighted to obtain what is then called a "degradation factor." Specimens are graded as "weather-resistant," "general-purpose," and "interior-application." Specific details of the grading system and mathematics involved are purposely omitted in this discussion inasmuch as the test procedure has not as yet been formally adopted by the Council.

RP/C panels are marketed along with other building products through mass merchandisers, cooperatives, building-supply chains, wholesaler-retailer channels, and building contractors. The 1973 market is estimated at approximately 200 million sq ft. The two largest definitive end uses are garage doors and greenhouse covers.

Excellent manuals which provide specific installation details for translucent architectural paneling in all types of construction are supplied by the manufacturers and their distributors.[2] These include instructions for glazing, sidelights, skylights, elevated skylights, patio roofs, partitions, and many other items. Excellent handling recommendations and lists of necessary installations and maintenance accessories are also provided.

Table III-1.8 Light-Transmission Values for Various Colors in Corrugated Translucent Paneling

Colors	% Typical Light Transmission	Colors	% Typical Light Transmission	Opaque Heavy Duty Panel Colors	% Typical Light Transmission
Clear	90	Awning white	45	Redwood brown	Completely opaque
Skylight green	65	Awning blue	25	Asbestos grey	Completely opaque
Sky blue	58	Awning yellow	50	Industrial green	Completely opaque
Sunlight yellow	80	Surf green	35		
Light ivory	66	Brown	10		
Coral	45	Black	0		
Forest green	46	Desert sand	10		
Awning red	9	Desert rose	10		

Source: Corrulux Corp., Houston, Tex.

References

1. Registered trademark, E. I. duPont deNemours & Co.
2. Glass Fiber Reinforced Polyester Structural Plastic Panels, NBS Voluntary Product Standard PS-53-72, April 1972, U. S. Dept. of Commerce.

III-2

Translucent Sandwich Panels

INTRODUCTION

Translucent fiber glass RP/C sandwich panels have reached solid commercial status as a primary element for architectural wall and skylight systems. Grids of aluminum I-beams form the structural core to which the skins are pressure-bonded. The resultant light-transmitting panels are structurally strong, weather-resistant, and sound-absorbent, with good insulation values and excellent decorative effects.

A unique feature of these panels is the inclusion of a fine glass fiber veil between the translucent skins. This increases both heat and sound absorption while maintaining high light transmission. At the same time it serves as a method to color individual grid sections together with standard color inserts.

The useful properties of translucent sandwich panels combined with engineered installation and accessories produce building enclosures particularly suited to public structures such as churches, schools, airports, and industrial plants.

MATERIALS

The translucent sandwich skins are pre-fabricated flat sheet produced by continuous processing of acrylic-fortified polyester plus chopped glass fiber mat (see Chapter III-1). Thicknesses range from 0.030 to 0.060 in. Special flame- and weather-resistant types are available in grades which closely parallel the corrugated sheet standards. All have smooth, easy-to-clean, glossy surfaces.

Cores are made from type 6035-T5 aluminum I-beams which are heliarc-welded and mechanically interlocked to form rectangular or decorative patterned grids. Bonding adhesives used are of the synthetic rubber types. These are formulated to be both pressure sensitive and thermosetting and become a non-rigid, high-strength polymer on curing. Because it outlines the grid sections, black adhesive is used for a contrasting effect.

Additional thermal insulation is obtained by packing grid-core spaces with glass-fiber veil mat which does not block out but merely diffuses the light transmitted. Thermal insulating "U" factors of 0.27 equivalent to 30 in. of concrete are obtained. Decorative, colored grid effects are made by coloring the glass fiber insulation in combination with neutral white or colorless skin sheets.

Although panels for custom applications may be fabricated using weather-resistant film surfaces such as duPont Tedlar, most panels are coated on the exterior face with an acrylic lacquer. This is the same coating recommended for long-term maintenance in other types of RP/C architectural paneling.

EQUIPMENT AND TOOLING

The processes for production and assembly of translucent RP/C sandwich panels are essentially proprietary. For general reader interest, however, they may be resolved into three basic steps, described below with the major equipment and tooling components necessary for each.

1. *Continuous line for fiber-glass resin, flat-sheet formation.* The main component for flat-sheet formation is a belt moving continuously through the forming, gelling, and curing stages of the sheet. Film, resin, and glass feed stations are provided as well as equipment for post-cure cut-off, cleaning, inspection, and handling. The major difference between this equipment and tooling and that for continuous panel production (see Chapter III-1) is that no means are necessary for corrugating or providing any surface treatment other than smooth and flat. The film surface determines and dictates the molded surface of the flat panel, and thicker film assures a high quality molded surface.

2. *Equipment for grid-core fabrication.* Again, a continuously moving assembly line is engineered for preparing the aluminum-grid core. The main steps are: milling the I-beams to thickness and parallelism, applying the adhesive to surfaces, cutting to length, notching (ends), assembling the grid, and providing a heating means to "tackify" the adhesive immediately prior to assembly.

3. *Finished panel assembly and production.* A continuously moving conveyor belt carries skins and cores assembled automatically or semi-automatically through heating and compression stations. Trim saws plus a lacquer-spray coating and drying facility comprise the final equipment station in the operation.

MOLDING METHOD—PROCESSING DETAILS

The method of panel fabrication becomes fairly obvious after consideration of the equipment and tooling explanation and Figure III-2.1. The salient features of the three major process stages for maximum control of finished panel properties are:

Flat RP/C translucent-panel production. Paneling for skins of the translucent-sandwich construction must be flat, uniform in thickness, and of good surface quality to be usable and to promote maximum contact and adhesion to the grid. Other process requirements for the surface skins are freedom from surface imperfections and from wide variations in the glass/resin ratio, and freedom from internal voids.

Grid-core fabrication. Grid-core thicknesses must be maintained to ±0.004 in. in order to guarantee skin contact and adhesion. The adhesive quality and uniformity of application must be consistent. The total bonding area represents only 8% of the total panel area.

Finished panel assembly and production. On the final moving assembly-belt conveyor, the bottom flat panel is laid down, the grid core with reactivated adhesive is carefully centered and applied, the diffusing layer of glass veil mat plus panel section color inserts (if any) are added, and the top flat RP/C panel is applied—all without pressure.

The temperature of the composite structure is raised to 240°F in a continuous, in-line through-put oven, immediately after which the material moves through pressure rolls carefully set to an adjusted, uniform thickness gap. The face is cooled and the bonded panels are trimmed to size.

As the final step in panel fabrication, an acrylic lacquer for promoting the best possible outdoor weathering is sprayed on what will be the exterior sandwich-panel surface. Setting of the lacquer and solvent removal are accomplished with blower air. Final inspection, sorting, and packaging for shipment complete the cycle.

PRODUCT DATA

Configurations. Many varied patterns and decorative grid configurations are available. The major structural outline of the grid sections, usually 8 in. × 20 in., normally corresponds to the facing of the aluminum grid extrusion to which adhesive is applied for sealing. Decorative embellishments or line-motifs may be added within the faces of the individual grid sections for further attractiveness of the panels.

The panel-outline configurations are patterned primarily after conventional bricklaying techniques: the square, oblong staggered and oblong roman styles and variations with a close double row of supporting grid along the lengthwise or major axis.

For the sake of uniformity and effectiveness over a widely viewed area, such as the side of a building in which the panels are installed, the grid sizes or individual panel dimensions are not varied once they are set throughout the entire assembly. Interest and accent are provided by use of the standard color inserts and by fabricated trim and panel-outline accessories. Transparent glazing is added to panels at eye level as a built-in feature if required.

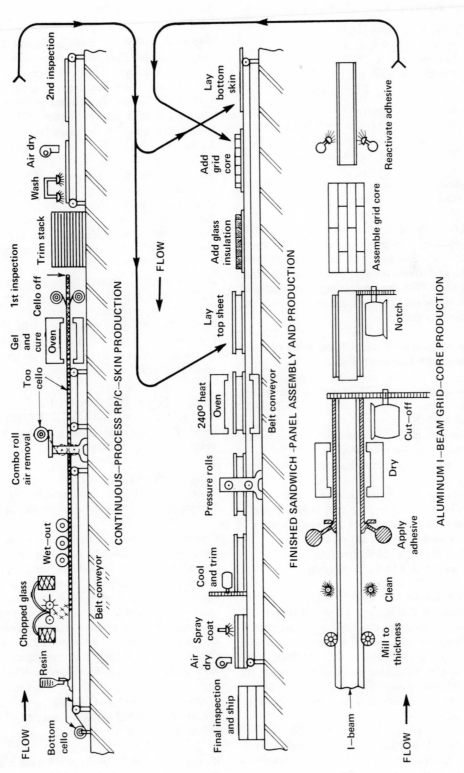

Figure III-2.1 Translucent sandwich panel manufacture.

Standard Sizes. Translucent sandwich panels are available in the following sizes:

Curtain wall panel sizes:
Widths   40 in. and 5 ft
Lengths   to 35 ft
Thicknesses $5/8$ in., $1\,9/16$ in., $1\,3/4$ in., $2\,3/4$ in., and 3 in.

Skylight panel sizes:
Widths $14\,1/4$ in. to 3 ft. $10\,3/4$ in.
Lengths $14\,1/4$ in. to 19 ft. $10\,3/4$ in.
Thicknesses $1\,9/16$ in. and $2\,3/4$ in.

Glazing panel size:
Thickness $5/8$ in.

Colors. Manufacturers offer approximately five opaque, architecturally compatible colors which are combined with several translucent colors to form the many decorative combinations used. With the contrasting fiber-filled grid areas an almost limitless number of color combinations is possible.

Representative Physical Properties of Translucent Sandwich Panels:

Light Transmission. The range of light transmission varies from 0 to 80%. Control is obtained by:
Skin selection: opaque 30%, translucent 40%, transparent 60%.
Glass-fiber angel hair inserted in the core: 0 to 40%.
Recommended light transmission range:
Walls: 20 to 28% generally; lower on south and west elevations in southern climates.
Individual skylights: 28% generally.
Small skyroofs: 20% generally.
Large roofs: 8 to 12% generally.

Thermal Insulation. The "U" factor of standard panels is 0.4 Btu per hr per sq ft per °F, allowing a wind velocity of 2 mph (perimeter aluminum not included) at a mean temperature of 8°F. Panels insulated with glass fiber angel hair have a "U" factor range of 0.27 down to 0.15, whereas panels with a "U" factor of 0.4 are generally satisfactory for general-purpose application.

Acoustical Insulation. Standard panels have a sound transmission loss of 29 to 32 db (average over the normal range of frequencies).

Coefficient of Thermal Expansion. 12.5×10^{-6} in. per in. per °F.

Impact Resistance. Panels made with 0.06 in. acrylic-modified polyester skins absorb 137 ft lb per sec impact (6.22 lb free-falling steel ball, $3\,3/8$ in. diameter, from $7\,1/2$ ft). Panels with 0.045 in. acrylic-modified polyester skins absorb 112 ft lb per sec. impact (6.22 lb free-falling steel ball, $3\,3/8$ in. diameter, from 5 ft). Impact strength of faces by ASTM D256 and Federal Specification 406-1071 is 15 ft lb per in. width (Izod unnotched flatwise). The combination of 0.045 in. thickness for interior skins and 0.60 in. for exterior faces is satisfactory for normal conditions. Above a second story, 0.045 in. is satisfactory for exterior. For gymnasiums, 0.06 in. interior is recommended.

Bond Strength. Accelerated aging—ASTM D-1037 (Forest Products Laboratory Cycle Test) caused no delamination of the bond after six cycles. The tensile strength of resilient bond tests to a minimum of 400 psi dynamic tensile. Humidity exposure has no effect on the bond (ASTM D1735).

Spans. Tables III-2.1 and III-2.2 measuring from sill to point of attachment (rough opening), show the allowable span for various panel mounting constructions.

Weight. Standard panels less than 14 ft 9 in. in length weigh less than $1\,1/2$ lb per sq ft. Longer spans weigh slightly more per sq ft.

Chemical Resistance. Panels are resistant to all chemicals commonly found in the atmosphere and in normal building uses, including polluted air, atmosphere of swimming pools, seaside atmospheric acid, atmospheres in paper mills and steel towns. Special protection may be required for the aluminum-clamping system in chemically corrosive atmospheres. (*see also* Chapter II-1.) (Chemical resistance tests: ASTM D543; Federal Specification 406-7011.)

Weathering. Outdoor-exposure tests in the Florida sun resulted in only a very slight color change in one year.

Water Absorption. Water absorption is about 0.13% by weight (ASTM D570).

Table III-2.1 Span Table for Vertical Panel Mounting[a]

Panel	Joint and Stiffener	Allowable Span
2¾ in. thick panels for walls (4 ft wide, standard grid)	2 in. batten	14 ft 9 in.
	#2 short span splice edge	15 ft 8 in.
	2 in. structural tee batten	16 ft
	#4 short span splice edge	16 ft 6 in.
	4½ in. fin batten	19 ft 3 in.
	#1 long span splice edge	22 ft 4 in.
	silhouette stiffener	27 ft 2 in.
2¾ in. thick panel unit walls (4 ft wide, standard grid)	#2 short span splice edge	11 ft 6 in.
	#4 short span splice edge	14 ft 4 in.
	4½ in. fin batten with #2 short span splice edge	15 ft 2 in.
	#4 short span splice edge plus structural tee	16 ft
	#1 long span splice edge	20 ft 8 in.
	silhouette stiffener	27 ft 2 in.

[a]20 lb per sq ft design load

Table III-2.2 Maximum Roof Span for Typical 2¾ In. Thick Sandwich Panels[a]

Panel and Stiffener[b]	Maximum Span	
	Load = 20	Load = 40
Panel	11 ft 5 in.	8 ft
Panel with #2 short span splice edge	12 ft	9 ft 6 in.
Panel with #4 short span splice edge	13 ft 3 in. cm[c]	10 ft 6 in.
Panel with double I-beam #2 short span splice edge	14 ft 6 in. cm	11 ft 3 in.
Panel with double I-beam #4 short span splice edge	15 ft 9 in. cm	12 ft

[a]3/8 in. minimum to 12 in. pitch.
[b]2 in. double tee batten is used.
[c]cm = cross mullion.
Notes: Rule of thumb: 8 ft for 40 lb load with stiffener 10 ft.
Recommendations: Check with specification or architect on any deflection limitations because spans are based on safe performance.

Table III-2.3 Fire Resistance in Panels

Panel	Flame Spread	Fuel Contributed	Smoke Density
Standard panel	200	64	123
Panel with fire-retardant resin	30–75	5–15	Over 200

Abrasion Resistance. The loss per cycle is 0.0602g (Federal Specification 406-109.1).

Maintenance. In order to insure satisfactory lifetime service, all panels must be properly maintained—a procedure that varies according to the location of the installation. Preventive maintenance is recommended every 10 to 15 yr on north and east vertical exposures in northern climates; 5 to 8 yr in average locations; and 3 to 5 yr in areas of excessively strong sunlight, such as roofs in Florida or California. Basically, maintenance merely involves washing the panels clean and re-finishing them with specially formulated acrylic lacquer.

Fire Resistance. A comparison of Standard panels with fire-retardant panels in ASTM E84 tunnel test yields the results shown in Table III-2.3.

Installation. Installation of pre-fabricated translucent sandwich panels requires highly en-

Figure III-2.2 A typical application of white translucent face sheets on a 12 in. x 24 in. aluminum grid core. This office and warehouse used 20% light-transmission panels 5 ft. wide and 17 ft. high. A 0.27 "U" factor for thermal insulation was obtained by packing low-density, colorless glass fiber batting between faces. Black decorative areas were obtained by packing with black glass batting. (Courtesy of Kalwall Corp.)

gineered accessories. These components form a system which two men can handle on almost all jobs. Jambs, sills, battens, sash inserts, stiffeners, and louvers required for erection comprise clamp-type extended, prefinished aluminum assemblies fitted with sealing tapes and fastened on the site with stainless-steel screws.

Installation details are available from panel producers. Although some imbedding caulk is applied on the job, most components are joined by removing release tapes from factory-applied sealing beads and then assembling with hand tools.

Figure III-2.2 shows an actual representative installation of translucent sandwich panels with RP/C skins.

Costs, Advantages, and Applications. The cost of architectural curtain-wall and roof-system paneling depends upon the specific job, because a good portion stems from the metal grids and framing. Panels in volume sell as low as $2.10 per sq ft, while an installed wall system would start at $3.00 per sq ft. The panels offer many advantages: low maintenance cost; good thermal insulation; low sound transmission; controlled light transmission; high impact strengths; light in weight; beauty in wide choice; fire retardancy, if needed; large panel sizes; easy, low-cost installation.

It is not difficult to see why large quantities of paneling are chosen for banks, plants, churches, schools, shopping centers, interior office walls, office buildings, airports, and private homes. Annual production is estimated at 2,000,000 sq ft per yr.

References

1. Panel Structures, Inc., East Orange, N.J.
   Patents: 2,874,653; 2,969,618; 3,014,561
   Kalwall Corp., Manchester, N.H.

   Patents: 2,981,382   2,931,468
            3,082,848   3,082,849
            D196,778   D198,259
            D199,524   D199,525

III-3

Opaque Panels

Opaque panels are available in several types: shaped and flat panels, sandwich panels (contact-molded and press molded). Each type will be covered in detail in the following discussion.

OPAQUE SHAPED AND FLAT PANELS[1]

Industrial RP/C paneling is both contact-and press-laminated to produce flat, moderately contoured, or shaped sheet which is used principally for surfacing applications. Press laminating is employed to produce panels up to 4 ft × 12 ft with various corrugated or other type configurations, mostly from ½ to 1 in. deep. Larger sheets (up to 8 ft × 200 ft in rolls)[2] are made by contact laminating. Panel thicknesses range from ⅛ to ¼ in. because thinner gauges can be more economically produced by continuous laminating (see Chapter III-1). Some very thick sheet (up to 1 in.) is produced by pultrusion (see Chapter V-2).

The principal applications of opaque shaped and flat panels include flat sheets for electrical insulation; flat sheets for lining tanks, trucks, trailers, rail cars, ducts, and other surfaces; ribbed configurations for truck and rail-car refrigerator panels, containers, and bulkheads. Although a few manufacturers offer standard flat sheet (electrical grades are made to NEMA specifications), production is usually tailored to customer end-use requirements.

Panel costs to fabricators range from $0.65 per sq ft for ⅛-in. thick sheet to $1.60 per sq ft for ⅛ in. thick refrigerator panels.

Materials. For low-cost sheet, chopped-mat fiber glass reinforcement is used with a general-purpose polyester resin. Higher strengths are obtained with woven-roving reinforcement. Specially selected resins are used for electrical materials and severe corrosive environments.

Extra surface resistance is provided by the standard gel coats or a high-resin surface obtained with surfacing veil mats. Embossed molds or film also add a weather resistant surface resin.

Typical formulations and mechanical properties for several opaque shaped and flat panels are presented in Table III-3.1.

Equipment, Tooling, and Molding Methods. Process and equipment for production of contact laminated sheets closely follows methods described below under Equipment and Tooling for opaque sandwich panels.

Press laminating in matched-metal dies follows methods described in Chapter IV-2. Press molding of sheet-molding compound (SMC) is expected to expand some of these markets because it can produce more complicated shapes. Excellent adhesion between SMC and plywood is possible.

Product Data. Shaped structural RP/C sheets are made in many configurations from simple lap-edges for wall uses to shallow trays for hatch covers.

Perhaps the major shaped sheet product today is the ribbed refrigerator (reefer) panel used as truck and railroad car liners for surfacing where impact and mechanical abuse are severe. Typical refrigerator panel configurations include rib liner panels, liner panels, convolute, and so on (consult manufacturers[3]). In order

Table III-3.1 Formulations for and Properties of Representative Opaque Shaped and Flat Panels

	Press-Molded		Contact-Molded	
	Structural	Electrical	Structural	Structural
Formulation				
Resin, parts	73–Polyester	80–Polyester	78–Polyester	70–Polyester
Fiber glass	27–Chopped mat	20–Chopped mat	22–Chopped mat	30–Woven roving
Filler	37–Calcium carbonate	30–Hydrated Alumina	20–Clay	20–Clay
Pigment paste	4–White	4–White	4–White	4–White
Catalyst	1–BPO	1–BPO	0.75–MEK peroxide +0.02–Cobalt (6%)	0.75–MEK peroxide +0.02–Cobalt (6%)
Properties				
Tensile strength, psi	14,000	11,000	9,000	28,000
Flexural strength, psi	27,500	20,000	18,000	24,000
Flexural modulus, psi	1.29×10^6	1.10×10^6	1.00×10^6	1.30×10^6
Izod impact ft lb (unnotched)	18.0	13.5	8.0	25.0
Barcol hardness	53	50	45	50

Gas	Effect
Methyl bromide (CH₃Br)	very slight effect
Carbon disulphide	no effect
Carbon tetrachloride (CCl₄)	very slight effect
Ethylene dibromide (1,2 dibromethane)	very slight effect
Ethylene dichloride (1,2 dichloroethane)	very slight effect
Phosgene	no effect
Hydrogen cyanide	no effect
Malathion	no effect

Source: Kemlite Corp., Joliet, Ill.
Notes: As aerosals, fumigating had no effect on structural RP/C panels in each case.
As concentrated vapor or condensing vapor, effect was very slight, e.g., loss of gloss over extended exposure.

to assist cooling systems to circulate conditioning air without special ducts, these panels incorporate air grooves which also serve as structural corrugations.

Refrigerator panels are combined with thermal insulation placed between the impact-resistant panel and vehicle wall of the structure. This improvement provides both mechanical and thermal advantages.

One patented refrigerator panel is made with integral air ducting.[4]

When panels are used in food processing plants, food coolers, or food vehicles, fumigation is occasionally required. Table III-3.2 shows the effect of fumigating gases on structural opaque shaped and flat RP/C panels.

OPAQUE SANDWICH PANELS

Structural RP/C sandwich panels have been used in the aerospace industry for some time. These include high-strength structures where the cores are relatively expensive, such as RP/C or metal honeycombs. However, low-cost, commercially important RP/C sandwiches have been slow in developing. Competitive composites such as aluminum and plywood-skinned foam-cored structures have led this market. One example is the exterior panels for mobile homes.

Recently, RP/C-skinned sandwiches, cored with urethane foam but especially with plywood, have become an important commercial material. Urethane-cored RP/C-skinned sandwiches give the unique combination of corro-

sion resistance, easy-to-clean surfaces, color, and inherent thermal and electrical insulation. Refrigeration applications are an accepted and growing use for this construction. Plywood-cored RP/C-surfaced sandwiches have become an established product for shipping containers, truck bodies, and other end uses where non-corrosive, high-modulus, impact-resistant, low-maintenance, easy-to-repair panels are needed.

Along with these product advantages plus economy and ease of fabrication, the industry's capability to produce very large seamless panels up to 10 ft wide and 200 ft long is an additional advantage that is especially important for the containerized freight modules which now enjoy worldwide acceptance. This market alone used well over 5 million sq ft in 1972.[5]

Two principal production methods are used today. These are contact-molding and press molding. The former accounts for perhaps 80% of the output.

Contact-Molded Sandwich Panels

Materials:

Resins. General-purpose gel coat and laminating resins are used. Both are catalyzed for rapid room-temperature gel and cure. Formulations employ standard pigment and filler additives (see Table III-3.3).

Reinforcements. Although some panel production employs the spray-up method using conventional chopped fiber glass roving, most fabricators take advantage of higher strength woven rovings. They use 24-oz materials available in the full, wide widths and long, continuous lengths required for panels. One ply of woven roving per face gives the required strength for most uses, and the cost is only 20% more than roving alone. Ease of processing the woven roving makes up for most of this difference.

Cores. Core material is primarily structural, no. 1-grade plywood which is C-C plugged, unsanded, exterior-grade, rough-cut, and ¾ in. thick. Exterior-grade quality is a major requirement, because panel edges are not surfaced and weather exposure is severe. Rotary cutting provides a rough plywood surface which is used without sanding to assist RP/C skin adhesion.

For less impact-resistant panels, other core materials are used, such as 2 lb per cu ft urethane foam. These produce better insulating

Table III-3.3 Formulations for Gel Coat and Laminating Resins for Room-Temperature Cure of Contact-Molded Sandwich Panels

Gel Coat Formulation	Parts	Laminating-Resin Formulation	Parts
General-purpose polyester resin	100	General-purpose polyester resin	100
Cabosil	2	ASP 400 clay filler	20
White-pigment paste	5.0	Dimethyl analine	0.1
Cobalt octoate	0.20	Benzoyl peroxide	1
MEK peroxide	0.60		

properties and also lighter weight panels suitable for refrigerator and architectural applications.

Molding Method/Processing Details. A typical process for fabricating contact-molded core panels is based on the use of woven roving on the panel facings. Following is an example of a contact-process cycle producing four 8 ft × 40 ft panels from one mold:

	(min.)
Clean and wax mold	15
Lay gel coat	10
Cure gel coat	45
Lay bottom laminating resin	10
Lay bottom woven roving	5
Lay core sheets	10
Lay vacuum bag and seal	10
Evacuate and cure	60
Lay top laminating resin	10
Lay top woven roving	5
Roll-out and cure	45
Demold and trim	20
Inspect and stack	20
Total	265

The long cure times indicated dictate the use of a minimum of two molds for efficient labor utilization. As in many other panel-producing operations, rather large floor spaces are required. Large sandwich-cored panels particularly require an ample processing area because the foam cores are in panel form and storage prior to fabricating is essential. A 10,000 sq ft per shift panel output requires at least 50,000 sq ft of floor space.

Equipment and Tooling:

Molds. Molds are deceptively simple because they are merely flat surface. They do, however, need to be quite large (10 ft × 220 ft), nonporous to insure good release, seamless, rugged enough to withstand accidental scratching, rea-

sonable in cost, and they must have surfaces good enough to produce easily-cleaned, attractive products.

Although some work has been done with metal and surfaced concrete, glass-reinforced plastic itself has been the preferred mold material to meet these requirements.

Cleaning and Waxing. Powered floor waxing equipment carries out this function very well.

Resin Application. Although sprays can be used, most systems are too costly when the required elaborate corollary exhaust equipment is added. One manufacturer,[6] however, does offer a reciprocating airless spray unit for panel manufacturing.

Simple hand coating is laborious, although, with assistance from moving hoppers or tanks, doctor blades, and rollers, it does have the advantage of little maintenance, short clean-up time, and low cost.

Woven-Roving Application. Reinforcements such as woven roving are simply unrolled onto the resin coating and hand-rolled to wet out.

Core Application. Both plywood and foam cores are generally limited in width (4 ft). Consequently, they need to be accurately positioned with little or no gap between them. Tracked vehicles move stacks of core sheets over the mold and lay them in place with the long dimension (usually 8 ft) across the sandwich-panel width.

Vacuum Bag. Film or a rubber bag is unrolled onto the core, sealed to the mold edges, and evacuated with conventional high-volume vacuum pumps.

Trimming. Panel trim involves moving the panel through side-trim saws and starting and stopping for length cut-off. Although conventional saws (diamond-faced) are used, the equipment requirements dictate heavy, well-controlled design. Side-trim control, for example, must be both accurate and capable of rapid correction, because the amount of mate-

Table III-3.4 Mechanical Properties of Contact-Laminated Sandwich Panels with Plywood Cores

Construction	Type Joint	Flexural		Tensile		Edge Compression		Face Impact
		Str.[a]	Mod.[b]	Str.	Mod.	Str.	Mod.	Str.[c]
Douglas fir—¾ in. Fiber-glass—1.5 oz sprayed chopped strand	Butt	3.6	0.715	2.2	0.547	4.1	0.311	10.3
—both faces	None	3.4	0.703	4.8	0.731	4.2	0.341	19.8
Douglas fir—¾ in. Fiber-glass—24 oz. woven roving	Butt	4.2	0.988	1.9	0.843	3.1	0.318	10.0
—both faces	None	4.2	1.07	6.0	0.912	4.9	0.371	21.7

Source: Owens-Corning Fiberglas Corp., Toledo, Ohio.
[a]Strength (psi $\times$ 10³).
[b]Modulus (psi $\times$ 10⁶).
[c]Strength ft/lb/per in. width.

rial allocated to trim is expensive and necessarily small (½ in.). Also, standards on squareness (¼ in.) are stringent.

Inspection and Shipping. Inspection, which is deceptively simple, can present difficulties when the problem is merely to get a look at the bottom side of an 8 ft $\times$ 40 ft panel. In order to turn over panels of this size and weight, a special panel flip unit is used which can stop the panel in the vertical position for inspection and then deliver it with either surface facing upward.

Moving panels up to 40 ft long for storage and shipping requires no special equipment except that components be oversize.

Product Data:

Advantages and Disadvantages: There are several advantages in using the contact-molding process for opaque sandwich panels:
- Low tooling and equipment cost.
- Very large panel sizes possible.
- Good skin bonding, promoted by the low-temperature cure.
- Weatherability, promoted by gel coating.
- Easily varied panel sizes.

The disadvantages of this process are:
- Smooth surface is possible on one skin only.
- Long cycle time.

Product Properties and Performance. Contact-laminated RP/C skins with plywood cores yield panels with the mechanical properties shown in Table III-3.4. Table III-3.5 presents a stiffness comparison of plywood directions. Table III-3.6 gives a comparison of the impact resistances of various container materials.

Table III-3.7 shows bolt-bearing loads for RP/C-surfaced sandwich panels with plywood cores.

Table III-3.8 gives additional mechanical properties for a few selected sandwiches made with other core materials.

Table III-3.5 Comparison of Stiffness of Panels with Different Plywood Directions

Specification	Flexural Modulus of Elasticity, EI, in lb 0 in.² $\times$ 10⁶	Maximum Moment of Inertia, in lb-in.
Standard ⅞-in. thick panel made from ¾ in. plywood with plywood face grain parallel to the test span	0.99	14,400
Standard ⅞-in. thick panel made from ¾ in. plywood with plywood face grain perpendicular to test span	0.77	11,900

Source: Lunn Laminates, Inc.
Conditions: Panels are made from fir-cored plywood with RP/C skins of 24 oz. woven roving. Determined on 12-in. wide spans, uniformly loaded. Values are average of several tests with spans up to 7 ft in length.

Table III-3.6 Comparative Impact Resistance of Various Container Materials

Material	Comparative Rating
18-gage steel	1.0
¾ in. plywood by itself	0.5
¾ in. plywood core with skin of laminated chopper fiber-glass on each surface	1.0
¾ in. plywood core with skin of 24 oz. woven roving on each surface	2.0

Source: Lunn Laminates, Inc., Wyandanch, N.Y.
Conditions: A comparison was made between the various materials listed using the identical test device for both. This test comprised a typical fork-lift tyne applied with the same force in each case. The relative force necessary to puncture each material is reported.

Lastly, Tables III-3.9 and 3.10 supply typical specifications for panels used for cargo containers.[8]

The thermal K factor for plywood-cored panels is 0.762 Btu/hr/sq ft/°F/in. thickness. A typical[7] 3-in. thick, urethane-foam-cored panel with nominal 3/16 in. thick skins has a U factor of 0.057 Btu/hr/sq ft/°F.

Applications of RP/C-surfaced structural sandwich panels include cargo containers, elevator cabs, hockey-rink walls, curtain walls, portable housing, portable schoolroom walls, basement paneling, walk-in cold rooms, truck and trailer bodies, railroad-car lining, marine deck panels, and commercial shower and bath partitions. (Panels may be fabricated as molded, or easily sawed, etc. to comply with dimensional requirements.)

The cost of ¾ in. plywood-cored panels is approximately $1.00 per sq ft to fabricators and the 1973 projection for cargo containers alone is 15,000,000 sq ft.

Reference Specifications:
MIL-P-17549C—Plastic Laminates, Fibrous Glass Reinforced.
MIL-R-7575C—Resin, Polyester

Requirements:
- The woven roving used shall be ± 10% of the stated weight per, sq. yd. on each side of the core. Roving will be continuous in both directions.
- The physical properties of the laminate shall conform to MIL-P-17549, Grade W, as modified by adding pigment.
- Plywood: The plywood core material shall be C plugged C grade, exterior type, all plies group 1 species, or better. Plywood core will be butted. All plywood must bear the DFPA Grade-trademark of the American Plywood Association, manufactured in accordance with PS 1-66.
- Tolerances: Finished panels
- Surface Appearance:
  Outer Surfaces: Outer surfaces shall be a white gel-coat surface with a minimum thickness of 0.010 in. backed with one layer of woven roving with white-pigmented resin. The surface will be smooth with no surface discontinuities of more than ¼ in. maximum size. Defects may be repaired with gel coat and sanded smooth as required.
  Inner Surfaces: Inner surfaces shall be made as a single layer of woven roving and white-

Table III-3.7 Bolt-Bearing Loads for RP/C-Surfaced Sandwich Panels with Plywood Cores

Specification	Ultimate Bolt-Bearing Load (in lbs)
¾ in. panel—5/8 in. plywood with 24 oz woven-roving skins, with ¼ in. bolt in double shear, ¾ in. edge distance	800
7/8 in. panel—¾ in. plywood with 24 oz woven-roving skins, with ¼ in. bolt in double shear, ¾ in. edge distance	1200
7/8 in. panel—¾ in. plywood with 24 oz woven-roving skins, with 5/16 in. bolt in double shear, 1 in. edge distance	2300
7/8 in. panel—¾ in. plywood with 24 oz woven-roving skins, with 5/16 in. bolt in single shear, 1 in. edge distance	2000

Source: Lunn Laminates, Inc., Wyandanch, N.Y.
Conditions: Skins are 24 oz. woven-roving reinforcing polyester resin.

Table III-3.8 Mechanical Properties of Contact-Molded Laminate Sandwich Panels Made With Various Core Materials[a]

Material Description	Direction of Test		Tensile Stress (psi)	Compression Stress (psi)	Flex Moment in #/ft of Width	E.I. Factor # in²×10⁶	1/4 in. Bolt Bearing lb
	Face Grain Parallel	Warp Perpendicular					
⅞ in. Plywood panels, CC-plugged ext. 5-ply.	(‖)	(|)	4,900	7,250	14,400	1.198	1,300
Same with 24 oz W.R.[b] both faces 3.2 lb/sq ft			‖	6,600	4,960	11,900	0.887	1,125
1⅛ in. end-grained balsa, with 24 oz W.R. both faces, 1.34 lb/sq ft	N.A.[c]	‖ or		2,600	1,320	12,000	0.192	37.6
1⅛ in. PVC foam core 2 lb density, with 24 oz W.R. both faces, 1.20 lb/sq ft	N.A.	‖ or		2,600	1,340	4,000	0.0008	27.1
1⅛ in. P.U. foam core 2 lb density, with 24 oz W.R. both faces 1.01 lb/sq ft	N.A.	‖ or		2,570	1,310	2,700	0.0003	27.1
⅞ in. paper honeycomb, with 24 oz W.R. both faces, 1.23 lb/sq ft	N.A.	‖ or		3,300	1,650	7,500	0.120	41.6
1⅛ in. lightweight concrete core, with 18 oz W.R. both faces, 2.42 lb/sq ft	N.A.	‖ or		1,860	960	17,500	0.301	24.2

Source: Lunn Laminates Inc., Wyandanch, N.Y.
[a]These test results are taken on the basis of a 2 ft span.
[b]Woven roving.
[c]Not acceptable.

Table III-3.9 Length and Width of Contact-Molded RP/C Sandwich Panels

Panel Size (Approx.)		Tolerances	
Width (ft)	Length (ft)	Width (in.)	Length (in.)
8	8[a]	±⅛	±⅛
8	20[a]	±⅛	+¼ −⅛
8	40[a]	±⅛	+¼ −⅛
10	40[a]	±¼	+¼ −¼

[a]Difference between panel diagonal measurements not to exceed 1/4 in.

Table III-3.10 Thickness of Contact-Molded RP/C Sandwich Panels

Nominal Thickness (in.)	Plywood Thickness (in.)	Actual Panel Thickness[a] (In.) Weight of Woven Roving (Each Side of Panel)	
		18 oz	24 oz
⅜	¼	0.340	0.350
½	⅜	0.465	0.475
⅝	½	0.590	0.600
¾	⅝	0.715	0.725
⅞	¾	0.840	0.850
1⅛	1	1.090	1.100
1¼	1⅛	1.340	1.350

[a]Tolerance on all these thicknesses .050 to −.050 in.

pigmented resin. The surface appearance will have a predominant woven cloth appearance.

Quality Assurance: The following inspection checks will be performed based on sampling plan techniques: inspect incoming plywood, roving, resin, overall size of panels, maximum thickness of panels, and surface appearance of laminate.

■ Weights
    ¾ in. panels (⅝ in. plywood) 2.7 lb/sq ft
    ⅞ in. panels (¾ in. plywood) 3.0 lb/sq ft
    1¼ in. panels (1⅛ in. plywood) 4.1 lb/sq ft

Press-Molded Sandwich Panels. Two methods are in competition with the contact-molded sandwich panels: the standard closed matched-die process and an open-sided step process.

Materials. Both methods use exterior-grade plywood cores plus reinforced plastic RP/C skins. While the press-molded, matched-die method can produce panels up to 4 ft × 12 ft and with encapsulated edges, the step process adds the capability to produce continuous panels of any required length. Although surface-skin materials for both processes can be the older mat or woven-roving reinforced-polyester resins, sheet-molding compound (SMC) is also used, especially in the step-cure process. SMC is a high-impact pre-formulated molding material containing long chopped glass fibers, fillers, pigments, release agents, and catalysts in a thickened polyester resin. It offers several process advantages (see Chapter IV-2).

Equipment and Tooling. Equipment and tooling are similar to hydraulic press equipment and tooling for fabricating translucent architectural panels (see Chapter III-1).

Molding Method. Pressing SMC in molds and curing it on both sides of core sheets in the step process are procedures that are currently challenging the older contact methods. In this process, the press, capable of applying a minimum of 300 psi, has a cooled-end section allowing the sheet to be stepped or indexed through the mold with a minimum visible cure line between steps. In this way, sheets up to press-platen width can be made as long as required.

Press-curing time for both SMC and wet-press-molding processes are approximately 2 min. for skins ¹⁄₁₆ in. thick. SMC requires 300 psi and 300°F for proper cure. Wet-process press-molding can operate at the lower pressure of 100 psi and 250°F temperature. Cycles are typically 3 min. for SMC and 4 min. for wet resins. Both have economically high output rates. Table III-3.11 presents a detailed comparison of the two processes.

Obviously, the handling, trimming, inspection, stacking, and packing of these panels in large sizes have the same requirements for equipment and floor space as in the contact process.

Product Data:

Advantages and disadvantages: The advantages and disadvantages of the matched-die and step-press processes are compared as follows:

Table III-3.11 Comparison of SMC and the Wet Process in Press Molding Opaque Sandwich Panels

SMC Process	Wet Process
1. Clean mold	1. Clean and lubricate mold
2. Load bottom SMC	2. Load bottom reinforcements
3. Load core	3. Apply liquid resin
4. Load top SMC	4. Load core
5. Close press and cure	5. Load top reinforcements
6. De-mold	6. Apply liquid resin
7. De-flash	7. Close press and cure
8. Inspect	8. De-mold
9. Pack and ship	9. Inspect
	10. Pack and ship

Matched-die press process:
- Advantages:
  - Smooth surface both sides
  - Core completely encapsulated
  - Core joints pressure-bonded
  - Fast cycle and high output
  - Capability for molded-in detail

- Disadvantages:
  - High equipment and tooling costs
  - Panel sizes controlled by possible mold sizes
  - Heavy gel coatings are not practical

Step-press process:
- Advantages:
  - Smooth surface both sides
  - Core-joints pressure-bonded
  - Fast cycle and high output
  - Capability for producing long sheets
  - Capability for producing molded-in detail

- Disadvantages:
  - High equipment cost
  - Heavy gel coatings are not possible

Product properties and performance. The application of glass-reinforced polyester skins to plywood produces a substantial improvement in stiffness. A deflection comparison of plywood with and without an RP/C skin is shown in Figure III-3.1.

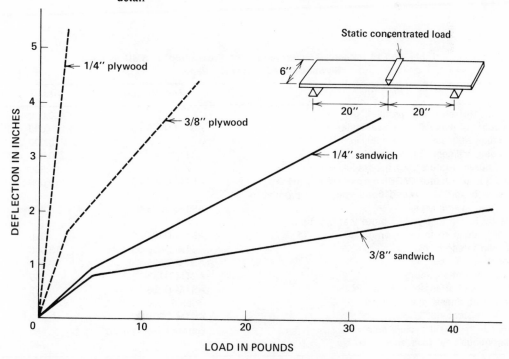

Figure III-3.1 Comparative deflections of plain plywood and sandwich panels: skins are 1½ oz. glass-fiber mat and polyester. (Courtesy of National Vulcanized Fiber Co.)

Table III-3.12   Composite Thicknesses of Press-Molded SMC-Skinned, Plywood-Cored Sandwich Panels

Ply	Jointed Overlaid Specimens (in.)	Unjointed Overlaid Controls (in.)	Plain Plywood Unjointed
Plywood core (⅝ in.)	0.613	0.603	0.602
Entire sandwich	0.762	0.749	

Source: American Plywood Association for Brooks and Perkins, Inc. material.

Table III-3.13   Flexural Strength Properties of Plywood Overlaid Both Sides with SMC

Specimen Description	Average Stiffness[a] EI, Lb.-In.^{2} (12-in. width)	Average Load-Carrying Capacity (12-In. Width)	
		Moment (lb.-in.)	% Joint Efficiency
Face grain perpendicular to span:			
RP/C continuous face and back butt-jointed plywood core	578,800	6,265	69
RP/C face and back unjointed	603,900	9,034	
Plain plywood	133,500	2,338	

Source: American Plywood Association for Brooks and Perkins, Inc. material.
[a]Average of two sets of specimens submitted to test.

Table III-3.14   Mechanical Properties of Press-Molded, SMC-Skinned, Plywood-Cored Sandwich Panels

Property	Test Method	Result
Modulus of elasticity, parallel, psi		1,400,000
Modulus of elasticity, perpendicular, psi	ASTM-D805-63	1,200,000
Bending stiffness (EI) parallel, in.^{2}		410,000
Bending stiffness (EI) perpendicular, in.^{2}		250,000
Modulus of rupture (MOR) parallel, psi		14,500
Modulus of rupture (MOR) perpendicular, psi	ASTM-D805-63	10,000
Plastic facing to plywood bond test grain parallel to force, average, psi	ASTM-D805-63	400+
Plastic facing to plywood bond test grain perpendicular to force, average, psi	ASTM-D805-63	320
Tensile strength, psi	ASTM-D805-63	13,000
Compression, psi	ASTM-D805-63B	7,500
Ball penetration, lbs	ASTM-D143-52	330
Izod impact strength, ft-lb per in.	ASTM-D256	18.0
Barcol hardness, minimum		65
Moisture content, %	ASTM-D805-63	2.5
Wear test, 10,000 cycles, mils	Gardner heavy duty	4–5
Continuous heat resistance		300

Source: Koppers Co., Inc., Pittsburgh, Penna.
Note: Properties are minimum values for 1/2 in. exterior-grade plywood Douglas fir, Grade C-C or better with approximately 1/16 in. thick fiber-glass reinforced-plastic facing on both sides.

Table III-3.15 Environmental Performance of Press-Molded, SMC-Skinned, Plywood-Cored Sandwich Panels

Test Type	Test Description	Times Cycle Repeated	Performance
Soak-dry	Soak 3 in. × 6 in. sample in ambient-temperature water for 8 hr. Follow with force dry at 145°F for 16 hr.	25	No surface cracking visible to naked eye. No apparent degradation of the bonding of the coating to substrate. Good appearance and color retention.
Boil-dry	Immerse 3 in. × 6 in. sample in boiling water for 4 hr. Follow with force dry at 145°F for 20 hr.	25	No surface cracking visible to naked eye. No apparent degradation of the bond of the coating to substrate. Good appearance and color retention.
Hot water-dry	Immerse 3 in. × 6 in. sample in 180°F constant-temperature water for 4 hr. Follow with force dry at 145°F for 20 hr.	25	No surface cracking visible to naked eye. No apparent degradation of the bond of the coating to substrate. Good appearance and color retention.
Soak-freeze-dry	Soak 4 in. × 6 in. sample in ambient-temperature water for 8 hr. Follow with freezing at 0°F or lower for 16 hr. Follow with force dry at 145°F for 24 hr.	10	No surface cracking visible to naked eye. No apparent degradation of the bond of the coating to substrate. Good appearance and color retention.

Source: American Plywood Association for Brooks and Perkins, Inc. material.
Note: Panels tested consisted of 5/8-in. exterior-type, C-D grade, plywood overlaid with 50 to 80 mils of SMC.

The mechanical properties of press-molded, SMC-skinned plywood-cored sandwich panels are given as shown in Tables III-3.12, III-3.13 and III-3.14.

Environmental performance is shown in Table III-3.15.

Applications and markets. The end uses of press-molded sandwich panels are similar to and compete with the contact-process products. Press-molded panels could, however, expand into applications where molded-in features would be an advantage, especially if a highly resistant gel coat is not a requirement. The cost to fabricators is currently competitive to that of contact-molded panels.

References

1. Produced by Lunn Laminates, Inc., Wyandanch, N.Y.
2. *Ibid.*
3. Molded Resin Fiberglass Co., Division North American Rockwell, Ashtabula, Ohio.
4. Koppers Co., Inc., Pittsburgh, Penna.
5. American Plywood Association.
6. Venus Products, Inc., Kent, Wash.
7. Molded Resin Fiberglass Co.
8. Lunn Laminates, Inc.

SECTION IV

HIGH-TEMPERATURE CURING PROCESSES

IV-1

Premix and Compound Molding Including BMC and Injection Molding of Thermosets

INTRODUCTION

Premix and compound molding involve co-mixing resin, filler, and reinforcement in a mechanical device; extruding or otherwise shaping the mixture for convenient handling; and charging it to a shaped mold in a compression, transfer, or injection press. The application of force and the behavior of the charge are unique in plastics processing because the molding material is puttylike in consistency. The charge resists pressure and hence is forced uniformly to every extremity of the mold, with the fillers lending sufficient plasticity to the compound to preclude separation of the three component phases.

The resultant parts generally are categorized by the following:

- Rigidity and dimensional stability. Bosses are molded in.
- Low cold-flow or creep.
- Hard surface and high modulus of elasticity.
- High impact and compressive strengths.
- Moldable up to 2 in. thick.
- Can have variation of thicknesses in same part.
- Wide range of colors possible.
- Excellent, smooth, paintable surfaces available with advent of low-profile resins and bulk-molding compound thickeners.
- Parts almost 100% finished out of mold (very little flash).
- Highest production rates and lowest part cost in RP/C field.

Some examples of applications of premix molded parts are electrical gear, such as structural elements used in heavy-duty apparatus: switchgear, arc chutes, insulator rods, and miscellaneous equipment for resistance to high temperature, vibration and arcing conditions; chemical equipment, such as humidity appliance fan housings resisting corrosive environments of hard water at 90°F. and other media; and also pipe flanges and fittings. Premix is used in transportation for automotive heater housings, other automotive structural elements, and distributor caps with molded-in contacts. General utility parts in premix include washtubs, trays, tote boxes, instrument and equipment housings. For the military, premix is molded into rifle-butt stocks and handguards.

The first premix compound was formulated in order to compensate for glass fiber washing which was occurring when experimenters attempted to adapt preform molding to small, complicated shapes with critical contours. It was determined that incorporation of fillers and shorter lengths of reinforcement into the resin permitted uniform deposition of a homogeneous compound into all areas and extremities of the mold without the undesirable material separation.[1]

While approximately 80 to 90% of the premix material molded is formulated by the processor using his own in-plant mixing facilities, a considerable amount of material is prepared and sold by compounders. These specialized premixes include flake, bulk-dried reinforced compound, pliable bulk, and continuous rope or slugs. Exceptionally high, well-controlled properties such as arc resistance, high dielectric strength, and others are the benefits gained by using specialized compounds.

The materials and processing and molding equipment for all forms of premix are included in this discussion. Special processing equipment for the specialized forms is designated as such. A detailed discussion of product properties and end-use performance concludes the chapter.

MATERIALS

Materials fall into four groups: resins, fillers, thickening agents, and reinforcements.

Resins. Standard vinyl-toluene-modified resin is the main type of polyester resin employed.

Acrylic-modified low-shrink, DAP, and styrene-modified types are also used.

Vinyl-Toluene-Modified Resin. The vinyl-toluene type is used for lower monomer volatility, higher hot strength out of the mold, better surface, and higher heat-deflection temperature. Table IV-1.1 presents the properties of this liquid resin; Table IV-1.2, the curing characteristics; Table IV-1.3, the properties of cast, unfilled resin.

Low-Shrink Systems for Low-Profile Molded Parts. Low-shrink or low-profile resin systems consist usually of a styrene-type base resin and an acrylic-type resinous modifier that contributes the low-shrink characteristic. Preferential chemical thickening agents MgO or $Mg(OH)_2$ may be further added for bulk-molding compound. From 10 to 40% modifying resin as modifier on the total resin mix in a low-profile premix system may be used with the optimum

level at 30%. From 0.50 to 1.50% of chemical thickener may be used.

There are many benefits of low-shrink resins in producing low-profile systems. The total micro in. variation over a 2-in. span is 375 for a cured low-shrink system, as opposed to 2750 for a conventional premix resin system.[2]

Table IV-1.4 lists the physical properties of styrene-type, low-shrink base resin; Table IV-1.5, the physical properties of ⅛ in. casting; Table IV-1.6, the properties of the resinous modifier; and Table IV-1.7, the gel time of a blend of base resin and modifier.

Table IV-1.1 Typical Properties of Liquid Vinyl-Toluene Resin

Property	Specification
Viscosity @ 77°F poise	25
Specific gravity @ 77°F	1.09
Weight lb per gal.	9.1
% polymerizable	100
Color, APHA, max.	150
Stability, uncatalyzed, in dark @ 70°F, months	4
Stability, catalyzed (2% benzoyl peroxide paste), in dark @ 70°F, days	4

Table IV-1.2 Typical Curing Characteristics of Liquid Vinyl-Toluene-Modified Resin

Gel time	3 min. 29 sec
Peak exotherm, °F	394
Time interval, 150°F to peak exotherm	4 min. 38 sec

Notes: SPI Gel Time Test, 1% benzoyl peroxide, 180°F bath. Recommended catalyst and concentration for molding = 2% BPO paste (1% actual BPO).

Table IV-1-3 Typical Properties of Cast, Unfilled Vinyl-Toluene Type Resin[a]

Property	Specification
Barcol hardness	38
Ultimate tensile strength, psi	7,730
Ultimate elongation, %	2.9
Ultimate flexural strength at room temperature, psi	18,500
Ultimate flexural strength at 160°F, psi	7,680
Modulus of elasticity in flexure at room temperature psi	0.67×10^6
Modulus of elasticity in flexure at 160°F, psi	0.17×10^6
Water absorption, 24 hr, %	0.49
Specific gravity	1.18
Heat distortion temperature, °C	99

[a]Determined on a 1/8 in. casting.

Table IV-1.4 Physical Properties of Styrene-Modified Low-Shrink Base Resin

Property	Specification
Viscosity @ 77°F, cps	1,200
Acid number	14
Color, APHA	100
SPI gel time (1% BPO, 180°F):	
gel time min.	7
gel to peak, min.	1½
peak exotherm, °F.	465
Specific gravity @ 77°F	1.12

Table IV-1.5 Physical Properties of ⅛ in. Casting

Property	Specification
Flexural strength	10,000
Flexural modulus	480,000
ASTM heat distortion, 264 psi, °C	137
Barcol hardness	50

Table IV-1.6 Typical Properties of Resinous Modifier

Property	Specification
Viscosity @ 77°F, cps	3,000
Specific gravity @ 77°F	0.95
Weight, lb per gal.	7.90

Table IV-1.7 Typical SPI Gel Time of a 70:30 Blend Base Resin to Modifier

Gel time, min.	7
Gel to peak, min.	2
Peak exotherm, °F	450

Note: Several reliable one-component, low-shrink, low-profile resins are currently becoming available.

Table IV-1.8 Typical Chemical Composition of Calcium-Carbonate Filler

Ingredient	%
$CaCO_3$	98.20
$MgCO_3$	1.20
Al_2O_3	0.12
Fe_2O_3	0.06
SiO_2	0.25
Copper	None
MnO	0.0035
Moisture	0.20

Table IV-1.9 Typical Physical Properties of Calcium-Carbonate Filler

Property	Specification
Particle size: range	Up to 25 μ
Particle size: %−10	80% minus 10 μ
Particle size: mean	5 μ
Dry brightness, %	95.0 (Minimum)
Index by refraction	1.59
Hardness (Mho)	3.0
Oil absorption (rubout)	9.0–10.0
pH	9.2–9.5
ᵃWater-soluble salts (determined by conductivity)	0.06% (as Na_2CO_3)
Specific gravity	2.71
Bulking Value:	
1 lb bulks, gal.	0.0443
Weight per solid gal., lb	22.57

The following catalyst systems for low-shrink premix can be used: 1% benzoyl peroxide, tertiary butyl perbenzoate or tertiary butyl peroctoate based on the total resin weight in the formulation.

Fillers. Fillers are used in premix as low-cost inert diluents, viscosity modifiers, plasticizers and flow-control agents, heat-sinks to lower exotherm, pigment extenders, and reducers of thermal expansion and contraction. Fillers also increase molded hardness, tensile and flexural moduli, and compressive strength and they generally enhance corrosion resistance and dimensional stability. New techniques involving low-shrink resins and chemical thickeners have added important dimensions to the use of fillers.[3]

Calcium Carbonates. Calcium carbonates comprise the prime powdered filler in premix formulating. Table IV-1.8 outlines the typical chemical composition of this filler. Table IV-1.9 presents its physical properties.

Clays. Fine-ground kaolin clays are the second most desirable filler in polyester premix. They produce more discoloration than $CaCO_3$ but color well in molded parts. They lend more plasticity to the mix but cause slightly higher shrinkage out of the mold. The best clays for premix have organic coating for low oil absorption in the range of 10 to 30. Table IV-1.10 gives the chemical composition of clays. Table IV-1.11 presents the physical properties.

Table IV-1.10 Typical Chemical Composition of Clays

Ingredient	%
Al_2O_3	38.5
SiO_2	45.5
Na_2O	0.1
TiO_2	1.3
CaO	Trace
Fe_2O_3	0.2
MgO	0.1
K_2O	Trace
Ignition loss	13.7

Table IV-1.11 Typical Physical Properties of Clays

Property	Specification
Fineness (residue on 325 mesh), %	0.1
Particle size (below 2 μ), %	35
(ESD μ)	3.5
Brightness (457 μ)	83
Oil absorption	32
pH	4.5
Free moisture, %	1
Specific gravity	2.58
Bulking, gal. per lb	0.0465
Refractive index	1.56

Aluminum Hydrate. Aluminum hydrate is a filler used for fire ratardancy in polyster premixes. The chemical composition is as follows:

Formula: $Al_2O_3 \cdot 3H_2O$
Approximate composition:

Al_2O_3—65%

Water of hydration—35%

Table IV-1.12 provides the physical properties of aluminum hydrate. Table IV-1.13 presents the improvements in physical properties of a standard aluminum hydrate-filled premix formulation compared with those of $CaCO_3$ filled premix.

In brief, aluminum hydrate provides filled premix molded parts that are stable upon extended heat-aging at 280°F. The material also releases water of hydration (on flame exposure) which acts as a flame snuffer, providing freedom from the toxicity of chemical flame snuffers. Water resistance is also improved.[4]

Talc. When used as a filler in premix talc provides specific properties such as water resistance and improved electrical strengths. It also permits greater facility of finishing (sand-

Table IV-1.12 Typical Physical Properties of Aluminum Hydrate

Property	Specification
Form	Fluffy, white powder
Particle shape	Plate like
Particle size	Many available, all through 200 mesh preferred
Refractive index	1.566
Specific gravity	2.40
Oil absorption	Low

Table IV-1.13 Comparison of Physical Properties of Standard Aluminum Hydrate Filled Premix Formulations[a] versus $CaCO_3$ Filled Premix

Property	Aluminum Hydrate Filled Premix	$CaCO_3$ Filled Premix
Barcol, 3 months in H_2O @ 180°F	44	41–28
Water absorption	0.19	0.2–0.5
Weight loss, 48 hr @ 480°F	7.4	7.6–8.4
Weight loss, 3 months @ 280°F	1.6	1.7–2.3
Flame resistance	nonburning	burns
Electrical properties	Arc resistance-above 200 sec	

[a]When molded, standard premix formulation with vinyl-toluene-modified resin (30%), aluminum hydrate (33%), 1/4 in. glass fibers (15%), and zinc stearate (1.4%) showed the improvement in properties shown here as compared with a $CaCO_3$ filled premix.

Table IV-1.14   Chemical Composition of Talc[a]

Ingredient	%
Silica (SiO$_2$)	59.14
Aluminum oxide Al$_2$O$_3$)	0.91
Ferric oxide (Fe$_2$O$_2$)	0.59
Calcium oxide (CaO)	1.26
Magnesium oxide (MgO)	30.98
Nonsilica residue	1.22
Loss on ignition	5.40

[a]Molecular composition — Theoretical: 3MgO-4SiO$_2$-H$_2$O.

Table IV-1.15   Typical Physical Properties of Talc

Property	Specification
Type of particle	Flat
Particle-size sieve fineness	
through 325 mesh, %	100
Hegeman fineness	5½
Melting point, °F	Approx.–2200
Ohms resistance	20–22,000
Physical constants (typical):	
Color	White
Reflectance, green stim.	
filter	97
Wettability	Fair
Suspension in water	Good
Ion concentration as pH	8.9
Specific gravity:	2.77
True density, lb per gal.	23.07
gal. per lb	0.434
Tapped bulk:	
20 g—500 times	56 cc
Apparent density,	
lb per cu ft	22.3
Loose bulk:	
20 g	110 cc
Apparent density,	
lb per cu ft	11.38
Relative abrasiveness	Soft

Table IV-1.16   Chemical Properties of Asbestos

Property	Specification
Chemical name	Hydrous Magnesium Silicate
Color	Gray
Texture	Very fine, flexible, silky
Length	Quebec Standard Test 0-0-0-16
Combined Water	12–15
Acid Resistance	Readily attacked by acids

Source: Johns-Manville Asbestos Fiber Division.

ing, drilling, and so on) out of the mold. Table IV-1.14 gives the chemical composition of talc; Table IV-1.15, the physical properties.

Asbestos. Asbestos helps prevent separation of the ingredients of polyester premixes during molding. It provides the best control of premix plasticity in a mixed state, even with bulk-molding compound (thickener). Asbestos also eliminates pre-gelation and spotting in low-profile systems but has the disadvantage of imparting a grayish-green color. Table IV-1.16 provides the chemical properties of asbestos; Table IV-1.17, the chemical composition; and Table IV-1.18, the physical properties.

Powdered Polyethylene. Powdered polyethylene is used as filler to provide an improved surface when molded and to assist in mold release if necessary. It melts or softens during the molding cycle. Flexural and impact properties are improved by LD polyethylene. Electrical and heat aging remain unchanged.[5] Table IV-1.26 presents the properties of polyethylene powder.

Additional Fillers for Fire-Retardant Premixes. For production of fire-retardant premixes, use the following chemicals as resin additives or modifiers: antimony trioxide, powder or paste; chlorinated or brominated waxes.

Chemical Thickening Agents. Chemical thickening agents are employed to selectively increase the viscosity of polyester BMC premix resin systems and to induce a superficial gel. The value of creating a thickened or BMC system out of a regular polyester premix lies in better molding qualities and more uniform molded physical properties. Almost all BMC systems employ low-shrink resins.

Table IV-1.17   Chemical Composition of Asbestos

Ingredient	%
(A) Silicon dioxide	38–42
(B) Magnesium oxide	40–42
(C) Water	12–15
(D) Iron oxide	Trace–6
(E) Iron trioxide	Trace–6
(F) Aluminum trioxide	Trace–3
(G) Calcium	0–3

Source: Johns-Manville Asbestos Fiber Division.

Table IV-1.18   Physical Properties of Asbestos

Quebec Standard Test (Guaranteed Minimum)	0-0-0-16				
Grade Designations	7T15	7T05	7T02	7TF1	7TF02
Rotap Screen Analysis (100 grams—30 mins.)					
Plus 6 mesh (%)	0	0	0	0	0
Plus 14 mesh	0.5	0.5	0	0	0
Plus 20 mesh	1	4	3	0	0
Plus 28 mesh	3	21	25	0.5	0.5
Plus 35 mesh	8	24	33	1	17
Plus 65 mesh	44	16	21	44	64
Minus 65 mesh	44	35	18	55	19
Wet Classification Bauer-McNett Test					
Plus 14 mesh (%)	1	1	0.5	3	2
Minus 200 mesh	90	74	83	93	95
Penetration Efficiency (%)	59	108	140	50	75
Color	53	58	60	53	57

Source: Johns-Manville Asbestos Fiber Division.

Many combinations of thickening agents may be formulated, such as MgO or $Mg(OH)_2$ alone, or $CaO/Ca(OH)_2$ or MgO/CaO combinations. The main purposes of selectively combined thickener systems are to control the time-rate of thickening after mixing, degree of final thickening, and so on.

Under the most desirable conditions it is possible to (1) add thickeners to the resin-filler mix without occurrence of any immediate viscosity increase prior to the completion of blending, and then (2) permit the full thickening action to take effect in less than 8 hr so that the thickened, reinforced mix may be molded as soon as possible after formulating. This eliminates a costly and undesirable long-term aging period.

Other variables are resin acid number, molecular weight, moisture content of resin, chopped glass or filler, impurities, type filler, glass finish, other additives, catalysts, accelerators and low-shrink additives, temperature, and refrigeration. The entire technology of superficial resin thickening is in its infancy at this writing.[6] Gaining consistency in mixing and molding is the problem of greatest magnitude that requires solution.

Magnesium Oxide and Hydroxide, Calcium Oxide and Hydroxide. Magnesium oxide and magnesium hydroxide have been found to exert a long-term chemical thickening effect and pseudo-gel (leather-hardening) at room temperature in paints. They are now used in polyester-premix bulk-molding compound as flow controller and pre-thickener to gain improved properties, both in the molding process and in finished premix parts. The magnesium and calcium compounds become effective by partially reacting with the unsaturation bonds in the basic resin. Patent numbers concerning chemical thickening are as follows: U.S. Patent Nos. 2,628,209; 3,431,320; 3,465,061; German

Table IV-1.19   Chemical Composition of Magnesium Oxide (MgO)

Ingredient	%
Magnesium oxide (MgO)	93.10
Carbon dioxide (CO_2)	0.46
Combined water (H_2O)	4.40
Calcium oxide (CaO)	0.79
Silicon dioxide (SiO_2)	0.27
Chloride (Cl)	0.17
Sulfate (SO_3)	0.68
Iron oxide (Fe_2O_3)	0.03
Aluminum oxide (Al_2O_3)	0.10
Manganese (Mn)	0.0017
Copper (Cu)	0.0002
Acid insoluble	0.05
Ignition loss	4.86

Note: Molecular weight 40.32.

1,131,881; and British 949,869. Calcined natural and CaMg ores are preferred over calcined precipitated materials. Tables IV-1.19 to 1.25 present the chemical composition and physical properties of magnesium oxide, magnesium hydroxide, calcium oxide, and calcium hydroxide thickening agents.

Table IV-1.20   Physical Properties of Magnesium Oxide (MgO)

Property	Specification
Appearance	Clean, white, odorless powder
Refractive index	1.64
Specific gravity	3.32
Weight lb per gal.	2.8
Bulk (loose)	21 pcf
Screen analysis, %:	
Through 100 mesh	100
Through 200 mesh	100
Through 325 mesh	99.5
Particle size:	
Average, μ	
Distribution, %	0.086
0 −0.05 μ	20.9
0.05–0.10 μ	48.8
0.10–0.15 μ	20.2
0.15–0.20 μ	7.1
0.20–0.25 μ	3.0
Surface area, m^2/g	185
Iodine number	135
Solubility:	
Water	Insoluble
Organic solvents	Insoluble
Mineral acids	Decomposed

Table IV-1.21   Physical and Chemical Properties of Magnesium Hydroxide (Mg(OH)₂)

Property	Guaranteed Specification
Physical characteristics	White or practically white, essentially odorless powder. Clean and free from visible evidence of contamination
Assay	Min. 94.5% Mg(OH)₂
Loss on drying	Max. 2.0%
Calcium oxide (CaO)	Max. 1.50
Bulk (loose)	20–30 pcf
Color	White or practically white
Mesh	All (100%) passes No. 100 Min. 98.5% passes No. 200 Min. 97.5% passes No. 325

Note: Molecular weight = 58.33.

Table IV-1.22   Chemical Properties of Calcium Oxide (CaO)[a]

Ingredient	%
CaO	97.0
MgO	0.9
SiO₂	1.3
Al₂O₃	0.8
Fe₂O₃	1.0
Sulfur as SO₃	0.05
LOI	0.50

Note: Soluble in acids, slightly soluble in water to form Ca(OH)₂.
[a]Dead-burned, high-calcium lime.

Table IV-1.23   Physical Properties of Calcium Oxide (CaO)

Properties	Specification
Appearance	White, greyish powder; sometimes yellow or brownish tint; odorless; generates heat upon contact with water
Screen size	Grade for thickening— 95% through 325 mesh
Specific gravity	3.40
Melting point	2570°C.
Boiling point	2850°C.

Table IV-1.24   Chemical Properties of Calcium Hydroxide

Ingredient	%
Available Ca(OH)₂	95
Available as CaO	75
MgO	0.60
SiO₂	0.20
Al₂O₃	0.20
Fe₂O₃	0.06

Note: Soluble in glycerine, syrup, acids; slightly soluble in water, insoluble in alcohol; absorbs CO₂ from the air.

Table IV-1.25   Physical Properties of Calcium Hydroxide

Properties	Specification
Appearance	Soft, white crystalline powder; bitter alkaline taste
Screen size	Grade for thickening— 98% through 200 mesh
Specific gravity	2.34
Bulk density, loose packed	20–30 lb/per cu ft
Melting point	Loses water at 580°C

Internal-Mold Release Agents. Many internal release agents are used in polyester premixes. Stearates are the most popular type. They should melt at temperatures slightly below molding temperatures. If the latter are greatly in excess of the stearate melting temperature (20%), surface imperfections are likely to result in the molded parts. Table IV-1.27 lists the common release agents used.

Coloring Pigments. Any desired color can be molded. Lighter colors require extreme care in maintaining cleanliness of the mixing and molding equipment and in duplication from part to part. Polyester pigment colors, both inorganic and organic in DAP monomer, are the most desirable.

General-purpose inorganic powdered types include iron oxides for blacks are preferred at 1% or less for freedom from reactivity with resin. Carbon blacks reduce pot life and cause variability of cure. Manganese dioxide is also used as a black colorant at a low percentage.

Reinforcements. Chopped glass fibers produced from hot-melt bushings with appropriate organic binders affixed constitute the major and

Table IV-1.26 Typical Properties of Extra-Fine Low-Density Polyethylene Powder

Property	Specification
Melt index, g per 10 min.	5
Density, g per cu cm	0.924
Bulk density, lb per cu ft	17–20
Vicat softening temperature, °C	97
Particle shape	Spherical
Average particle size, μ	<30
Volatiles, max. %	0.1
Color	White

Table IV-1.27 Internal-Mold Release Agents

Material	Melting Point (°F)
Stearic acid	157
Lead stearate	221
Aluminum stearate	239
Magnesium stearate	270
Zinc stearate	272
Calcium stearate	302
Barium stearate	321

Table IV-1.28 Properties of Glass-Fiber Chopped Strands

Property	Specification
Glass fibers	"E" glass composition preferred "A" glass composition sometimes employed
Lengths	⅛ in., ¼ in., ½ in. and longer. 2 in. usual max.
Ignition loss	1.8–2.2% preferred
Bulk-density	115 max.[a]
Fuzz	0% preferred
Water repellency	Sizing to be water repellent
Styrene solubility	Minimum filamentation due to immersion in monomeric styrene is desirable
Resistance to Degradation during mixing	Fibers should have maximum possible resistance. Mixing times should be limited to lowest possible: 2–5 min. in 100 mixer preferred

[a]For strand integrity, see Figure IV 1.1, J-M Bulk Density Tester and method for integrity of fiber glass chopped strand.

most dependable type of reinforcement for premix.

Significant parameters for classifying the glass fiber chopped strands are listed in Table IV-1.28. Uppermost in importance in processing glass fibers for premix is the preservation of strand integrity. Bulk density and plasticity or spiral flow measurements accomplish this before and after mixing respectively.

Sisal. Sisal is natural vegetable fiber desirable as a reinforcement because it is a monofilament. It resists degradation and retains impact strengths over longer mixing times than glass fiber. Sisal is undesirable for electrical and other parts to be used outdoors where water absorption will result from exposure to the elements.

Asbestos Reinforcement. Asbestos fibers for premix reinforcement possess the same chemical composition and properties as those listed under *Fillers.* However, fibers processed to conform to length and bulk requirements of grades 4 and 5 are the most desirable (approximately ¼ in.). Some special "cleaned" grades are available which impart less grayish-green color to the molded premix.

The most important factor in using asbestos is the necessity of "opening" the fiber to as nearly complete single-entity fibers as possible. Asbestos fiber as reinforcement for premix is beneficial for bulking, providing maximum mix plasticity and good surface appearance. Asbestos does not, however, supply the maximum possible electrical properties nor the highest desirable impact strengths.

PVA Fibers (Polyvinyl Alcohol). PVA thermoplastic monofilament fibers for RP/C end uses are not water soluble as is normal chemical polyvinyl alcohol. However, the moisture pick-up of the fiber itself is 4 to 5%; of a molded polyester-premix compound reinforced with the PVA fibers, 0.5%; of the same composition reinforced with glass fibers, 0.2%[7]

Advantages:
- Excellent toughness; good impact strength; no deterioration of fibers during mixing, extrusion, or molding processes; good abrasion resistance.
- Good flow properties. Decrease of modulus of PVA fiber at molding temperature means increase of flexibility of fiber (this phenomenon is reversible). When molded compound returns to room temperature, fiber regains stiffness.
- Smooth molded surface without using low-profile polyester resins.
- Density is one half of glass fiber.
- Good weatherability. Small differences of thermal expansion coefficient between resin and fiber—no hair cracking. PVA fiber itself has also excellent weatherability.

Disadvantages:
- At a softening point of 390°F PVA fiber does not melt sharply but decreases in strength above its softening point.
- Difficulty of flash removal because of good impact strength. Recommend high-speed grinder or 570°F heated iron.
- Low modulus at high temperature. Not recommended for composites used at high temperatures.
- Contains fines and differential fiber lengths.

Glass Plus PVA Blend Premix. The optimum fiber cut length and fiber content of PVA fibers makes for suitable entanglements with glass fibers. Such an optimum fiber network will deform appropriately during the molding process, preventing segregation between resins and fibers. In other words, the result is an improvement of the flow properties of premix compounds. Optimum entanglements depend upon the composition of the premix (viscosity) and the molding conditions (rate of deformation and shape of mold). It is necessary to find the optimum fiber-blend ratio in each case. In one example, the compression molding of circuit-breaker parts, ¼ in. PVA fiber constitutes 5%, ¾ in. glass fiber—10%, polyester resin—30%, $CaCO_3$—55% (by weight). Flexibility of PVA fibers at molding temperature and good affinity to glass fibers cause good entanglement of the two. PVA fibers are compatible with glass fibers and mixtures compensate for the disadvantages of each individual fiber. Molded products have good mechanical properties and good appearance.

Chopped or Diced Fabric. Chopped or diced fabric, usually nylon tricot (but may be glass), is another reinforcement for premix. The maximum used may approach 15% by weight.

Advantages:
- Low cost
- Exceptionally smooth, non-porous molded surfaces
- Excellent stain and water resistance
- Good electrical properties imparted to finished premix

Disadvantages:
- Produces a high-bulk, light, fluffy premix which requires special handling facilities and resists extrusion
- Causes high molding shrinkage

EQUIPMENT AND TOOLING

The following discussion treats presses and molds required for compression, transfer, and injection molding of premix; it includes batch-mixing equipment; extruders; and other tooling necessary for premix processing.[8]

Presses for Compression Molding of Premixes. Several classes of presses for compression molding are available to the premix processor.

Open Rod or Column Presses. In open rod or column presses, usually four side or corner posts (variations from two to six or more) are prestressed and accurately machined for platen travel. Separate top and bed plates provide maximum working area. The platens move up and down, guided by close-fitting bushing retainers. Hydraulic units are mounted in the press base (into the floor if necessary) or on top of the press unit to provide up-acting or down-acting platens respectively. Designs are available for 1 to 20,000 tons.

The column presses are preferred for accessibility. A 15% design overload should be allowed when severe or unusually high off-center thrusting is anticipated.

Welded Rigid-Frame Presses. In rigid-frame presses, rigid fabricated elements constitute the side risers or separating press members which are welded to the top and bottom masses.

Slide or travel is accomplished by a beveled guide shoe bolted to the moving platen which rides against a dry-lubricated phenolic-laminate liner adhered to the risers. Both the guide shoe and stationary bottom platen are water-cooled to maintain dimensional accuracy. The accessible working area is limited to the front and back only.

Hydraulic units are top-mounted and usually incorporate two single-acting hydraulic cylinders for positive fast advance and one double-acting main cylinder for infinitely adjustable pressing and opening speeds.

The capacity of the welded rigid-frame press is limited to 500 tons. This press is one of the best available for eliminating the sidethrusting caused by off-center loading which generally is present in the premix molding of an eccentric or non-symmetrical part. Injection and transfer units may be easily incorporated with the hydraulic unit.

Housing-Type Presses. Uprights in housing-type presses are comprised of a boxlike member accurately aligned to the top and bottom platens by keying. A prestressed tie rod passes through each upright. Construction may be either open or closed-side.

For platen surface travel, a gib system with either four or eight contact guides controls sideways and front-to-back movement. Contact is made between a dry-lubricated phenolic cross-grained laminate and a replaceable hard-ened-steel wear plate adhered to the frame or upright boxlike ways.

The working area is not quite equivalent to that available in the column presses and is larger in the open-housing than in the close-housing construction. Hydraulic units are usually top-mounted. Open-housing presses are adaptable to jobs requiring molding pressures of 250 tn. and over. Closed-housing presses are available in a wide variety of tonnages, bed sizes, strokes, day-light openings and closing speeds. Both types are extremely rigid and capable of heavy production work. They are desirable for eliminating side-thrust associated with forming solid-phase premix or granular plastics.

Side-Plate Presses. In side-plate press design, tension members also become the side uprights separating both platens. Guide plates on the moving platen travel along ways on the side plates to control and prevent misalignment. However, the press hydraulic cylinder is also made to correspond as closely as possible to the size of the movable platen in order to resist side-thrusting. Side rib reinforcing may be added for extra rigidity.

The working area is naturally limited to front and back accessibility. Hydraulic units may be top-mounted for the down-acting conventional type of press or bottom-mounted for the box or multiple-opening (2 to 20) press, both of which are included in this classification.

The pressing capacity of side-plate presses is limited to 3000 tons. These presses are economical to build, fall midway between column and side-housing presses for rigidity, and are very adaptable to many general and specialized requirements of the plastics industry.

Loop-Frame Hydraulic Presses. A cast or welded "J" forms an oval or loop which confines the working area of the loop-frame hydraulic compression press and actually incorporates perfect parallelism into the design by absence of any fabricated joints.

The hydraulic cylinder is mounted within the loop. Travel of the moving platen is guided against bevel surfaces machined into the sides or beveled corners (four) of the loop frame.

Working area is accessible only from the front and back. Hydraulic units are either top or bottom-mounted. Multiple-opening styles may also be designed into this press construction.

Although stroke is limited to 15 in. within the loop, two resistance platens can also be included for greater adaptability.

In addition to the obvious advantage of rigidity of alignment, loop-frame presses have automated controls, speed, and a variety of tonnage capacities. The latter are limited only by physical size and economics.

These presses are adaptable to molding components (printed circuit boards for example), of such extremely uniform and controlled thickness that post-molding machining is not required.

C-Frame Presses. The C-frame press presents the ultimate in accessibility, being open on three sides. It is probably the least desirable, however, for maintaining complete rigidity. Hydraulic units are usually top-mounted. These presses are available only in limited capacities.

Guided-Platen Presses. In order to obtain more substantial rigidity in press frames and minimize even further any deflection upon mold closing, systems have been devised for extremely accurate guidance of the platen from an external loading position to the location in the press for normal mold closing.[9]

Required Press Tonnage Versus Press Design-Standards. A pressure of 1000 psi on a projected part area constitutes a starting point for determining the press tonnage required to mold a given premix part size. As much as 3000 or 4000 psi may be required for high lateral material movement, eccentric part design, or deep draw. A molder who limits himself to less than 500 psi calculated on a projected part area may find himself wanting in the event of job changes or emergencies.

A minimum press stroke of 36 in. (48 in. full, open daylight) is desirable. The platen size should be 30 in. L-R × 24 in. F-B for low-tonnage and 48 in. L-R × 36 in. F-B minimum for larger tonnage.

Regarding deflection, a moving platen or a press bed should not deflect in bending and shear more than 0.002 in. per ft, when considering the maximum press tonnage uniformly distributed over two-thirds of the platen area.

The unloaded moving platen should show a parallelism of 0.001 in. per ft when resting on four corner supports. Pressure build-up time from 0 to maximum tonnage should be a maximum of 5 sec.

Controls Required for Hydraulic-Compression Presses. The capability of rapid advance of at least 300 in. per min. and adjustable slow close of 0 to 10 in. per min. should be provided for in premix presses. The required control elements are hydraulic oil-temperature gage, hydraulic pressure gage, main ram tonnage control valve, pullback (breakaway) or press-opening tonnage control valve, intermediate speed control valve, pressing-speed control valve, pullback-speed control valve, decompress hand valve, speed build-up hand valve, adjustable timers, switches limited to control shift from maximum closing speed to throttled slow-close and assure gentle press opening, electrically controlled push buttons, hand selector switches, and indicating light stations.

Variations of Compression Molding. Injection molding units may be fabricated for mounting into a compression press, which then becomes the clamping device.

Transfer Presses. The transfer-press method is similar to molding with compression presses, using four-post presses with platens large enough to accommodate the required molds. The hydraulic transfer cylinder may operate at an angle from the side or through a hole in the top mounting and the platen of the press directly over the mold.

The tonnage of the press represents the clamping pressure of the mold and must exceed any force which may open the mold along the seams or joints during the pressing cycle.

Many advantages accrue in polyester premix molding from this method. One of the most beneficial is the ability to mold around and encapsulate studs or inserts pre-set into the cavity prior to injection of the compound.

Variations of Transfer Molding:
High-Speed Automation. By using a pre-heating cycle geared to start during the molding of the previous charge, the total cycles of transfer molding for a ⅛ in. thick part were shortened from 30 sec. to 17 sec. (actual cure time 9 sec.).[10]

Cold-plunger molding. Cold-plunger molding received its name because of the rope, slugs, or cut extruded premix, placed automatically on a sliding ramp or track from which they are forced by a ram (at room temperature) into heated molds. The molds are mounted on a rotating turret and follow a four-or more stage molding cycle for rapid fill, fast cure, and rapid removal from molds. It is mostly applicable to smaller size parts such as electrical and electronic components, many with molded-in inserts.[11]

Vacuum-mold transfer molding. In molding thermosets, better surfaces and elimination of voids are accomplished by exerting a vacuum during the molding cycle. The vacuum effectively removes solvent vapors, condensation products, and so on.

The process requires a two-part mold containing an outer annular groove into which a soft, high-temperature-resistant (Viton)®[12] ring has been placed. An inner annular groove is connected to the vacuum line.

In closing the mold, the Viton ring first engages the upper mold half forming an airtight seal. The continuously running vacuum pump exerts a vacuum and puts the mold cavity under negative pressure. The charge is transfer-injected into the mold cavity and cures under vacuum.

The pertinent factors and dimensions of a typical mold are as follows: multiple cavity, 0.020 in $\times$ 0.250 in. gate, 0.003 in. $\times$ 0.250 in. vent enlarging to 0.020 in. $\times$ 0.250 in., 0.250 in. trapezoidal runners from the center to the transfer-entry chamber in the mold.

Presses for Injection Molding of Thermosets. Variations in the usual functioning of thermoplastics injection presses have been brought about to take advantage of their high-speed production cycles in premix molding. Thermosets in general and polyester premix in particular are adaptable to injection molding. The size of the injection press required should be decided upon after discussions with reliable manufacturers of this type of equipment.

The differences between molding functions of injection-type presses using thermoplastics

Figure IV-1.1 Bulk-density tester for determining strand integrity of chopped glass fiber for premix molding and BMC. (Courtesy of Johns-Manville Fiber Glass Reinforcement Division, Engineering Drawings Available on Request.)

and thermosets have been ably treated by E. W. Vaill in a discussion which is reprinted here for direct reference.[13]

Molding Cycles. The basic reason for this rapid growth in the injection molding of thermosets is the shorter molding cycles which can be obtained from the fast and uniform homogenizing and temperature control of the material being plasticized in the screw barrel. The cost and time savings over injection molded thermoplastics has been conclusively proven. Stock temperatures in the range of 220° to 260°F. are normally obtained. Since injection of the material into the mold seldom exceeds five seconds, high mold temperatures in the order of 375° to 410°F may be used. As a result, extremely fast cycles or cure times are obtained. In thickness over ⅛″, the cycles in many cases are faster than those for similar thermoplastic molded parts. Thermosetting materials cure or polymerize in the mold and are discharged hot with no cooling required, while thermoplastic parts must be cooled in the mold to become sufficiently rigid to be removed without distortion.

Screw Design. Most thermoplastic injection machines can be converted for the molding of thermosets by the installation of a new screw and barrel especially designed for these materials. Although the L/D ratio of screw barrel (length to diameter) for thermosets is normally the same as for thermoplastics, the compression ratio (screw base diameter, back to front) seldom exceeds 1.1 to 1 for thermosets as compared to compression ratios of approximately 2.5 to 1 for thermoplastics. This lower screw compression ratio results from the difference in fluidity when hot of the thermoplastic and thermosetting materials. Much of the heat is caused by surface friction and the shear of the material passing through the screw flights. Most thermoplastic materials become a molasses-like fluid while the thermosetting materials become a soft, flexible putty.

Some of the injection machines for thermosets now being produced have hollow screws using water "bubbles" to maintain more accurate screw temperatures and closer control of the plasticizing temperature. Machines using solid screws must rely on accurate control of barrel temperature, screw back pressure and rpm. for control of plasticizing temperature.

Controlled screw temperatures are recommended for the more heat sensitive thermosetting materials, such as the one-step resin phenolics and melamines.

Screw Rotation. The rotation of the screw in revolutions per minute is a very important factor in the proper plasticizing of the material passing over the flights. The rpm. of the screw is determined by the plastic being molded and the capacity of the screw versus the actual charge required for each shot. Screw speeds between 50 and 90 rpm. are normal.

Rotation of the screw occurs only during the back stroke or plasticizing stage of the molding cycle which can be either electrically or hydraulically driven. We will not discuss here the relative merits of the electric motor versus the hydraulic motor drive. Both have their advantages and disadvantages. Injection machines using either of the two methods have proven very satisfactory in the molding of thermosets.

Barrel Temperatures. Thermoplastic screw injection barrels are normally heated to temperatures between 350 and 600°F, using electric heater bands. Thermoset materials, on the other hand, require barrel temperatures of about 150°F on the back or feed end, approximately 200°F in the middle of the barrel and 190°F at the nozzle. The heating media is usually hot water or hot oil circulated through channelled metal jackets around the barrel.

Screw Back Pressure. The rotation of the screw forces the molding material over the screw flights and accumulates it in the front of the barrel. Screw back pressure is the pressure developed at the tip of the screw on the charge of material. The amount of back pressure of the screw tip on the material is determined by a guage on the injection hydraulic cylinder. The pressure is controlled by a flow valve governing the rate of hydraulic oil flowing from the cylinder back into the oil reservoir.

If the flow valve is wide open, the rotating screw easily forces the material forward, little frictional shear is developed and the material is not highly compressed in front of the screw. As the flow of oil is restricted through the flow valve, pressure is built up on the material, frictional shear becomes greater, the temperature of the material increases and the oil pressure in the cylinder becomes greater. The higher the back pressure, the slower the plasticizing or back travel time of the screw will be after the injection shot has been made.

Screw rpm. and back pressure should be so coordinated that the plasticizing time should not exceed 30 sec., otherwise the hot metered charge accumulating ahead of the screw may precure. This would cause plugging of the barrel nozzle and prevent the injection of the material into the mold at the start of the injection stroke or forward motion of the screw.

Insufficient back pressure will result in low material temperatures which can also prevent the proper flow of the material into the mold on the injection stroke if the material has not reached the proper plastic state.

Normal back pressures range from 500 to 1,500 psi on the front end of the screw. The guage pressure is usually calibrated to read the injection cylinder pressure in psi. If the cross section area of the hydraulic cylinder is 50 sq in. and the cross section area of the screw barrel is 5 sq in., the ratio is 10:1. Therefore, if the gauge reads 80 psi, the back pressure on the material in the barrel is 800 psi.

Injection Speed. Injection speed is the time required for the forward stroke of the screw ram to inject the material into the cavities of the mold. Optimium injection times should range between 5 and 10 seconds. Too fast cavity fill or injection time may cause entrapped air and gas in the cavities which results in porous and/or blistered parts. Too slow injection time will cause the material to set up or precure before the cavities are filled to proper density. The result will be porous and/or non-filled parts.

Injection speed can be controlled by either a flow valve on the hydraulic line, or by controlling the injection pressure or a combination of both.

Injection Pressure. Injection pressures will vary greatly according to the type of material being molded, the geometry of the runners and gates and the geometry of the molded parts.

Normally, for phenolic and melamine materials, the injection pressure will range between 12,000 and 20,000 psi on the material at the front of the barrel. A secondary or holding pressure of between 5,000 and 10,000 psi follows the injection pressure stroke to maintain sufficient pressure on the sprue and runners until the material becomes cured or rigid. This secondary pressure tends to reduce sticking or "packing" of the material in the sprue and runners.

Injection Mold Runner and Gate Design.
Runner and gate design of injection molds for thermosetting materials are very important factors in obtaining fast and uniform cavity fill without gas entrapment and/or precure.

Table IV-1.29 may be used as a guide for determining optimum cross sectional areas of runners and gates according to the weight of the molded parts and the weight of the material to be transferred through the runners and gates. Areas and weights are based on general purpose phenolic materials. Larger gates and runners are necessary for fibrous filled improved impact materials. The data sheet, Table IV-1.30, may also be used for quick reference.

For multiple cavity injection molds in which one main runner feeds branch runners, each runner should be of a cross-sectional area con-

Table IV-1.29 Data for Injection Mold Runner and Gate Design

Cross-Section Area of Runner or Gate (sq in.)	Round Diameter (in.)	Rectangular Section (in. × in.)	Weight of Material Transferred Through Runner (g.)	Weight of Material Transferred Through Gate (g.)
0.005	0.115	0.125 × 0.080	—	5
0.020	0.160	0.160 × 0.125	—	65
0.040	0.226	0.233 × 0.172	40	150
0.060	0.276	0.267 × 0.224	85	235
0.080	0.320	0.297 × 0.270	130	—
0.100	0.356	0.330 × 0.300	175	—
0.120	0.390	0.375 × 0.320	220	—

Table IV-1.30 Data Sheet for Areas of Circles and Shot Weight

Diameter		Area (sq in.)	Shot Weight (g.)	
Fraction	Decimal		Gate	Runner
1/16	0.0625	0.0030		
5/64	0.0781	0.0048	4	
3/32	0.0937	0.0070	14	
7/64	0.1094	0.0094	22	
1/8	0.1250	0.0122	36	
9/64	0.1406	0.0155	48	
5/32	0.1562	0.0191	64	
11/64	0.1718	0.0232	82	
3/16	0.1875	0.0276	100	10
13/64	0.2031	0.0324	120	20
7/32	0.2187	0.0376	144	30
15/64	0.2344	0.0431	168	43
1/4	0.2500	0.0490	200	58
17/64	0.2656	0.0554		73
9/32	0.2812	0.0621		88
19/64	0.2968	0.0692		105
5/16	0.3125	0.0767		124
21/64	0.3281	0.0845		142
11/32	0.3437	0.0928		160
23/64	0.3593	0.1014		180
3/8	0.3750	0.1104		200

sistent with the material weight that passes through it.

Assume: Two cavities are fed by one base runner

Weight of each molded part	50 g
Weight of branch runner	10 g
Total	60 g

Then the total weight of material going through the base runner to the two cavities is 60 + 60 = 120 grams.

From Table IV-1.28 we determine that the cross section area of the branch runner should be 0.046 sq in. for a 50-g part and the cross section area of the base runner should be 0.076 sq. in. for the 120 g of material passing through it.

This would hold true for runners up to 3 in. long. For longer runners, the area should be proportionally larger to compensate for frictional losses during the flow of the plastic material.

Geometry of Runners. There are three basic types of runners: rectangular, half round and full round.

1. Rectangular; This is usually machined into only half of the die plate and is the least expensive to mill, but harder to polish. A minimum radius of 1/16 in. is essential at the corners to obtain best material flow and mold release. This design requires higher injection pressures and causes slower cavity fill times than the half round or full round runner design.

2. Half Round: This is machined into only half of the die plate. The runner is easy to polish and because of the semi-circular contour, the molded runner releases easily from the mold.

This is probably the most popular type of runner even though higher injection pressure and slower fill times are normally required as compared to the full round runner.

3. Full Round: A semi-circular runner groove is machined into each half of the die plate. It is important that the runner halves match closely to obtain the best material flow. It has been definitely proven that the full round runner will result in faster cavity fill time with lower injection pressure than with either the half round or the rectangular runners. Full round runners are particularly recommended for transfer molding by the screw plunger and screw injection methods.

Gate Design
1. Rectangular Gate: The main advantages in this gate design are cleaner breakoff of the gate at the edge of the molded part and the fact that the gate can be "fanned" out to obtain better surface finish on the molded part, particularly on large flat surfaces. This gate design is commonly used with half round and full round runners.

2. Half Round Gate: There is no particular advantage in this gate design except ease in grinding and polishing the gate if it must be enlarged.

3. Full Round Gate: This design of gate does not lend itself to edge gating on a flat piece at the mold parting line if the cavity is located in only one half of the mold. However, it should be used wherever part design will permit.

The data included in this report are intended primarily as a guide to better mold design. However, the actual runner and gate areas and their design are dependent on the following factors:

1. Geometry of the molded part.

2. Location of the gate or gates.

3. Molding pressure available.

4. Number of cavities.

5. Plasticity and heat sensitivity of the material selected for the application.

6. Physical properties required in the molded part.

Detailed plunger molding pressure tests have been run to determine the effect of the runner and gate design on pressure required to all the mold cavities. The full round runner and gate system was taken as a standard.

The rectangular runner and gate system required 35% higher injection pressure to fill the cavity in the same time.

The half round runner and gate system required 15% higher injection pressures to fill the cavity in the same time.

It should also be borne in mind that, in general, the higher the injection pressure, the greater the runner and gate wear or erosion. Therefore, the full round runner and gate design should be used wherever possible.

Mold Vents. No discussion on the subject of runner and gate design would be complete without mention of the importance of mold vents. The venting of an injection mold for thermosets is extremely important because of the volatile gasses developed during the plasticizing of the thermosetting material and the extremely fast fill time of the mold, generally in the order of 6 seconds. Unless the gasses and air in the mold can be readily expelled, severe burning or oxidizing of the material and staining of the surface of the mold in the gas entrapment area will occur. Vents are very small grooves in the parting line of the mold extending from the lip of the cavity to the outer edge of the the mold and furthest from the gate. Vents are normally 0.001 to 0.003 in. deep and from 0.025 to 0.125 in. wide. These vents may be enlarged and added vents placed in the cavity, depending upon the particular flow pattern of the material in specific molds. As many as 4 or 6 vents may be required in intricately shaped cavities. These vents can readily be ground in the mold after initial shots have been made and the entrapped air or gas areas as determined.

Conclusion. It should be emphasized that the data included in this discussion are to be considered as a bench mark or a starting point in designing molds for screw injection molding of thermosets. Naturally, gates in multiple cavity molds will have to be balanced according to the flow pattern and molding conditions for a particular type of material, as well as the geometry of the molded part. The runner and gate cross section areas recommended in this paper may be classed as being minimal. This is done for two specific reasons. First, the frictional heat developed during the injection of the material into the mold causes considerable wear in the gate and runner system, particularly in the use of mineral-filled and glass-filled materials. These will ultimately cause the runners and gates in particular to become larger. Secondly, after a mold has been built, it is much easier to grind gates out to a somewhat larger cross section area if they are too small than it is to weld up gate areas if they are found to be too large. The data given on molding conditions may also be considered as optimum conditions which may vary according to the particular make of injection machine and the materials being processed.

In their fifty-six year history, thermosets have been constantly improved. New resins and modifications have been developed, many of which have found application in markets formerly dominated by wood, glass, ceramics and metals. Similarly, press equipment and molding techniques have been improved to effect greater production economies and make possible design features previously thought to be impractical, if not impossible.

We look forward to a great and profitable future in the injection molding of thermosets.

Process Control for Injection Molding of Polyester Premix. Control points or monitoring for process control of the injection molding of polyester premix should include the following: stock temperature at the nozzle area; material pressure at the nozzle; screw displacement or shot size; hydraulic injection pressure; pressure of plastic at sprue, in runners and gates, and in multicavities for pressure balancing; mold temperatures especially at knock-out pins; material flow rates; clamping pressure on mold; mold hold time; ram travel speed; screw-turning speed; screw forward time.

Factors Associated with Injection Molding of Glass-Reinforced Polyester-Premix Compounds. Molding polyester premix by the injection method represents a fairly new and revolutionary technique. It is not yet on a 100% stabilized basis. Not only has sufficient testing not been completed, but also very few material suppliers have reciprocating screw machines in their testing stations where desirable data should be generated. However, technology to date has accumulated a good deal of information, summarized below.

- Glass degradation of "long" (¼ in.) lengths may result from passage through screw, runners, mold, or gating system. Suitable enlargement of contricting openings on an experimental basis with diligent analysis of results should resolve these shortcomings. Starting points: gates—0.250 × 0.093, + 0.002 land; runners—¼ in. trapezoid; sprues — ⁷⁄₃₂ in. diameter.

- Production of high quality parts at an extremely desirable economic advantage is possible and should give impetus to expanded use of the process. Defects occur at top molding speeds, but with a 25% cutback in molding time, the premix injection cycle is still 10 sec faster than compression molding.

- Feeding of material to the in-line reciprocating screw has been problematical. It has been improved, however, to include hydraulic stuffers that pressurize the material to apply a positive force to fill the shot chamber while the screw is rotating. Cylinders accommodating one weighed premix charge per cavity in a rotating "six-shooter" mount have been successfully used to feed the correct-size charge automatically to the stuffer. The presence of the stuffer eliminates the necessity for a screw-head valve. Cylinder pressures up to 1000 lb psi may be required. The stuffer cyclinder may be supplied with mild heaters to raise the fluidity of materials such as high-viscosity alkyd molding compounds. The temperature should obviously be kept substantially below the gelation point.

- Proper molding of reinforced thermosets requires a compound that has superior flow properties, internal lubrication, and freedom from separation of the batch components. This includes powders as well as dough-type premix. The molding of reinforced-polyester powders has already been evaluated.

- Injection of premix material from the barrel into the mold may be either directly in line through the screw and barrel or through the side of the mold at the parting line.

- Polyester premix requires "cold" runners so that design of the thermal pattern of the mold can prevent heating of the compound until it is in the actual part cavity. Heating the barrel to 150°F provides some likelihood of material kickover (gelation). Swivel barrels and quick-disconnecting nozzles, however, have been designed to preclude loss of production time if gelation of one charge does occur.

- Proper or optimum screw design for premix injection molding will probably be a thing of the future. Proper L/D ratio is not yet confirmed. Screws tapered from the middle towards each end show considerable promise. Very short flights and close tolerance to barrel ID are desirable to prevent flow-through and glass degradation. The flights must be filled all the way and not have voids on the rear side or contamination, which requires mechanical cleaning. Chamber wear also results from glass-reinforced compounds, caused partially by corrosion as well as abrasion.

In the final analysis, a special screw design has been found necessary or desirable in the past, for each individual reinforced material or resin.

Some preferences exist for a bullet-nosed screw. A ram may be preferred in other cases. Also, a spring-tensioned front plunger may be used to prevent back-flow.

- To transfer the premix in one machine a 15 lb premix shot was injected through a ⅝ in. tip into a mold cavity in 15 sec. This required a 20,000 lb injection force.

- With polyester premix, a correct balance of duration of screw-injection pressure must be maintained. Persistence of high pressure after completely filling the mold may cause separation of the clamped mold, which in turn causes warping of the molded parts. Conversely, slackening the screw pressure prematurely may cause poor surface, an unfilled mold, and other problems.

- To overcome pressure difficulties, a 10 ft tapered or wedge-shaped lock designed into the mold on one machine increased the locking pressure by a factor of 2.5.

- A reciprocating or shuttling-screw feed system has also been devised to feed two side-by-side presses. Each side may contain the mold for a different size premix part.

- The results of variation of molding conditions on premix-part ultimate strengths are as follows:

An increase in cavity pressure resulted in lower tensile and lower impact strengths. A lengthening of fill time resulted in higher tensile strengths.

An increase in mold temperature caused higher part shrinkage and lower impact strengths.

- Optimum molding conditions follow: injection pressure—850 psi, cavity fill time—1 to 5 sec, mold temperatures—375 to 410°F., time in cavity (cure time)—5 to 30 sec., range of total molding cycle—17 to 40 sec.

Compression, Transfer and Injection Molds for Premix. Services of a specialized moldmaker should be consulted for the design and construction of compression and transfer molds. Premix molds for injection presses generally should follow the requirements for sectionalized thermoplastics injection molds, but special treatment is necessary because of the higher mold temperatures required.[14] A detailed discussion of the mold requirements for standard compression, transfer, and injection molding of premix follows. (*See* also discussion of mold fabrication in Chapter IV-2.)

Mold Materials for Compression and Transfer Molds. The best mold materials for long runs are forged and machined from billets of tool steel. For shorter runs, kirksite, meehanite, aluminum alloys, and some plastic tooling (300 psi maximum permissible pressure) are used. These materials can be cast and are therefore cheaper to produce than those which need forging and machining. Many molds are also fabricated from slab stock of tool steel which lowers cost.

Molds should not distort during the molding cycle and development of the required pressure. If 1000 psi maximum premix pressure is required for intricate design, optimum performance material (tool steel) should be used.

Heating Methods:
- Steam is the cheapest heating method and the best for temperatures up to 350°F and for larger, multi-contoured molds. It requires coring. Coring should provide uniform heating of all mold surfaces and uniform flow-through, preventing accumulation of any condensate.

- Electric heat is best for smaller or flat molds. It is the cleanest type heat with the least complicated equipment, but it is most expensive. It requires coring for heating elements. The thermal controls should be of the proportioning type, not the direct on-off type in order to prevent wide temperature variations.
- Oil is the best and safest heat for higher mold temperatures above 350°F. No high pressure is required to achieve higher temperatures.
- Insulation involves the insertion of transite layers between the mold and press platens.

Design of Joining Surfaces of Plug and Cavity for Premix Compression and Transfer Molds.
- Telescoping requires no shear-off. The mold halves should telescope at least ¼ in. to prevent squeeze-out.
- A gap or tolerance of 0.003 to 0.004 in. is required at joining edges but may be 0.002 in. for smaller molds. The same is true of molds for low-shrink or bulk-molding compound premixes.
- Knock-out pins should be designed as large as possible to prevent fracture of the part. They should be tapered to seat properly in a "down" position, and the gap should be small as possible (0.001 in. preferred).
- Flashing requires 0.002 in. maximum at the point to be broken loose and 0.020 in. in the center of the slug or piece to be removed.

Mold Design to Accommodate Part Shape and Size:
- The sidewall draft should be 2.0 degrees optimum.
- The part thickness should be 0.090 in. optimum but may vary from 0.080 in. to ⅜ in. or from ³⁄₁₆ in. to 1¼ in. in the same part Sections thicker than ¼ in. are subject to internal cracking unless they are highly filled.
- The surface finish should be completed with 400 and 800 grit. Chrome plating to provide 125 mμ. in. finish on molded parts is also used.

Injection Molds. Injection-mold planning requires engineering data and decisions regarding part design, shot capacity, melting capacity or throughput capacity for premix, clamping pressure, and mold dimensions.

The injection-mold parts, in order, are: locating ring, sprue bushing, front clamping plate, front cavity plate "A," leader pins, leader pin bushing (0.001 in. fit), rear cavity plate "B" to prevent flash, support plate, actual mold cavity, spacer block, ejector retaining plate, return pin, ejector plate (may be hydraulically driven), knock-out pins, rear clamping plate, sprue pillars, support pillars with a tight fit of 0.001 in. to 0.002 in., stop pin, and heating means usually consisting of electrical-resistance elements.

Molds may be either single or multiple cavity or family molds—(2 up to 100 part sizes in same cavity all of the same general size and shape). The services of a qualified mold designer should be retained for best results.

Mixing Equipment:

Sigma or Spiral Blade. The sigma or spiral blade mixer is efficient, fast, and thorough. Close tolerance between the blades and sidewall (¼ in. preferred), however, contributes to undesirable degradation of the glass fiber reinforcement and concomitant loss of physical strengths. The mixing chamber is jacketed for cooling water (120°F maximum operating temperature).

The two mixing blades are geared to rotate at differential speeds of 20 and 40 rpm respectively (lower speeds for larger 100 to 500 lb mixers) and the blades are reversible for dumping and cleaning. The mixing chamber is pivoted around one blade axis so that it may be rotated through 90° for dumping with the blades running (see Figure IV-1.2).[15]

Muller Type. The muller-type mixing equipment is less efficient and not as thorough as the spiral blade, but it results in less glass degradation. Some units are furnished with a rubber-surfaced kneading roller and a rubber-lined wall. Larger units have automatic discharge.

Continuous-Mixing and Extruding Equipment. The reciprocating screw of the continuous-mixing and extruding types combines resin, filler, and glass into a well-mixed compound. The mechanical agitation was formerly regarded as too severe for glass fiber reinforcement but the path of the mixing blade has since been shortened. Well-designed, automatic continuous systems for premix to accommodate glass, resin, and fillers are commercially available (see Figure IV-1.3).[16]

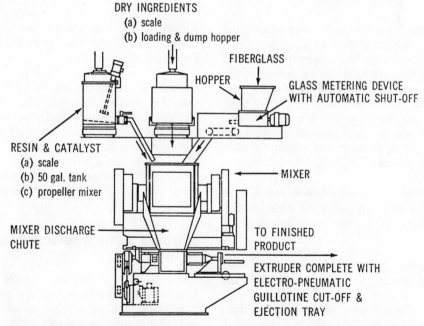

DRY INGREDIENTS
(a) scale
(b) loading & dump hopper

FIBERGLASS

HOPPER

GLASS METERING DEVICE WITH AUTOMATIC SHUT-OFF

RESIN & CATALYST
(a) scale
(b) 50 gal. tank
(c) propeller mixer

MIXER

MIXER DISCHARGE CHUTE

TO FINISHED PRODUCT

EXTRUDER COMPLETE WITH ELECTRO-PNEUMATIC GUILLOTINE CUT-OFF & EJECTION TRAY

Figure IV-1.2 Automated glass, resin, and filler feed system built around a spiral or sigma blade mixer for compounding batch-type premix and BMC mixing equipment. (Courtesy of Baker Perkins, Inc.)

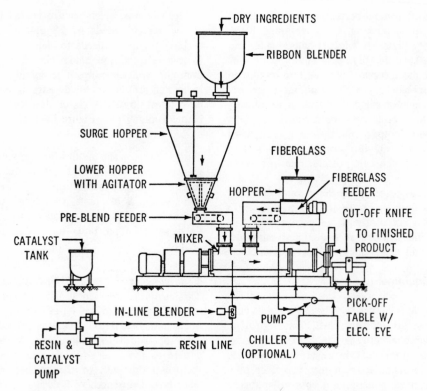

Figure IV-1.3 Continuous-feed automated system for feeding glass, resin, and filler to a continuous, in-line mixer for preparation of premix and BMC compounds while keeping glass degradation to a minimum. (Courtesy of Baker Perkins, Inc.)

Belt Mixing. For low-profile resin systems and thickened mixes (bulk-molding compound), some advantages are to be realized in introducing glass fiber reinforcement as it is chopped or by dropping it into a resin-filler slurry carried on film on a moving belt (similar to SMC preparation). Rotating wheels or rollers force the glass into the resin, thereby avoiding the shearing action of other mixers which results in glass degradation.

Gear Mixing. Mixtures of resin and filler or resin plus filler and glass may be made by passing them over or through the surfaces of meshing gears.

Impregnation Type. Continuous roving may be drawn through a resin mix with the ratio controlled by a limiting die. The extrudate may be either chopped when wet to form a dough-type compound, or "B"-staged and chopped to form a granular or pelletized reinforced-molding compound.

Kneaders. Kneaders contain two lens or oval-shaped agitators which compress the reinforcement into the resin mix.

Extruders. Extruders are desirable to compact the premix compound, eliminate excess air, and put the compound into an easy-to-handle form for charging the press. Extruders are available with a vacuum cabinet to further eliminate entrained air.

Automatic premix slug sizers and loading trays for multiple-cavity compression molds are advisable.

MOLDING METHOD/PROCESSING DETAILS

Three facets of premix molding are discussed here with corresponding pertinent procedures:

(1) Standard or general-purpose premix, involving normal polyester or alkyd resins and used for most non-critical industrial problems. This type can be modified for specific proper-

ties such as chemical durability or high electrical strength.

(2) Low-shrink system premix used for improved molded surfaces.

(3) Bulk-molding compound mixes with low-shrink resins plus a chemical thickener used for improved surfaces and better molding capability. These are basic formulations and may be modified to suit specific requirements or property demands.

Formulation and Mixing Procedure for Standard General-Purpose Premix

Mixing Procedure. The mixing procedure for standard, general-purpose premix includes the following steps:

1. Weigh all powdered filler ingredients (asbestos, zinc stearate, and calcium carbonate) into the same container and charge to the mixer. Mix for 5 min.
2. Weigh the catalyst into the resin in a separate container and mix for 20 min. Add the resin to the mixer and mix for 5 min.
3. Weigh the glass fiber into a separate container and charge (do not sift) to the mixer all at once for 1, 2, or 5 min. as specified (keep the time as short as possible). This completes the mixing cycle.
4. Empty the mixer by rotating the hopper 90°. Clean with solka-flock and methylene chloride only. Do not use rags or towels.
5. Extrude if desired and weigh into correct-size charges.
6. Mold premix parts at 270 to 350°F with 500 to 1000 psi pressure for 45 sec. Provide cooling fixtures. If necessary, punch-drill and trim dies for finishing *after*

Table IV-1.31   Formulation of Standard, General-Purpose Premix[a]

Ingredient	% by Weight
Vinyl-toluene resin	30.0
Paste catalyst (50% BPO)	(1.6% of resin)
Zinc stearate	1.0
7TF-1 asbestos floats	7.0
Calcium carbonate or clay or combination	47.0
¼ in. glass-fiber or sisal reinforcement	15.0
Total	100.0

[a]Using standard vinyl-toluene resin.

cooling or while parts are held in cooling fixtures.

Variations of Standard, General-Purpose Premix. It is recommended that the premix molder give earnest consideration to the use of special filler and wetting-agent materials which have been found to greatly ease the preparation of the mix and to improve the molding performance of standard polyester-premix compounds.[17]

Fillers. The use of finely divided calcium silicate as filler in a standard, general-purpose polyester premix at a level range between 0.90 and 1.92% by weight and molded for 45 sec. at mold temperatures of 315°F top and 310°F bottom with 800 psi contributes the following benefits:

- Greatly improves the molded surfaces by imparting excellent gloss and extremely low waviness.
- Increases the Barcol hardness from 35 to 48.
- Increases the flexural strength in a sisal premix by 20%.
- Greatly reduces internal porosity, and improves the homogeneity of the resin-filler-reinforcement mix.

Wetting Agents. The use of an anionic surfactant, which is basically di-octyl sodium sulfosuccinate, incorporated into the resin-catalyst component at a level ranging from 0.04 to 0.15% by weight in a sisal-reinforced compound accomplishes the following:

- Reduces mix viscosity and eliminates the undesirable dryness and granularity of the mixed compound, improving handleability and eliminating undesirable molding characteristics caused by the compound dryness.
- Contributes to improved surface finish and gloss and the elimination of post-molded surface waviness.
- Results in no loss of physical strengths.
- Provides a synergistic effect by combining with the precipitated zinc stearate to greatly improve release from the mold surfaces.

Formulation and Mixing Procedure for Low-Shrink-System Premix[18]

Mixing Procedure. The mixing procedure is the same as for standard premix.

Molding Comments. Low-shrink-system premix may be molded at higher temperatures

Table IV-1.32 Formulation of Low-Shrink-System Premix

Ingredient	% by Weight
Polyester resin	19.0
Low-shrink modifying resin	13.0
T-butyl perbenzoate catalyst	(0.32 parts added to resin)
Calcium-carbonate filler	47.0
7TF-1 or suitable asbestos floats	5.7
Phosphate-type internal release	0.3
¼ in. chopped-glass fiber or sisal	15.0
Total	100.0 (excluding catalyst)

Table IV-1.33 Formulation for BMC Premix with Low-Shrink Resin and Chemical Thickener

Ingredient	% by Weight
Polyester resin	18.0
Low-shrink modifying resin	7.0
T-Butyl perbenzoate catalyst	0.25 (1% of resin)
Calcium-carbonate filler	48.0
Zinc stearate	1.0
Finely pulverized polyethylene	5.0
Magnesium oxide	1.0
7-TF-1 asbestos floats	5.0
¼ in. chopped-glass fiber	15.0
Total	100.0 (excluding catalyst)

and shorter molding cycles than standard mix. It does not have a propensity toward driving the filler and reinforcement to the extremities of the mold as well as would chemically thickened systems. Therefore the ratio of filler to resin may be increased to aproach 2:1.

Asbestos floats assist in producing better surface properties as molded and tend to eliminate blotchiness caused by the pre-gelation characteristic of the low-shrink resin system.

Formulation and Mixing Procedure for Bulk-Molding Compound with Low-Shrink Resin and Chemical Thickeners:

Mixing Procedure. The mixing procedure for bulk-molding compound mix comprises the following steps:

1. Add catalyst and magnesium oxide to blended resins in the same container and stir or mix for 20 min. Allow to stand for removal of voids.
2. Add dry fillers to the mixer and blend for 5 min.
3. Add resin blend into the mixer and mix for 5 min.
4. Add reinforcement and mix for the shortest time possible (maximum 5 min.).
5. Extrude or leave in bulk as desired.
6. Allow 24 hr for chemical thickening to take full effect.
7. Mold at 300 to 350°F. in the shortest possible molding cycles with pressures of 500 to 1000 psi as required.

BMC Molding. MgO permits maximum thickening of the premix compound to the leather-hard state and during molding, intensifies penetration of the resin, filler, and reinforcement to all mold extremities. Shapes with more complicated contours and detail are therefore possible.

Finished parts possess low surface profile to 125 mu in. and also receive and hold painting in a manner superior to standard premix formulations. Fiber pattern, surface waviness, and expensive finishing processes are eliminated.

This BMC mix may be made to have higher impact strengths because bulk-molding compound systems can be mixed more simply by impregnating strands of glass fiber rather than by mixing in a spiral-blade mixer, with its accompanying degradation of the glass filaments. Dimensional stability and freedom from warpage and distortion out of the mold are improved. The occurrence of internal voids is lessened.

One difficulty, however, lies in the removal of low-shrink parts from mold at the completion of the cycle.

In automated molding, it was found that compaction in a screw-injection machine generally improved the strength of glass-reinforced thermoset molding materials. A ⅛ in. fiber length favored the best combination of lower degradation and highest molded strengths.[19,20]

Summary. While compression molding is the method used for the largest percentage of premix, any of the three types of premix discussed may be used for compression, transfer, or injection molding.

Quality Control. The trouble-shooting guide to polyester molding listed in Chapter IV-2 is applicable to premix molding. In addition, there are several other checks that should be made.[21]

Incoming-Material Checks:

Fillers. Chemical and physical properties that may affect the operation such as particle size, moisture content, pH.

Resin. Gel time, peak exotherm, viscosity, monomer content, acid number.

Catalyst. Active oxygen content and peroxide percentage.

Fiber-glass reinforcement. Bulk density and binder content. Spot-check fiber length.

In-Process Evaluations:

Batches. Gel time and peak exotherm after mixing.

Cure. Spiral-flow distance.

Mold temperatures. Regular evaluations during shift.

Finished parts. Impact and flexural strengths, wrinkles, surfaces, voids, and other specific requirements.

Part dimensions. Continual inspection.

Scrap or reject quantities. Weigh quantity and use information to immediately correct trouble areas.

Maintenance. Preventive maintenance of process and equipment is recommended.

Premix Molding Plant Layout. The manufacturing plant illustrated in Figure IV-1.4 might be described as a typical custom premix-molding plant. The products of this plant are custom-molded premix parts for the automotive and appliance industries. The largest parts are not more than 4 ft × 5 ft. The major production areas as shown in the layout are the mix room, press room, and paint room.

The mix room has four premix blenders which allow four different types of mixes to be prepared at the same time. In a custom-molding plant as shown, there are always several different jobs running which require separate mixes. The mixer capacity is 50 to 100 gal. units. Resins are supplied from the resin room by meters.

The press room has fourteen presses, eight

which are 400 to 600 ton capacity approximately 5 ft × 10 ft of floor space. The 200 to 300 cover approximately 4 ft × 8 ft of floor space. An 18 ft × 15 ft work area in front of each press has been allowed to give sufficient area for finishing and secondary operations that must be done at the press. Such operations are cooling and shrinking each part on a shrink fixture, deflashing, drilling, studding, riveting brackets, and preparing mix. Aisle ways have been allowed at the back side of each press to permit molds to be changed and presses to be serviced without disturbing the operations of the adjoining presses.

Because the illustrated plant is manufacturing automotive parts, it is necessary that there be facilities to paint these parts. Not all automotive parts are painted, but those that are must be done to specification. This requires that the plant facility be capable of handling a variety of cure temperatures and paints. A good size paint room and preparation area have been set aside for this operation.

The percentage of the total plant space allotted to each area shown in the layout is listed below:

	%
Mix room	3
Press room	18
Paint room	11
Resin room	3
Dry materials storage—receiving	11
Boilers-molds-maintenance	12
Hardware and packaging storage	9
In-process storage	2
Secondary finishing	6
Finished goods—shipping	14
Offices and lunchroom	11

The plant illustrated can employ 150 to 200 hourly employees and 15 to 20 salaried employees.

The power and service utilities required consist of steam for mold heating; gas for cure ovens, plant heat, and boilers; air for fixtures, drills, clamp equipment, and stamping fixtures; electricity for lights, press, and equipment motors; water for paint booths, boilers, and coolers. Air, steam, and water services supplied to the press lines are best provided through covered floor trenches with all pipe well insulated.

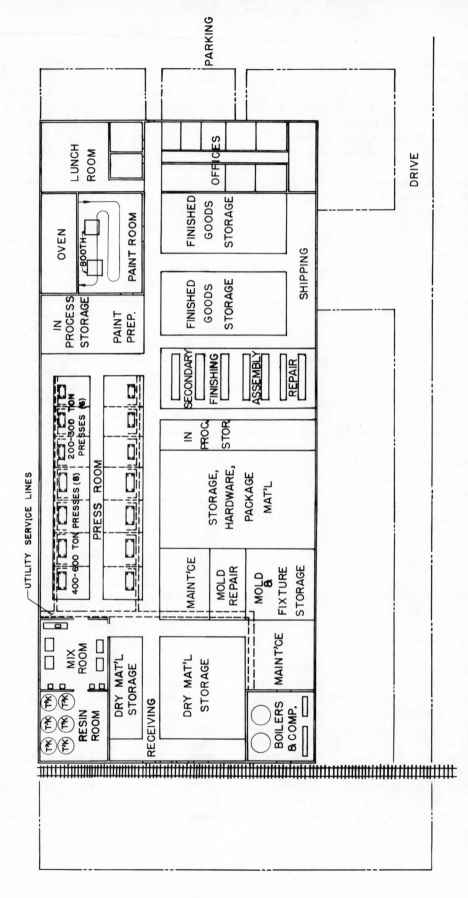

Figure IV-1.4 Typical plant layout for a premix molding operation. Especially prepared for this volume by Owen Schofield and Richard Mahan. (Fabricon Products Division, Eagle Picher Lead Corp., Grabill, Indiana.)

Because of the hazardous materials being used, it is necessary that an adequate safety program be implemented. An adequate supply of CO_2 extinguishers well located throughout the plant is a necessity. Extremely hazardous areas such as the resin-room storage area, mix room, and paint room should have explosion-proof electrical wiring and equipment. These rooms should also have blow-out panels on the exterior walls. Small mixers and hoists used in these rooms should be pneumatic-type devices.

The materials and processes create a quantity of dust in the building. It is therefore necessary that an adequate ventilation system be installed. Dust collectors and exhaust systems will help reduce dust in the mix room and at the de-flash and drilling stations at the presses and in the secondary finishing areas. These systems will also help remove fumes. Because of the dust, the paint room should be isolated and sealed from the rest of the plant and a positive pressure maintained in order to reduce dust contamination of the parts being painted.

In bulky parts produced, the size and weight of materials received and shipped, and the size and weight of molds and secondary finishing equipment create the need for strong handling equipment. It should consist of a 6000 lb fork truck and several 2000 lb units, all with high-lift capacities. Overhead lifting equipment is essential in the mold-repair area. Smaller pallet trucks and material carts are also a necessity.

PRODUCT DATA

This section discusses contributory part and mold design parameters, product properties, premix advantages and disadvantages.

Design Rules for Premix.[22] *Mold Design and Draft.* A side-wall draft of only 2 degrees is considered adequate except when only one vertical edge is required, in which case the side opposite should be machined to a minimum 3 to 4 degree draft. Also, in the latter case it is advisable to cock the mold cavity and tilt the plunger. Knockout pins should have a minimum tolerance of 0.002 in. per in. of diameter.

In setting up the speed of closing of the mold, both the charge weight and the type fillers are the governing factors and probable errors in weight must be taken into account. For a charge less than 1 lb. in weight, the rapid mold close may be made to within 0.003 in.,

while for charges greater than 1 lb., the rapid mold close should stop at 0.010 in. Twenty sec of full closing time should be allowed for parts in which the minimum recommended thickness occurs.

Part Thickness. Allowable thickness variations within the same molding are: 0.080 in. minimum to ⅜ in. maximum, and 3/16 in. to 1¼ in. In special cases, parts with thicknesses up to 2 in. may be molded. The parts with thicker sections naturally require longer cure times.

Part Shrinkage. Nominal shrinkage of premix molded parts varies between 0.003 and 0.006 in. per in. and is influenced by type of resin, type of fillers, uniformity of mixing, exotherm, part thicknesses, and mold temperature. Predictable contraction tolerances for design purposes in premix molded parts have been determined as:

Thickness of length, in.	Shrinkage, in. per in.[a]
less than 0.1	0.005 over-all
1–4	±0.010
4–10	±1/64 to ±1/32
over 10	±1/32 to ±1/16

[a]Except bulk-molding compound.

Additional design rules are stated in Table IV-1.34.

The following description presents the most reliable information available regarding actual physical design of premix parts to obtain maximum rigidity and broad distribution of unavoidable deflection. Such treatment is intended to solve the problems not of breakage in use, but of breakage during assembly, installation, intermediate processing and shipping, where it is probable that extreme stresses due to impact, elongation and angular deflection will be encountered.

1. For parts whose surface area is greater than 15 sq in., nominal wall thickness should never be less than .090 in.

2. Avoid large flat sections. Use convolute, rib, or gusset where possible.

3. Peripheral edges of parts should be flanged or made heavier than nominal wall thickness whenever possible.

Table IV-1.34 Design Rules for Premix

Minimum inside radius, in.: 1/32	Maximum size part to date, sq ft: 25
Molded-in holes: Yes	Limiting size factors:
Trimmed in mold: Yes	Press capacity, flow
Built-in cores: Yes	Metal edge stiffeners:
Undercuts: Yes	Yes
Minimum draft recommended, deg.: 1	Bosses: Yes
	Fins: Yes
Minimum practical thickness, in.: 0.060	Molded-in labels: No
	Raised numbers: Yes
Maximum practical thickness, in.: 1¼, 2 possible (see above)	Gel coat surface: No
	Shape limitations: Moldable
	Translucency: Possible
Normal thickness variation, in.: ±0.007	Finished surfaces: Two
Maximum thickness build-up: As desired	Strength orientation: Random
Corrugated sections: Yes	Typical glass loading, % by wt: 0–35
Metal inserts: Yes	Sisal, max. %: 20
Surfacing mat: No	

4. Reinforce flange mounting holes by means of gussets when mounted against a flexible gasket. Plastic directly beneath head of mounting bolt should be in compression whenever possible. Where compression mount is not possible, increase the thickness of the flange between the reinforcing gussets.

5. Gussets should extend beyond the center of the mounting hole.

6. Avoid sharp breaks along the outer edges which interfere with rapid deflashing of the parts.

7. Avoid wide or long peripheral flats on any part in order to increase strength and reduce the possibility of warpage.

8. Warpage of long flange areas or of rectangular openings can be reduced (a) by thickening these areas, (b) by introducing additional angular bends as close to the edge as possible.

9. Where openings in side walls are required, if one can use the kiss-off or mash-off construction to obviate the need for sliding cores, it is highly desirable to do so. Not only is the original cost of the tool reduced but die set-up costs and die maintenance costs are markedly reduced.

10. When a large opening in a side wall is situated close to the edge of a part and a kiss-off construction can be utilized, it is often desirable to use a sliding core. The movement of the sliding core should be operated on a time delay and should come home after the die is closed to reduce knitting problems at the edge of the part.

11. Molded holes should be at least one diameter away from the edge of the part. Unfortunately, this important rule is one which is most flagrantly violated.

The eleven points just discussed result in adding to the weight of a premix part. However, these steps will not increase the cost of premix parts but, in the long run, will reduce cost because of the reduction in scrap at the molder's plant and at the user's assembly plants.

Performance. The general range of properties which may be expected from the various formulations for premix molded parts is presented in Table IV-1.35.

Advantages and Disadvantages of Premix Molding.[23] Premix has the lowest cost of all the molded, engineered, high-performance RP/C products. The low costs are due to the high filler content and the rapidity of molding cycles, while still maintaining the percentage of reinforcement required to provide acceptable engineering properties.

The most prominent disadvantage is that premix uses short-length fibers which cause a loss of general strengths. Also, present mixing methods result in serious fiber degradation with concomitant loss in desirable impact strengths in the molded parts.

The Present and Future of Premix. Premix is presently in a fast-moving state of fluid change in which older, original handling and processing methods are being challenged by newer ones. The resins, fillers, and reinforcements are being swiftly improved. This is illustrated by low-shrink resins, chemical thickening, improvements in molding methods and in the integrity and mechanical and chemical durability of glass fiber reinforcements.

The major consumer of premix parts is currently the transportation (automotive) industry. New applications are evolving annually at a rate that justifies the hard work and effort expended by all concerned. It makes one glad to see this excellent example of free enterprise and American ingenuity at work.

Table IV-1.35 General Range of Properties of Premix Including BMC

Property		Low	High	ASTM Test Method
Physical Properties:				
Specific gravity		1.4	2.0	D792
IZOD impact, notched		2.0	8.0	D256
Flexural strength, psi		6,000	26,000	D790
Flexural modulus of elasticity psi $\times 10^6$		1.5	2.5	D790
Tensile strength, psi		3,000	10,000	D638
Tensile modulus of elasticity psi $\times 10^6$		1.5	2.0	D638
Compressive strength, psi		15,000	30,000	D695
Water absorption, %		0.05	0.25	D570
Deflection temp. °F, 264 psi load		300	>570	D648
Resistance to heat, continuous		250	400	
Flammability, 1/8 in. thick		SE-0	Non-burning	UL Subject 94
1/16 in. thick		SE-1	Non-burning	UL Subject 94
Flame resistance IT/BT, sec		95/90		FSTM 406
Thermal conductivity[a]	A	0.11	0.14	
	E	1.32	1.68	
	C	4.5	5.8	
Barcol Hardness		35	70	
Machining Qualities		Good	Excellent	
Electrical Properties:				
Arc resistance, sec		100	400	D495
Track resistance, min.		600	>800	D2303
Electric strength, VPM, ST		320	400	D149
S/S		300	330	D149
Dielectric constant, 60 Hz		5.7000	5.100	D150
Dissipation factor, 60 Hz		0.015	0.004	D150
Dielectric constant, 1 MHz		4.600	2.100	D150
Dissipation factor, 1 MHz		0.0128	0.008	D150
Power factor, 60 Hz, %		2.400		D150
Chemical Properties:				
Resistance to acids		Good	Excellent	
Resistance to alkalis		Fair	Good	
Resistance to solvents		Good	Excellent	

[a]Explanation of thermal conductivity data:

Code	System	Units
A	American	Btu-ft/ft²/°F/hr
E	English	Btu-in./ft²/°F/hr
C	CGS	Cal-cm/cm²/°C/sec

(This information from C. L. Ward, Premix, Inc., North Kingville, Ohio: Personal communication.)

Conversions:

$A = E/12$
$E = 12A$
$C = E \times 3.45 \times 10^{-4}$
$E = C \times 0.29 \times 10^4$

References

1. "Premix Molding," R. B. White, New York, Van Nostrand Reinhold, 1964.
2. Determined using Bendix "microcorder", Ann Arbor, Mich.
3. SPI 23rd RP/C Div. Proceedings, 1968, sec. 12-D, Geiber, C. A. & Waters, W. D.
4. SPI 20th RP/C Div. Proceedings, 1965, sec. 11-B, Connolly, W. J. and Thornton, A. M.
   SPI 20th RP/C Div. Proceedings, 1965, sec. 4-E, B. D. Pratt.
5. "Modern Plastics," Aug. 1968, p. 137.
   SPI Premix Committee Minutes 15th Seminar, Apr. 2, 1969.
   Bulletin No. PR 85-570, U.S. Industrial Chemicals Co.
6. SPI 20th Proceedings, 1965. sec. 6-A.

SPE Retec, Cleveland, Ohio, Oct. 4-5, 1965, J. P. Walton.

SPI Premix Seminar, Apr. 2, 1969, W. H. Deis.

SPE Retec, Cleveland, Ohio, Sept. 25, 1969, pp. 55 and 85.

SPI 25th RP/C Proceedings, 1970, sec. 6-D, F. Fekete.

7. SPI Premix Committee Mtg., Grand Rapids, Mich., Apr. 2, 1969.

8. *Plastics Technology*, Sept. 1969, p. 17.

SPE Journal, Feb. 1970, Vol. **26**, p. 21, C. J. Olowin.

Plastics Technology, Aug. 1968, pp. 33–42.

"Plastics Technology Processing Handbook," Oct. 1968, p. 185, B. Blanchard.

9. Minutes SPI Press Molding Committee, Oct. 21, 1970, Nashville, Tenn., C. J. Olowin.

10. SPI 21st RP/C Proceedings, 1966, sec. 11-F, C. A. Marsczewski.

SPE Journal, June 1963, p. 557.

11. H. E. Dutot, SPE Journal, Sept. 1968, p. 46.

G. B. Rheinfrank, SPE Retec, Cleveland, Ohio, 1963, p. 1.

R. L. Tracey, British Plastics, Jan. 1966, p. 40.

M. Sears, SPI 22nd Preprint, 1967, sec. 12-A.

U.S. Patent 2,738,551, Mar. 20, 1956.

12. E. I. duPont Co.

13. E. W. Vaill, SPI RP/C 23rd Proceedings, 1968, sec. 11-A.

See also:

SPE RP/C Retec Proceedings, Cleveland, Ohio, 1965, p. 27, J. M. Grigor.

SPE Retec "Promise of Thermosets, Chicago, Ill., Dec. 10, 1968.

SPI 23rd RP/C Proceedings, 1968, sec. 11-B, J. M. Grigor.

SPI 23rd *Ibid*, sec. 11-C, R. Paci.

Plastics Technology, Jan. 1968, p .15.

Ibid, Dec. 1967, p. 15.

E. E. Moneghan, Stokes Equipment Co., Personal communication.

SPI 25th RP/C Proceedings, 1970, sec. 1-B, D. H. Stone.

14. "SPI Handbook Reinforced Plastics," First Edition, Oleesky-Mohr, New York, Van Nostrand Reinhold, 1964, p. 530.

"Plastic Mold Engineering," Laslo Sors, New York, Pergamon Press, 1967.

"Plastics Mold Engineering," J. H. DuBois, and W. I. Pribble, New York, Van Nostrand-Reinhold.

15. For notes on the low-degradation mixer *see* SPE RP/C Retec, Cleveland, Ohio, Sept. 25, 1969, p. 79, W. J. Foley, Littleford Bros.

16. Plastics Technology Processing Handbook," Oct. 1968, p. 85.

17. Information released especially for this volume by Smiths Industries Pty., Ltd., Sydney, Australia.

18. Personal Communication, Johnson, R. B., Reichold Chemical Corp.

19. G. Kaull, SPE Retec, Cleveland, Ohio, 1965, p. 1.

SPI 25th Proceedings RP/C Div. 1970, sec. 6-E, Nussbaum and Czarnowski.

20. *Plastics Technology*, Aug. 1970, p. 50, Hoffman, K. R., and Vellturo, J. W.

SPE Journal, Vol. **27**, Sept. 1971, p. 30, Reinfrank, G. B.

21. J. Potrubacz, SPE Retec, Cleveland, Ohio, 1967. (Added paper not in index.)

22. The following section is adapted from S. S. Oleesky and J. G. Mohr, "Handbook of Reinforced Plastics," New York, Van Nostrand Reinhold, 1964, p. 530.

23. "Premix Molding," R. B. White, New York, Van Nostrand Reinhold, 1964.

IV-2

Preform and SMC Methods for Matched-Die Molding

INTRODUCTION

As this book is going into print, an amazing American industrial and technological success story is also being written—that of the prodigious growth of matched-die press molding in RP/C brought about by the combined developments of chemical thickening of polyester resins to make possible sheet-molding compound (SMC) and of the science of adding thermoplastic resins to the same polyesters to negate shrinkage and improve the molded surface profile.

In prior technology, almost as much money was spent sanding, filling, and re-sanding matched-die molded components, particularly to provide surfaces suitable for painting and generally matching automotive standards for smoothness. With the advent of SMC and low-profile resins applied to RP/C, we now have a sheetlike material that can be taken directly off a roll and fed to a press in a manner similar to the handling of sheet metal for stamping and drawing. Molding and process efficiency can be greatly improved, and parts can be molded and sent directly to an automotive assembly line without elaborate surface-finishing operations.

Another most fortuitous circumstance is that the mushrooming of this SMC-LP technology in the first few years of the 1970s represents an increase of almost 100% in new business for RP/C. In other words, RP/C press molding using SMC and low-profile resins is not really taking away from the old established matched-die molding business which continues to supply parts for less-critical surface applications. What it is doing is forging ahead to supersede metals in many diverse applications. The transportation industry has been the first significant commercial activity for the material while it is currently gaining acceptance in appliance and other markets. This is all new business.

Preforming in RP/C implies pre-shaping glass reinforcement into a form simulating the shape of the article to be molded. It is usually done by chopping glass into an air stream and drawing the air through a shaped, perforated screen. The glass is then deposited and a liquid organic binder is applied which, when heat-cured, bonds the deposited fibers together to preserve the shape. The preform must fit the portion of the mold usually the plug over or onto which it is to be placed for molding. Shaped preforms are combined with a matched-die molding mix at the press and then compression-molded to form the finished part.

Preforming also refers to pre-cutting and/or stapling chopped-strand molding mat over a prepared solid form prior to pressing. In addition a slurry preform process is used in which glass fibers, cellulosic fibers, and an organic bonding ingredient are suspended in a water-mix, deposited on a screen, and retrieved for heat-curing and molding. The slurry process is more or less proprietary and is not in widespread commercial usage, although good molded surfaces are produced because of close fiber packing and the use of cellulosic fibers together with the glass fibers.

As stated the SMC method was applied to RP/C when it was determined that the method

175

in a 1951 patent for accelerating polymerization was also effective in thickening polyester resins.[1] By use of the thickener, resin, filler, and reinforcement are combined into a flat sheet which becomes leather-hard, and handleable for weighing and charging at the press, thus eliminating the necessity for separate preparation of these 3 ingredients and individual combination at the press-molding operation. The accompanying and timely development and use of low-shrink or low-profile resins (first patent in 1968) has strongly tended to eliminate some of the surface defect problems originally encountered in the preform process and the subsequent need for elaborate post-molding finishing operations.[2] The possible opportunities for automation in matched-die molding are made far more practicable using SMC.

Many normal utilitarian products and industrial intermediates, particularly in the automotive and transportation markets, are produced by the matched-die molding processes. Parts with compound curvatures in sizes up to 150 sq ft in area and up to 17 ft in length or longer are possible in flat molded-sheet form. Some examples are shell chairs, flat molded-sheet stock, automotive body and structural components, housings for applicances and electrical gear, safety equipment, and one-piece molded boats up to 17 ft in length. SMC molding has been substituted for preform molding in all small- to medium-sized parts, while the preforming method persists in larger parts.

The entire preform and SMC processes are similar to that used in premix (Chapter IV-1), except that longer fibers and higher percentage of reinforcement make possible superior mechanical and other engineering properties.

In order to eliminate or avoid at the outset any potential misunderstandings concerning terminology and proper classification, we have established that the two main categories in matched-die press molding remain as preform and SMC methods. It is possible to use either general purpose or low-shrink modified polyester resins in either process.

Because of their relative individual importance, each method merits a separate discussion of processing, except for the information regarding molding equipment and mold design common to both. This later information is presented more easily as a unit immediately following this introduction, thus diverging slightly from the usual chapter outline.

EQUIPMENT AND TOOLING FOR PREFORM AND SMC METHODS

Presses. Compression presses are the major type utilized in matched-die molding of RP/C. Many varieties are manufactured and different sub-types are available. It is the responsibility of the molding-shop owner or engineer to pick the type of press most suitable to his needs, based on part and mold size, production rate, and other parameters.

Purchasing a press requires a major capital expense. It therefore must be selected to provide most all-around performance and prompt the greatest return on the initial investment.

It is impossible to specifically detail all considerations in purchasing presses for matched-die molding. The following list presents press construction and size and operational variables that have to be categorized and numerically rated on the basis of job requirements prior to purchasing any unit.

- Cost: Depends upon the total tonnage and physical size.
- Flexibility: The equipment should be adaptable to many molding jobs; show as wide a variability as possible in adjustable daylight; have fast and slow approaches, automatic loading, a wide range of adjustment of hydraulic pressure, an unobstructed platen area, and simple controls.
- Design: The press design should be simple and rugged but dimensionally perfect. It should be built to be free of deflection under maximum loading, to withstand a 15% overload, to assure easy maintenance and lubrication, cleanliness and elimination of dust and molding offal on exposed parts (tie rods, switches, etc.). Maximum electrical and mechanical safety should be provided for.
- Press variables: The correct sizes or values for break-in, warm-up, running, maintenance, and safety must be agreed upon by the press manufacturer and molder when selecting a press to fulfill the desired function. Recommendations of the press manufacturer should be rigidly followed for the above-mentioned factors. (See also the discussion of types of molding presses for RP/C in Chapter IV-1.)

Mechanical Performance
(A) *Type*
    1. Upacting or Downacting—determined by

the size of the machine and the space and building construction that is available.

2. Strain Rod or "A" Frame Type—
   (a) Capacity total tonnage or pounds per square inch unit force on the platen area;
   (b) Pull-back or breakaway force in tons.

(B) Press dimensions should cover the following items:
   1. Shut height.
   2. Daylight opening.
   3. Left to right dimension between rods or frame.
   4. Front to back dimension between rods.
   5. Length of stroke, main ram.
   6. Work height, floor to top of lower heating plate.
   7. Size and area of hot plates.
   8. Maximum allowable deflections.
      (a) Under uniform loading conditions.
      (b) Under concentrated loading.

(C) Operating cycle determined by the products being projected and the production requirements.
   1. Fast Close—adjustable in inches per minute. Slow Close—fully adjustable in inches per minute.
   2. Pressing speed; decompression time.
   3. Slow-Open or breakaway in inches per minute—adjustable as required.
   4. Fast-Open speed in inches per minutes and can be adjustable.

(D) Two pressure hydraulic system with high volume low volume high pressure with full range adjustability of speed and pressure on both units.
   1. Designed for the minimum HP requirements to accommodate the desired cycle.
   2. Power source available.

(E) Type of Fluid—
   1. Oil, water, oil emulsion, waste water system or synthetic fire resistant fluids.

(F) Type of Hydraulic System—
   1. Self contained or central hydraulic distribution system. Each of these systems has its own merits and careful consideration should be given to each one to determine the most flexible operation for the particular requirements of the operation.

(G) Heating System—
   1. Maximum heating capacity required by
      (a) Steam
      (b) Super Heated Water
      (c) Circulating Hot Oil System

(H) Heating Controls —
   1. To be either indicated or recording type with thermo-couples inserted in the heating platens or molds.
   2. Controls to be 110-volt service arranged to supply adequate power for all cycle operations without failure or interruption, or low voltage which is detrimental to solenoid coils.
   3. A 2KVA transformer is advisable to be installed on the control panel, operating from 220/440 to 110 to assure an adequate supply of control power. This then assures a fail-safe operation in the event of a power failure.[3]

Mold Design. Prior to proceeding with mold design and construction, the following questions should be considered and answered regarding the particular part involved. (*see also* Part Design Rules, this chapter.)

Let us assume we are about to design a new RP/C part or panel. We first must arrive at a definite understanding of the purpose or function of this part. What do we and the user expect this part will do? Have there been other parts like this made before? What were their good and bad points? Why and how will our part design be better? Only by good design, styling, and engineered use of the material can we expect to produce a better product.

Reinforced plastic is a relatively new material, so experience and experienced designer are somewhat limited. But we must make good use of that experience, for our critics will not permit us to make the same mistakes twice. RP/C does, however, offer the designer a freedom in design, strength, and styling which he has not had before.

Once we have firmly established the part's function, we must calculate or estimate both the normal and extreme abuse loads it will be expected to withstand. While wind or humanly applied loads on a fender or car door panel may be relatively light, the impact from stones or other car doors and the loads from teenagers standing on the fender or swinging on the door may well be our most critical load conditions. These are not normal, but if the panel is to be a public success, it must withstand these abuse loads well.

Now, to determine the nominal part thickness, we may use simple stress analysis or past experience. Using our estimated loads, we can calculate the thickness using a safety factor of 2 to 5 depending on the part's use and our sureness of the loads to be applied. On the other hand, we may choose to rely on the free experience of the custom molders' engineers or use the general "rule-of-thumb" of increasing the RP/C thickness to 2 or 3 times that of the previous steel parts. While a nominal RP/C thickness of 0.100 in. has been used to replace standard #20 gauge (0.035 in.) sheet steel, it is possible to go as low as 0.060 in. on panels which will have no heavy or abuse loads. Conversely, a truck fender which may be stood on, or impacted with large tire-thrown stones may have to be .250 in. thick to remain good looking after years of such abuse. It is best to add a little extra in the small high-stress areas and shave a few thousandths off the nominal thickness to save weight and money.

Having decided on a nominal thickness, we must look at the critical loads and areas. Failures will generally occur around loading points or in hinge, mounting, or attachment areas where the stress load levels are high. Applying the highest load conditions, we should make a simple stress analysis at these points to determine the highest stresses. The part thickness should then be increased at these points to withstand these stresses or a metal reinforcing and load-spreading plate should be added. In most cases, these high-stress areas are relatively few and the stress analysis is limited unless you are working a special military aircraft or automotive applications on which you are trying to cut weight to an absolute minimum.

Fortunately, RP/C is a resilient material and tends to spread and absorb its loads over a larger area. Then, too, it is relatively easy to reinforce an RP/C panel locally if your later calculations or test failures prove there is a need.

The thickness may be increased where necessary by easily removing metal from the matched metal mold. Thus, RP/C materials prove to be kind to the designers' errors.

Besides determining the part thickness to take care of high-stress loads, there are many other factors the designer must consider such as the following:

1. Is this part moldable with no undercuts or impossible shear edges?

2. Are the draft angles as large as possible for easy part removal from the mold (1° minimum, 3° normal, 6° for textured sides)?

3. Have we used generous corner radii wherever possible to prevent resin-rich shrinkage cracks?

4. Have we designed this part to cover as much area, function, or purpose as possible?

5. Could the complete assembly be molded in fewer pieces?

6. Have we calculated part shrinkage and allowed sufficient overage of the mold dimensions so that the part will seek equilibrium at the proper dimensions at room temperature after molding? (See Table IV-2.1).

7. Have we taken advantage of the deep draw moldability of RP/C?

8. Have we molded in as many openings and holes as possible?

9. Have we molded in color or texture, if desirable?

10. Have we designed in extra stiffness by good use of contours and/or style lines?

11. Have we remembered that RP/C is more flexible than metals?

12. Can this part be easily molded, drilled, assembled and finished?

13. Is this design as simple as possible or have we needlessly complicated it?

14. Will this part do its job successfully for the life of the assembly?

15. What is wrong with this RP/C design?

Other Mold Design Criteria[4,5]. Wherever possible, all critical openings should be molded in one piece, leaving extra stock on the inner edge for final dimensioning. Even though the initial tooling cost will be larger in order to accomplish this, it will result in savings and elimination of extra costs in final assembly and assembly fixtures.

Stress cracks will develop across a bond line, especially a short one. It is therefore advisable to avoid a design where tensile stress concentrations may occur across a bond line and to make bond areas as large as possible wherever they are to be used.

Because RP/C does not possess the stiffness of steel, concentrated loads must be spread

Table IV-2.1 Part Shrinkage Standards for Preformed and SMC Matched-Die Molded Parts of RP/C

Conditions	Cold Mold to Cold Part Shrinkage (in. per in.)
Filled general-purpose polyester resin without reinforcement	0.004
Preform or SMC parts—general-purpose polyester resin + filler containing 35% fiber glass reinforcement	0.0015
Preform or SMC parts—low-shrink polyester resin plus filler and 35% fiber glass reinforcement	0.00000 (May be controlled between −0.00015 expansion and 0.00014 shrinkage)
Thermoplastics, unreinforced (for comparison)	0.004–0.020

over a wider part area than would be necessary with steel. Although it is more costly per part, it may be advisable to design into the part extra rigidizing materials, such as a welded, extruded aluminum frame to eliminate cold flow or to otherwise mechanically increase stiffness in the RP/C.

Mold Construction. A good set of matched-metal tools or dies is a prime requisite for turning out high-quality finished molded parts. A properly designed part should precede mold design.

The essential elements of a mold are: proper type steel or other material (steel is the most widely used general-purpose mold material); a heating means such as an electric platen (shallow or flat molds), steam-heated platen, mold metal cored for electric heat (flat molds) or cored for steam (larger molds); guide pins to properly seat mold halves; a shear edge or pinch-off edge to cut off the excess, contain the resin, create back pressure, define the extremities of the molded part; a mold finish which dictates the primary finish of the molded part; can eliminate post-finishing, and contributes to proper release of the part from the mold.

The steps necessary in detailing the proper construction of a suitable matched-die mold follow:

Preliminary Planning. Many mold programs start with inadequate information and part prints lacking the vital finer points necessary to plan and build a solid functional mold.

A clear and accurate print is the number-one step toward that goal.

Other items to be finalized are:
1. Shrinkage factor
2. Draft angles
3. Press limitations and data
4. Part tolerances
5. Mold surface specifications

The responsibility of the dimensional accuracy, finish, and general appearance is the builder's. However, without the close cooperation between the molder and mold maker, many problems will arise in the crucial "deadline period". These can be avoided with common understanding at the start. This cooperation is vital and will have a bearing on the ultimate success of any mold.

Duplicating Models. The ideal situation in the mold-building industry would be that a wood master model be furnished with proper allowed shrinkage and be in "die position" at the beginning of any mold program. However, if this master is not available, then ample lead time must be allowed to fabricate it. This can be done while the basic mold design is drawn up.

Every precaution should be taken to insure that the master model be carefully inspected. Both molder and mold maker should join in a mutual responsibility in this respect.

After shear-edge details have been added to it, the master is used to cast the epoxy laminate male and female Keller duplicating models. These will be the identical surfaces that will be duplicated into the male and female steel by the Kellers. The duplicating models must have a sturdy reinforced frame to keep them from warping and to withstand the normal shop abuse.

A final inspection of the models is necessary in order to guarantee that they conform to the master.

Mold Design. After an agreement has been reached on the basic items, the preliminary mold layout can be drawn up. As many points as possible must be incorporated into the design at this stage. The body of mold must be heavy enough to prevent deflection during molding.

The following are items to be considered:

1. Materials. (See Table IV-2.2). The mold-material selection must be suited to the size and shape of the part. No. 1045 or 4140 steel plate, forged billet, or castings can be used. Castings are best suited for larger complex shapes and contours. Machining stock of only ¼ to ½ will result in a minimum of machining time. However, the cost of casting patterns may offset any savings gained using forged billet or plate.

A detailed study of casting versus plate in several orders showed that "rough" machining a billet was more economical provided that large and adequate machine tools are available. The primary advantage of roughing plates or billets is the lesser danger of porosity in the grain structure of the steel as compared with the casting.

2. Heating Channels. There are no set rules for location of steamlines. Contour and shape will dictate this. Common sense should prevail in the placement of the lines or chamber stressing of a balanced pattern. A heat distribution of plus or minus 5° should be maintained. Inlets and outlets must be drilled in the mold opposite the position of the press operator.

3. Shear Edges. The right-angle shear condition with a by-pass or "pinch-off" of the male and female of $\frac{1}{16}$ in. is recommended. A clearance of .002 to .004 maximum between shear edges is very important to maintain. An open shear will result in defective parts because resin will not penetrate to the pinch-off periphery.

Shear edges should be machined as an integral part of the parent steel. Professional mold-fitting and heat-treating procedures will insure a solid cutting edge.

The carbon content of the 40-50 carbon steel mentioned in Item #1 allows shear edges and shut-off areas to be flame hardened to 54-58 R.C.

4. Guide Pins. Incorporating the proper size guide pins relative to the cavity size is a critical point. Guide pins are the main guiding factor insuring the close alignment of the male and female halves of the mold. Always use pins "husky" enough to take load and "side thrust". Commercially purchased pins and bushings are precision-fitted and should be used rather than shop-made items.

If the mold has an off-balanced cavity, supplementary wedge bars should be used. These interlocking bars will relieve wear and tear on the guide pins as well as the shear edge by taking up the side thrust in the molding operation.

5. Part Removal. A professionally built mold with ample draft and proper polish will release the part without any ejector aids for the majority of matched die-molded parts. A comparison of component requirements for standard and SMC molds for RP/C is shown in Table IV-2.3.

FABRICATION

The actual cutting of steel should commence only after both parties are in agreement on all requirements of a solid producing mold. Too often the desire to cut steel too soon results in errors and delays instead of improvements in the progress.

Each mold program will require a different machining schedule depending on the size and contour of the part to be produced.

An orderly procedure of machining should be planned and followed by a competent mold follower or leader. His duty is to supervise the various stages in the building of the mold and he will give it the personal attention it needs.

A typical machining procedure on a RP/C part follows:

Machine Operation
1. Grind and square 1045 Carbon steel forgings.
2. Drill eyebolt holes for handling.
3. Rough Keller core and cavity leaving approximately ¼ in. of stock.
4. Regrind forgings square.
5. Finish Keller leaving approximately .010 stock for finishing operations. Finish Kellering should include as much machine detailing as possible. The precise Kellering using the duplicating models as a guide will eliminate a great deal of hand finishing.
6. Drill 23⁄32 dia. steamlines. With contour in cavities established, locations of the lines can

be checked to insure that best judgment was exercised in the original design.
7. Miscellaneous Machining.
   Final milling and drilling operations are completed. Included is the drilling of the guide pin and bushing locations. This is one of the critical phases of mold. Precise drilling will allow proper alignment of the male and female shear edges.

The mold is now ready for the fitting and polishing operations in the following order.
1. Disc Sander with # 24 Grit
   This removes Keller ridges. Follow up with #80 Grit to remove the rest of the tool marks.
2. Stone #180 Grit
   This step is to form or "true up" the contour by removing waves or depressions that are machined into the cavities due to machine deflections.
3. Stone #320 and #400 Grit
   This smooths out the roughness of the previous stone and prepares the surface for the "papering" stage. The direction of stoning should be applied in opposite directions to insure complete coverage.
4. Flame-harden female shear to hardness of 54-58 R.C.
5. Spotting Press
   The mold is now mounted into the spotting press for fitting. The core or male shear is still "soft" and, during the fitting operation, is literally sheared into the cavity, metal to metal. When the proper depth has been reached, the shear relief to a clearance of .090 is achieved.
6. An RP/C Part is molded and checked for an actual part thickness. This shot will also establish the shear condition accuracy.
7. Aluminum Oxide Cloth #180 Grit
   The final polishing phase. Again, a changing diagonal pattern of polishing should be used.
8. Aluminum Oxide Cloth #240 to #320
   This phase should be stroked in one direction. It is also the range required for RP/C parts used by the automotive field.
   In an actual test, a cavity surface was finished with #240 Grit cloth and another with #320 Grit. The molded part was inspected by a surface profilometer. No appreciable difference in the surface finish of the part was noticed.
9. Felt Buffer
   Use of a #1800 Grit buffing compound is the final step in the polishing phase. This is the operation which applies the "high luster" to the mold. It should be applied in a cross-directional pattern to remove the fine surface marks. A word of caution—a high luster

Table IV-2.2   Review of Materials Used in Fabricating Molds for RP/C Molding

Material	Suitability of Use and Comments
Cast plastic and plastic-faced	Suitable for short runs and prototypes (see chapter II-4).
Meehanite—cast (semi-steel)	Extensively used in early RP/C molding, but rarely used today. The material is not durable and develops too much porosity due to casting. Requires continual patching in service.
Kirksite—cast	Has greater expansion than steel when used in combination, and also "grows" in continual service at molding temperatures. Not 100% suitable.
1040 and 1045 steels (50 carbon type)	Good machineable mold material. Suitable for parts where post-molded surface finish is not critical.
4140 steel	Suitable for parts where appearance is a prime requirement. Has some porosity.
P-20 steel or equivalent	Best mold material available. Almost no troublesome porosity, and has good polishability. Better grade of steel and is preferred for extremely long runs. Cost is approximately 2.5 times that of 1045.

can be achieved on any mold but it is of no use if the initial steps mentioned earlier have not been properly executed.

10. Tryout Press

A final tryout of the mold with the molder's representative present should be made. After the tryout satisfies everyone concerned, the mold is covered with a silicone base grease to protect it while in transit to the molder's plant.

CHROME PLATING

This is a decision the molder must make. A 0.0005 to 0.001 thickness of hard chrome does have some advantages, such as improving mold release and eliminating the use of release agents. It will also help protect die surfaces. However, chrome does have some disadvantages. It will not improve the mold surface finish; in fact, it will actually degrade it slightly.

The mold surface must be prepared in the pre-described manner before chrome can even be considered. Another point to consider is the problem of repairing the mold after it has been chromed. This would require that it be dechromed before any major repairs can be made and then re-chromed. These are hidden costs which are not originally considered.

Summary. A molded RP/C part will be no better than the mold from which it was produced. Mold making is an art and it should not be entrusted to anyone on the basis of low price alone. Consider the qualifications, the experience, and the facilities of the mold builder. The following critical points will insure a successful mold:

1. Adequate preliminary planning.
2. Clear understanding of the mold design.
3. Accurate duplicating models.
4. Utilize a professional mold maker's experience and skills.
5. Complete satisfaction of finished mold before acceptance.

The proper physical and visual finishes required of RP/C parts by many industries can be achieved in matched metal molding. However, only with the teamwork of the molder and mold maker in the fabrication of the molds will our industry attain this goal.[6] See Table IV-2.3.

THE PREFORM PROCESS IN MATCHED-DIE MOLDING

The preform process entails a method for preparing a shaped reinforcing fiber glass or other "mat" in precisely the shape of the article to be molded and combining it with matrix ingredients at the press. The actual steps in the preforming process comprise:

- Formulating and mixing polyester resins, catalysts, fillers and so on.
- Preforming the reinforcing fibers over a prepared screen by air suction and maintaining the preformed shape with use of a binder material spray applied and heat-cured.
- Readying press equipment.
- Molding.
- Trimming and finishing.
- Inspecting and packing for shipment.

The technical and engineering essentials for carrying out the complete preform process for matched-die molding are particularized herein.

**Table IV-2.3 Comparison of Major Structural and Operational Elements
for Standard Vs. SMC Molds for Matched-Die Moulding
of Reinforced Plastics**

Type or Method of Molding	Matched Metal Dies for Preform or Mat Molding	Matched Metal Dies for Molding SMC
Mold Component:		
Surface finish	High polish satisfactory; can be chrome-plated if desired; non-chromed surface hides "laking"	Chrome plating preferred over 325–1200 grit finish, buffed and polished
Shear-edge	Flame hardened to resist pinching and dulling due to glass	Flame-hardened or chrome-plated to reduce abrasive wear
Guide pins	Required 0.001 in. clearance on diameter	Extra-strong and accurate guide pins required to resist sideways thrust due to off-center charge or assymetrical mold; must protect shear edge
Ejector pins	Not necessary for most matched-die molding	Generally required for SMC; air-blast ejection preferred; cellophane preferred to cover ejector head during molding
Telescoping at shear edge	Travel should be 0.040–0.050 in.	SMC requires 0.025–0.8 in. telescope for developing proper back pressure and best mold fill-out
Clearance at pinch-off	0.002–0.005 in.	0.004–0.008 in.
Landing or molding to stops	Needed to properly define part thickness	Not necessary; part thickness determined by weight of charge
Optimum part thickness	0.090–0.125 in. optimum = 0.100 in.	0.125 in.
Molding temperature	235–275°F	1 sec, per 0.001 in. at 275°–280°F Range = 260–290; 340°F for thin parts
Molding pressure	200–500 psi	500 for flat to 1000 for deep draw; slow close required for last ¼ in. travel.

Materials. As in the case of premix, the materials for matched-die molding fall into 3 major categories:

Resins supply the matrix function, bonding the other ingredients. They must be catalyzed and subjected to heat and pressure in order to cure.

Catalysts are materials which supply oxygen to the resin network to initiate polymerization and effect a complete cure.

Reinforcements consist mainly of glass fibers which fill the major function of supplying superior mechanical strength when combined into a resinous structure.

Fillers supply the auxiliary properties of improved flow during molding, dimensional stability, and improved surface. They also lower cost.

Resins. There are three major polyester types of resins: orthophthalic, isophthalic, and the acrylic or other thermoplastic-modified low-profile or low-shrink type. A breakdown of their properties are shown in Table IV-2.4. Mention is also made here of additional resin improvements resulting from new monomers, one-component low-shrink types, and advanced isophthalic resins for improved craze resistance.

Catalysts. Properties of the catalysts most commonly in use in a matched-die molding system are presented in Table IV-2.5.

Table IV-2.4 Properties of Polyester Resins for Matched-Die Molding

Property	General-Purpose Orthophthalic Type	Isophthalic Type[a]	Low-Profile or Low-Shrink Resins
Description	Rigid high viscosity, good release from mold, high gloss, less crazing, low exothorm	Resilient, medium viscosity, higher hot-strength, excellent craze resistance[b]	Usually, series of 2 resins combined to give low-shrink resin in molding-125 M.I. (Mixed properties)[c]
Viscosity poise	25–35	26–30	12–13
Specific gravity, liquid	1.14	1.17	1.67
Monomer	Styrene	Styrene	Styrene-acrylic
Deflection temp. 264 psi (casting), °C	71	83	205
Flexural strength, psi	14,000	22,000	12,000
Flexural modulus, psi	580,000	541,000	1,650,000
Tensile strength, psi	9,900	10,900	4,700
Tensile elongation, %	1.8	4.0	—
Barcol hardness	50	45	60
SPI gel time, min.	5–7	5–7	—

[a]*See also* Resins Based on O-Chlorostyrene *t*-Butyl Styrene and Vinylester Types, *SPI 20th RP/C Preprint,* 1965, sec. 2-C, 18-A.
[b]Additional craze resistance claimed for molded polyesters subjected to several post-cure bake cycles at 280°F (23rd SPI Proc., 1968, sec. 19-F).
[c]One-component low-shrink resins are also available. See SPE RP/C Retec, Sept. 25, 1969, p. 67; and *SPI RP/C Division Proceedings,* 1972, sec. 12-A.

Table IV-2.5 Properties of Catalysts for Polyester Resins in Matched-Die Molding

	Benzoyl Peroxide[a]	t-Butyl Perbenzoate	t-Butyl Hydroperoxide	t-Butyl Peroctoate	Diperoctoate[b]
Available as solid liquid, paste	S,P	L	L	L	L
Peroxide, %	S = 96–99 P = 50	98	72.1	97	90
Active oxygen, %	S = 6.5	8.1	12.7	7.18	7.45
Decomposition temp., °F in benzene at life of 1 min.	272	331	354	266	253
Activation energy, kcal mol.	30.0	34.7	—	31.2	33.6
Kick-off per temp., °F	200	245	270	198	190
SPI gel time, for general purpose rigid resin, 1% peroxide concentration 180°F, min.	4.4	9.5 (212°F)	8.8 (212°F)	3.8	2.7
Recommended concentration peroxide	0.07–1.0	1.0	1.0	1.0	0.3–0.5

[a]BPO not recommended for use with low-shrink resins. *See also* "Curing Effects of 2,2—Azobisiso Butyronitrile"—0.4% concentration equivalent to 1% BPO, *SPI 20th RP/C Preprint,* 1965, sec. 2-E.
[b]Actually a difunctional perester, 2,5 dimethyl 2,5 di(2-ethyl hexanoyl peroxy) hexane.

Table IV-2.6 Properties of Glass-Fiber Roving for Preforming

Property	Specification
Type	Silane preferred; chrome desirable for low static.
Strand integrity	Medium hard, minimum fuzzing.
Ignition loss, %	0.5–0.8
Ribbonization	3–5
Stiffness	3.6–4.2
Color	True white
Static	Minimum (under 3000 v.)
Yd per lb nominal	221
Chopability	Should be excellent with minimum filamentation for good fiber locking and conformity to sharp radii without bridging. Sometimes desirable to use combination of hard- and medium-integrity types for better control of fiber-locking and distribution on preform screen.

Source: Data Courtesy Johns-Manville Fiber Glass Reinforcements Division.

Reinforcements. Properties of the glass-fiber roving types required to fabricate preforms are shown in Table IV-2.6. Table IV-2.7 compares the properties of mat-type reinforcements used in preform molding.

Fillers. A great deal of basic technical information is available regarding fillers. Chemical and physical properties are sufficient however for procuring materials suitable for use in resin mixes intended for matched-die molding. Please refer to the discussion of fillers for premix in Chapter IV-1 because the same types, principally clay and calcium carbonate, are also applicable to matched-die molding.

Other Materials:
Mold-Release Agents. The following agents are desirable to assure separation of the molded part from the mold at completion of the cycle:
- *Mold break-in.* Heavy silicone grease or petroleum jelly is smeared on the mold for each of first 10 or 12 resin plus glass-molding shots. The purpose is to fill the pores of the mold and avoid major nonrelease problems during the life of the mold.

- *Internal-release agents.* Internal-release agents are added to molding mix. They come to the surface and usually melt to form a lubricating layer or to create a high-contact angle, preventing sticking. They are usually added at levels from 0.5 to 1.5%. The solid-type are phosphates, lecithin, paraffin oils and stearates.
- *External-release agents.* External-release agents are added by rubbing, brushing, or a spraying application to mold surfaces prior to the molding cycle. Several kinds of agents are available: silicone waxes, fluids, and lubricants; paraffin oils and waxes; polytetrafluoroethylene and other filming types.

Preform Binder Materials. A polymeric binder is applied to preformed fiber glass by spray and cured by heat to a fibrous glass structure built up on a screen. Most preform binders are based on emulsifiable polyesters.
- *Emulsion type.* Emulsions can be easily prepared in any matched-die shop. The formula, suitable resin, and recommended methods may be procured from several of the larger resin suppliers.
- *Emulsifier-containing resins.* These resins may also be procured and developed into a suitable preform binder by a much easier process than preparing the complete emulsion from start.
- *Solvent types.* Several shops use a solvent type preform binder, that is resin in an alcohol vehicle. Much more rapid curing is possible but toxicity and fire hazards exist.
- *Catalyst.* The resin must be catalyzed during preparation of the preform binder. Pot-life problems therefore exist.
- *Solids content.* Preform binders in use contain from 5 to 10% resin-binder solids based on requirements.

Colorants.[7] Colorants for addition to either general-purpose or low-shrink, low-profile resins for RP/C matched-die molding consist of either organic or inorganic dry powders or pigments blended into a suitable vehicle using a 3-roll paint mill. The vehicle constitutes an alkyd-base resin usually with a diallyl phthalate or equivalent monomeric system to provide good uniformity of mix and percentages and practically unlimited shelf life of the resultant pigment paste. The amount of color pigment used in a paste varies from 15 to 75% and is a function of the oil absorption or the particular pigment.

The dry pigments vary widely in cost, and this is directly reflected in the price of the finished pigment paste. Factors affecting the selection of the proper pigment are the hue, intensity, cleanliness, and brightness of the color to be matched; the need for special conditions such as resistance to exposure to ultraviolet rays in outdoor exposure, high-humidity conditions, electrical field, and high temperature in service; the need for unusual or specific chemical resistance; the effect of the catalyst in the system (BPO catalyst is not compatible with organic phthalocyanine pigments); the temperature attained in the mixing process, molding and pressure; the tightness of the mold with either preform or SMC molding; the range of color variation that can be tolerated by the purchaser of the molded parts.

The color manufacturer should be depended upon to match the specific color required and to provide or point out probable tolerances that will occur. He should set suitable color standards for his own production batches and inform the molder in advance that variations will occur. This should be decided in advance even though it may be necessary to provide deliberately altered pigment paste or molded coupons or panels to illustrate the probable range of variation. Once this range is established, the molder should also acquire agreement from the purchaser to accept his molded parts so that complete understanding is reached by all parties concerned. As in any other quality-controlled industrial process, a range of variation is certain to occur. Standards should not be set close to either of the probable extremes of variation.

Pigment pastes are used in quantities as required that vary from 0.1 to 7% for colorants, and from 0.25 to 12% for whites. In most cases, it is advisable to incorporate some quantity of white pigment for aid in color clarity, stability and consistency.

Finally, it must be stated that the coloring material itself will not hide or even partially mask the inadequacies of a molding mix, nor disguise the malfunctions of a mold or of the molding process itself. This is especially true of low-shrink or low-profile resin systems.

Types of Preform Machines. The mechanisms for collecting the glass fibers in preforming them preparatory to molding constitute a specific class or genus of equipment. The several types which have been generated or evolved are described here together with their particular *raisons d'etre*.

Single-Station Plenum Type. Appearance: This type includes an enclosed housing with an upper chamber, lower screen and blower to hold the glass onto the screen. The blower air is ejected through an external waste-glass collecting chamber. *Opening*: The opening is cut through the rotating base of the plenum and provided with either a cast rim, or spaced lugs for loading and removal of the screens. *Chopping*: A fiber glass roving chopper and spreader are mounted at the top of the closed plenum. *Collection*: A perforated screen simulating the part shape is mounted on the horizontally rotating turntable to collect the glass in the proper thickness. *Blower*: A high horsepower motor drives the shaft of a blower to draw air through the screen from below, causing the fibers to be held onto the screen. *Vents*: Horizontal vents are placed in the sides of the plenum at the top and also lower screen level. Diagonal vents are also used. Vents may be adjusted in width of opening to provide side drafts and thereby control the deposition and distribution of chopped glass on the screen. *Preform Binder*: Spray heads are located at appropriate levels within the plenum and are activated during the cycle to completely cover the screen area with binder spray. *Static Control*: Humidity sprays mounted in the top of the plenum hood dispense water to control or remove static electricity generated by the glass being chopped. *Automation*: Rotation, chopping of glass, delayed binder application, and shutoffs are all automatically controlled from a master panel with times and solonoids. *Curing*: Curing of the preform binder is accomplished by removal of the screen containing deposited glass and transferring it to a separate oven. *Advantages*: Major advantage—control of fiber distribution is better than in open-spray type. *Disadvantages*: Compaction is lost when blower suction is relaxed from the glass during transfer of the screen to the oven; many screens are required for a single part, and fabricating errors are possible: bulkiness of the preforms may result in washing during molding.

Shuttle Type Machine. Appearance: Similar to the single station plenum type in appearance,

Table IV-2.7 Types of Reinforcements and Reinforcing Mats for Matched-Die Molding

Type	Description and Advantages
Preforming	Requires open-spray or plenum for fiber collection to shape of male. Fibers (roving) chopped and directed at screen. Low solubility powdered or liquid polyester binder applied and cured at site. Preform plus resin mix charged to mold at pressing operation. Part shape exactly duplicated in preforming, but requires 3 or 4 manual steps.
Chopped-strand mats, low-solubility type	Glass chopped to 1 and 2 in. lengths, usually mixed. Combinations of powdered fiber-bonding resin plus surface-liquid spray binder applied and cured to form flat-sheet mat. Low solubility binder required so that no washing occurs. Provides good general-purpose molding mat but is limited in drape over compound curvatures.
Chopped-strand mats, mechanically bonded	Chopped fiber-glass strands in 1 to 2 in. lengths are laid down randomly on belt and needle-punched for bond. No chemical binder applied. Mat has excellent drape over curvatures in molding, low I.L., and low tendency to wash. Has lower molded strengths due to bent glass strands during needling.
Continuous-strand	Continuous, unchopped glass fiber laid down on moving belt in swirl pattern. Nonsoluble polyester binder applied as liquid and cured. Mat has excellent drape and permits high flow-through of resin mix in molding. High binder content tends to cause molded surface fiber pattern; requires veil mat overlay in molding.
Drum-wound expanded mats	Glass fiber drawn onto 4-ft diameter rotating drum from reciprocating fiber-melting unit. Binder applied during or after winding. Mat cut longitudinally and drawn transversely. Mat has excellent drape, bulk, and flow-through, but provides lower molded strengths than chopped-strand mats.
Surfacing mats	Fabricated from either drum-wound and expanded or blown fiber. Either veil or overlay mats used as a molded surface protector and equalizer to eliminate fiber pattern associated with chopped reinforcing glass fibers.

except that two screen holders shuttle, one each from each side of the plenum. *Operation*: Operation is the same as in the single station plenum type except that screens are mounted on a sliding shuttle so that one screen is inside the plenum (collecting fiber) while the second screen is shuttled to the side for preform screen removal and makeready for the next cycle. *Openings:* Two vertical-rise doors, automatically operated, are stationed on the shuttle-track sides of the plenum hood. A manually operated door is placed on the front of the plenum for cleaning, maintenance repair, and so on. *Advantages*: The operation of this machine is faster than the single station type. *Disadvantages*: Relaxation of the suction on the screen before curing induces undesirable bulkiness in the preformed glass fiber.

Self-Curing Shuttle Machine. Appearance: This machine is larger than the original shuttle machine, and consists of a plenum hood, an extra-large base for housing the heating unit, and a large, external, screened waste collection chamber. *Operation*: This machine was developed to provide larger preforms than were possible with the rotary plenum type, and also to make possible curing of the preforms directly on the screen without relaxing the fibers held by blower-induced suction. *Chopping*: Four roving cutters are mounted in the top of the plenum and operate automatically according to timed cycles. *Screen Size*: This machine accomodates screens up to 60 in. diameter or larger which are much larger than those possible with the rotary plenum type. *Blower*: A streamlined air intake, and a tunnel effect in-

Mat-Forming Binder or Treatment	% Ignition Loss	Mat Thickness (in.)	Mat Weight (oz per sq ft)
Low-solubility polyester	5–10	As received (bulks considerably)	4–6
Both powder and liquid binder insoluble in styrene when cured to prevent fiber-washing during molding	5–10	0.040 0.090	¾–3
None	0.5–1.0 (glass-fiber sizing only)	0.060 0.300	1½–10
Polyester, insoluble type	5–14	0.040 0.090	¾–3
Polyester, urea borate, or other insoluble type	5	0.030 0.090	¾–3
Polyester compatible, but nonwashing	Veil = 5–12 Overlay = 0.9	10–30 mil	—

duced by the sidewalls of the forming chamber produces better fiber distribution than is possible on the open-spray type machines. Also, air dampers in the blower system preselect and recirculate an amount of heated air back through the preform to conserve heat. *Heaters*: Extra-large capacity gas-fired heaters are included for high volume short time period heating. This effects a rapid cure rate for the preform and greatly reduces the total preforming cycle time. The oven section and heating units are self contained. *Advantages*: This machine will operate either as a closed chamber type or as a directed fiber type, whichever is the most efficient for the part involved. In general, it provides the desirable combination of automatically controlled fiber distribution on the screen together with large sized preform capa-

bility, plus extremely rapid cure. *Disadvantages*: The cure time and cycling are not as rapid as in the automatic rotary plenum type machine.

Automatic Rotary Plenum Type. *Appearance*: A four-stage preform screen holder or assembly is mounted on a central pivotal shaft which permits the assembly to rotate in a vertical plane akin to a ferris wheel. The plenum housing is at the top of the machine and is accessible from a platform via steps or ladder. *Main Functions*: Chopping, collecting glass on the turntable screen station, binder spray, humidity control, etcetera, are the same as for the single station plenum type machine. *Cycling*: Positions: collect glass fiber at the top station with the screen rotating in the plenum chamber;

cure the preform binder in the rear side and bottom stations (quadrants); (blower suction is never released from the deposited glass fiber throughout the entire preforming cycle and in both oven curing stations, and not even during advancement of the ferris wheel from one station to the next); finally, take off and screen cleaning and makeready are accomplished at the front floor-level position. The screens are not removed from the base housing and hence must be cleaned and maintained in place. *Automation*: The entire operation is fully automated. One operator only is required to tie in glass roving strands, remove preforms, clean screens, and clean the plenum chamber every hour. *Advantages*: This method represents the most uniform and rapid preforming rate available. It provides the lowest cost of preform per pound, is automated as fully as possible, and the suction is never relinquished from the deposited glass thus permitting maximum compaction and resultant freedom from associated molding defects. *Disadvantages*: This type preform machine is limited in size to a 30 or 40 in. turntable diameter at each station. It is also limited to fairly symmetrical shapes and to a screen size which fits the turntable, making it uneconomical for screens much smaller than the turntable, hence limits machine adaptability. Also, the versatility of this machine is limited to running only one size or two very similar size preforms at any given time.

Open Spray or Directed Fiber Type: Single Station. Appearance: The turntable is mounted slightly above the floor, and rotates circumferentially in a plane approximately 10 degrees off vertical. The glass roving is chopped and sprayed by hand from a separate unit. *Screens*: Screens are mounted on a ledge, lugs, or other non-clamping means equidistant from the center of the turntable. The screens do not require clamping and are held on by force of the blower air. Different sized screens may be alternated for intermittent preforming of different shaped parts, each size requiring only a common base. Hence, one preform machine can supply preforms for two or more presses molding different sized parts. *Turntable rotation*: This is set at 20 to 30 rpm for smaller open-spray machines, and 10 to 16 rpm for the large machines. *Chopping*: Glass roving is chopped using a separate assembly on a frame or wheeled cart located in front of but not physically connected with the preform turntable housing. This unit provides the function of holding the roving, chopping and ejecting onto the screen. One to four or more roving strands are fed to the cutter and the stream of chopped glass is aspirated by blower air into a 4 in. diam. hand-held flexible tube, 1, 2, 3, or 4 ft long. The glass is propelled by the high velocity blower air onto the rotating screen. Distribution of the glass fiber on the screen is manually controlled, and is a function of the operator. For exceptionally large parts to be performed, two chopping units may be employed. *Preform binder*: The emulsified polyester preform binder is delivered from a pressure to a spray gun mounted on top of the glass delivery tube. The binder is hence controlled by the operator, and is usually admixed with the glass during chopping and application, but may also be applied separately for such requirements as enriching the outside surface of the finished preform with binder solids. *Blower*: A large HP blower is mounted with motor in a duct behind the turntable, and moves air through the turntable opening to hold the chopped fibers on the screen. *Heating and Curing Means*: In the operation of small open-spray machines (single station), screens are removed after preforming and placed in a separate oven. In machines for larger parts 8 to 17 ft diameter, a gas-fired oven unit is mounted on the top of the preformer doors closed over the turntable (still rotating) following preforming, and heat from the oven circulates rapidly through the glass fiber on the preform screen to cure the binder without permitting it to relax and become bulky. *Automation*: The timing of the preforming cycle is automatic, and may be intermittently varied for sequential preforming of various sized parts. Glass spray and preform binder application are manual. The operator starts each individual cycle with a foot switch provided. *Advantages*: Different shaped parts may be preformed intermittently; exceptionally large parts may be handled on this equipment; compaction of fibers may be constantly maintained in larger units possessing attached curing ovens; the binder spray is directed into the fibers and is controllable, thereby preventing excessive losses; extra glass and binder may be selectively placed for increased thickness and reinforcement where required, such as on outside radii and corners to prevent resin-rich areas in molding, and differential part thicknesses. *Disad-*

vantages: Glass distribution is a function of and subject to inconsistencies and vagaries of the operator. This factor plus tendency for the glass to fall off the screen results in more variability and waste of glass for this system than for the closed-chamber automated systems. The external choppers require constant maintenance, and there is a higher possibility of contamination than in a closed system.

Open-Spray Rotary Type. Appearance: This machine consists of a four-stage vertical plane rotary ferris wheel assembly with capabilities for open-spray preforming at the front station (floor level), and with curing carried out in the other three stations. *Main Functions*: Chopping, binder spray, rotation and venting for glass control are controlled the same as in the single-station open spray type machine. *Blower*: The heating and curing means are similar to that outlined for the rotary plenum type machine. Cooling means are sometimes employed using a separate blower in the fourth or take-off stage for the purpose of cooling preform and binder solids and stabilizing the preform before handling. *Advantages*: Although glass deposition and binder spray are manual operations in this equipment, the curing is automated. Hence, preform compaction is induced by blower air suction, and dense preforms result. Larger parts are preformable than with the rotary plenum type machine. *Disadvantages*: This process requires slower preforming cycles. A higher glass waste results, and there is a greater chance of drawing and entrapping dirt and foreign matter into the preforms.

Choppers. Choppers are required to break the glass roving into finite lengths for required processing. There are several types of choppers available. The type most widely used has the blade impressed against the elastomeric roll. It is mounted on top of the plenum or in open spray units. Blades may be milling or razor type. The advantages of this chopper are that it gives the cleanest cut, preserves glass-strand integrity, and has an adjustable speed. The disadvantages are the high maintenance cost—it is subject to rapid blade and rubber wear—and the lengthy downtime required to change either the rolls or the length of cut.

In the guillotine type of chopper, glass roving is exuded through a hole in the carbide bed plate and fractured off in finite lengths by a revolving, spring-loaded carbide-tipped cutter blade, with the tips strongly impressed against the bed plate. Among the advantages of this type of chopper are the easily adjusted speed and length of cut and the low maintenance. It is difficult, however, to preserve the initial strand integrity.

Plenum-type preform machines require a spiked spinner or beater just below the chopper for several reasons. It directs the fiber, prevents uneven accumulations, and controls the bulk of the preform. A higher spinner speed produces greater filamentation and fiber interlocking to hold the glass on the screen. Interlocking is also partly helped by combining hard- and soft-finish (integrity) glass fiber roving types. If the preforms are too thick (low bulk), glass washing plus thin spots and wrinkles may occur during the molding. If the glass is too highly compacted, it is difficult for the resin to penetrate during molding.

Fabrication of Preform Screens. These screens are made from perforated metal shaped to match the outside surface of the force or plug side of a matched-die set. Air drawn through holes in the screen during preforming causes the chopped glass to be held on the screen. In completing the cycle they are charged with resinous binder, cured, released and use in press molding.

The technology developed for screens involves shaping, fabricating and acquiring optimum performance.

Screen Materials. 16-gage mild steel with $1/8$ in. openings, rows staggered on $3/16$ in. centers, 45 to 50% open.

18-gage mild steel with $3/16$ in. openings, holes staggered on $1/4$ in. centers, 45 to 50% open.

Hand lay-up RP/C shaped into a cast of the mold plug, ribbed for rigidity, and drilled to establish the proper air flow.

Preparing Form for Screen. Shape the plaster of Paris like the mold plug minus the thickness of the preform-screen metal. Use a wood tool with metal strips where screen segments must be welded. No inside radius in screen is to be less than $1/8$ in. Do not shape by hammering but by bending.

Cutting, Shaping and Fabricating:
1. Cut and bend the screen metal to form butt-joints.
2. Cut the metal to be formed preferably

along flat sections of part, not curvatures or corners. Some joining at corners is unavoidable, however.

3. Tack-weld along seams with sections clamped firmly against the screen form.
4. Gas-weld to join all sections and parts.
5. Mechanically grind and file to remove burrs.
6. Allow ¾ in. of preform screen material beyond the actual periphery of the molded part size so that the accumulation of extra glass will provide material for the glass pinch-off when the mold closes.
7. Weld finished screen to rim sized for proper mounting on preform machine turntable.

Baffling. Baffling comprises the placement of either perforated or solid sheet-metal panels parallel to the surface but inside the screen and usually stood off ½ in. to 1½ in. or more. The purpose of baffling is to provide equal or uniform air flow through all areas of the screen, thus establishing the most uniform glass deposition possible. Baffling is almost always necessary in screens for plenum-type machines but is not so critical in the open-spray type. It is also less critical in symmetrical shapes than in non symmetrical. Some need for baffling may be eliminated by varying the hole sizes in various areas of the preform screen. More or less glass may be deposited on critical areas of the screen to either compensate for or permit resin-richness, respectively. Success in proper air flow through the preform screen requires trial and error. Case history accounts and experience help to solve problems.

Treatment for Release:
1. Treat the external surface of all preform screens for release of the glass fiber shape after application of the binder and curing. Material with a high-contact angle is required.
2. Spray and bake on teflon emulsions.
3. Wax surface. Many types of waxes are available.
4. Consider chromium plating. It has not been 100% successful in use with preform screens.
5. Apply phenolic primer by dipping and baking on. It provides good release and does not flake off. It may be easily cleaned or chemically removed for general clean-up and then reapplied.

Screen Cleaning. Screens should be brush-cleaned frequently during preforming-every 10th cycle. A hand or motorized brush with wire bristles is used to remove fibers and caked binder. An air blast from a hand air gun can remove fibers from under the screen and between the screen and baffle. A caustic (oakite) dip-bath is valuable when truly necessary to remove caked fibers and binder and to open holes. The screen should be chemically cleaned, washed, and recoated for release afterwards.

Preform Operation in Relation to Equipment Involved:

Glass Fiber Storage and Use. Store glass in a dry area at a constant temperature (70 to 80°F) and humidity (50 to 60% rh). Maintain these conditions throughout the year if possible.

Glass handling at the plenum-type machine requires 35-lb packages of roving placed on a rack or elevated supports. They may be near floor for ease in maintaining and replacing. The glass either has tie-in tails or is in a continuous creel package for continuous feed. For fiber guides for 2 to 4 or 5 strands the 1 in. diameter ceramic type on wire stand-off is used with the open-spray machine. Glass-roving packages are arranged on a cart or are nested on the floor near the guides for feeding to the chopper, which is nearer to the floor than in the plenum type. The original strand integrity and ribbon of glass roving should be protected because too much bulk and soft glass produces wrinkles in the molded parts.

Fiber guides should be located so as to prevent whip and the generation of static. A flat plate over a roving ball with a 1- to 2-in. diameter smooth center hole for the strand to run through prevents whip, gobs, and snarls. The strand should not be run through small-diameter tubing on the way to the chopper because this generates excessive static.

Excessive glass ribbon is broken up when the strand passes through guides, especially if it is permitted to pass over right-angle bends. This assures complete glass break-up in chopping.

Preforming Efficiencies. The loss of glass for plenum hoods is 5 to 8%; for open spray 10%. Glass loss of 2 to 3% is due to rejected preforms which are nonuniform of the wrong bulk, and so on. (See Table IV-2.8.)

There is a 2 to 3% loss of binder in the

tank due to poor stability-spillage, and other factors. The loss by spraying through the screen is 4 to 5%. The binder makes up 5 to 15% of the total weight of the preform. The binder makes up 5 to 15% of the total weight of the preform. There is some loss of binder up the oven stack.

Cleanliness. Many preforms are lost due to dirty shop conditions. The shop should be kept clean and be frequently inspected. Preforms must have clean air, preferably from the outside and filtered. The air which flows to the preform screen should be air-conditioned and kept at a constant temperature and humidity. Positive pressure should be maintained in the room if possible. Machinery and preform should be kept well lubricated. Grease must be cleaned from areas which the preforms or glass strand are likely to contact because it shows up worse than bugs and other dirt in molded parts, especially in white or light colors. The chopper, blade rolls, and rubber rolls should be frequently inspected. Shredded rubber in preforms causes molding defects and tramp iron in a preform may cause mold damage.

Curing. The secret in creating preforms of the proper bulk and best molding properties is to keep suction applied during both the preforming and the cure. The best preforms do not have the strand relaxed or the blower vacuum released in changing from preforming to curing. The binder assists in drawing a better pressure head on the deposited fibers. Overcuring is undesirable because it results in brownish discoloration and therefore molding rejects.

Preform Automation. The plenum-type preform operation most closely approaches the possible degree of automation. It yields the highest number of preforms per man-hour of labor.

The complete preform delivery system to the presses may be automated. Major requirements are that (1) preforms be dried en masse immediately before traversing the press line of the area and (2) the whole preform delivery line be covered to protect against dirt and other contaminants.[8]

A highly automated and specially constructed deviation from regular preforming equipment and process possesses the following engineering features:[9]

- Filtered air is brought in from outside and passed at high velocity down a completely enclosed stack over the preform screen. The pipe length is 17 ft. This length is 5 times the diameter from the nearest bend to establish minimum air turbulency. Glass losses are thereby reduced and preforms are more consistent in weight.
- Air passes through the screen and is directed outside to avoid disturbing the room air. This arrangement makes for easier control, cleanliness, and maintenance.
- Higher air velocities are possible to accomplish optimum glass compaction.
- Screens are mounted on a rotary turntable and pass underneath the downward-directed stack, thereby lessening the time elapsed between each cycle.
- Glass is chopped externally and is fed into the downward air stream.
- Screens are baffled to better control the airflow, that is, there is less airflow through flat facing surfaces and more through vertical surfaces. Preforms are thereby made more consistent in shape and weight.
- A cone-type construction for the turntable apron surrounding the screen (dished downward toward the center of the screen) forces the glass fibers to the center, reducing glass-fiber losses. The fibers also tend to build up on the shear edge, providing more uniformity in molding. The tapered cone also permits a tighter fit between the screen and its mounting.
- Preform binder is delivered from a pressurized pot system rather than an atomized air system. The glass pattern dictated by the main airstream is therefore not disturbed.

Although an open-spray preformer cannot by definition be fully automated, the nearest thing to it has probably been accomplished.[10] A 60 in. turntable open-spray machine is in operation which permits the production of 40 to 50 preforms per hr., continuously holding the air pressure for both compaction and cure. The main fan pulls 38,000 CFM (150 h.p., 2650 rpm). The burner is capable of 4,000,000 Btu. to maintain oven temperature and raise preforms up to 500°F. The process speed is also increased by action of a 7.5 h.p. cooling fan at the takeoff station.

Preform Binder Preparation and Delivery. Large pressure tanks are needed for binder storage and delivery to spray guns. If the binder emulsion is to be prepared at the operation, emulsification equipment and testing devices will be necessary for control. If, however, emulsified resin is received, a vigorous mixing to let down to proper solids level (usually from 50 to 10 or 5% solids) is all that is necessary. Satisfactory performance is realized from atomizing spray guns (50 to 60 lb of air pressure), with pressure exerted on the resin tank to force the emulsions to the gun, are satisfactory.

Solonoid-activated spray guns for the plenum-type machine are arranged to turn on and off at the appropriate points in cycle. For the open-spray type the gun is held in the hand of the operator or is affixed to the top of the glass-delivery tube for intermittent manual operation.

Plenum Preform Hood. The plenum-preform hood may have anti-static humidity sprays in the top or it may pass glass over an anti-static solution or chemical. The hood baffled at two or three levels to assist in directing the glass to the desired area of screen. Baffles are air gaps which are adjustable for changes of product or external conditions. The binder should not be turned on until the screen is covered with a light layer of glass.

Resin Mixing Equipment for Match-Die Molding Mix. The mixing operation consists of storing materials; making them accessible to the mixing operation; and weighing, blending, mixing, conditioning and delivering them to the molding operation.

If the preform and molding operation is large enough in scope bulk storage of materials, particularly of resins, may be used. Piping may be installed to deliver the resin to the weigh station or to meter it directly into the main mix container. For a smaller operation resin may be kept in 55 gal. drums or below at room temperature. Catalysts must be stored in a concrete vault removed 150 ft from the operation.

Bulk storage of powdered fillers (clay, for example) is not as desirable because of fines separation. Bag storage on pallets is most commonly used, although some losses occur. The bags are transferred to the weigh station on hand or motorized fork-lift trucks.

Scales for weighing materials include a floor platform scale ranging to 500 lb and tare-weighted with printweight facilities. Also used are 100-, 50-, 10- and 5-lb scales and a flat-pan gram scale for weighing minor ingredients and catalysts.

Mixing equipment consists of table-top mixers with elevatable housings for minor-sized blends (speed—200- to 300 rpm) and large floor-mounted propeller-type mixers with the motor on a sled moving on vertical tracks. For a 500-lb batch, mixer blades should be 8 in. in diameter, mounted (2 sets) to rotate at the bottom and middle of a 500 to 1000-lb batch can with a speed of up to 200 rpm. Batch cans must have a valve at the bottom for dispensing the resin mix. Another necessary piece of equipment is the small homogenizing mixer with speeds up to 5900 rpm for intense mixing or suspending thixotropic agents, coolants, or other materials. Perfect uniformity is the secret both in small and large operations. For transfering the mix, wheeled carts or an overhead monorail system from the mixing room to the pressline are used.

A well-organized program of safety, fire prevention, and housekeeping must be implemented in the entire mixing system in order to maintain the best possible order and freedom from undesirable accidents.

Engineering Data for Preform Machines (Applies to all preform machines)

Chopping Rate. A standard 3½ in. diameter (10-in. circumference rubber-and-blade roll type cutter operating at an optimum speed of 450 rpm passes 375 ft or 125 yards (0.6 lb) of a single roving strand per minute. An average size 5 sq ft screen for a part to be preformed and molded with glass coverage of 6 oz per sq ft would require a total weight of 2 lbs. Consequently, the required amount of glass would be delivered in 3.6 min. using one roving strand, and 1.8 min. using two roving strands. The chopper speed could be increased to 750 or 800 rpm to multiply glass deposition, but inefficiency would result due mainly to the increased chopper maintenance that would become necessary.

Preform Weight Versus Thickness. The relationship of the weight of the cured preform versus corresponding thickness is of interest in control of the quality and uniformity of preform production. Thickness determinations are

Table IV-2.8   Cured Preform Weight versus Thickness

Glass Preform Weight (oz per sq ft)	Nominal Preform Thickness (in.)
2	0.068
3	0.090
4	0.112
5	0.134
6	0.156
7	0.178

made by measuring in a deep-throated thickness gage test stand using a dial gage with a 1⅛ diameter foot and with a 10 lb load applied to that area. Nominal thickness values are presented in Table IV-2.8.

Weight variations in adjacent areas of a preform of greater than ± 35% may be expected. Molded part weights and thicknesses corresponding to 35% glass content are:

A preform weight of 6.0 oz ft molded to 0.125 in. thickness weighs approximately 17 oz. per sq ft.

For a preform weight of 4.5 oz sq ft molded to 0.100 in. thickness, the finished part weight would be aproximately 13 oz sq ft.

Turntable Size—Plenum Type Machine. The optimum turntable size is one that is just large enough to accommodate the screen. The falling and deposition of glass after chopping requires a streamlined flow without turbulence. Therefore, the screen should adequately fill the turntable. Unnecessary glass loss will result from using a small screen on a large turntable in the plenum type machine. Plenum type machines have been made with turntable diameters up to 72 in.

Turntable Size—Open Spray Type. Turntable size in this type machine is not critical. As stated, many different size screens may be adapted to a single turntable size. Open spray machines in excess of 200 in. (17 ft) turntable diameter are in operation.

Preform Screens. Optimum metal thickness and type is 16 gage mild steel for fabricating preform screens. Two designs for perforation of the screen are used commonly: ⅛ in. holes in staggered rows on ³⁄₁₆ in. centers (45% open area), and ³⁄₁₆ in. holes in staggered rows on ¼ in. centers (51% open area). Baffling

to control air flow through the screen by placement of vanes, etc, behind the screen is not recommended unless absolutely necessary. It is used most frequently in non-symmetrical parts. Preferential glass buildup can be established by varying hole size and frequency on a custom built basis if needed.

Blower Size and Air Flow Through the Screen. All commercial preform machines are designed to provide passage of 1000 cfm per sq ft of area through the screen. This is true for both plenum and open spray types. A minimum static pressure of 9 in. water prior to glass deposition is necessary. Experimental conditions of 3000 cfm/sq ft and 15 to 25 in. water static pressure were found to reduce glass fiber loss. As soon as screen resistance exceeds blower capacity, fibers will no longer be held on the screen. Automatic dampers are beneficial. These may be installed and made to open and permit passage of more air through the screen as the preforming cycle progresses, and the glass fiber buildup for a given part increases. By this means, the proper and desirable static pressure head may be maintained across all parts of the screen through the complete cycle.

Horsepower of Blower Motors: Power requirements necessary to induce the 9 in. static pressure minimum across the screen on various size preform machines are shown in the following sub-tabulation:

Plenum type—36 in. machine	40 hp
54 in. machine	75 hp
72 in. machine	100 hp

Open-spray type machines of from 60 to 200 in. diameter turntables require 150 to 200 hp motors. Physical dimensions of the blower (diameter and width of fan) are commensurate with the motor hp to provide the required air flow. It is possible to use larger motors, but results are of doubtful value unless the screen area is exceptionally large.

Binder Application. Binder application may be made using either air atomizing or airless-spray type spray guns.

Oven Temperatures and Capacity Required for Curing Preforms. The optimum preform curing temperature is 400°F. Higher temperatures will discolor the preform and degrade the binder solids. The 36 in. rotary plenum type

Table IV-2.9 Number of Preforms Produced by Rotary Plenum Type Preform Machines

Turntable Size (in.)	Number of Preforms per hr
36	60–80; up to 120 under special conditions
72	10–30; depending upon size and weight
48	20–30
72	10–30
greater than 100	3–5

preform machine requires oven capacity of 300 Btu/hr, while the 72 in. machine will require a 600 Btu/hr. oven. The larger open spray type machines may require ovens with as much as 4 to 8 million btu/hr heat capacity if the exceptionally rapid curing rate is desired.

Number of Preforms Produced Vs. Machine Size. Table IV-2.9 provides this data.

Glass Loss During Preforming. Any molder is extremely vulnerable to watered-down or lost profits if excessive glass loss in preforming is allowed to persist. Glass loss in preforming is influenced by static pressure, glass type and loading in weight per sq ft on the screen, and ratio of surface area of the screen to the size and shape of forming chamber and degree of suction (plenum machines). It is essential to maintain a uniform static pressure across the screen. Automatically adjustable dampers have been used. The static pressure may be increased if 9 in. water is inadequate to minimize or prevent losses. In plenum machines, glass distribution may be improved and losses reduced by the use of diagonal venting in the

hood sides used together with the horizontal vents top and bottom. Table IV-2.10 presents results of a measurement of glass losses at varied static pressure.

If the glass used possesses excessively high integrity without at least some degree of filamentation, losses from the preform screen will result. If the preform binder spray is too thin in solids contained, or if the spray is directed so as not to cover or impinge properly, glass will be lost from the screen. Improper or no baffling will cause glass losses. Unclean screens will cause glass loss by permitting sticking and subsequent ruination of otherwise passable preforms. The open spray type of preforming equipment is subject to the highest percentage of glass losses due to its innate reliance on manually controlled application of the chopped glass. Proper selection, training and surveillance of skillful, conscientious workers is essential. Electro-mechanical development of means for automatic fiber deposition in open-spray preform machine types is drastically needed. The use of venturi-type hoods or sheet-metal channeling tubes with slot-shaped cross sections to direct and distribute the chopped glass properly over the surface of an open-spray preform screen have been used with great success. These and other improvements yet to come would combine the advantages of the plenum type operation with those of the open-spray machine, and allow the existing benefits of adaptability to unsymmetrical shapes to remain.

MOLDING METHOD/PROCESSING DETAILS FOR PREFORM MOLDING

Formulations for Preform Molding. The formulations and variations associated with

Table IV-2.10 Glass Losses vs. Static Pressure in a 30 in. Plenum Preform Machine[a]

Glass Buildup oz/sq ft	Static Pressure, in. Water	Corresponding Air Velocity in fpm	Glass Fiber Loss in %
4	5	400 to 600	20
	10	600 to 800	5
	15 to 25	750 to 1000	4
8	5	100 to 250	30
	10	250 to 350	17
	15	350 to 400	15
	20	400 to 500	12

[a]Conditions: Preform area was 4.23 sq ft in the 30 in. plenum machine.
Scrap loss was less from a 5 sq ft area preform screen than from a 4 or 2 sq ft screen.

Table IV-2.11   Formulations for the Preform Method of Matched-Die Molding

Ingredient	Resin Mix	Combining Ratio as Mixed at Press	% Final Composition
1. Resin			
Polyester resin, 35 poise viscosity	68.0		
2. Monomer			
Styrene (to lower viscosity to approx. 8 poise)	7.0		48.7
3. Catalyst		65.0	
Benzoyl peroxide (0.8–1.0% of resin)	0.6		
4. Fillers			
Clay, calcium carbon, or combined fillers	25.0		16.3
Internal mold release	0.1		
Color pigments			
(If required) Paste dispersions in DAP preferred over dry pigment additions	0.5–5.0		
5. Reinforcement			
Glass fiber as preform or mat	—	35.0	35.0

preform molding are shown in Table IV-2.11.[11,12,13,14]

1. *Resins.* Resins may span the range from rigid to resilient in order to fill the requirements for molded properties. These properties vary from a higher modulus of elasticity to higher impact strength at each end of the scale respectively.

Orthophthalic and the tougher isophthalic types possess the best general-purpose properties. Both have high shrinkage, however, and require the use of veil mat to provide a resin-rich surface as molded and to eliminate surface waviness, porosity, and fiber prominence. High-shrink resins require filling and sanding after molding.

Vinyl-ester resins reputedly lower the cure time in preform molding by as much as 10 sec. per cycle and lessen molding shrinkage, although they are not completely free of contraction or cooling. Epoxy resins also exhibit less shrinkage.

Acrylic or other thermoplastic-modified, low-shrink or low-profile polyester resins substituted for resin component in the above formula do not contract on cooling directly out of the mold and therefore greatly reduce molded surface waviness and fiber prominence. They do require internal mold release.

2. *Monomers.* In addition to styrene, ortho and monochlorostyrene provide faster molding cycles. Costs are higher, however, and some mold scumming may result. The material *t*-butyl styrene monomer improves surface smoothness.

3. *Catalysts.* Benzoyl peroxide (BPO—50%) in paste form is the standard catalyst workhorse for use with general-purpose polyester matched-die molding resins. BPO is not recommended for use with low-shrunk resin systems, however. Granular BPO dissolved in monomeric styrene will represent a considerable cost saving. Normal molding cycles for BPO are 235 to 265°F. *t-butyl* perbenzoate and *t-butyl* hydroperoxide are employed when higher molding temperatures and longer room-temperature pot-life of the molding mix are desirable. To supersede BPO at ⅓ to ½ required amount, 2,5-di-peroctoate is employed. A fire-retardant BPO compound is also available.

4. *Fillers.* Clays treated for oil-absorption values of 28 to 32 are preferred. The average particle size of clays and $CaCO_3$ is 4.8 μ. The range is 0.4 to 28 μ.

The range of the amount of filler in matched-die preform molding mix may be extended to 40% of the resin mix. The optimum is 25%. Above 25%, the mix is too viscous, and filtering of the filler material when passing through the reinforcement occurs. As a result physicals, particularly impact strengths, are lowered.

Unlike clays, calcium carbonate is nonporous and does not require treatment for low-oil absorption. $CaCO_3$ provides better resistance to crazing and cracking after molding. It provides higher molded impact strengths, but causes a 10 to 15% higher laminate weight because of greater specific gravity.

5. *Reinforcement.* As stated, reinforcement may be preformed glass fiber chopped from roving into 1, 1½ or 2 in. lengths or into a mixture of lengths. It may also be low-solubility-type chopped-strand mat or swirl mat (*see* Table IV-2.7). Chopped glass should be as static free as possible.

The optimum glass content required is 35% for preform and mat matched-die molding, although it may range from 20 to 50%. Glass content has the greatest influence on and is directly proportional to molded physical strength. Smooth surface finish is inversely proportional to glass percentage. Medium glass content with resilient resin favors better impact strength at the expense of surface finish. Glass content is usually varied at the expense of the total resin/filler shown in Table IV-2.11 held in constant proportion unless higher filler content is desired.

Mixing Procedure for Preform Molding.
The steps in the mixing procedure for the preform method of matched molding follow:

Batch-Mixing Procedures for Preform:
Preblending. Resin is piped from bulk storage or delivered to the weigh station out of drums. (For equipment, see Mixing Equipment, page 4.). Weigh the catalyst and disperse it in the portion of styrene, making sure the catalyst is dissolved if granular material is used.

Mixing Method
1. Add resin to main mix tank and start mixer.
2. Add monomer and stir until dissolved—requires 3 to 5 min.
3. Add preblended catalyst-monomer mix and stir until dispersed—requires 5 to 10 min.
4. Add filler and mix until dispersed— requires 20 min. Keep temperature from rising.
5. Remove and clean stirring blades.
6. Allow mix to stand 1 hr. for bubble rise prior to molding.
7. Mix is to be used within 8 hr. after mixing to prevent gelation in the container.

Control Viscosities:
- Original resin viscosity—35.0 CPS
- After monomer-catalyst addition—800 CPS
- After filler, addition, and mixing—15,000 CPS
- Ready to mold at press—15,000 to 30,000 CPS

Readying Press Equipment. Prior to molding, the press and mold should be inspected for cleanliness and proper functioning of the moving parts, the hydraulic system, and the control buttons. Safety for both operating and surrounding personnel should be kept foremost in readying the press for operation.

The mold should be checked for the production of proper part thickness, alignment, and register of pinch-off or telescoping edges. This can prevent damage which would immobilize the mold, produce defective parts, or interfere with the highest desired possible production rates.

Molds can be cleaned with compressed air, high-flash or nonflammable solvents on paper or rags (descummed), brass wool, or frequently resharpened brass chisels. Harder substances will damage mold surfaces.

Preform Molding Procedure. The molding procedure for preform matched die molding involves preparing the glass form and resin mix, materials to be molded, charging them to the press, molding, part removal, finishing, inspection and handling for packing and shipment.

1. In preparing the equipment, clean the mold by air blast and by rubbing off adhered resin spots using brass wool and brass bar or rod chisels. The flash from the previously molded part must be completely cleaned off the shear edge and all chips removed from the mold surface.

2. To prepare the charge, weigh the preform on a suitable scale and adjust the weight by carefully removing or adding thin layers of glass fiber as required. Add fiber glass or premix in putty form at the corners or edges if extra reinforcement has been found necessary. Cover the entire preform with veil mat if resin mix is to be poured onto the opposite side.

3. Weigh or meter the resin mix and pour it onto the preform. A pour pattern should be predetermined through practice and rigidly adhered to for each part to be molded, and for the entire run of the particular part. Pour-

ing the resin mix onto the preform may be carried out either in the mold or prior to placement of the preform into the mold. The pour pattern should be made essentially in the form of an "X" or cross from corner to corner so that when the press is closed, the resin mix will flow evenly throughout the entire reinforcement mass without trapping air and producing "islands" or air voids. Much can be learned about the proper pour pattern by trial and error with a specific part, or by drawing on experience gained with prior parts molded. The ultimate pour pattern depends upon the shape, total size, or area, and complexity of curvature of the part to be molded. The necessity for a long gel-time in the mold may be eliminated by making the pour outside the press. If the resin-pour side is the finished part side, place veil mat over the top before closing the press.

4. Fast close of the pressmold is made at 300 in. per min. Slow close for the last ¼ in. is made at 3 in. per min. (total slow close range provided for in the press should be 1 to 10 in. per min.). The slow close is made to avoid jetting of the fluid resin mix and also the pulling of fibers out of position at the pinchoff.

5. In molding, a pressure of 250 to 400 psi is exerted against stops built into the mold. Back pressure is created by expansion of the mass of resin and glass suddenly increasing in temperature, and also by the exotherm when it occurs.

6. Allow a cure time of 1.0 to 3 min. for small parts and up to 20 min. for parts in the range of 100 sq ft or larger in area.

7. When the press opens automatically at the end of the timed cycle, the part may be removed with assistance of an air blast, brass chisels, suction-type lifters and the like. If the pressmold is run at a temperature in the range 220 to 300°F, the part, for greater ease in removal from the mold, may be made to stick to the mold half which is run 5 to 10°F cooler than the other half.

8. Automate if desired or feasible by introducing resin through a rotating jet for deep draw parts or by a T-bar pipe or manifold with jets 2 in. apart for flat parts.

9. Devise supporting fixtures for loading the preforms and unloading the molded parts. It is more judicious to design loading fixtures for large rather than small parts. Small parts cure quicker and require less handling.

10. Preform molded parts out of the mold may be satisfactory as molded, but may also require cooling in a fixture to prevent warpage. Surface finish and repair are necessary if the part is to be painted, but not usually if low shrink resins are used. Trimming, deburring and drilling may be combined into one operation to save time.

11. Carry out full inspection and testing.

Inspection and Testing. Quality control sheets should be made up from the job specifications originally agreed upon and prepared for estimating the job. These sheets should include checklists for the following:

- *Surface Finish.* Absence of voids in surface, and the condition as visually inspected and as measured with a profilometer (if required).
- *Condition of Shear-edge.* Removal of flash; absence of loose fibers, torn laminate, and voids.
- *Part Shape.* Measurements of height, width, and length using a surface plate and gages; thickness and distribution using a jaw-type fixture with dial gages, warpage using prepared dimensional-tolerance fixtures, preferably step gages.

Laminate parts may be cut and destructively tested for any of several specific parameters as outlined in a specification for the part. Consult ASTM and U.S. Government tests applicable to the testing of RP/C.

Table IV-2.12 Variations of % Glass Reinforcement Vs. Glass Weight per Sq. Ft.

% Glass Concentration	Glass Weight (oz) per sq ft for 0.100 in. Molded Part (1.0 lb per sq ft)	Glass Weight (oz) per sq ft for 0.125 in. Molded Part (1.3 lb sq ft)
20	3.2	4.2
35	5.6	7.3
50	8.0	10.2

Table IV-2.12 presents the normal variations of weight of glass reinforcement per sq ft of laminate for various glass concentrations.

A trouble shooter's guide for the correction and elimination of defects in preform and mat molding is presented in Table IV-2.13.

Table IV-2.13 Trouble Shooter's Guide to Polyester Matched-Die Molding

Defect	Description	Possible Causes	Suggested Remedies
Blister	Round elevation on the surface somewhat resembling a blister on the human skin. (Delamination within the molding.).	*Undercure.* Blisters extend over considerable area.	Increase molding time. Increase concentration of catalyst.
		Expanding vapor. Moisture, solvents, or entrapped air.	Dry pre-form thoroughly. Avoid surrounding pockets of air with resin.
Crazing	Fine cracks in resin. May extend in a network over the surface or through the molding.	*Highly reactive resin.* During polymerization, the reaction mixture develops a peak exothermic temperature in proportion to the degree of unsaturation of the base polyester resin. Resins having a high degree of unsaturation tend to cure rapidly, developing high exothermic temperatures. In rigid systems, then, the stresses caused by the shrinkage accompanying polymerization, and the expansion due to heat, are likely to produce crazing even in areas which are uniformly filled with glass fibers.	Reduce catalyst concentration. Curing time is increased slightly, but the reaction period is lengthened, thus decreasing the exothermic temperature developed. Reduce molding temperature. Rate of polymerization is reduced, decreasing the exothermic temperature. Add inert filler. The quantity of reactive material per unit volume is thereby decreased and, with it, the exothermic temperature. Add flexible resin. Such resins are generally less reactive than are rigid resins, hence have the same effect as an inert filler on the exothermic temperature. Toughness of the curing resin is increased, also. Dilute with less styrene. Addition of styrene tends to make the resin more reactive, hence raises the exothermic temperature. Shrinkage is also increased by addition of styrene, which contributes to crazing.
		Resin-rich (glass-starved) areas. Resin-rich (glass-starved) areas are likely to show crazing even with resins which are only moderately reactive because of the stresses set up by the concentration of resin.	Improve uniformity of preforms. If the preform has thin sections, adjustments should be made to the preform screen and/or the plenum chamber to provide better fiber distribution. If distribution of glass fibers cannot be improved, the procedures indicated in connection with *Highly reactive resin* may help to eliminate crazing. Resin-rich areas are frequently the result of washing of the preform. See notes on *Washing of Preform* for details.
		Undercure (see *Odor*).	

Odor	Styrene odor, which differs from the normal odor of completely cured polyester resin. Incomplete cure is generally accompanied by low hardness, poor strength, etc.	*Undercure.*	Increase curing time. Increase concentration of catalyst. Increase molding temperature.
		Inhibition leading to undercure.	Examine pigments for inhibitory effect. If they act as inhibitors, they should be eliminated or additional catalyst should be added. Examine extenders: they may act similarly.
	Odor of benzaldehyde, i.e., sweet, cherry-like odor. Often confused with styrene odor, although the two are distinctly different.	*Side-reaction involving the oxidation of styrene to benzaldehyde.* The odor varies with the nature of the base resin. Less reactive resins, in general, tend toward the benzaldehyde odor; more reactive (greater unsaturation) resins posses only a very slight odor of benzaldehyde.	Reduce concentration of catalyst. Use a more reactive resin. Lower the mold temperature. Post-bake in air oven at 250°F. to get rid of residue odor.
Pitting	Regular or irregular holes on the surface of the molding.	*Trapped air.*	Improve fit at cut-off. A close fit—0.002 to 0.004-inch—at the cut-off restricts the flow of excess resin from the mold. A close fit increases the hydraulic pressure in the system, which reduces bubbles to an insignificant size or forces them into solution.
		Air in resin mix.	Allow mix to stand before molding. Reduce viscosity of mix by adding styrene (provided other properties are not altered). Degas the resin.
		Improper molding temperature.	Raise or lower temperature until best results obtained. Use differential temperatures. A 5 to 20°F. differential between halves of molds will reduce fiber patterns and increase gloss on hotter side.
Prominent fiber pattern	Excessive prominence of glass fiber pattern on surface of molding.	*Character of resin.*	Resin may distort at too low a temperature. Such resins generally display fiber patterns more prominently. Lower molding temperature may help.
		Coarse preform.	Use a surfacing mat. The coarse pattern of glass fiber obtained with the standard glass mat or preforms can often be so reduced.

Table IV-2.13 Trouble Shooter's Guide to Polyester Matched-Die Molding (Cont.)

Defect	Description	Possible Causes	Suggested Remedies
Resin-starved areas	An area over which the reinforcing material has not become impregnated or an area containing very low concentrations of resin—usually due to an excess of reinforcing material.	*Poor flow.* When pressure is applied to the resin as the press closes, the resin takes the path of least resistance, flowing into the lowest pressure areas.	Lower viscosity. Resin mixes of low viscosity offer less resistance to flow and, therefore, are more likely to flow into high pressure areas, i.e., areas containing more glass fibers. See *High viscosity* under *Washing.*
		Early gelation. Occasionally, very reactive or highly catalyzed resin will gel before the resin has wet out sections of the glass fiber preform.	Reduce concentration of catalyst, lower mold temperature, or add inhibitor. (Either of these practices will reduce the speed of reaction and thus give the glass more time to wet out.)
		Excess glass fibers. Areas in which there exists an excess of glass fibers produce high-pressure areas, which are slow to wet out. Even if they do wet out completely, fibers or crushed glass might become prominent.	Use more uniform preforms.
		Poor cut-off. A reasonably close fit at the cut-off in a matched metal mold is required so that back-pressure will cause the remainder of the preform to fill out when the resin has flowed to one extremity of the preform.	Improve fit at cut-off. Examine distribution of resin on preform. Use an excess charge of resin.
Resin-rich areas	Area filled with resin mix having little or no reinforcing material. (May craze during cure and fail in service.) (See *Resin-rich* section under *Crazing.*)	*Poor design.*	Improve design. The simplest design for easy molding has walls of uniform thickness and no sharp corners. When variations in wall thickness are required, they should either be gradual or so designed to provide for additional glass fibers in the preform. Otherwise, additional glass fibers must be added to resin-rich areas.
Voids	Numerous, small air bubbles trapped in the resin.	See *Pitting.*	See *Pitting.*

Defect	Cause	Remedy
Distortion of molding from the mold dimensions.	Unbalanced Construction. Since the thermal coefficient of expansion of polyester resin is about 10 times that of glass fiber, non-uniform distribution of resin and fiber will tend to warp the construction toward the resin-rich side.	Balance the construction by more uniform glass distribution. Use cooling jig to hold the molded piece to the desired dimensions. Use less styrene to reduce shrinkage due to polymerization. Use inert filler to reduce overall shrinkage due to polymerization. Use a lower molding temperature to reduce thermal shrinkage after curing.
Warping	Uneven cure. If a molded piece is not cured at the same rate from both sides of the construction, it tends to warp toward the side which cures first.	Adjust both of the mold surfaces to the same temperature. Eliminate hot spots on the mold surface.
	Design. Any curved surface of a molded piece tends to shrink to a curvature of shorter radius as it cools. This phenomenon accounts for the sides of box-shapes bowing in, and for pipe closing when split longitudinally.	Use cooling jigs. Use resin with lower heat-distortion temperature. Use as large radii as possible and/or stiffen sides of boxes with extra thickness or metal reinforcement.
Abnormal tearing or displacement of reinforcing material during molding.	High-viscosity resin mix. A high-viscosity mix offers too much resistance to flow, which results in displacement of the glass fibers in the preform.	Slow down press closing speed. Add styrene. Reduce amount of inert filler. Use filler of lower oil number.
	Early gelation. Occasionally, a very reactive or highly catalyzed resin will gel before the press has completely closed and before the resin has filled out the preform. Result: washing.	Reduce concentration of catalyst. This increases gel time, thus allows for a bigger period of flow. Add inhibitor. Small quantities will increase the gel time without seriously affecting the curing rate after gelation.
Washing of preform	Poor preform. Preforms having an excess of loose glass fibers, resulting from insufficient binder, will display washing. Washing also will occur if the preform binder is soluble in the resin system to the extent that it dissolves before the resin flow has ceased.	Adjust binder distribution. Where washing has resulted from loose fibers, additional binder is required. Where a fluffy preform is demanded, additional binder should be applied to the surface fibers only.

Source: Rohm and Haas Co., Philadelphia, Penn.
Source: Paraplex, P-Series Polyester Resins Technical Data, Rohm and Haas Co., (August 1956).

Post-Molding Finishing Operations. Finishing methods or processes for preform matched-die molded RP/C parts generally include machining (drilling, punching, and so on), surface finishing (sanding, filling, priming, painting), bonding (mechanically and chemically fastening or assembling molded parts together).

Machining. A complete treatise on the mechanical post-treatment of matched-die parts and RP/C in general is presented in the *Handbook of Reinforced Plastics of The SPI* (Oleesky and Mohr. New York: Van Nostrand Reinhold, 1964, p. 381). Additional information has also been published: *Plastics Design and Processing*, September, 1968, page 24, and *Plastics Technology*, June, 1968, page 49. Very little new technology has been added, except for eliminating the necessity for finishing.[15]

Surface Finishing. The advent of more detailed and accurate methods of surface-profile measurement have brought about an increased ability to study surfaces and evaluate the effectiveness of various finishing methods. Surface variations across 0.100 in. span in typical matched die molded parts are listed in table IV-2.14.

There are several types of surface profiles that must be considered in matched-die RP/C molded parts:

- Short term roughness is measured across 0.010 in. increments. It is caused by mold marks, tooling scratches, veil mat fibers, pits, and other imperfections.
- The average surface roughness is measured across 0.100 in. increments. The causes are glass prominence and molding deficiencies (laking, for example).
- Average surface waviness is measured across increments greater than 0.100 in. and is caused by differences in thermal expansion, differential thermal expansion between glass and resin, part warpage, untrue mold surface assembly induced stresses, and so on.

Assembly methods and techniques. Reinforced plastic parts may be assembled to themselves or to metal parts by any one or a combination of the three common methods, namely riveting, bolting or bonding. Each has its advantages and disadvantages as noted below.

Riveting. Rivets are used for permanent fastenings where the joint has only limited tensile or peeling loads, but is primarily working in shear. Rivets are also used with a bonded

Table IV-2.14   Surface Variations in Typical Matched-Die Molded Parts (0.100 in. span)

Matched-Die Molded Part	0.100 in. Increment Range Microinches
Unsanded matched-die laminate	300–1250
Exposed glass strands on laminate surface	1000–2500
Properly sanded with hard block backing for 120 grit paper	125–250
Finished grade laminate or part after sanding, painting and buffing	25–50 (no peaks)
Low-profile resin, as matched-die molded	25–50
Automotive grade steel (unfinished)	250–500
Premix molded with conventional resin	200–500
Premix molded using low-profile resin	50–200

Notes: Priming and painting requires a starting surface of not over 250 microinches to provide an acceptable final finish of the automotive type. (1 microinch = 1×10^{-6} in.)

After establishing surface smoothness under 250 microinches, many varied finishes may be applied, such as graining, or painting. Priming and painting require bake cycles up to 300°F, which may disturb an improperly filled surface.

Also available is a well-developed system for optical evaluation of surface waviness in molded parts, in which observation is made of the distortion of an illuminated grid reflected from the molded part surface. Selection of control standards makes possible classification into several grades for comparison and evaluation (see SPI 23rd RP/C Proceedings, 1968, Sec. 1-C, Kralovic, W.).

joint to increase joint strength and at bond ends to prevent the start of peeling failures. Aluminum rivets of $\frac{1}{8}''$ to $\frac{3}{16}''$ diameter with large $\frac{3}{8}''$ to $\frac{9}{16}''$ diameter heads are generally used since high-shear steel is not effective in RP/C. Aluminum forms easier and the large heads spread the head load. Although the head may be formed by hammering, it is best formed by a rivet squeezer because of better control and less chance of damaging the assembly. Be sure the rivet hole is not too large, the large head or a washer is against the RP/C panel and the rivet squeezer is properly set for a good formed head. Rivets are commonly used to hold and locate thick metal reinforcing nut plates.

Bolting. Bolts are used for take apart joints

wetting was accomplished in a large molding operation. Although fast-curing resins shortened curing cycles from 90 to 30 sec., press-open times were not essentially shortened. An analysis of the entire 8-stage press cycle was therefore made (see Table IV-2.15) and steps made to establish improvements.

Automatic wetting of preforms outside the press was carried out. Unloading jigs and fixtures were developed and constructed next to each press unit. Automatically wet mat was positioned on the mold and molded parts were automatically removed. The results were a reduction of the mold-open time to 25 sec. per cycle and production rates of 60 parts per hr. from each single mold cavity.

Variations Benefiting Preform Matched-Die Molding. Many interesting material improvements, solutions to difficult molding problems, and innovations in the molding processes have been accumulated over the years. These novel methods have all resulted in improvements to the process and to the quality and performance of the finished products molded. The methods include resin improvements, hot gel coating, and improvements in the molding processes themselves.

Resin Improvements. Polyester resins in which the styrene monomer was substituted 100% by orthochlorostyrene showed approximately 10 times the reactivity in curing (same catalyst content) as normal styrene resins in matched-die molding.[30] The use of orthochlorostyrene-modified polyesters in matched-die molding shows potential for:

- Extremely rapid molding cycles with resultant increased production rates and economic benefits.
- Considerably higher production volumes requiring 2 sets of matched-die tools.
- Greater physical strengths and modulus of elasticity and therefore lower formulation cost.
- Dimensional stability, resistance to cracking and improved part surface finish resulting in less warping out of the press and lower assembly and finishing costs.

The effect of using this material on mold scumming, mold release, and a full range of molding conditions has not yet been determined.

The addition of complex halogenated anhydrides plus selected fillers such as hydrated alumina $(Al_2(OH)_3)$ to polyester resins or matched-die formulations renders them acceptably nonburning or fire retardant, when used as matched-die molding resins for preform molding.[31, 32]

Hot-Mold Gel Coats. A gel coat formulation based on a polyester resin plus catalyst and fillers is sprayed (airless spray preferred) on a hot matched-die mold and rapidly cured. Resin-reinforcement composite is then added, the press closed, and the molding cycle is completed.[33] The gel coat forms a nonporous film and protective coating by welding or adhering itself to the laminate during the molding cycle.

Hot gel coats are used mostly for large, flat parts. The use of SMC shows some potential for precluding the necessity for hot-mold gel coats.

The major steps for application of hot-mold gel coats follow:

- Hot gel coats are supplied as proprietary formulations. However, they generally comprise 50 to 60% resin, 0.5 to 0.8% BPO catalyst, 10-12% white pigment, and 30 to 40% inert filler. Fillers should be designed to prevent resin shrinkage and eliminate fiber pattern. An internal mold-release agent should be included for high-contour parts.
- Application viscosity should be 3000 to 6000 CPS (Brookfield, #4 Spindle).
- Preferred mold temperatures are 230° to 270°F. Temperatures of 280° to 310°F. require 0.3 to 0.5% BPO.
- A good mold polish is desirable and chrome plating is preferred.
- It is advisable to mask peripheral off-mold areas to protect against overspray.
- The improved technology of airless spray is responsible for making hot-mold gel coating systems available to matched-die molders. For effective atomization, airless spray requires a fluid pressure of 3600 lb per sq in. This is accomplished with a 40:1 pump ratio and 90 psi air pressure. The pump system must be well filtered to prevent clogging of gel coat particles. For best results the fluid pressure exerted impinges the material onto a 0.009 in. nozzle orifice with a 20 to 60° spread or fan. The spray rate is variable from fractional amounts up to 6 gal. per min. at 1500 psi.
- The film laid down should be uniform and 0.015 in. thick (gaged frequently).
- The rate of gel coat material delivered should not exceed the capability of sprayer to cover

the mold. Approximately 5 sec. of spray time per sq ft will be required.

- Cure-time requires 5 to 30 sec. before the molding may proceed. Two similar molds should therefore be reciprocally worked.
- Normal cure time and part removal may proceed as usual. Some extra time may be required to clean up the overspray and change the masking.

There are many advantages in using hot-mold gel coats:

- It eliminates necessity for internal resin coloring.
- It provides higher gloss and lower fiber pattern.
- It allows the use of dirty or discolored molding mat or preforms. This is important for pastel shades.
- It eliminates 15 to 20 cents per ft. which would otherwise be required for filling, sanding, and painting. In some cases, hot-mold gel coating may be automated—large flat parts, for example.
- Chemical resistance may be imparted to an entire molding or panel by using chemically-resistant resins in the hot gel coat only. Weathering resistance is also improved. A resin may be used which would serve as a base or primer to be painted over, if required.

Among the disadvantages of hot-mold gel coating are the following:

- Over-catalyzation results in tearing, cracking, blistering, and poor adhesion.
- Under-catalyzation causes washing and a poor, incompletely covered surface.
- The gel coat thickness must be absolutely uniform. Tearing and cracking results in thick areas and tearing and washing (loosening) may result in thinly applied areas.
- It is difficult to spray and get uniform coating in molds that are highly contoured and have inaccessable portions.
- The cost of 3 to 7 cents per sq ft for hot-mold gel coating is expensive and higher than internal coloring.
- The initial equipment outlay is a minimum of $1000.
- It requires masking, an abnormal clean-up of overspray, and a frequent change of mold masking.
- It hides fiber pattern partially, but not completely. Some post-finishing prior to painting may therefore be necessary.

Molding with Retractible Stops. Considerable work has been carried out to attempt an improvement in the molded surface character of matched-die molded parts. The porosity in molded parts was thought to be caused by the loss of pressure on the molding in the press; loss of pressure results in nonfilling, pits, voids, blisters, and variations in thickness within the part.

An improvement was accomplished by retracting the press stops or lands after the press had closed and gelation of the resin had taken place.[34, 35.] The theory behind this method was to maintain positive pressure on the molded part at all times, even after gelation. Surface evenness and freedom from porosity were positively improved, but at a great expense due to the equipment cost.

The advent of low-profile, low-shrink resins plus preparation and exploitation of SMC materials may have filled the gap that existed when consideration was being given to the possible benefits of this molding method.

Vacuum-Compression Molding. The elimination of air from a material charge during or immediately prior to molding has the benefits of minimizing trapped air, lowering necessary molding pressure, eliminating the need for antioxidants from some resins, and reducing stress-cracking. Denser, more uniform parts result.

Vacuum-compression molding units have been devised and put into operation for specific applications where densification and elimination of voids were an absolute requirement.

With this type of tooling, molding cycles are somewhat longer. Both thermoplastics and thermosets may be molded using the technique.[36, 37.]

Cold-Press Forming. Reinforced phenoxy, polypropylene, styrene-acrylonitrile and PVC (all thermoplastic and also cross-linked thermoset PVC) resin compounds (moldable sheets) have been developed. These materials may be molded as follows:

1. Place material in a heated chamber.
2. After arriving at the proper molding and flow temperatures, place material immediately into a cold die in a punch or compression press.
3. Stamp or close press with the alacrity and force necessary turn out satisfactory molded parts.
4. Finish or trim to size. Molds are not al-

ways built for pinch-off capability.[38, 39, 40].

(*See also* Chapter V-3, Cold-Forming).

Typical reinforced, moldable PVC sheet contains 25% 2 in. chopped fiber glass reinforcement. The PVC used has approximately 6 min. of thermal life (before molding) at 390°F. and up to 50 min. at 320°F. The sheet is made up in thickness of 0.025 in. Multiple-thickness layers are used to achieve part-thickness requirements.

A material flow of 300% at 2850 psi forming pressure and 390°F is possible. Optimum flow conditions occur at lower pressures and temperatures. A fairly good finished-part appearance can be attained at 2000 psi and 374°F. The heat deflection temperature of the PVC is approximately 212°F. Hence, 18 sec. are needed after press close for a 0.150 in. part originally heated to 356°F and 7 sec for a 0.075 in. part.

The resultant physical properties are slightly lower than those obtained with reinforced polyesters. Some advantages are lower press and mold costs and faster cycles, despite the requirement for post-finishing.

Summary of Preform Molding Process. It is fitting to conclude the discussion of the preform process for molding in matched dies by illustrating a typical and representative plant layout (See Figure IV-2.1). Information and data concerning product performance are presented at the conclusion of the chapter following the discussion of SMC.

THE SMC PROCESS IN MATCHED-DIE MOLDING

SMC is made possible by the interaction after mixing of certain specific chemicals with an unsaturated polyester resin to increase its molecular weight without gelation but produce a 1000-fold viscosity increase that greatly enhances material handling and also ease and efficiency of molding. Hence, SMC brings together the advantages of handleability and ease of molding that characterize premix plus the higher strengths of preform molding and improved surface properties especially when used with low profile resins.

Briefly, the steps in the formulation and molding of RP/C using SMC comprise:

- Selection of specific resins, catalysts, fillers (plus thickeners), and reinforcements.

- Precombining the resin and fillers and delivering them onto a moving belt or carrying membrane.

- Chopping glass fiber reinforcement into the resin mix, covering with a top film and kneading to eliminate entrapped air.

- Allowing time to pass for the resin to thicken to a moldable consistency.

- Weighing, removing the film, charging the SMC to the press, and carrying out the complete molding cycle as for preform molding.

General advantages such as consolidation of parts, low tooling costs, molded-in ribs and bosses, greatly improved and paintable surface quality plus automated processing and other economics in material handling and in press-molding time are some of the benefits derived from using SMC. The process is fairly new and is undergoing continual change and improvement. The prognosis is good and use of the material should flourish in coming years.

MATERIALS

Resins. Although general purpose resins are usable with SMC, most of the present action and future potential centers around low shrink or low profile resins. Hence, it is of interest to discuss briefly the technology of this important new concept. Low shrink resins are a combination of two or more but usually two different components or phases—a liquid thermosetting type like a high-reactivity polyester, and a finely dispersed or dissolved thermoplastic which may include polyethylene, polystyrene, polymethylmethacrylate, copolymers of styrene, copolymers of various methacrylates, polyvinylacetates, polycaprolactones, PVC, cellulose acetate butyrate, cellulose acetate propionate, and others. The preferred molecular weight is in the range of 25,000 to 500,000.[41] The resins are usually sold in a two part system to be mixed at the operation. One-component systems are also marketed. The thermoplastic phase is the discontinuous one and exists following cure as fine particles dispersed in the polyester or continuous phase. While in the uncured state both phases must have a common monomer, not necessarily styrene. The 2-part systems provide more flexibility because the ratio of each can be controlled in mixing for better control of post molded shrink. From 5 to 40 parts of thermoplastic resin in the poly-

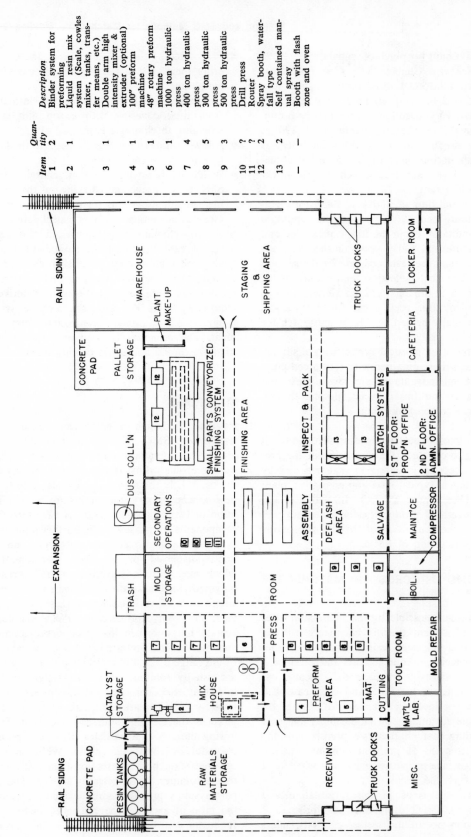

Item	Quantity	Description
1	2	Binder system for preforming
2	1	Liquid resin mix system (Scale, cowles mixer, tanks, transfer means, etc.)
3	1	Double arm high intensity mixer & extruder (optional)
4	1	100" preform machine
5	1	48" rotary preform machine
6	1	1000 ton hydraulic press
7	4	400 ton hydraulic press
8	5	300 ton hydraulic press
9	3	500 ton hydraulic press
10	?	Drill press
11	?	Router
12	2	Spray booth, waterfall type
13	2	Self contained manual spray
—	—	Booth with flash zone and oven

Figure IV-2.1 Typical plant layout for the preform process in matched-die molding. (Drawing prepared by R. W. Meyer, courtesy of General Tire and Rubber Co., Akron, Ohio.)

ester phase represents the total range. Two component systems were necessary at the outstart due to incompatibility and the necessity to keep both phases under positive agitation after mixing to prevent separation prior to molding.

Early problems with low shrink systems were efficiency of resin usage, exuding, or phase separation after addition of thickening agents, scumming during molding, and poor paint adhesion. The thickening problems have been solved by incorporating a modifier for the thermoplastic phase similar to that for the continuous polyester phase so that both resins take part in the molecular weight readjustments when the thickener is added.[42]

By what subtle mechanism does the phenomenon of low shrink take place in the co-curing of the 2 phases in these resins? By a very interesting one. For comparison, general purpose polyester resins, when studied in an adiabatic system with temperature and pressure controls to simulate molding conditions, and means for determining expansion and contraction at the exact time they occur, show a sudden thermal expansion when polymerization and accompanying exotherm occur. This is followed immediately by an almost equal and opposite thermal contraction or shrinkage, so that when the system finally cools to room temperature, the final dimensions are *7% less* than the initial volume.

In the case of the two-phase low shrink resins, the same initial strong thermal expansion occurs except that the resin remains essentially distended and does not contract in any significant degree until a delayed polymerization is complete, and cooling commences. As room temperature is approached, the two-phase resin remains at approximately *3% greater* than the initial dimensions.

In explanation of this anomalous but welcomed behavior, it is postulated that when the 2-phase system is subjected to heat under curing conditions, the continuous phase (highly reactive unsaturated polyester) cross-links rapidly and, expanding, liberates considerable exothermic heat. The dispersed droplets of the thermoplastic resin also polymerize but do so more slowly because the principal reaction is the homopolymerization from the suspending monomer. This means that the continuous phase gels and becomes crosslinked in the early stages of cure, but the thermoplastic

droplets expand due to the exothermic heat and arrive at a suspended state, being locked or frozen into the matrix of the continuous phase. The droplets continue to polymerize and give off exothermic heat to support the distended state, but at a slower rate than that of the continuous phase.[43,44]

There are three observable facts which strongly support this analysis. The first is that the resin droplets of the discontinuous phase, readily visible in the microscope at 500 to 600 diameters, appear to have internal voids or a foamlike structure, indicating that after they had done the work of permanently distending the entire resin structure, they have attempted to shrink when the mass cooled to room temperature, but were inhibited by their already rigid boundaries. The second supporting fact is two-part: first, when these 2-phase low shrink resins are cured in hand lay-up room temperature cure systems, there is no evidence of anything but normal general purpose resin type shrinkage because the rapid initial exotherm and sudden thermal expansion are not present; secondly, in areas of a part to be press-molded using the low-shrink resins classed as ribs and bosses, complete freedom from contraction of the resin in reaching room temperature is never fully realized, and sink marks result. It is believed that the resin mass within the central area of a deep rib is insulated from the high initial heat shock provided by the pressmold, and the sudden necessary high rate of polymerization and rapid expansion of the continuous resin phase are masked. The third supporting fact is the positive evidence of the sustained expansion and distended condition of the low shrink resin mass in forcing the pressmold slightly open soon after press-closing and arrival of the resin mass at the mold temperature and start of the protracted polymerization.

In the case of the one component low shrink resin systems (a more recent development), the ultimate result after curing is the same as the resins supplied in two components. There are two major differences: (1) the two incompatible phases are made permanently miscible with common monomers, wetting agents, or other means, and (2) upon curing they perform in a manner identical to their first generation cousins in extending the polymerization and limiting initial shrinkage, except that, after curing, the incompatible phase exists as very minute droplets that are substantially smaller in size and require more sophisticated means for observation.[45]

Catalysts. Catalysts for use and best results with SMC molding systems remain the high temperature forms, *t*-butyl perbenzoate and *t*-butyl hydroperoxide. The preferred molding temperature range is 295 to 315°F.

Other catalysts have been evaluated at lower molding temperatures including *t*-butyl peroctoate (260 to 280°F), 2, 5-dimethyl peroxy hexane (260 to 280°F), and *t*-butyl peroxy pivalate (220 to 240°F). Questionable cures and sink marks plus poor surfaces resulted in each case, and reduced mixed batch potlife resulted in each case but the latter.

Contrary to anticipation, benzoyl peroxide catalyst is not recommended for use in SMC systems with low shrink resins. Not only is the mixed pot life unduly short, but in molding, sink marks are exaggerated, and surface profile is worsened.

Some improvement in and rounding out of catalyst function and performance is expected by blending of top-performing catalysts, such as *t*-butyl perbenzoate and *t*-butyl peroctoate.

Fillers. Calcium carbonate fillers have proven most effective as resin extenders and molding and surface enhancers in SMC. The 2-micron and 10 to 40 μ particle sizes are the most commonly used, with the 2-micron size preferred to resist settling and avoid separation during molding.

The use of aluminum hydrate, 325 mesh grade, renders an SMC molded part satisfactorily self extinguishing. Small quantities of a 0.4 μ filler such as bentonite clay or calcium carbonate assist in keeping the aluminum hydrate in suspension during processing. This material is approximately three times the cost of calcium carbonate.

Some applications such as resistance to specific acids are better serviced by use of kaolin clays as filler instead of calcium carbonate. Almost all fillers are coated to prevent high absorption of the resin.

Thickeners. The effect of certain alkaline earth oxides and hydroxides and many other specific chemicals in inducing a real and workable thickening in polyester resins is not fully understood. What is understood is the visible physical changes that occur to form an improved molding compound. The difunctional earth oxides or hydroxides react with the terminal carboxyl groups on the polyester chains resulting in doubling or tripling the average molecular weight of the resin. This bonding is not permanent, but diminishes only slightly when the thickened compound is subjected to press temperatures in molding. The thickened resins have their own built-in back pressure, and fill the mold cavity at a uniform rate, driving air and volatiles out to prevent voids and porosity. The result of this behavior is that the molding material is held securely in place until catalysis takes over and the resin gels, and the sustained thermal expansion due to the low shrink components occurs. The result is a hard compact and dense molded structure with superior molded physical properties.[44]

The viscosity changes are measurable by viscosimeter and production monitoring should be set up on an inflexible time schedule through the course of mixing and use of the thickened compound batch or rolls. The viscosity changes pass through ranges of from 1000 to 3500 centipoise to anywhere between 10 to 60 million centipoise after final thickening. These are measured respectively by #6 and #T-F T-bar spindles in a Brookfield HBT viscosimeter run at 5 rpm and 1 rpm (see Appendix I). Samples of each resin-filler-thickener mix should be taken and stored in a jar together with the resultant SMC rolls, and viscosity determinations taken up until the time the material is used in molding. Penetrometer measurements to determine viscosity are also made on the finished SMC mixes (without reinforcement) but Brookfield viscosity measurement of the resin batches in beakers or jars should also be relied upon.

The real goal in formulating SMC mixes is to establish a rapid thickening rate as soon as the material comes off the end of the machine, but not before the fiber glass reinforcement has been incorporated, and also to be able to exert a definite controllable end point to proper thickened state for optimum moldability without having the material continue to thicken beyond the usable state. Aging is the only reliable recourse, because this technology is not sufficiently advanced to properly coordinate low working viscosity with immediate establishment of high molding viscosity. Whereas the development of molding viscosity may take from 1 to 2 weeks at room temperature, it will occur in 2 to 3 days at 90°F.

Although many chemicals have been observed to induce thickening as described in the patent literature, the effects of 3 systems only will be presented here. Table IV-2.16 illustrates the action of oxides and hydroxides of Ca and Mg, and Table IV-2.17 for cyclic acids and anhydrides in combination with MgO. Table IV-2.18 lists changes induced by LiCl in combination with MgO. The Ca and Mg oxide plus hydroxide systems are in widest present-day usage. Undoubtedly other effective proprietary systems are known and in use. The Ca and Mg materials increase the rate of thickening, but do not achieve maximum possible viscosity, while acids and anhydrides permit attainment of higher ultimate viscosities in the compounded sheet.

The viscosity values presented in the tables resulted from resin plus thickener only with no filler added. Actual viscosity guidelines for mixing and feeding resin paste to an SMC machine are: starting resin—1000 to 3000 cps; resin plus filler and thickening agent immediately after mixing—20,000 cps; starting resin-paste viscosity at the SMC machine at the point of adding chopped glass—20,000 to 70,000 cps; Optimum molding viscosity after aging of the compound—20,000,000 cps (range 10 to 60 million cps); maximum measurable viscosity—160 million cps. Compounds with viscosities up to 160 million cps have been molded, however, with increasing difficulty. The higher the viscosity the higher the molding pressure required.

Many "gremlin" type of problems exist with thickeners for polyesters used in SMC. It is necessary to wrap finished rolls of SMC in a non-permeable membrane and tape closed to prevent loss of volatile monomer, and contact with moisture-ladened air. The moisture level of fillers, thickening ingredients, film, and other materials coming in contact with the sheet must be controlled at a low minimum. Amounts of as little as 1% of polar compounds such as free water, alcohol, glycols, and acids affect the thickening response of polyesters. In addition, the following are also critical: type resin and manufacturing control, types of mixing equipment and thoroughness of mixing (without overheating), resin paste temperature at the doctor blade, and the SMC storage temperature (see discussion of Equipment and Tooling.)

While SMC technology has made possible a real growth within the RP/C industry, it has created a few problems. Some of the early experience with defects such as molded surface porosity, non-filling, etc., showed that presses with low-pressure molding capability were not adequate, and that it was necessary to acquire presses with tonnages up to 4000 to provide molding pressures in excess of 1000 psi. It is the present goal of pioneers and researchers in this field to bring to light materials which will not thicken polyesters into a leather-hard state but into a softer synthetic gel so that, in molding, only 500 psi pressure is required, voids are eliminated from the fabricated sheet and molded components, and lower pressures are possible to cause the material to perform in the same manner as currently in creating its own back-pressure and adequately filling the mold.[46]

Carriers for Thickening Chemicals. In order to better stabilize the action of chemical thickeners, keeping out undesirable moisture accumulation, and avoiding extreme viscosity variations, use of carriers for the thickening agents has been found effective. Acrylic and styrene syrups and other non-thickening, unsaturated polyester resins without monomer are used, and all have been found inert to the alkaline earth oxide or hydroxide thickening agents.

A typical carrier mix would consist of carrier—55%, 2-micron calcium carbonate—40%, and magnesium oxide thickener—5%. This mix would be blended at whatever concentration required in order to introduce the correct amount of thickener.

Mold Release Agents. Zinc Stearate at 1½ to 2% of the total SMC batch weight provides the most satisfactory mold release agent. It does not interfere with painting of molded parts, but has a tendency for foaming during mixing.[47]

Calcium stearate has been found effective at one-third the amount of zinc stearate required. It is limited in the upper percentage range, however, as 1½% causes mold scumming during molding, and reduces molded surface gloss.

Phosphate based internal release agents are also used effectively.

Colorants for SMC. Color problems for SMC are slightly more critical than for preform matched die molding. Whereas the same

Table IV-2.16 Thickening Effect of Ca and Mg Oxides and Hydroxides and Combinations in General-Purpose Polyester Resin

Combination No.	Material	% Added to Polyester	Resin Viscosity Change cps (Initial = 1650 cps)				Remarks
			1 hr	24 hr	1 wk	2 wk	
1	CaO	2.8 (6.0 same)	1650	1650	1650	1650	No thickening action
2	Ca(OH)$_2$	6.0	1650	—	14.7×10^6	—	Too rapid initial thickening
3	MgO	3.6	6500	1.9×10^6	10.2×10^6	—	Too rapid initial thickening
4	Mg(OH)$_2$	5.2	1650	—	0.5×10^6	0.6×10^6	—
5	CaO Ca(OH)$_2$	3.0 2.0	1500 1600 (3 hr)	0.65×10^6	8.6×10^6	—	Slow thickening
6	CaO Ca(OH)$_2$	1.0 4.0	1850 68000 (3 hr)	0.55×10^6	5.1×10^6	—	Fast thickening
7	CaO MgO	2.5 1.8	3320	1.1×10^6	4.20×10^6	—	Heaviest thickening
8	CaO Mg(OH)$_2$	3.0 2.6	1650	—	—	0.38×10^6	—

Table IV-2.17 Thickening Effect of Cyclic Anhydrides in Polyester Resin Containing 2% MgO

Combination No.	Type Anhydride Material	% Added to Polyester	Viscosity Change, cps			Remarks
			24 hr	1 wk	2 wk	
1	-0- (MgO only)	0	30,000	1.4×10^6	13.6×10^6	—
2	Maleic	1.0	9700	0.03×10^6	0.2×10^6	—
3	Maleic	3.0	4900	5800	8000	Inhibits
4	Tetrahydrophthalic	3.0	52,000	25.3×10^6	64.0×10^6	—
5	Phthalic	3.0	42,500	16.4×10^6	64.0×10^6	—
6	Hexahydrophthalic	3.0	12,000	—	33.9×10^6	—
7	Succinic	3.0	15,400	—	28,800	Inhibits

Table IV-2.18 Thickening Effect of Combinations of LiCl and MgO in Polyester Resins (at room temperature) (no filler)

Combination No.	Material or Combination	% Added	Resin Viscosity Change, cps			
			4 hr	24 hr	4 days	8 days
1	MgO	1.5	6,200	67,000	1.84×10^6	6.53×10^6
2	MgO LiCl	1.5 0.7	5,800	40,000	13.76×10^6	46.00×10^6
3	MgO LiCl	1.5 1.0	5,800	49,500	19.20×10^6	58.00×10^6
4	MgO LiCl	2.0 0.7	5,400	51,000	26.90×10^6	64.00×10^6
5	MgO LiCl	2.0 1.0	5,000	70,000	31.40×10^6	64.00×10^6

Table IV-2.19 Basic Properties of Glass Fiber Roving as Reinforcement for SMC

Roving Property	Explanation and Significance in SMC	Typical Values	
		Hard Type	Soft Type
Strand Integrity	Measures how well filaments in the strand are bonded together and resist filamentation and degradation during handling and chopping. High integrity strands chop and disperse well in SMC paste, remain separate and mold easily. Low integrity strands show filamentation, mix with more difficulty and mold poorly.	Maximum— no filamentation—	Some filamentation—
Sizing Solubility (drape test, in.)	A measure of how rapidly the sizing on the glass strand is attacked or solvated by the monomeric or resinous ingredients. Either low or high solubility strand must wet well. Low solubility indicates resistance to degradation and permits strand breakup and restricts flow in molding, but provides higher physical strengths.	Low 3 to 4	Medium to high less than 2 in.
Ignition Loss, %	Quantitative determination of the amount of organic solids applied to and contained on the roving. For SMC, high ignition loss favors high integrity and a good crisp strand for best chopping properties and quickest assimilation into resin paste. Low ignition loss causes higher filamentation and poorer mixing.	1.8 to 2.2	1.0 to 1.3
Ribbonization	A numerical interpretation of the degree to which a roving strand has been bonded to itself for proper handling and transporting through guides to the chopping equipment. Highest possible values = 7, Ribbonization of roving for SMC should be in the low range to assure rapid breakup and easy dispersion.	2 to 3	2 to 3
Choppability	A visual estimation of the behavior of fiber glass roving when chopped. Chopping behavior of all types of roving for SMC should be excellent with greatest possible freedom from filamentation and excessive clumping.	Excellent	Good to excellent— some filamentation
Stiffness, in.	A measure of the resistance of dry roving to flexure under its own weight. Related to SMC, stiffness value results from the combined degrees of strand integrity and ignition loss, and is valuable as an additional control.	5.4 to 6.0	4.5 to 5.0
Yards per Pound	This is a measure of the yield of the glass fiber roving product, and is due primarily to the count of the number of strands (ends) in the roving strand. Yardage should be controlled within fairly narrow limits in order to control the percentage of glass reinforcement in the SMC sheet.	221	221

Source: Johns-Manville Fiber Glass Reinforcements Division, Waterville, Ohio.

classes of pigment materials may be applied, each problem in molding colored SMC requires its own separate solution.

Color success and reproducibility in SMC are influenced by the following: color intensity, original choice of pigment, degree to which monomer loss is prevented, degree of thickening of the SMC, proper temperature of the mold within close limits and elimination of peaks and lows, adequate molding pressure, tightness of the mold to contain and build back pressure without large shearedge gaps, and others. Success with one color or pigment in an SMC mix does not guarantee success with another, or with that color in another type of SMC compound.

It is recommended that surface textures between different materials be coordinated in assembled systems involving SMC and others. It is wise to select a different RP/C color to contrast with or complement color of other materials in a coordinated system. Deliberately selecting a different shade or hue is more judicious than attempting an accurate match. A variety of surfaces and textures will produce different shades regardless of all precautions. The molder who must use color to sell his products is warned of these limitations, as well as of the advisability of selecting colorants and shades which will perform well in the middle

of a known range of variation, instead of being close to what may be a statistically disastrous limit.

Reinforcements. Requirements for fiber glass to be used in SMC are possibly more stringent and demand closer tolerances than in most other RP/C operations. The correct combination of properties must be met. The main variables include strand integrity, sizing solubility, ignition loss, ribbonization, choppability, stiffness and yards per pound. The different extremes of compounding and molding SMC require the correct combination of glass fiber properties. Glass fiber manufacturers have fortunately been able to answer the call and provide a full range of glass fiber product types with a full range of properties.

Table IV-2.19 lists the basic properties of glass fiber as received from the manufacturer and relates the significance of each to formulating SMC. Table IV-2.20 compares the performance of the three integrity types of glass reinforcement in molding and end-product properties of SMC.

As regards the function of sizing solubility and related properties to performance in SMC, it should be pointed out that a type of organic coating (sizing) for glass fibers with improved performance was recently introduced to the

Table IV-2.20 Performance Comparison for Various Integrity Types of Glass-Fiber Reinforcement for SMC

Performance Characteristics	Glass Type		
	Hard	Medium	Soft
Compounding			
Chopping	Good; static not a problem, fiber distribution good	Good; static not a problem, fiber distribution good	Fair; static can be a hindrance, fiber distribution fair
Mat saturation	Good	Good	Fair
Fiber wet-out	Poor	Fair	Very good
Molding			
Flow	Very good	Good	Fair
Fiber orientation	Little	Little	Some
Molded Part			
Physicals	Low	Med. (20% > hard)	High (40% > hard)
Chemical resistance	Fair	Fair	Good
Electrical properties	Fair	Fair	Good
Knit-line strength	Good	Fair	Poor
Sink marks	Least obvious	Apparent	Prominent
Waviness	Little	Some	Some

market. This material makes possible the best properties of high and low strand integrity and solubility types in one single product. The particular sizing is insoluble in organic monomers and resins, but permeable to them. This means that good wetout and assimilation of the glass fiber in SMC compound pastes may take place without degradation of the fiber bundles and, due to the penetration through the sizing material, molding efficiencies and molded physical properties remain exceptionally high.*

Chopped fiber lengths used in compounding SMC vary from ¼ or ½ in. to 1 and even 2 in. Increasing the fiber length improves molded efficiency. In molding, the shorter length fibers have been found to provide better compound flow, to increase molded strengths across weld-lines, and ¼ in. fibers have been used effectively in reducing sink marks over ribs and bosses. Actually, part design and molding criteria will be the principal regulators of glass lengths required.

To summarize the role of glass fiber reinforcement in SMC:

1. For most SMC applications, use either an insoluble/permeable sizing type fiber glass product, or one that is intermediate in integrity and solubility type for the most fortuitous balance between efficient compounding and best molding behavior plus molded product properties.

2. Where high molded part strength is of prime importance, and part design is relatively simple, use a roving with a soluble, low integrity product.

3. If the part design is extremely complex, use a roving product with high integrity and an insoluble sizing.

4. Optimum glass fiber content is 30 ± 2%.

EQUIPMENT AND TOOLING

The choice is left to the molder to either purchase commercially made SMC for his in-plant use or manufacture his own. The decision depends upon the size of the molder's operation and the investment he is prepared to make. The most basic SMC machines cost from $15,000 to $75,000 and require four to six operators. Many types of SMC are available commercially.

Source: Technical Product Data Sheets—PDS 750F and PDS-752-E, Johns Manville Fiber Glass Reinforcements Division, Waterville, Ohio.

Based on the success of SMC to date, the estimate is that the material will become a part of almost every matched-die molding operation.

SMC Machine Characteristics. The following specification is presented for the reader who will (1) want to understand the preparation of SMC, (2) purchase an SMC machine, or (3) build his own SMC machine.

There are seven basic elements to consider in the construction of an SMC machine:[48]

1. Frame
2. Conveyor system—belt or film carrier
3. Polyethylene film stations, usually 2
4. Doctor-blade mechanisms, usually 2
5. Choppers across the full machine width
6. Wet-out and compaction means and devices
7. Take-up section

Three additional support systems are necessary for machine operation:

1. Glass roving creel, either over or to the rear of the machine
2. Resin-paste preparation and feed system
3. Glass static eliminators

There are four types of SMC machines:[49]

1. Mat impregnator, the original type of machine used. It was superseded because of the high cost of mat.

2. Belt-type impregnator. This machine has resin spreaders in 2 locations and deposits chopped glass reinforcement by gravity where it may be effectively coalesced into the resin. Because the belt runs along a flat-plane course, disc rolls and a perforating roll are required for wet-out.

3. Roll-type impregnator. This machine contains no belt, the resin and glass are carried by the polyethylene membrane. Increasing tension on the sheet as it courses through serpentine rolls accounts for kneading and elimination of air voids. Serrated rolls may be incorporated to assist in wet-out. Belt- and roll-type machines are approximately equivalent in cost. The rolls may be heated for increasing the wet-out speed.

4. Large roll impregnator. In this machine, a large drum replaces the smaller drums of the roll-type. Planetary compaction rolls are located around the periphery of the larger roll to assist in wet-out. As in the roll-type, the large drum may be

heated or cooled. There is probably some space saving with this machine.

Figure IV-2.2 (upper part) illustrates an actual belt-type machine for fabricating SMC. Figure IV-2.2 (lower part) illustrates a roll-type SMC machine.

SMC Machine Operation. The technical aspects of SMC machine operation are discussed below.

Resin Mix and Resin Feed. The size of the mixing facility should be geared to the amount

Figure IV-2.2 (*Top*) A roll-type SMC machine with rolls through which the film carrying the combined resin and reinforcement is threaded for impregnation. The heater system is also shown. (*Bottom*) A belt-type SMC machine capable of turning out up to 10,000 lb. per hr. of moldable mat with an average thickness of 0.100 in. Courtesy I. G. Brenner Co. (top) and Engineering Technology, Inc. (bottom.)

of compound to be produced. This may vary from 1200 lb per hr. for a 25 in. machine at 10 ft per min. to 7200 lb per hr. for a 50 in. machine at 30 ft per min. The weight of the compound may be estimated at 0.75 lb per sq ft for a sheet thickness of approximately 0.070 in.

The location of the resin-feed system should be as close as possible to the point of application onto the moving belt, with some use of gravity fall if possible.

Batch-type mixing is limited to a 30 min. maximum pot life because the resin starts to thicken causing difficulty in wetting chopped-glass fiber strands. The recommended batch-mixing methods consists of mix in the main part of the resin-filler batch and adding the MgO thickening mix through a metering and mixing pump. This results in uniformity of thickening time of the resin batch laid on the belt and corresponding uniformity in both thoroughness and rate of wet-out of the glass fibers by the resin mix.

Batch mixing, practiced in the early days of SMC formulation and use, was found to cause too much variation of the resin-paste viscosity, with the major increases coming just at the finish of amount compounded, and creating difficulties of fiber wetting. Therefore, continuous mixing was developed. In continuous mixing, two separate batches are prepared and pumped by metering devices through a static mixer. Batch A (mixed) contains resin, catalyst and filler, while Batch B contains the inert polyester or other carrier, thickener, and a small amount of filler as a suspendor. The static mixer intimately separates and brings together small streams of the 2 batches metered together in the correct ratio, and delivers a homogeneously-mixed resin paste which is nearly as possible consistent in viscosity onto the SMC machine at the chopping station.

The viscosity of the resin-filler mix is 15,000 cps at the start. After introduction of the thickening mix it increases to 30,000 cps. The maximum tolerable viscosity after 30 min. is approximately 50,000 cps. Measurements are made using Brookfield viscosimeter samples of checks on the liquid mix (penetrometer measurements also on the liquid). Viscosity samples should be taken every ½ hr. or for every 2000 lb. molding mat produced. (See the discussion of SMC formulations and molding methods.)

Application to the Moving Belt. The resin mix (thickener included) must be applied onto 0.002 in. polyethylene film laid on the belt from a roll. The resin is applied so that a 3 in. minimum gap exists between the edge of the resin band and the edge of the polyethylene film. Resin should be applied using an air-pressure control feed device requiring a sealed resin system. If a metering and mixing pump is used, the feed is continuous and under pressure. Although means are provided for eliminating voids in the sheet, care should be taken to avoid inclusion of air in the mixing stages. An adjustable doctor blade should be applied 1 to 2 ft downstream from all points at which resins is applied onto the film. Resin thickness should be ⅛ in. to end up with ⅛ in. SMC mat.

The purpose of all-automated mechanical handling of film at this stage is to eliminate air entrapment which, in excessive amounts, causes deleterious voids in molded parts.

Figure IV-2.3 shows an automatic mix and feed system for delivering resin, catalyst, and fillers (including thickeners) to an SMC line.

Chopping. Chopping the glass fiber roving should be carried out in the hood placed over the full width of the machine preferably with slowly rotating choppers whose blade width and back-up rubber impressor rolls extend across the full width of the belt. Fiber glass roving strands are fed in every 1, 2, or 3 in. 1 to 4 choppers may be used.

Roving is delivered to the choppers from pallets or creel packs stationed on the floor at the end of the machine. The humidity-temperature level should preferably be controlled at 50% RH, 70° to 80°F at all times to eliminate static and assist in controlling the glass distribution to a uniform, even pattern with a minimum of variation from one area unit area to another.

The gap between the rubber impressor roll and the chopping blade roll should be adjustable to compensate for rubber wear, dulling of blades, variance in roving hardness, and any emergencies.

One or two additional or spare chopping units should be provided for easy replacement in the event of failure of a chopping unit in service. The roving should have low to medium ribbon and should break and disperse easily.

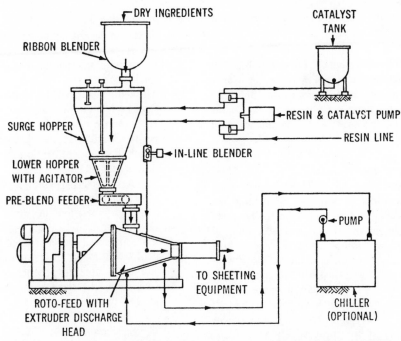

Figure IV-2.3 Schematic diagram of equipment available for automatic mixing and feeding of resin, catalyst and fillers to an SMC line. (Courtesy of Baker-Perkins, Inc.)

Blades in the chopping arbor or mandrel should be staggered in 2 or 3 groups along the length of the arbor with a uniform distance between each blade. By this means, there will be no violent impact during revolution because some of the blades will be in contact with the rubber roll at all times.

Wet-out. Devices for initially forcing chopped glass fibers into the resin mix comprise (1) rows of longitudinal, stationary rolls with 3 in. diameter discs, staggered so that all areas of the mat are covered; (2) reciprocating transverse rolls that impress glass fibers into the resin mix; (3) no wet-out devices at all, depending on mechanical manipulation of the material between films for wetting out.

Impregnation Techniques. As a further means of accomplishing the most perfect possible union between the resin mix and the glass reinforcement and of eliminating or suppressing voids, additional steps for impregnation are desirable. A critical stage is reached at this point because the resin mix has started to thicken and complete fiber wet-out and impregnation must be finished before the thickening end-point is reached.

Two methods for impregnation are possible: (1) compression or kneading, which is principally accomplished along a flat stretch of the SMC mat in the direction of the moving machine; (2) bending and stretching, which involves passing SMC mat over, under, and around different-sized rolls in a serpentine course so that the top and bottom skins are laterally moved with respect to each other. The advantages and action induced by each method are obvious. Perforating the top film sheet is sometimes associated with the compression and kneading method.

Wind-up. Pinch-rolls should be provided as the last stage before the SMC mat leaves the machine and proceeds to the wind-up mechanism. These rolls are necessary to define sheet thickness and to provide constant tension in the mechanism. The wind-up device itself should be the Hobbs Turret type or an equivalent with a minimum of 2 creels. Wind-up may be made on 4 in. diameter kraft tube of the required width. The mat line does not stop when changing wind-up creels.

Machine Start-up. Machine start-up must be done by hand. It is a difficult task and a bait should be used. All safety precautions should be taken.

Manpower Requirements. For a 50 in., full scale production machine 6 men are necessary: 1 to supervise, 2 to mix the compound and feed the resin, 1 to watch over the chopping operation and tend the creel, and 2 at the wind-up and take-off station.

Aging. It is usually necessary to age SMC material from 24 to 36 or 72 hr before molding in order to obtain optimum thickening. An extremely valuable step would be to have the SMC already thickened and moldable directly off the machine without aging. To accomplish this end, organic anhydrides used together with MgO, Mg(OH)2, and Ca(OH)2 thickeners and also irradiation by radio-active materials on the finished sandwiched sheet have been evaluated. (See Tables IV-2.16, 2.17, 2.18). Long-term aging remains as the most widely used technique, however.

In-plant Controls for SMC Manufacture.
It is beneficial at this point to itemize and tabulate factors that have become time-proven parameters for the control of SMC in 4 separate areas: raw materials, consistency of resin compounding, SMC sheet preparation, and molding performance.[50]

Raw materials control:
1. Variations in resin properties constitute the most critical raw-material problem.
2. The degree of reactivity of each resin batch should be available and the thickener level adjusted so that the molding viscosity is known throughout the usable life of the SMC batch.
3. The resin supplier should send preshipment test data including thickening curves for each resin batch to the RP/C manufacturer.
4. The viscosity of the resin mix both after mixing and while on the machine after thickeners are added should be measured and controlled by a Brookfield viscosimeter.
5. A complete log should be maintained on every batch and should include viscosity details during mixing, after thickening, through maturation, and in storage.

Compounding:
1. A record of the input weight of all materials should be accurately kept and balanced with the output quantity of every final batch.
2. The temperature of the resin mix prior to thickener addition must be maintained at room temperature to control the viscosity. Variations in viscosity seriously affect glass content, wet-out, and final mat weight per sq ft.
3. Mixing times, particularly when adding thickener, should be rigidly controlled to eliminate variation in molding characteristics and shelf life.
4. Viscosity should be controlled by a Brookfield viscosimeter to 2 million centipoise and by a laboratory bench-top precision penetrometer after thickening above 2 million cps.

SMC Preparation:
1. The glass-to-resin ratio should be controlled to within ±2% glass fiber. The use of acetone extraction or washout is satisfactory for this determination. The efficacy of chopping should be checked frequently and long fiber lengths should be avoided unless required.
2. SMC mat weight per sq ft should be controlled to within ± 0.02 lb per sq ft to insure mold charge consistency and reduce material waste.
3. Improper densification or poor static control can effect fiber wet-out. This parameter should be checked visually with the microscope.
4. As soon as rolls come off the SMC machine roll-up device, they should be completely encapsulated in cellophane or aluminum foil with the edges tape-sealed to prevent loss of volatiles which would deleteriously affect molding characteristics.
5. Samples of each SMC resin batch should be procured at the time of thickening and be checked in the jar for the rate of viscosity increase using a Brookfield viscosimeter and a penetrometer as described.
6. Viscosity changes should be referred to a standard control graph. The optimum plus upper and lower limits of the range of moldability should be established. The plot should show the penetrometer needle

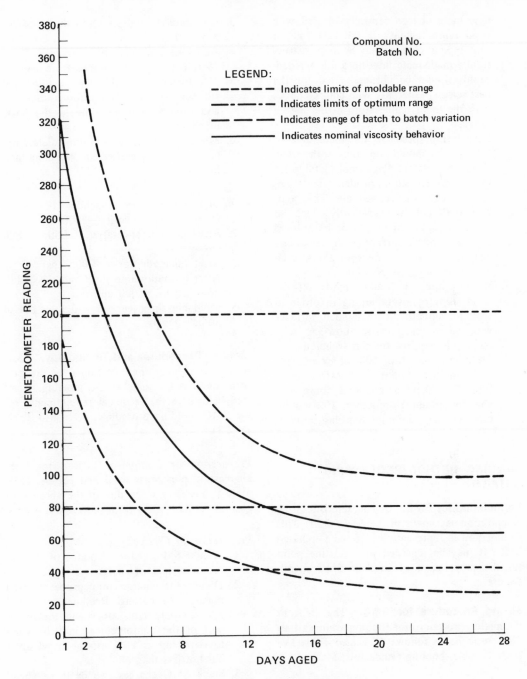

Figure IV-2.4 SMC viscosity control curve.

penetration in 0.1 mm. versus aging time in days[51] (see Figure IV-2.4).

Molding Performance:
1. Molding performance should be evaluated for every SMC batch produced.

2. Various SMC performance factors are sometimes peculiar to individual parts and should be determined.
3. The charge pattern into the mold is critical and should start with the simplest, most logical procedure for a given part.

One basic charge centrally located with good reproducibility is the best rule of thumb. Split charges may be used if structurally inadequate flow lines are avoided.

4. Molds should be chrome-plated for the best release, especially with colored SMC, with the mold temperature in the 300°F range.

5. Fast-closing speeds to the material contact point should be run with slow, steady, slow-close for complete fill prior to gelation, reduction, of fiber orientation, and removal of trapped air. The compound should be sufficiently thick to avoid development of pin or blow holes in the molded surface in postmolding heating and processing operations, such as bonderizing and painting.

6. It is desirable to mold slightly off the mold stops by overcharging in order to establish smooth, dense molded surfaces with minimum porosity. Low-gap telescoping is required for this technique.

7. Molding pressures of 300 psi are possible with SMC on some parts. However, pressure of 1000 psi or greater is more suitable for attaining smoother, 100 mu in. surfaces and minimum finishing costs.

MOLDING METHOD/PROCESSING DETAILS

Formulation for SMC. A characteristic SMC composition is shown in Table IV-2.21. This formulation is representative of the purchased SMC. It may be regarded as a starting point for the preparation of in-plant SMC manufacture.

Mixing Procedure for SMC. The steps in the mixing procedure for the preparation of an SMC resin batch follow: (*See also* Table IV-2.22. Trouble-shooting Guide for SMC).

Preblending. Same as for preform molding. The liquid catalysts may be dispersed in styrene prior to batch addition.

Mixing Method:
1. Add resin to main mix tank and start mixer.
2. Add monomer and stir until dissolved for 3 to 5 min.

3. Add catalyst and stir until dispersed for 5 to 10 min.
4. Add filler and mix until dispersed, about 20 min. Keep temperature from rising.
5. Add thickeners or thickener solution and stir for 20 min. Add through metering and mixing pump on line to the SMC machine.
6. Mix may start to thicken immediately unless care is exercised in weighing and handling.

Control Viscosities (See Appendix I)
1. Original resin viscosity—3500 cps
2. After monomer-catalyst addition—800 cps
3. After filler addition—15,000 cps
4. After formation into sheet and glass addition—30 to 50,000 cps
5. Moldable sheet viscosity—0.5×10^6 to 50×10^6 cps

Molding Procedures and Technology. SMC represents a great simplification over the preform method for matched-die molding. Itemized below are the technical requirements for the SMC molding procedure: (See also Table IV-2.3)

Preparation of Equipment. Clean mold by air-blasting plus brass wool and chisels. SMC mold is easier to clean due to the absence of glass in the flash. Only resin squeeze-out occurs.

Preparation of Charge:
1. Weight SMC charge. Volume of part $\times$ 1.7 = weight of charge.
2. Divide the charge into properly sized portions by cutting. Rectangular shapes are generally satisfactory and the major part of the reinforcement may be sized to cover up to 80% of the mold area, but not less than 40%.
3. Stack SMC charges two or three ahead of the mold cycle. Keep them covered with film to prevent contamination and hardening from monomer loss.
4. The charge placement in the mold should be made to eliminate air entrapment and "islands," and also to avoid "weld lines."
5. The charge placement should also be made so as not to off-set the mold alignment when the press closes. This is es-

Table IV-2.21ᵃ   Formulations for SMC Used in Matched-Die Molding

Ingredient	Resin Mix	Combining Ratio for Sheet Formation	% Final Composition
1. Resin Low-shrink polyester resins preferred (total resins plus carrier mix)	40.0	} 70.0	} 28.0
2. Catalyst % t-butyl perbenzoate as % of resin	1.0		
3. Thickeners Ca(OH)₂, Mg(OH)₂ or MgO (See Table IV-2.12)	1–1.5 or 2–3		1.4
4. Filler CaCO₃ preferred	57.0		40.0
5. Internal mold release Zinc stearate	1–1.5		0.6
6. Reinforcement Glass-fiber roving ¼ to 1 in. lengths	—	30.0	30.0

ᵃSee also Table IV-2.24, Trouble Shooting Guide for SMC.

pecially critical and necessary with unsymmetrical parts.

6. The charge may be preheated to 100 or 120° to improve the flow. This is advisable for deep-draw parts.
7. Cutting the charge to shape is preferable to folding.
8. Utilize the scrap and cut the edges for material savings and economy.

Mold Close:
1. Fast close: 600 in. per min.
2. Intermediate close—40 in. per min.
3. Slow close: at 10 in. per min. This permits proper compound flow and should last only 4 sec.

Molding Pressure. 500 to 1000 psi on material. Stops should be built into the mold, but are not always required or used. Back pressure is created. The pressure should build up to a maximum in 2 sec.

Mold Temperature. 275 to 290°F for 0.100 to 0.125 to 340°F for thin parts.

Cure Time. Approximately 1.0 sec. for each 0.001 in. part thickness.

Automation. Automate by rig to place the charge accurately in the same location in the mold each time. SMC makes it possible to

traverse unbroken sheet through the mold to stamp or mold it intermittently (step-molding).

Part Removal. Mechanical ejector pins are built into the mold for part removal.

Typical Cycle Time:

Open mold	7 sec
Part removal	10 sec
Load SMC	20 sec
Close mold	10 sec
Cure	73 sec
Total	120 sec

For purposes of test molding, a test part was designed and working mold fabricated for laboratory use in evaluating component SMC ingredients and molding performance. The part was used to great advantage in the development of SMC and in the selection of optimum combinations of resins, reinforcement, fillers, and thickeners for specific applications and for technical service. The part may be used to evaluate mold sticking, surface smoothness effect of fiber integrity and lengths, compound shrinkage and sink marks, mottling and virtually all other molding parameters preparatory to designing and releasing the compound for use in actual production. It may also be cut up and destructively tested for compliance with physical-strength requirements. Figure IV-2.5

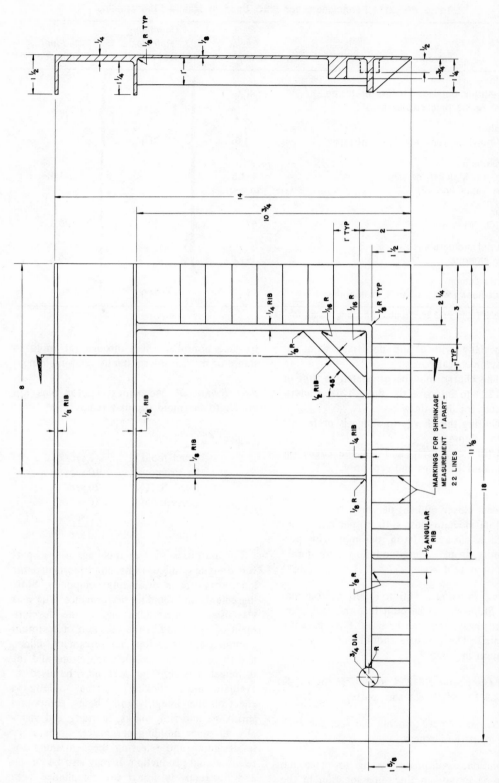

Figure IV-2.5 Test part for evaluating SMC formulations and their molding performance. The part has differential flat thicknesses, vertical grooves both narrow and wide and shallow and deep, plus markings on one inch centers for determining shrinkage. (Courtesy of Johns-Manville Fiber Glass, Inc.)

presents an engineering drawing of this test part.

Table IV-2.22 provides a trouble-shooting guide for dealing with difficulties encountered in regular molding production using SMC. Finishing operations for postmolded SMC parts are the same as those described for preform molded parts (see "Postmolding Operations"

Table IV-2.22 Trouble-Shooting Guide for Molding with SMC

Problem	Description	Possible Cause	Remedy
Mold not filled	The mold is not filled at the edges.	Charge weight too low.	Increase charge weight until the material appears at the telsecoping edge.
		Temperature too high, SMC hardens before the mold is filled.	Lower temperature.
		Closing time for the press is too long, the SMC hardens before the mold is filled	Shorten closing time.
		Pressure too low.	Increase pressure.
		Area of the charge is too small.	Choose a charge with larger area.
	The mold remains unfilled at the edges only in a few spots.	Charge weight too low.	Increase charge weight until the material appears at the telescoping edge.
		The charge escapes before the mold is closed.	Place the charge more carefully.
		Clearance of the telescoping edge is too large or length of telescope is too short and allows the SMC to escape before the mold is filled.	Make clearance smaller or telescope deeper.
			If the fault is minor a higher temperature or excess material may help.
	Mold is not filled in some spots, although the entire edge is filled.	Charge weight too low.	Increase charge weight until the material exudes from the telescoping edge.
		Air cannot escape from the mold.	Arrange the charge in such a manner that air cannot be trapped and so that the mat pushes the air ahead during the flow.
		Blind holes or pockets make it impossible for air to escape.	Deaeration of the enclosures by a 3 part construction of the mold or by bleeding air past ejector pins.
			If the fault is small, increase of pressure may help.

Table IV-2.22 Trouble-Shooting Guide for Molding with SMC (Cont.)

Problem	Description	Possible Cause	Remedy
Burning	Dark brown or sooty surface in places where the part is not completely filled.	By compressing trapped air and styrene vapors the temperature is raised to the ignition point.	Choose a charge which will not trap air, but pushes the air with it as it flows.
			If these brown spots appear on blind holes or pockets they have to be deaerated by 3 part mold construction or by ejector pins.
Blisters	Round elevations on the surface of the cured part.	Air entrapped between the layers of resin mat.	Remove trapped air from the charge by prior compression.
			Decrease the area of the charge so that air can escape better.
		Too high a mold temperature (monomer vapors).	Lower mold temperature.
		Curing time too short (monomer vapors).	Increase curing time.
	Round elevations on the surface of the cured part of heavy section.	Only with heavy wall thickness. Internal stress tears the laminate between the individual layers.	Decrease the area of the cut piece so that the glass fibers of the various layers mesh better.
			Lower mold temperature.
		Weak spot along a knit line.	Shape the charge in such a way that no knit line can form.
		Decrease in strength in one direction in spots with extremely long flow paths (glass fiber orientation).	Shorten the flow path by increasing the area of the charge.
		Damage during removal from the mold, caused by: (a) Undercuts (unintentional). (b) Ejection pins have too small an ejection area. (c) Insufficient number of ejection pins. (d) Sticking to the mold. (e) Incomplete curing.	(a) Remove undercuts. (b) Increase ejection area. (c) Increase number of ejection pins. (d) See "sticking" (e) Increase curing time or temperature.
		Only with heavy wall thickness. The laminate	Decrease area of the charge so that the glass

Table IV-2.22 Trouble-Shooting Guide for Molding with SMC (Cont.)

Problem	Description	Possible Cause	Remedy
Internal cracks		cracks because of strong shrink stress between the individual layers.	fibers of the various layers mesh better.
			Lower mold temperature.
Sticking	It is hard to remove the finished part from the mold. In some spots the material sticks to the mold.	Mold temperature too low.	Increase mold temperature.
		Curing time too short.	Increase curing time.
		SMC was unpacked too long. With rolls of SMC open only the outer layers.	Keep rolls sealed in foil until used.
		Mold not broken in. The mold is new or has not been used for a long time.	Use a mold release on the first few moldings.
		Mold surface too rough.	Polish surface.
	The cured part is hard to remove. In spots material sticks to the mold. At the same time pores and scars show at the surface.	Area of the charge too large. Air on the surface cannot escape due to short flow path. Trapped air delays cure.	Decrease area of charge. Add small charge on top of larger charge.
Surface porosity	If these pores are numerous, the part is difficult to remove.	Area of the charge is too large. Air on the surface cannot escape because the flow path is too short	Decrease area of charge. Add small charge on top of larger charge.
Mold abrasion	Dark to black spots on the surface of the cured part.	Abrasion from the mold.	Chrome plate the mold.
Warpage	The part is slightly warped.	Warpage due to shrinkage during hardening and cooling.	Cool the part in a jig. Employ low-shrink or zero shrink resins in compounding.
		One mold much hotter than the other mold.	Reduce temperature differential of molds.
	The part is badly warped.	Warpage is due to glass fiber orientation caused by particularly long flow path.	Shorten flow path by increasing the area of the charge. Employ low-shrink or zero-shrink resins in compounding.
Wavy surface	Waves are found on long, vertical, thin walls at a right angle to the direction of flow. Also with other adverse flow conditions (large dif-	Complex design interrupts uniform flow.	In most cases cannot be eliminated completely. Improvement can be obtained by: (1) Increased pressure. (2) Change design of the mold.

Table IV-2.22 Trouble-Shooting Guide for Molding with SMC (Cont.)

Problem	Description	Possible Cause	Remedy
	ferences in wall thickness) an irregularly wavy surface may occur.		(3) Alter position of charge. Employ low-shrink or zero-shrink resins in compounding.
Sink marks	Laking (shiny and dull spots) on surface, or opposite rib or boss.	Non-uniform shrinkage during molding.	Employ low-shrink or zero-shrink resins in compounding. Increase temperature of one half of the mold. A difference of 10 degrees F is usually sufficient.
			Increase pressure. Shorten length of chopped fibers. Change mold design. Alter position of charge. Narrow the clearance of the telescoping edge.
Erosion of cut off on mold	Metal breaks off in the direction of applied thrust.	Inaccurate or weak guide pins.	Provide accurate mold guidance.
			Strengthen guide pins.
			Place the charge to minimize side thrust.
Dull surface	Surface not shiny enough.	Pressure too low.	Increase pressure.
		Mold temperature too low.	Increase mold temperature.
		Unsatisfactory mold surface.	Chrome plate mold.
Flow lines	Local waviness on surface.	Mold closure is improperly designed or damaged.	Follow recommendations on tool design.
		Mold temperature too low.	Increase temperature.
		Glass fiber orientation in places with extremely long or adverse flow paths.	Shorten flow paths by increasing the area of the charge.
		Mold shifting causing excessive pressure drop at one edge.	Improve mold guidance.

Source: Courtesy of Marco Chemical Division, W. R. Grace Co., Bulletin No. 710-B15.

in Chapter IV-2). By way of summary, a recommended plant layout for a typical SMC molding operation is illustrated in Figure IV-2.6.

Variations in the Use of SMC. Several variations in the use of SMC in the actual matched-die molding process and in related applications are extant. They are included here

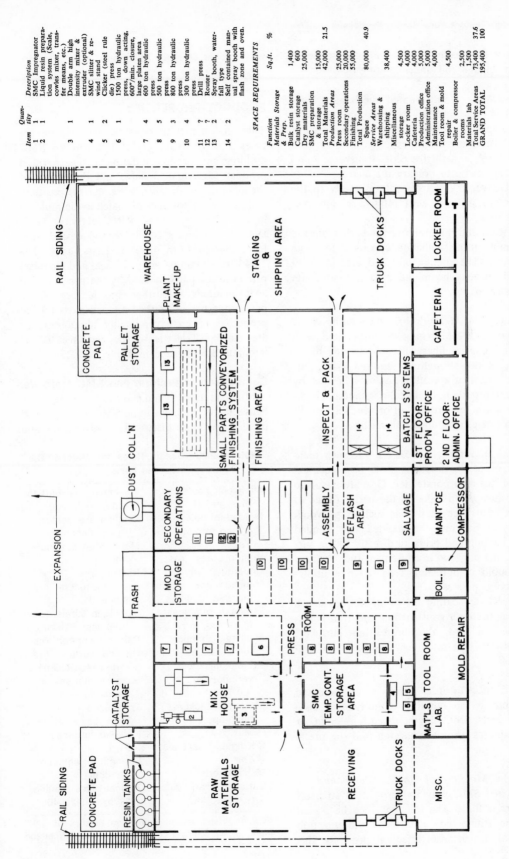

Figure IV-2.6 Typical plant layout for SMC preparation and molding. (Drawing courtesy of General Tire and Rubber Company, Akron, Ohio, prepared by R. W. Meyer.)

Item	Quantity	Description
1	1	SMC Impregnator
2	1	Liquid resin preparation system (Scale, cowles mixer, transfer means, etc.)
3	1	Double arm high intensity mixer & extruder (optional)
4	1	SMC slitter & re-wind stand
5	2	Clicker (steel rule die) press
6	1	1500 ton hydraulic press, down acting, 600"/min. closure, large platen area
7	4	600 ton hydraulic press
8	5	500 ton hydraulic press
9	3	800 ton hydraulic press
10	4	300 ton hydraulic press
11	?	Drill press
12	?	Router
13	2	Spray booth, waterfall type
14	2	Self contained manual spray booth with flash zone and oven.

SPACE REQUIREMENTS

Function	Sq.ft.	%
Materials Storage & Prep.		
Bulk resin storage	1,400	
Catalyst storage	600	
Dry materials	25,000	
SMC preparation & storage	15,000	
Total Materials	42,000	21.5
Production Areas		
Press room	25,000	
Secondary operations	20,000	
Finishing	55,000	
Total Production	80,000	40.9
Service Areas		
Warehousing & shipping	38,400	
Miscellaneous storage	4,500	
Locker room	4,000	
Cafeteria	4,000	
Production office	5,000	
Administration office	5,000	
Maintenance	4,000	
Tool room & mold repair	4,500	
Boiler & compressor rooms	2,500	
Materials lab	1,500	
Total Service Areas	73,400	37.6
GRAND TOTAL	195,400	100

229

to assist the reader in enlarging the scope of his knowledge and understanding of SMC.

Automation. Uninterrupted mechanical continuity in molding SMC has been proposed and the components assembled and found workable.[52] These are three steps:

1. Unroll and slit the SMC transversely and lengthwise to prepare the charge.
2. Deliver the charges intermittently to the mold using a zero relative-motion machine.
3. Mold and then remove the parts from the mold using a suction cup on a movable arm. The suction cup and arm arrangement may be modified to accommodate any shape.

Variations in Methods of Formulating SMC Components. SMC may be formulated with reinforcing glass fibers oriented in one direction for added strength as required. Also, particular customer specifications may be met by using fire-retardant, chemical-resistant and high-electrical-property resins, all of which are available in low-profile, no-shrink resins.

Molding of Large, Flat Sheets. Known techniques, especially involving SMC, have been expanded to incorporate RP/C materials into composites which enhance the performance of otherwise ordinary materials such as plywood. SMC material may be molded and bonded under heating and pressure directly to the surface of the plywood in equipment such as large single-mold compression type or in multiple-layer book presses.

The RP/C coating on plywood provides a material for both construction and transportation with a hard, abrasion-resistant surface[53] that may be fabricated very easily. (See Chapter III-3).

Diced SMC. In order to provide an increased flow during the molding of intricately contoured parts, SMC may be diced into ¾ in. or 1 in. squares prior to charging into the press-mold.[53]

Injection Molding of SMC. Compound or formulated SMC may be rerolled to strip off the carrier films. The roll may then be placed into the stuffer box that feeds the injection screw of a thermoset injection press. In this way, higher strength injection-molded thermosets are possible.

PRODUCT PROPERTIES

This section first presents comparative advantages and disadvantages of preform vs. SMC molding and product differences, design rules; product performance data; and discussion of the advantages, disadvantages, market position, and relative importance of matched-die molding in the whole gamut of RP/C activity. Table IV-2.23 provides a set of general design rules for either preform or SMC parts. Table IV-2.24 lists the comparative physical and chemical properties of parts molded by the preform and SMC methods. In further consideration of the part properties of finished molded parts typical properties of an exceptionally high grade of fire-retardant, electrical-grade, flat-sheet RP/C laminate are itemized in Table IV-2.25.

Comparison of Preform and SMC Methods.

Major processing and end-product differences between the preformed and SMC methods and molded parts are discussed as advantages and

Table IV-2.23   Design Rules for Matched-Die Molded Parts

Minimum inside radius, in.: ⅛	Maximum size part to date, sq ft: 200
Molded-in holes: Yes	Limiting size factor: Press dimensions
Trimmed in mold: Yes	Metal edge stiffeners: Yes
Built-in cores: Yes	
Undercuts: No	Bosses: Yes
Mimimum draft recommended, degrees: 1	Fins: Not recommended
Minimum practical thickness, in.: 0.030	Molded-in labels: Pre-from-yes; SMC-no.
Maximum practical thickness, in: 0.250	Raised numbers: Yes Gel-coat surface: Pre-form-yes; SMC-no.
Normal thickness variation, in.: ±0.008	Shape limitations: Moldable
Maximum thickness build-up: 2 to 1 Max.	Translucency: Pre-form-yes; SMC-no. Finished surfaces: Two
Corrugated sections: Yes	Strength orientation: Random
Metal inserts: Yes	
Surfacing mat: Yes (except SMC)	Typical glass loading, % by wt: 25–40

Source: "Handbook of Reinforced Plastics of SPI," Mohr, J. G., Oleesky, S., New York, Van Nostrand Reinhold, 1964, p. 536. *(with additions)*

Table IV-2.24   Physical and Chemical Properties of Preform and SMC Molded Parts

Property	Preform Molding	SMC Molding
Tensile strength, psi	15,000–25,000	12,000–20,000
Tensile modulus, psi, $\times 10^{-6}$	0.8–1.8	0.9–2.0
Flexural strength, psi	25,000–40,000	25,000–40,000
Flexural modulus, psi, $\times 10^{-6}$	1.25–10	1.25–10
Compressive strength, psi	18,000–30,000	15,000–30,000
Impact strength, ft lb per in. notch	10–20	8–20
Specific gravity	1.4–1.7	1.7–2.0
Heat-distortion temp., °F	300–400	350–450
Resistance to heat, continuous, °F	300–350	300–350
Burning rate	Slow to self-extinguishing	Slow to self-extinguishing
Hardness: Barcol	40–70	40–80
Rockwell	H40–H105	H50–H112
Moisture absorption, 24 hr, %	0.1–1.0	0.1–1.0
Chemical resistance		
Weak acids	Good—excellent	Good—excellent
Strong acids	Fair	Fair
Weak bases	Fair	Fair
Strong bases	Poor	Poor
Organic solvents (except methylene chloride)	Good—excellent	Good—excellent
Nominal glass content, %	35	35

disadvantages of SMC as follows:[51, 54, 55, 56, 57, 58, 59, 60, 61, 62, 63]

Advantages of SMC:

- Controlled impregnation of glass reinforcing strands is possible as well as intimate, long-term contact of the resin and glass before molding. These factors eliminate variations caused by preform nonuniformity and resin-pour at the press.
- A uniform rate of mixing resin and adding glass fibers can be achieved with no sudden, hurried pouring or other fast action at the press as in fiber glass preform molding.
- SMC offers greater freedom in part and mold design than does preform molding.
- SMC makes possible more complex matched-die molded (high-strength) parts with bosses, molded-in inserts, ribs, differential thicknesses in the same part (0.090 to 1.0 in.), sharp radii and corners, and other modifications which are not possible or practical with preform molding. This variability makes possible the elimination of subsequent machining in molded parts.
- SMC eliminates some of the costly resin mixing and transfer equipment and the multiple steps required in the preforming and molding operations.

- There is no flash to pinch off during molding as in preforming. The molding process is therefore easier and the tooling will have longer life.
- There is less material to inventory than in the preform process.
- Resin-rich and starved areas are eliminated, thus preventing cracking and crazing in molded parts.
- The average physical properties except tensile strength are superior to those of preformed parts at the same glass levels. The out-of-mold Barcol hardness is 20 to 50% higher.
- SMC charges may be weighed accurately with no loss of glass or binder resin at the preforming operation.
- All SMC scraps from corner trim and other adjustments made while cutting are moldable and may be used. This eliminates what could be a major source of loss.
- Using SMC, a smooth molded surface results without fiber pattern. Surface veil is not necessary. Smoother surfaces are available if low-profile resins are employed.
- The smoother molded SMC surface results in an improved wear layer and therefore superior chemical durability and resistance to weathering.

Table IV-2.25 Typical Properties of a Fire-Retardant, Electrical-Grade, Matched-Die Molded Flat-Sheet RP/C Laminate

Property	Grade 1582	ASTM Test Method
Perpendicular electric strength, S.T., ⅛ in. in air—VPM	400	D-229
Parallel electric strength, S. x S.—KV	40	D-229
Arc resistance (⅛ in. thickness)—sec.	185	D-495
Carbon Tracking Resistance—min.	> 600	D-2302-64T(8)
Power factor, 60 cycle—%	5.0	D-150-59T
Impact strength—Izod notched—ft-lb per in.	6	D-229
Flexural strength—psi	18,000	D-229
Flexural strength tested @ 130°C.—psi	9,000	D-229
Compressive strength—psi	30,000	D-229
Water absorption @ 23°C.—% 24 hr	0.24	D-229
Flame resistance—sec.	15	D-229 Method 1
Specific gravity	1.83	D-792
U.L. recognition number	E 23525	
N.E.M.A. grade resignation	GPO 2 & 3	
Standard color	Brown	
U.L. temperature rating	130°C. Elec./ 160°C. Mech.	
U.L. slow burning rate in. per min.	0	
U.L. subject 94 flame retardance	SE-1	

Source: Courtesy of Glastic Corp., Subsidiary Monogram Industries, Cleveland, Ohio.

- A higher back pressure is created during molding because of the elimination of lands or stops. Air voids are therefore largely eliminated.
- Production finishing costs are reduced by using SMC because filling, sanding, and finishing for painting are all eliminated.
- SMC molded parts may be painted directly out of the mold.
- SMC material is in an easily handable state and can be cut accurately and rapidly with a knife.
- SMC resin in the pseudo-gel state is leather-hard, non-tacky, and non-sticky.
- SMC mat is uniform in thickness and weight per sq ft and is also better suited to automated matched-die molding in parts of all sizes than are preforms.
- The glass-to-resin ratio may be varied in the SMC mat itself without requiring trial-and-error molding for good quality parts at the press. Preforms are bulky and not as readily suited to a wide range of glass-to-resin ratios, especially high glass, without part defects. High glass content in preform molding tends to produce dry, exposed-fiber areas, especially on corners.
- Weld-lines can be virtually eliminated in SMC molding by lapping during placement in the press. It would be difficult to mold for inserts and bosses by forcing or placing preforms over extensions from the mold surface.
- Faster molding cycles are possible using SMC.
- Undercuts can be molded if a split or retractible mold is used, whereas preparing preform with an undercut is difficult.

- The weight of an SMC charge can be directly calculated by multiplying the volume of the part by the specific gravity of SMC (approximately 1.7).
- SMC thickness may be accurately controlled by presetting the heavy limiting rollers during sheet manufacture.
- Air voids may be removed or greatly reduced by manipulating and/or kneading during sheet fabrication. This provides molded parts with fewer surface imperfections and voids, particularly if low-profile resins are used.
- A mild preheating of the SMC charge permits a better flow in the mold without altering the gel time.
- The longer cure time of SMC does not seriously affect part removal or ultimate properties.
- SMC materials molds successfully after startup and mold break-in without an external mold release applied and with only a 1.5 to 2.0° draft in the mold walls.
- SMC compounds may be hand-tailored for different job requirements.
- The fiber glass sizing or finish or type of glass fiber reinforcement may be varied to satisfy different job requirements. Hard to soft (high to low integrity) finishes and soluble to insoluble finishes are available.
- Resin-rich radii and bridging of reinforcement are eliminated by using SMC.
- There is no danger of filtering the filler out of the resin mix as happens when it passes through the preform during molding.
- Ejector pins are more effective with the thickened low profile resins of SMC and present fewer troublesome problems than with other types of matched-die molding.
- SMC combines the ease of moldability of the premix with the higher-performance physical properties of preform molding.
- Parts such as automotive components previously molded in metal drawing or stamping can be combined reducing two or three separate components into one unit when molded using SMC.
- Higher gloss in the molded surfaces is possible in SMC especially if the profile resins are incorporated. Surfaces may be improved and regulated by higher pressures or by slightly overcharging and molding "off stops."
- The possibility for complete molding automation exists with SMC.

Disadvantages of SMC:

It is virtually impossible to eliminate all the porosity and voids from in between the carrier films during SMC sheet formation. Porosity in molding must be eliminated by higher pressures requiring greater press tonnages.

- SMC requires ejector pins or buttons in the mold, preferably large in area.
- SMC material must be kept cool and wrapped in foil after forming in order to prevent pregelation and loss of the monomer which would make the mat unusable.
- SMC requires 4 to 5¢ per lb total cost to process, and is slightly higher in raw-material cost than preforming. It also requires costly non-reusable carrier film on both surfaces—approximately 0.7¢ per sq ft per side. The cost of 1 lb of SMC includes 1.8¢ for film. However, even though SMC is more expensive, the savings in molding equalize its cost with that of preforming because of the time and labor savings and the high cost of preforms.
- It is impossible to make changes once the mat is made up. SMC is not as adaptable as preform molding for changing the gel time of resin mix and other procedures or adjustments.
- The size of parts made with SMC is limited, unless the parts are fairly flat. Preform molding is better suited to large shapes with complex compound curvatures and deep draw. SMC may be molded in deep-draw parts having no baffles which would tend to create weld-lines.
- SMC requires higher molding temperatures than preform molding.
- SMC requires a more complicated die design with a heavier guide pins, and deeper telescoping. The gap at the shear edge, however, is less critical.
- SMC requires more careful, accurate and consistent placement of the charge for non-symmetrical parts.
- In SMC molding, the press must close rapidly to induce material flow prior to gelation.
- For proper SMC molded surface and release a more expensive, hard, chrome-plated mold surface is preferred.
- In SMC molding, the charge must be pyramided toward the center to avoid trapping air during mold closure. No trapped air is permissible in the pour pattern in preform molding either.

- The minimum pressure for molding SMC is 500 psi which is starting to extend out of the low-pressure range. Pressures up to 1500 or 2000 psi may be necessary to avoid and eliminate porosity, pores and pinholes.
- The thickener is difficult to control in fabricating SMC sheet and the viscosity of the resin mix is not always changed uniformly upon addition of the thickener. A better control is needed. Post-metering and mixing equipment is expensive.
- It is impossible to use veil mat with SMC for a better molded surface.
- The best set of conditions would be to make SMC usable directly off the end of the mat-forming machine. There is a wide-open area for improvement here because presently-known chemical thickeners require 24 to 72 hr to develop full molding viscosity.
- Colors in SMC cost more and control of shade, freedom from surface imperfections, variation in hue, bleaching out and so on are more difficult to control.
- In SMC, some orientation of fibers occurs because of the flow in molding. This is true because strengths vary more widely in a specific SMC molded part area even though the glass content remains uniform. SMC is weaker in tensile strength because the fiber pattern and directional fiber alignment are disturbed during molding. Pressure and die closure cause higher range of strength variation in SMC.

In decreasing order, molding methods are rated for uniformity and narrowness of the range of spread of physical strength values as follows:

1. Continuous swirl mat, as in large area moldings.
2. Preformed parts, such as deep draw.
3. SMC
4. Premix molding has the narrowest range of strength variation but is also the lowest in ultimate strengths.

- The largest SMC parts moldable range between 50 and 100 lb in weight, as compared with parts 2 or 3 times that size possible in preform molding.
- Excessive variance exists in the quality of SMC formulations supplied by different compounders. This has resulted in the occurrence of blow holes, or pin holes (small volcanoes) in post-molding heating operations for automotive and other applications, such as bonderizing or other afterbake processes preparatory to painting. Pin holes are eliminated by formulating a higher viscosity SMC material. It is possible to establish proper viscosity and other controls to gain a higher percentage of production recovery of good quality parts without pin holes or blow holes. Continual quality control and tight surveillance are required.
- Reactivity or sensitivity of resins to the thickening agents is inconsistent and must be improved. Thickening agents must also be improved for more reliable viscosity control.
- Higher integrity reinforcing glasses without filamentation or static and with higher physical properties are required.
- Thickened resin is substantially resistant to acetone and other common cleaning solvents used in polyester operations. This factor makes mixing equipment and machine clean-up more difficult once the resin has thickened.
- Low-shrink resins used in SMC present pigmentation problems. Colors, especially light pastel shades, are sensitive, easily mottled and inconsistent, making color matches difficult to establish and control. Color technology for SMC is improving, however.
- Sink marks occurring on flat surfaces directly opposite ribs and bosses are difficult to eliminate, even though low-profile resins are employed. Sink marks or lines inhibit design freedom when the design must be adjusted to hide the sink marks.
- Press parallelism is more critical in SMC than in preform molding and must be held within 0.005 in. corner to corner.
- The physical properties of molded SMC part are overly sensitive to part design, charge configuration, and placement, SMC formulation characteristics, and press closing speed.
- As in premix molding, SMC is sensitive to weld or knit lines. It is difficult to get reinforcing fibers to join across a butt joint, although the surface due to low profile resins looks normal and uniform. Physical strength weaknesses of course result.

Consumer Advantages of Matched-Die Molded RP/C:

1. Modern, smooth, continuous, flowing line design with fewer parts.
2. Lasting beauty and pride of ownership with higher trade in values.

3. Will not rust, rot, or corrode. No appearance disappointment.

4. Lighter than steel or aluminum. Less weight to lift, carry, and move.

5. Quieter. Insulates and absorbs sound.

6. Better heat insulation. Cooler in summer, warmer in winter.

7. Stronger; high impact, resistant to nicks and dents.

8. Easily repaired by cementing, filling, sanding, and repainting.

9. Better performance. Less weight equals higher pay loads, better gas mileage and faster automotive acceleration.

10. Safer. Absorbs more impact energy with less appearance damage.

Manufacturing Advantages:

1. Greater design freedom. Better limits on deep-draw and contours.

2. Lower cooling costs. Fewer tools at 50 to 80% less cost.

3. Shorter cooling time. Start-up and changes made in less time.

4. Fewer parts. Molding includes several former metal parts.

5. No rust-proofing required — eliminates deep-dip priming or galvanizing.

6. Easier handling. Less weight means quicker and lighter handling.

7. Lower manufacturing costs from fewer, lighter parts with less tooling.

8. Quicker changes at lower costs. Less tool rework.*

Market Position of Preform and SMC Molded Parts.

Matched-die molding represents approximately 25 to 30% of the total poundage of RP/C parts produced. Because it most closely approaches straight-line production and automation propensity, it has the potential for tremendous growth, particularly with the advent of improved systems such as low-profile resins and SMC.

Matched-die molding methods supply parts to many fields but the largest user by far is the automotive market. Hence it is fitting to examine the behavior of the material as compared with steel in manufacture of automotive parts.

Source: Harley, R., SPI 20th RP/C Proceedings, 1965, Sec. 14-D.

Advantages of RP/C versus Steel:

■ Lower Tooling Cost. The cost of reinforced plastics tooling is substantially below that of steel. Steel usually requires compound or series dies to produce one part. Reinforced plastic requires one mold per part, so that fewer tools are needed, and those are less costly since they require less material because of the lower pressures involved.

■ Shorter Tool-up Time. Reinforced plastic requires approximately 50% of the tooling time for steel, primarily because of the need for fewer tools. It permits greater flexibility in design changes because there are fewer tools to coordinate.

■ Fewer Parts. In some cases, greater designs flexibility permits substitution of one reinforced plastic part for several steel parts, reducing the number of assembly operations.

■ Possibility of Improved Styling. Reinforced plastic permits sharper lines for improved styling, but the design must be moldable and finish sanding minimized to preserve sharp lines.

■ Less Weight. A 25% reduction in weight is not unusual.

■ Will Not Rust or Corrode. Years of service exposure to all kinds of weather and road conditions have proven the durability of reinforced plastic. Although subject to more wear from stone pecking, it is inert to many metal-corrosive chemicals that may be expected in service.

■ No Yield Point. When steel is subjected to impact beyond its yield point, a permanently dented door or a crinkled fender results. Reinforced plastic cannot be dented. It can be fractured, but the force necessary to do this is much greater than the force needed to dent the standard steel auto body panel.

■ Strength. It is reported that the reinforced plastic body is undamaged in many minor collisions. Where damage does occur, it is claimed that it is localized, not transmitted through the body causing a sprung hood or a jammed door, as with steel.

Disadvantages of RP/C versus Steel:

■ Higher Material Costs. Cold-rolled body steel ready for forming costs about 7¼ cents per pound. Resin and glass ready for processing cost about 27 to 33 cents per pound, but about 25% less is required.

■ Higher Forming Costs. Labor content is considerably higher in the molding processes, due to the longer cycle time required to form and cure each piece. Methods are still relatively crude compared to steel stamping operations.

- Higher Assembly Costs. In assembly of reinforced plastic body components, the bonding adhesive requires a curing period which is not parelleled in spot welding of steel body components. Furthermore, direct bonding of reinforced plastic parts to metal parts has not yet proven feasible. Bolts, rivets or other types of mechanical fastening are required.

Reinforced plastic bodies produced thus far are unique designs employing unique metal reinforcement members. With present designs and assembly processing, it is not practical to use reinforced plastic unique body parts in conjunction with common steel parts of a higher volume vehicle. Processing such a reinforced plastic-steel combination low volume body in the same plant with a higher volume all steel body would require considerable development.

- Higher Finishing Costs. In surface finishing operations, more sanding is necessary to remove ripples and fiber patterns in the surface and putty-rub must be employed to fill in pits, etc. to produce an acceptable finish. When a painted reinforced plastic part is adjacent to a painted metal part of the car appearance differences are, of course, most noticeable. Under these conditions there have been problems matching the surface and appearance of the painted metal.

What Is Required to Make Reinforced Plastic More Competitive with Steel in Automotive Applications.
Following are some of the product and processing improvements the automotive industry would like to see reinforced plastic suppliers develop in order to give their materials wider industrial usage:

1. Better tooling, more automation, and resins that cure in seconds rather than in minutes.
2. Surface on parts from mold requiring practically no finishing.
3. Improved bonding methods.
4. Improved manufacturing techniques, so that it is practical to handle reinforced

plastic and metal panels through the same paint process.
5. Improved reinforced plastic panel appearance, so that it would be feasible to mix reinforced plastic and metal panels on the same vehicle during processing and with a satisfactory appearance match on the vehicle.

APPENDIX I
1. Procedure for measuring viscosity of high viscosity resin pastes for SMC
Equipment
Brookfield HBT viscometer mounted on Brookfield Helipath Stand
Brookfield HBT spindles
Brookfield T-Bar spindles and chuck
Procedure
A. Viscosity = 30,000-500,000 cps.
1. Carefully immerse No. 6 spindle into sample up to notch in spindle stem. Air bubbles must be excluded.
2. Attach spindle to viscometer.
3. Turn on viscometer at 5 rpm. Let it run until readings stabilize or reach a maximum. Record maximum reading.
4. Viscosity = reading X 16,000 cps.
B. Viscosity — 0.5-80 million centipoises
1. Attach round weight to T-Bar chuck.
2. Insert T-F T-Bar spindle (smallest one) into chuck and hand tighten chuck as tightly as possible.
3. Attach chuck to viscometer.
4. Lower viscometer by hand until lower tip of spindle is just above surface of test sample.
5. Turn on Helipath Stand to lower spindle into sample.
6. Turn on viscometer at 1 rpm after crossmember of spindle penetrates surface of sample completely (about ⅛" below surface). Note the position of the viscometer dial.
7. Record the dial reading after 2 complete revolutions (2 minutes).
8. Carefully loosen chuck, taking care not to put vertical stress on the viscometer shaft. Viscometer may then be raised clear of spindle.
9. Viscosity is calculated from dial reading by multiplying reading by 800,000 centipoises.
C. Viscosity = 80–160 million cps
1-5. Same as B.
6. Same as B, except 0.5 rpm is used instead of 1 rpm.
7. Record the dial reading after 1 complete revolution (2 minutes).
8. Same as B.
9. Viscosity = reading X 1,600,000 cps.

Source: Harley, R., SPI 20th RP/C Proceedings, 1965, Sec. 14-D.

Source: O-Reilly, J. T., SPI 20th RP/C Proceedings, 1965, Sec. 14-B.

2. Standard Test for Measuring the Response of Polyester Resins to Chemical Thickening—

Preparation of RVF Viscometer: Before any determination can be made, the Brookfield viscometer and stand must be set up properly. Detach the wooden handle by unscrewing the nut, then slide the handle back, leaving the metal holder rod visible. Clamp the viscometer to the stand and adjust meter to exact level position using the bubble level. Set the hexagonal knob located at the side of the upper case to twenty. Select spindles #5 and #7 and place them in a beaker of water at 100°F and place the beaker in the water bath at 100°F (±0.5°F) for at least 15 minutes before using for determination.

Test Procedure: Weigh 200 grams (±0.1) of sample into an 8 oz. tall form jar. Cover the jar with a lid having a small hole in the center through which a thermometer (30-150°F) is inserted. Place the jar in a 112°F (±2°F) water bath. A metal jacketed glas-col heating mantle (Catalog No. M-612) controlled by a variac at a setting of approximately 23 is suitable. Stir the sample with the thermometer occasionally and allow it to heat to approximately 102°F. While the temperature is adjusting, weigh 17.00 grams (±0.01) of calcium hydroxide (Mallinckrodt A. R. Grade No. 4195) onto a square of aluminum foil.

Remove the jar from the bath and clamp it to a stand which holds a high speed air stirrer (arrow) with a metal agitator. Insert the agitator blade to about ½ inch below the surface of the resin and adjust the jar to allow room for the thermometer. Stir the resin with the thermometer until the temperature has dropped to 100.5°F, then remove the thermometer. Recenter the agitator blade and start the air stirrer slowly to create a vortex but not cause a large amount of air bubbles. Simultaneously add the calcium hydroxide and start a timer. Begin timing when the first portion of calcium hydroxide contacts the resin. All of the calcium hydroxide must be added within 25 seconds. Use a wooden handled spatula to facilitate the complete addition of calcium hydroxide. After all of the calcium hydroxide has been added, lower the agitator blade to near the bottom of the jar and increase the stirring to maximum. After 60 seconds of elapsed time, quickly stop the stirrer, withdraw the agitator, remove the jar from the stand and place it in the 100°F water bath and cover it with the single-holed lid.

Using extreme care, attach spindle No. 5 to the Brookfield viscometer. Insert spindle No. 5 through the hole in the lid into the sample to the depth indicated by the groove cut in the shaft of the spindle. Set the rotation adjustment for 20 rpm. Engage clutch and permit motor to make enough revolutions to obtain a steady reading. This probably will take at least two revolutions.

After 2½ minutes elapsed time, engage the clutch and stop the motor when the pointer is in view. Record the reading indicated on the scale.

Viscosity in Centipoises (No. 5 Spindle at 20 rpm) = Scale Reading X 200.

Restart the viscometer and leaving it running between readings at 2½ minute intervals, i.e. 5, 7½, 10 minutes, etc. until a sudden large rise in scale reading is noticed. After the large rise in scale reading, but before the pointer goes off the scale, change to spindle No. 7. Take readings at 5 minute intervals for two readings then switch to 10 minute intervals. If the pointer approaches the end of the scale, reduce speed of the viscometer to 2 rpm. Take the final reading 30 minutes after the first scale reading which was greater than 50.

Viscosity in Centipoises (No. 7 Spindle at 20 rpm) = Scale Reading X 2,000.

Viscosity in Centipoises (No. 7 Spindle at 2 rpm) = Scale Reading X 20,000.

Plot the results on 5 cycle semi-logarithmic graph paper with viscosity logarithmic and time linear. Draw a smooth curve through the plotted points. Report the time recorded at 10,000 cps and the viscosity recorded thirty minutes after that time. Clean the spindles thoroughly by soaking in acetone and wipe dry.

Equipment—

> Brookfield RVF Viscometer
> No. 5, No. 7 Spindles
> 200 gms. Polyester resin
> 8 oz. Tall Form Jar & Lid
> 100°F, 112°F Water Bath
> 17 gms. Calcium Hydroxide
> 30-150°F Thermometer
> Stop Watch

Source: Espenshade, D. T. & Lowry, J. R. SPI 26th RP/C Proceedings, 1971, Sec. 12-F. (Rohm & Haas Co., Philadelphia, Pa. Reprinted by permission).

References

1. Fekete, Frank, SPI 27th RP/C Proceedings, 1972, Sec. 12-D.
2. *Ibid.*
3. SPI 16th RP/C Proceedings, 1961, Sec. 20-G, T. Kramer.
4. SPI 20th RP/C Proceedings, 1965, Sec. 14-D, R. E. Harley.
5. SPE Retec Preprint, Cleveland, Ohio, 1966, p. 92.
   "The Tool Engineer" July, 1955, Morrison,

R. G. & Martin, Morgan; Bulletin No. 710-B15, W. R. Grace Company.
SPI Preform Committee Summary Papers, 1961, A. J. Wiltshire.

6. Adapted by permission of SPI 20th RP/C Preprint, 1965, Sec. 18-E, Mendiola, N. J.

7. Blanchard, B., SPE 9th RP/C meeting Postprint, Cleveland, Ohio, 10/4/71, p. 117.
Blanchard, B., Personal communication, 3/29/72. Plasticolor, Inc., Ashtabula, Ohio.
Dunn, H., Personal communication, Koppers Co., Inc., Chemical Pigments Division, Newark, N. J., Feb. 6, 1972.

8. SPI 16th RP/C Proceedings, Preform Review Papers, 1961, Sec. 20-D, R. W. Meyer. See also *Modern Plastics,* Vol. 36, Feb. 1969. pp. 87 92.

9. SPI 24th RP/C Proceedings, 1969, Sec. 7-D, Weslock, E. A., et al.

10. Personal communication, I. G. Brenner, I. G. Brenner Co., Newark, Ohio.

11. *Plastics Technology,* Aug. 1969, p. 37, M. W. Reilly.

12. SPI 25th Proceedings, 1970, Sec. 6-E, Nussbaum and Czarnomski.

13. Minutes SPI Preform and Mat Die Molding Committee, Macon, Ga., Oct. 22, 1969.

14. SPI 25th RP/C Proceedings, 1970, Sec. 6-D, F. Fekete.

15. SPI 22nd RP/C Proceedings, 1966, Sec. 11-E, W. K. Glesner, et al.

16. SPI 19th RP/C Proceedings, 1964, Sec. 17-E.

17. Rohm & Haas-Low Profile Resins Technical Data PI-751.

18. SPI 19th RP/C Proceedings, 1964, Sec. 2-A.

19. SPI 20th RP/C Proceedings, 1965, Sec. 10-A.

20. *Modern Plastics,* Apr. 1969, p. 172.

21. SPE-RP Retec, Cleveland, Ohio, 1967, p. 21.

22. SPI 23rd RP/C Proceedings, 1968, Sec. 18-D.

23. SPI 20th RP/C Proceedings, 1965, Sec. 14-D, R. G. Harley.

24. SPI 23rd RP/C Proceedings, Sec. 11-E, E. Carley.

25. SPI 20th RP/C Proceedings, 1965, Sec. 10-D, G. I. Davis, et al.

26. SPE RP/C Retec Preprint, Cleveland, Ohio, 1964, p. 26.

27. SPI 20th RP/C Proceedings, 1965, Sec. 10-D, J. Deef.

28. SPI 26th RP/C Proceedings, 1971, Sec. 21-A (Preform Review), R. W. Meyer.

29. SPI 21st RP/C Proceedings, 1966, Sec. 11-C, R. W. Meyer and R. J. Savage.

30. SPE RP/C Retec, Cleveland, Ohio, 1964, p. 17, Thompson, et al.

31. Product Data Sheets, W. R. Grace & Co.

32. *Modern Plastics,* Oct. 1965, W. J. Connolly and A. M. Thornton.

33. SPE RP/C Retec Preprint, Cleveland, Ohio, 1964, p. 31, G. G. Hooper.

34. SPI 23rd RP/C Proceedings, 1968, Sec. 1-C, Kralovic, W.

35. SPI 22nd RP/C Proceedings, 1967, Sec. 3-C.

36. *Modern Plastics,* Sept. 1968, p. 200.

37. Product Data Sheets, Johns-Manville Products Corporation.

38. SPI 25th RP/C Proceedings, 1970, Sec. 6-B.

39. Product Data, Union Carbide Corporation.

40. Proceedings Brooklyn Polytechnic Inst. RP/C Conference, December, 1969, I. Muscat. See also SPI 22nd RP/C Proceedings, 1967, Sec. 8-A for fusible, moldable sheet reinforced with jute, nylon, fiber glass, steel and other metal fibers.

41. Fekete, Frank, SPI 27th RP/C Proceedings, 1972, Sec. 12-D.

42. Rabenold, Ronald, SPI 27th RP/C Proceedings, Sec. 15-E.

43. Kroekel, C. H., and Bartkus, E. J., SPI 23rd RP/C Proceedings, 1968, Sec. 18-E.

44. Espenshade, D. T., and Lowry, J. R., SPI 26th RP/C Proceedings, 1971, Sec. 12-F.
See also:
U.S. Patent No. 2,628,209 (U.S. Rubber Co./ W. R. Grace Co.)
U.S. Patent Nos. 3,434,320 and 3,465,061— Koppers Co.
SPE RP/C Retec, Cleveland, O., 9/25/69, p. 85, Fekete, F.

45. Personal Communications: Wright, F. M., and Walton, J. P., Marco Division, W. R. Grace Co., Linden, N.J., Apr. 6, 1972.

46. Personal Communication, Fekete, Frank, President Quality Controlled Industries, Inc., Ruffsdale, Penna., Apr. 6, 1972.

47. SPI RP Handbook, Oleesky-Mohr, Reinhold, 1964, p. 371.
SPI 26th RP/C Proceedings, 1971, Sec. 12-F, Espenshade, D. T., et al.

48. *Plastics Technology,* Dec. 1970, p. 23, Miller, E. R.

49. SPI 26th RP/C Proceedings, 1971, Sec. 21-B, Meyer, R. W.

50. SPI 26th SPI RP/C Proceedings, 1971, Sec. 15-B, Sheatsley, R., and Ring, E.

51. Personal Communication, E. Ring, Goodyear Aircraft Corp., Akron, Ohio, 2/16/71.

52. Plastics Design and Processing, Feb. 1970, p. 18, Carter, N. A.

53. Personal Communication, Rosenberger, F. A., Fiberite Corp., Jan. 4, 1970.

54. Personal Communication, Pollman, G. A., Johns-Manville Fiber Glass Reinforcements Div., Waterville, O., June 21, 1971.

55. SPI 27th RP/C Proceedings, 1972, Sec. 12-A, Wright, F. M.

56. SPE RP/C Retec, Preprint, Cleveland, O., Oct. 9, 1967, Humphrey, E. F.

57. *Plastics Design and Processing,* Jan. 1968, p. ,8 Morrison, R. S.

58. SPI 20th RP/C Proceedings, 1965, Sec. 6-A, Eastwood, N.

59. SPE RP/C Retec, Cleveland, O., 9/30/68, p. 63, Wright, F. M.

60. SPI 21st RP/C Proceedings, 1966, Sec. 7-A, Tiffan, A.

61. SPI 21st RP/C Proceedings, 1966, Sec. 11-C, Meyer, R., and Savage, R.

62. SPI 22nd RP/C Proceedings, 1967, Sec. 8-B, Walton, J. P.

63. SPI 22nd RP/C Proceedings, 1967, Sec. 8-C, Tiffan, A. J., et al.

SECTION V

MISCELLANEOUS THERMOSET MOLDING PROCESSES

V-1

Filament Winding

INTRODUCTION

The art of placing filaments over or onto structures to resist applied forces appears to be a common technique present in the earliest historical records, and even found in pre-civilization excavations. Upon their death, the pharohs of ancient Egypt were wrapped in strips of fabric interlaced with aromatic herbs and some rosin as an adhesive and, thanks to the dry climate, we are able to see the actual cocoon in museums in most large cities. In the twelfth century, Ghengis Kahn utilized filaments of bull tendon, rosin, and wooden strips, all overwrapped with a silk-fiber circumferential winding, to produce the excellent bows which his horse-mounted archers used to conduct the first blitzkriegs. Early cannons were wrapped with iron wire to hold the explosion within control. Later applications of filaments in pre-stressed concrete, and the hosts of patents recognizing the load-carrying ability of chains, ropes, and wires[1] laid the foundation for the term and substance of the new technique called *filament winding.*

The first classical RP/C filament-wound structure, which incorporated an understanding of the role of glass filament placement was completed in 1947 by Mr. R. E. Young, working under a consulting contract with the M. W. Kellog Co. and supported by the U.S. Bureau of Ordinance in its search to develop a strong and lightweight pressure vessel.[2] Many others were also working simultaneously on the art of oriented or directed filaments but to the author's knowledge, the record is quite poor.

In 1948, Mr. Young formed a consulting firm called Young Development Labs which started in Long Island, N.Y. and later moved to Rocky Hill, N. J. Much of the activity of this small company was devoted to the development of containers for solid rocket propellant in cooperation with the Allegheny Ballistics Laboratory. The propellant grain was encapsulated in fiber glass epoxy-resin cocoons, which made possible the Antares and Altair solid rockets.

Following the lead of Young Development Labs, several other companies began serious work on filament winding from 1948 through 1953. Notable among those pioneers were The Walter Kidde Co.; Lamtex Industries; Sharples Machine Co.; Spiral Glass Pipe Co.; Apex, Inc.; and A. O. Smith Co. In general, their products fabricated were either pipe or military hardware ranging from pressure vessels to torpedo-launching tubes.

In 1955, Young Development Labs completed the first analytical treatment of integral-end pressure vessels through the "netting analysis" and the development of the ovaloid dome. Young was later bought out by the Hercules Powder Co. and full-scale development of the third stage Minuteman and the second stage Polaris commenced in 1958.

Filament winding has since progressed with considerable acceleration. Some of the landmarks in the development of the concept can be appreciated in the following chronological listing:

1952—Patent on winding tubes, Francis, K. J., #2,614,058

1955—Winchester, model 59 semi-automatic shotgun with FW barrel

1955—Patent on vertical winding, DeGanahl, C., #2,714,414

1955—Naval Ordnance Labs (NOL) set up ring tests on initial development work on FW

1956—First FW motor flies on third stage of Vanguard missiles

1958—Patent on continuous FW on site, Imber, R., #2,822,575

1959—Large, continuous tube FW, Sweden

1960—Decision to use FW cases on Polaris missile

1964—First book on FW, Rosato and Grove

1964—Conical FW motor case for two-stage Sprint missile

1965—Contract completed for design and manufacture of equipment to build 22-ft diameter × 55-ft long FW motor case, #AFML-TR-65-436, November 1965.

1968—Patent on composite pipe, Bradley, R. C., #3,379,591

1968—Patent on flexible FW pipe, Skoggard, B. B., #3,399,094

1969—Contract for FW wing design and manufacture No. AFML-TR-68-378, March 1969

1969—Patents for sand-fiber pipe, Grosh, J. L. #3,483,896, #3,470,917

1970—Patent for on-site FW tank manufacture, Clements, H. R. #3,524,780

Reports are provided through the efforts of Mr. Charles Tanis, AFML, Dayton, Ohio.

Product and Mechanical Development. It is fortunate or unfortunate depending on viewpoint that the impetus of the great amount of work brought on by the need for highly efficient pressure tanks for the missile and space program in the years 1960 to 1969 made the industry think of filament winding almost exclusively in terms of the classical idea of winding as carefully located and filaments tensioned onto mandrels in sophisticated machines. This situation actually delayed development of commercial filament winding. This chapter will treat the military uses of filament winding, such as high-specific-strength tankage, but it will also cover the vastly larger and more practical aspects in which the greater part of the reader population is vitally concerned.

Commercial filament winding is concerned with the placement of directed filaments and other reinforcements in close relation to the loads that are to be opposed. Discussed here will be the machines and techniques that do this work and also the placement of the very largest amount of materials onto the very simplest of mandrels in a degree of precision that meets the specification of the ultimate customer. All forms of placement of filamentary materials will be considered.

Basic Types of Filament Placement. For introductory purposes Figure V-1.1 illustrates the basic types of filament placement. They are described below:

- *Classical Helical Winder*—Helical winding, back and forth on a mandrel is done with this machine (*A*).
- *Circumferential (hoop) winder*—In this method, also known as spool wrapping, the filaments are wound in a continuously advancing manner like thread on a spool (*B*).
- *Polar winder*—This machine provides a continuously advancing wrap from end to end, around the ends of the mandrel's axis (*C*).
- *Continuous helical winder*—This type of winding is done upon segmented mandrels which pass under the rotating winding heads, also upon an extruded mandrel, or upon a release film on a stationary mandrel (*D*).
- *Continuous normal-axial winder*—This machine functions in the same way as the continuous helical winder, except that it has axial and circumferential filaments (*E*).
- *With continuous rotating mandrel (Drostholm wrap)*—In this process, the mandrel rotates and the fibers are added from a stationary creel at a slight angle. The wrap advances axially and helically wraps the mandrel, which has a patented movement mechanism. The mandrel, very slightly tapered, merely rotates and protrudes well past the cure area or oven, so that the product (usually a round tube) is curved and is strong enough to withstand the pull after leaving the support of the mandrel (*F*).
- *Fiber-placement machine*—With the advent of exceptionally stiff (E=45 × 10⁶) and costly ($300.00/#) boron fiber-epoxy tapes for filament winding or placement, special machines have been and are being built to lay down such tapes in a precise manner from x, y, and z coordinates. The machine depicted Figure V-1.1 (*G*) is a tape dispenser with 6 degrees of freedom and is capable of placing the tape within ± 0.010 in. of the desired path, although not without

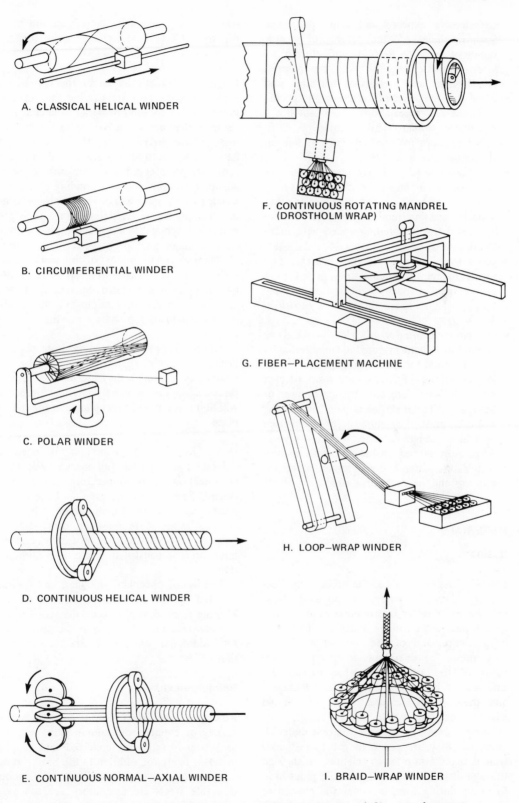

A. CLASSICAL HELICAL WINDER

B. CIRCUMFERENTIAL WINDER

C. POLAR WINDER

D. CONTINUOUS HELICAL WINDER

E. CONTINUOUS NORMAL—AXIAL WINDER

F. CONTINUOUS ROTATING MANDREL
(DROSTHOLM WRAP)

G. FIBER—PLACEMENT MACHINE

H. LOOP—WRAP WINDER

I. BRAID—WRAP WINDER

Figure V-1.1 Schematic representation of basic methods and types of filament placement.

considerable expense and some problems. Such machines are the parents of a new family of precision RP/C manufacturing devices and will enlarge the use of the material into aircraft and aerospace structures and other commercial areas where high specific strengths and repeatability are essential.

- *Loop-wrap winder*—This machine places fibers into loops for products such as tension straps for guy lines (*H*).
- *Braid-wrap winder*—Filaments or tapes are wrapped in a may-pole machine over a mandrel which can be plastic, glass, or other material needing reinforcement. It is saturated with resin, cured, wrapped with more braid, and so on (*I*).
- *Inside-wrap winder*—In this process, filaments of reinforcements are applied in a programmed manner to the inside of a rotating mandrel (see also in *Centrifugal Molding*).
- *Combination braid-wrap winder*—Axial or longitudinal filaments are added alternately with the braid for special properties.
- *Wound fabrics*—Filaments are wrapped over mandrels which are flat, round, square, or hexagonal. The material is then cut off the mandrel, press or vacuum molded, and used as a high-strength material for hand lay-up. Also, such fabrics are wound with incomplete fabric patterns to give screen effects for filtering and/or decorative effects.

MATERIALS

Resins:

Epoxy. Practically all of the military applications of filament winding use epoxies because they have far better interlaminar strength and less shrinkage than other materials. The costs of post-cure, and of the resin itself are of only minor consideration compared to the value of the article and what it can do. Epoxies can be easily prepegged and B-staged, and these handy features make the actual winding of the article much easier. For all these reasons, epoxies are the best material. However, they are prone to get thin if wet-wound (exotherm prior to gelation) as they go through their initial cure. Tanks and pipes have to rotate during cure, or suffer the chance of becoming unbalanced, that is, overloaded with

resin on one side and starved on the other. Figure V-1.2 presents typical epoxy formulations.

Polyesters. Practically all of the commercial applications use polyesters rather than epoxies. Polyesters are as good or better in chemical resistance, they are about half as expensive, and they require no post-cure. However, they do have to be carefully catalyzed and, although some are pre-pregged, the problems of getting adequate compression and bonding of the materials to each other without applying a vacuum or other procedure legislates against such general use. Chemical tankage and piping account for the major part of polyester consumption via filament winding. Most of this work is done with systems which cure in a few minutes after the laminate is completed. Smaller items may be cured in ovens, thus minimizing the chance of loss caused by too short a pot life.

Vinyl-ester resins are becoming quite popular[3] because of their impressive list of chemical resistances. Although they are more expensive, these resistances plus their physical properties will find them a ready acceptance in the market place.

Other Resins. The common bis-phenol-A polyesters still find[4] a full market. For water treatment at room temperature, the common isophthalic types are adequate and are used in great quantities. Most manufacturers have developed expertise in handling these different types of polyesters with equal facility and will quote on tank construction, etc. with no hesitation.

The use of phenolics, silicones, and furanes, is noted but the use of these resins in filament winding is considered outside the scope of this chapter. (see the references at the chapter end for additional and more specific information.[5,6,7,8,9])

Reinforcements:

Rovings. Fiber glass rovings come in a wide variety of counts from simple yarn or strand to 120 ends (strands) collected and supplied in cakes, ready for addition to the head through a tensioning device. In military applications, elaborate creels are mandatory because even-tension is necessary and the manufacturer's

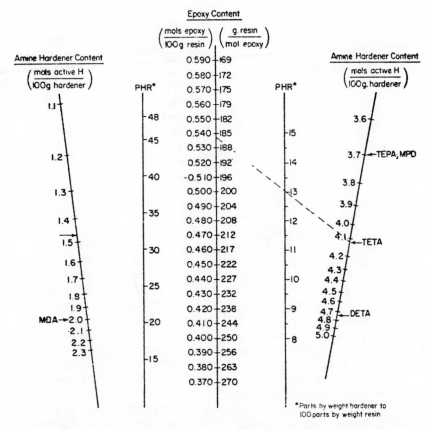

Figure V-1.2 Nomograph Calculator for Epoxy Resin/Hardener Ratios. (Stiochiometric Resin/Hardener Ratios.) To determine parts by weight of amine hardener required for 100 parts of epoxy resin, place straight edge so as to connect epoxy content with hardener content. For parts by weight of hardener for 100 parts by weight of resin, read PHR scale. (Data courtesy of CIBA Corp.)

Table V-1.1 Typical Properties of Glass Fiber Roving for Filament Winding

Property	Specification
Type of glass	E
Type of binder	Silane
Roving no. of ends	20 (exact end), ±0
Ignition loss	0.80%, ±0.10%
Yardage of fiber	13,500, ±10%; 7,500, ±10%
Roving yd per lb	20 (exact end) min.: 638, nom.: 675, max.: 722
Roving moisture content	0–0.05%, 0.05% max.
Average roving stiffness	4.00 in. + 0.3 in. − 0.3 in.
Roving ribbonization	Min.: 4, nom.: 5, max.: 6
Roving package hardness	40, +10, −10
Roving package types	3 in. ID × 10¾ in. length embossed tube
Roving package weight limits	Min.: 31.5, nom.: 35.0, max.: 38.5
Meter setting	24,000 yd per pkg. metered roving
Waywind	2½
Spacing	Min.: 1/16 in. max.: 3/16 in.
Catenary	Max.: ¼ in.
Roving chopability, clumping rating	Min.: no. 1, max.: no. 2

Source: Johns-Manville Fiber Glass Reinforcements Division.

specifications to the glass supplier are very tight and costly. In commercial use, the rovings requirements are less stringent. The finish must be compatible with the resin and the material must be clean and possess the proper low moisture content.

Preimpregnated Roving. Much roving is supplied in the preimpregnated or "resin-containing" state. A 20 end, 660 yd per lb roving strand impregnated with 20% B-staged epoxy or polyester will yield 528 yd. per lb in winding. The individual glass strands must have had absolute uniform tension applied in dipping to avoid catenary. The presence of the pre-impregnating resin assists in arriving at this condition. The roving must be refrigerated, and even under these conditions will show gel and cure time drift in 3 to 6 months. The pre-impregnated roving must be tensioned uniformly when applied in winding. Preheating to soften the resin and improve lay is almost a necessity. Even though more expensive, use of this product permits the winder to circumvent many of the mechanical and handling problems associated with resin-dip and wet winding.

An extraordinary type of "prepreg" roving has been generated for use in specialty filament winding to provide a stronger, non-conductive or corrosive, lighter weight material to supersede metals. This product consists of 5 ends of a 75 1/0 fiber glass yarn (66% by weight) and 4 ends of a 420 denier polypropylene fiber (33% by weight of the finished strand) "whirl-wind" twisted to form a balanced strand, 965 yd per lb, and purveyed on a large creelable package. In application, the glass-plastic strand is helically or otherwise wrapped under tension onto a suitable mandrel and heated to a temperature sufficient to soften the polypropylene fraction and cause it to coalesce with the glass fiber reinforcement. The result is a most interesting composite which, although limited in heat deflection temperature, is capable of many diverse applications. (data courtesy Frickert, P.J., Deering Milliken Corp., Laurens, S.C.).

Fabrics. Fabrics are not recommended for classical winding because of the problem of wrinkling on the ends of the wrap where the head reverses. On some tank applications, however, the reversing is not abrupt and tapes are used with no reported problems. In winding large cylinders and pipes, woven roving fabric, slit into tapes as wide as 12 in., has been doing a creditable job. With circumferential winding fabrics as wide as 140 in. are commonly applied. Such techniques have a long history in the electrical field. In the Drostholm process[10] fabrics in tape and wide forms can be applied to increase the through-put in combination with other reinforcing materials.

Chopped Glass and Mat. In large applications, it is now common to employ the glass-resin depositors (spray-up gun, see Chapter II-2) together with fabrics and rovings to give the item random rather than directional strength where it is required by the design. Chopped material is commonly applied to the mandrel immediately after the gel coat as a barrier. The mandrel is then wrapped with rovings and/or fabrics. In chemical tank fabrication, a final overspray of chopped glass is added as a barrier against outside attack. Surface veil mat is also applied to establish a resin-rich surface of finite, controlled thickness.

Other Materials. Common washed beach sand is alternately wrapped with filaments to make a composite water pipe. Other fillers are also used, such as microspheres, thermoplastic spheres, phenolic spheres, and the like as well as sisal fibers, veneers, and a host of products. They act as the core in the cylinder and enhance the structure, putting the glass in tension or compression rather than in a bending stress. A cored cylinder uses less of material than a solid wrap of glass and resin.

EQUIPMENT, TOOLING AND PROCESSING METHODS FOR FILAMENT WINDING

Because equipment, tooling and molding methods are inseparable in discussing filament winding, they are conjointly presented in this chapter. The subjects treated include mandrel types, machines and methods, the kinetics of chain drives, and processing variables.

Mandrels. In any RP/C application, a mold of some sort is required to suport and shape the laminate. This is also true of filament winding and molds are usually termed *mandrels.* Molds can usually be of somewhat weaker construction in the case of hand lay-up or spray-up, but mandrels for filament winding must be made with great care so that they will be strong enough to withstand the tensions built up by the filaments or strands as well as the strains caused by the oven-curing temperatures that are part of most of the processes. Moreover large mandrels must be strong and rigid enough to not sag during the process. The mandrel usually has to be removed and in some cases this is an engineer's paradox, because the hole for removal may be only slightly larger than the axle. A great deal has been written about mandrels and many materials have been investigated. The following, more popular types will be discussed in this section: water soluble or meltable salts and plasters; collapsible metal designs; plastics and foams; inflatable mandrels; low-melting alloys; mechanical break-outs; combination types: segmented-aluminum-pressurized, metal-substructure plaster-facing, and metal-segmented; reusable pipe mandrels, wing mandrels, house mandrels, blade mandrels, the mandrel machine, tape-wrap mandrel.

Water soluble or meltable salts and plasters. Sand-PVA is one of the best types for mandrels up to 5 ft in diameter. Mandrels should be reduced in weight by embedding large tubes or spheres in them. The PVA easily washes out with water so that removal is easy.

Eutectic salts are better than the low-melting metal alloys and are applicable in mandrels up to 2 ft in diameter.[11]

Soluble plasters are not recommended for mandrels because of their resistance to winding loads and the rather poor wash-out behavior. Sometimes break-out chains embedded at the time of casting will facilitate the wash-out. (They are pulled out thus creating a channel for the hot wash water.)

Collapsible metal designs. Segmented aluminum mandrels are suited for 3- to 5-ft diameters and for quantities over 25 units. Many complicated designs for segmented types have been built and used.[12] The largest segmented mandrel ever constructed was 22 ft in diameter and 55 ft long.[13]

Plastics and Foams. Foams are ruled out for use in mandrels because of their low strengths and poor performance at elevated temperatures. However, for certain low-cost applications with polyester resins and room-tempera-

ture cures, styrene foams coated with a suitable release (to mask the foam from the styrene in the resin) can be used successfully. The foam is washed out with acetone when the cure is completed. In many cases this type of mandrel works well because the tank is made in essentially two steps.. The first step amounts to the inner gel coat and a thin application of 1 or 2 passes of strands which are permitted to cure, thus reinforcing the mandrel. The second phase commences with the remainder of the winding applied onto the reinforced mandrel. After the cure, the mandrel is dissolved.

Laminated plastic mandrels are usually relatively thin laminates made over a mold and joined together to form the tank shape. After mounting on an axle, to provide perfect rotation, this type of mandrel is essentially the same as a foam mandrel with a thin skin of cured winding. An advantage of the laminated plastic mandrel is that it is constructed from materials that serve as the resistive surface and, as such, it is left in to become part of the tank structure.

Inflatable Mandrel. The inflatable mandrel is one of the most interesting and was used in the manufacture of a railroad tank car 8½ ft. in diameter and 50 ft. in length.[14] The choice of this type of mandrel was literally forced upon the builder because with a 50 ft. unsupported shape, the weight of a steel or aluminum segmented or collapsible type would generate enough sag to cause unacceptable strains in the laminate as it rotated. The inflatable mandrel was made over a long central axle and a series of circular spiders which helped to support the skin and serve as points for exerting the torsional loads.

In smaller tanks, present techniques utilize blow-molded or rotocast polyolefins as mandrels. They are mounted on a mandrel axle and overwrapped as required to take the interior pressures. Pressure is usually applied through a rotary joint, although in some cases, hot water is pumped into the shape and the cure can progress almost as fast as the winding. A good understanding of the products and of the behavior of the resins is a positive requirement.

Low-melting Alloys. Low-melting alloys are restricted to smaller mandrels and low-tension windings no larger than 1 ft. in diameter by 1 ft in length. Since these materials resemble lead, their high density and tendency to creep limit their size and applications. However, for large-volume applications involving smaller tanks and pressure bottles, they can be considered production tools. After melting out the alloy material is reusable.

Mechanical Break-outs. Plaster-chain mandrels are quite popular and with proper design and placement of the chain, fast production and low labor cost in mandrel removal are achievable. Many types of piping elbows and the like are made over plaster-chain types of break-out mandrels.

Combination Types. The segmented-aluminum-pressurized design is good for mandrels from 12 to 20 ft in diameter. It offers simple design, ease in fabrication and low weight. The control of pressurization and dimension present problems, however, especially during oven curing.

The metal-substructure, plaster-facing design is appropriate for mandrels which range in size from 12 to 20 ft in diameter and are produced in small quantities or one-of-a-kind designs. It shows advantages where small changes in diameter or provision for inner stiffening rings are required. The plaster can be screeded to many shapes, thus affording a low-cost way of generating unusual tank designs. Removal of such mandrels is a relatively easy process and is another reason for popularity. The TU-312 rocket motor case was wound over this type of mandrel.[15]

The reusable metal-segmented mandrel, 22 ft in diameter and 55 ft in length, was designed to withstand the compression load of the tensioned reinforcing filaments, the weight of the completed tank (30 tons), and also its own considerable weight (100 tons). In addition, it had to be removed in sections through an opening 100 in. in diameter. Mandrel strength and dimensional stability had to be maintained both during the winding and during the final cure at elevated temperatures. A study was made to compare the various types of materials and construction that could be used with this mandrel, and an axle-less, stressed-skin, sheet-steel fabrication was chosen.

This huge mandrel was made by a fabricator who had the proper equipment to roll 22-ft. diameter cylinders and weld them into a 45-ft

long cylinder. After proper clips, bulkheads, and lapping strips had been positioned within the cylinder, it was flame-cut into 9, equal-width (92½ in.) strip staves which could later be passed through a 100-in. diameter port hole. The end caps were made from hammered and forged pie-shaped steel sheet, welded into spheroids. After clips and splicing strips were installed within, the end caps too, were also flame-cut into 9 sections so that they could be later passed through the 100 in. porthole.[16] This mandrel was built, installed, and used frequently enough to prove that it was indeed a feasible way to construct such a huge item.

The mandrels described above have all been based upon the concept of the classic filament-wound product, i.e., pressure vessels with integrally wound heads. Now we will consider some of the wide variety of mandrels designed and built for filament winding in which there is easy access and in which a wound part can simply be slipped off the mandrel or the mandrel collapsed for easy removal. In this chapter discussion of *Machines and Methods,* several types of simple mandrels are described because they are so much a part of the machine. Below, we will describe special mandrels for applications for other than pressure bottles.

Pipe Mandrels. Pipe mandrels are exceptions to the types mentioned above. In most types of pipe manufacture, the resin shrinkage causes the laminate to tightly contract around the mandrel. The mandrel must therefore be made with a slight taper along its full length or with an absolutely smooth and truly cylindrical shape along the whole length. In both instances, strong mechanical or hydraulic forces must be applied to force the part off the mandrel. With helically wound pipes, it is best to place the end of the wound part against an anvil and arrange to pull the mandrel from the part. Such action generates forces in the pipe, tending to diametrically expand it and to put some shear forces into the resin matrix. Parts 4 ft. in diameter and 80 ft. in length have been removed using this technique.

An aluminum mandrel has been used by some manufacturers, despite its greater vulnerability, because it is made hollow and can be heated and cooled as required. In production, the mandrel is heated to enhance the cure. It is then chilled, a procedure which shrinks it away from the laminate and makes removal quite easy. In the larger sizes, it is common for the mandrel to literally "fall out" of the pipe.

Some firms use water pressure to do the removal job. This means laminating an enclosing end onto the wound pipe and also another containing end on the mandrel. Then, by pumping water between the 2 heads, considerable force is generated between piston and cylinder and the water leakage between the 2 elements serves to lubricate the area and also to dissolve the PVA if it has been used as a release.

Wing Mandrels. In 1968, the Air Force's Materials Laboratories issued a contract to North American Rockwell and Aerojet General Corp., to investigate the project entitled "Manufacturing Methods for Plastic Airframe Structures by Filament Winding.'" This project, a part of the Air Force's manufacturing-methods program, was one among many whose purpose was "To develop, on a timely basis, manufacturing processes, techniques, and equipment, for use in economical production of USAF materials and components." Such programs were important to our RP/C industry at the time, because they demonstrated the problems and the feasibility of a great deal of far-reaching thinking. Without government sponsorship, it is unlikely that any programs of this type would ever have gone past the proposal stage. A careful perusal of this program is recommended.

The project was based upon the manufacture of an analogue to the T-2B twin-jet aircraft wing structure using the filament-winding technique. The final tests of the product revealed that following data:

- The filament-wound wing was 40% lighter than a comparable aluminum wing.
- The filament-wound wing was 6% over the ultimate physical strength design-load goal.
- The filament wound wing had a bending stiffness/weight ratio that was 22% greater than that of a comparable aluminum wing.
- The filament-wound wing had a bending stiffness/weight ratio 97% greater than that of a comparable aluminum structure.
- The preliminary cost estimates at the time (1968) was only approximately 80% of that of a comparable aluminum wing.[17]

The wing was wrapped over several types of mandrels to generate the spars, the ribs, and the outer skin. The wing mandrel was made from a steel frame with a sand-acrylic mixture screeded onto the surface to give the proper contour. This mixture was oven-cured under a vacuum and was easily removed after winding using water and steam. Other techniques for holding the strands during winding can be properly classified as mandrels. Practically all of the wing structure was made upon a variety of mandrels. Space does not permit a complete discussion of the entire program.[18]

House Mandrels. Development of this concept was funded by the United States Department of Defense and conceived by the University of Michigan in collaboration with the Aerojet General Corp. It was developed in the search for low-cost housing and was based upon earlier work sponsored by the Hercules Co. and The University of Michigan. The mandrel is designed for easy portability in spite of its size, 8 ft x 20 ft x 36 ft. One of its features is the hinged sides supported by a series of jack screws which are used to move the hinged wall into a rectangular cross-section 36 ft. high. After the winding is completed, the jack screws are retracted, bending each wall inward to the center axis. The mandrel then becomes considerably smaller in periphery and the wound house section can be easily lifted off the mandrel.

For two story homes, two 8 ft x 20 ft x 36 ft units are placed upon a large turntable and a hoop winding is laid up over both units, securely bonding them together.[19]

Blade Mandrels. When the transonic-supersonic wind tunnel of the USAF Arnold Engineering Development Center was severely damaged by 1 or 2 stainless steel blades flying off the motor of the axial compressor, the center's engineers decided that RP/C blades would be a far better replacement. They would be 66% lighter, and the centrifugal loads would be less than 25% of those of the steel blades, greatly lessening the strain on the compressor. There would exist the possibility for higher speeds without any over-stressing of the rotor hub. Accordingly, the Parsons Corp. of Traverse City, Michigan, developed a filament-placing machine and a very special mandrel to make the blades.

The mandrel was composed of a series of trapezoidal cross-sectional bars of low-temperature eutectic alloy, individually wound with the material to form a box beam. After the winding, the units were reassembled in the correct order to form the chord of the blade, and using register pins to spot the necessary locations, provided, were then overwrapped with directional fabrics to provide the outer skin shape of the blade. The completed lay-up was then placed within a female mold and cured in an oven with the temperature raised, and the metal alloy was allowed to drain from bleed holes. The alloy was recast into RP/C molds for the manufacture of the next blade. In this case, the use of meltable mandrels and development of a unique winding machine represent the kind of new thinking that the author recommends. This may be summarized as the placement of filaments and directed fibers in line with and to resist applied forces.[20]

Mandrel Machines. The mandrel machine is based upon vertical winding technology that was developed for the manufacture of reinforced-concrete pipe. Generally, the mandrel structure is easily assembled with an outer skin of perforated screen or punched metal. The core is installed vertically, and the plaster extruder moves vertically upwards laying down a band of plaster in a circumferential wrap as the core rotates. The capability exists for advanced units to deposit 2 and even 3 rows of plaster per revolution. After the plaster has been dried and cured the plaster extruders are replaced with motor-driven high speed rotary files. These are set to the proper diameter and move up the rotating plaster-covered core mandrel shaving off the excess plaster and providing a smooth surface thus making ready for the application of the release agent and subsequent filament winding. Obviously, the spherical ends are made in much the same manner.[21,22]

Tape-Wrap Mandrel. The tape-wrap mandrel is capable of both winding and then moving the wound laminate smoothly forward in an axial direction. The mechanism amounts to an endless belt of steel strapping, $3\frac{1}{8}$ in. wide by about $\frac{1}{16}$ in. thick which is wound around a mandrel support composed of a series of beams parallel to the axis and are held in a cylindrical form by means of several discs with slots in them to accomodate the beams (see Figure V-1.3). The endless belt returns through the hollow center of the mandrel after unwrapping from the downstream end of the beam supports. It is then guided by idler rolls back around the end of the hollow axle to the re-

Figure V-1.3 Continuous, large filament winding pipe machine utilizing endless steel belt for the mandrel. Pipe up to 12 ft. in diameter has been produced by this method. (Courtesy of Drostholm Products.)

wind position. In this manner, the steel tape brings about a smooth mandrel surface, rotation, and a simultaneous advance in the axial direction. This mandrel can be quite easily changed from 1 diameter to another by the addition of spacers and other devices (see the discussion concerning the Drostholm Machine).

Machines and Methods. In the industry, the majority of filament winding machines have been built by the companies actually using the technique. However, several companies have sprung up devoted to the business of building a few types of rather standard design. They also have the engineering capability to design and build some very special and interesting machines. (See list of machinery builders.)[23]

The military's need for pressure bottles, solid rocket-motor housings, and fuel tanks in the 1960s provided virtually unlimited funds for several huge and complex winding machines. Some were primarily constructed to evaluate the various methods that might eventually be used, should production of large units become a necessity. They presently reside in various manufacturing complexes awaiting the requirement for more high-specific-strength tanks and the like.[24]

The machine picture is most interesting however, when commercial filament winding (placement) is considered. With the need for low-cost, high-strength, and chemical resistant tanks and pipes, machines to fill these needs have been produced in great proliferation. Almost all of them have been designed to do a specific job, and they have usually been built by the company requiring them. General practice indicates that a consultant is hired to design the machine and monitor its construction, completion, try-out, and eventually approve its operation.

A very few design concerns still remain which are capable of designing a complete system with the required sophistication for a full-blown production winding operation system. This is not to say that a good engineering firm cannot do the job. There are only a few firms, however, presently (1973) devoting their whole efforts to the design and construction of filament-winding equipment and systems.

Filament-placing machines may be classified under the following headings:
Helical winders (classical)
Helical circumferential winders

Helical circumferential dwell winders
Polar-wrap winders
  Plain polar-wrap, rotating mandrel
  Plain polar-wrap, tumbling mandrel
  Circumferential polar-wrap, plain and tumble types
  Polar-wrap, rotating mandrel
  Circumferential polar-wrap, race track
Continuous filament-placement machines:
Helical segmented mandrel
Helical stationary mandrel
Helical continuous mandrel
0-90° wrap
Strickland-B wrap machine
Continuous, on-site winding machines
  Composite type
  Boot-strap type
Vertical carrousel machines
Horizontal-tank machines
Continuous large-pipe machines
Elbow and tee machines
Box-winding machines
Filament-placing machines[25]

Within each type, there exist many variations, all designed to meet the requirements of the particular manufacturer and product.

Helical Winders (classical). Helical winders basically wind a helical pattern along a cylindrical mandrel. The machines look like a lathe with a traversing carriage upon which is mounted a pay-off head that can dispense roving strands, tapes, or bands of filaments called *ends.* An end of roving usually contains 8, 12, or 20 individual fiber glass strands. They are usually supplied from spools, tensioned evenly, saturated with a resin in a small bath, and fed onto the mandrel which rotates in positive relation to the traversing speed of the carriage.

These machines can be highly automated to make 20 to 30 ft lengths and to include automatic insertion and ejection of mandrels upon completion of the specified number of wraps. When weight might cause mandrel deflection, these machines are set with a set of vertical axis and the pay-off head travels up and down. Curing ovens for these machines are often vertical also.[26] Products such as pipe made in these winders have thickened ends because of the dwell time needed for the head to change directions. Unless pipe designers can use such a thickened area, the material is either cut off or ground down with considerable cost and/or scrap in either case. This factor is an advantage, however, in "Bermuda" and "Isotensoid" bottle windings.[27]

Classical helical winders are used throughout the industry. Some of the earliest were used to make electrical items such as hot sticks, fuse tubes, and stand-off insulators. As mandrels got longer and larger, pipes, tubes, ducts, and tank sections became popular products.

Helical Circumferential Winders. Helical circumferential winders are built like the classical helical type, with the additional capability of "shifting gears" which enables the head to make a very slow traverse in relation to the mandrel rotation, providing angles as high as 85° to the axis (similar to those in "hoop" wraps). These machines are set so the wrap on a pipe mandrel can be alternated between the helical and circumferential (hoop) mode.

Classical winders with the hoop-wrapping capability are an extension of helical winders and can produce a product with more resistance to internal pressure than is possible with only helical wrapping. Manufacturers found that the addition of hoop wraps could provide equivalent pressure capability with less material.

Helical Circumferential (Hoop) Dwell Winders. Helical circumferential dwell winders are the same as classical helical winders, with the "dwell" feature incorporated into the helical mode. Such a mechanism in the simpler machines often requires 2 extra sprockets in the head drive circuit. This permits the winding of tanks with either hemispherical or modified ellipsoid ends, and in which there may be various sizes of polar openings. More advance machines employ cams, tape-controlled electrical-servo-drives, and the like.[28] Again, for some larger or longer mandrels, the machines can be vertically oriented to minimize any strains caused by bending loads. These winders are quite versatile and are the most popular. Their transmissions or drive circuits permit combinations of helical wrap angles, different dwell times, and a wide variety of special features, such as automatic programming from start to finish. The head design on these winders is also quite advanced and can be made to thrust or retract as a function of the shape of the part and for the special guidance of the strand or

band as the ends of a tank are wrapped during the dwell phases.

This type of dwell winder is the type most frequently used for internally pressurized tanks and was the basis for development of the netting analysis for the structural design. Tanks wound on dwell machines have integral ends and are quite high in specific strength.

Polar-Wrap Winders. There are several types of polar-wrap winders. They range from those used to wind small high-pressure hydraulic accumulators from 6 ft in diameter to very large size machines that have wound tanks 22 ft in diameter and 55 ft long.[29] Machines to apply essentially polar winds are designed around the product to be produced. Rarely do we see a so-called universal machine. Most larger polar-wrap machines have a second capability for hoop wrapping and some can also do helical wraps.

Polar-wrap machines are generally used to build convex parts that will benefit from fibers being placed nearly parallel to the major axis. Internal pressures are resisted by a hoop wrap in the cylindrical portion so that the design is significantly simpler to accomplish. The ability of the polar-wrap style to apply more material per hr has obvious economic interest.

Plain Polar-Wrap, Rotating Mandrel. These machines are usually arranged to hold the mandrel vertically. The payoff head is attached to an arm that has a stroke equal to at least one-half the length of the mandrel, allowing room for clearance. Either the axis of the mandrel or that of the crank-shaped head-holding arm is supported at an angle that recognizes the size of the polar openings on the mandrel. The mandrel is geared to synchronize its rotation with that of the rotating payoff head, so that the bands of reinforcement can butt, lap, or be spaced, according to requirements.

Plain Polar-Wrap, Tumbling Mandrel. This is a much more complex machine, but the reinforcement tape or band can be fed directly onto the mandrel without passing through a series of pullies or guides. Thus very large or thick tanks can be wound without the need to stop and recharge the filament supply.

In such machines it is difficult to establish dynamic balancing and both sides of the supporting axes must be synchronized. Hence the mechanism must also include synchronized rotation of the mandrel as it is tumbling, so that the filament band can be butted or lapped or can produce whatever pattern is required.

Circumferential Polar-Wrap and Hoop— Plain and Tumble Types. These machines are identical to the plain polar-wrap types, except that they have a gear change where a second head is brought into play and the mandrel rotates about its longitudinal axis as the pay-off head traverses the article from end to end.

Polar-Wrap, Rotating Mandrel. This machine is best described as a device that supports the mandrel on 1 end, holds it at an angle that allows for the size of the polar openings, and then rotates the supporting axle around an axis located midway along the length of the rotating mandrel. Thus, it is actually rotating or tumbling the mandrel around an axis normal to the longitudinal axis. The mechanism also rotates the mandrel around its longitudinal axis, reversing the polar circumferential and hoop wrap method of wrapping as described above. The advantages claimed for this complex machine are the same as for the polar-wrap with the rotating mandrel in that the reinforcement can be continuous, is disturbed to a lesser degree, and thus will develop slightly higher physicals than plain polar-wrap, rotating mandrel.

This method of manufacture allows those interested in the development of winding geometry and the investigation of resin/fiber interface phenomona to adjust physical parameters in a wide range with but small changes in the machine itself.

Circumferential Polar-Wrap (Hoop), Race Track. The mandrel in these machines is supported on axles that are alternately raised and lowered, tipping the mandrel's longitudinal axis with respect to a race-track-shaped horizontal track which guides a motor-driven creel of rovings and impregnating resin. The creel is not unlike a small electric train on a track whose speed and position in the machine are directly related to the rotation of the mandrel. By displacing the mandrel axis to clear the polar openings, such a device is similar to the polar-wrap, rotating mandrel machine. One interesting feature is that several creel trains can run on this track, all simultaneously paying off rovings, thus affording enormous poundage possibilities. Another feature, gained by shifting gears, is that the creel trains can be made to move slowly on the straight portions or run forward and reverse, generating hoop and/or helical windings on the rotating mandrel. Such machines are truly universal and as long as the track encompasses the mandrel, all types of wraps can be accomplished. Small tanks can be handled by adding an inward extension onto the train's pay-off head.[30]

The race-track machine reached its zenith with the manufacture of a rocket-motor case 22 ft. in diameter and 55 ft. long for the Air Force's Materials Laboratories (1963-1965). Although no complete tank was ever made, the machine demonstrated the feasibility of the concept.

Continuous Filament-Placement Machines:

Helical Segmented Mandrel. This machine makes use of filament wraps fed by creels which are held in a rotating frame with a hole in its center through which the mandrel passes. At least two contra-rotating frames are used. Thus, as the mandrels move through the axis of rotation of each frame, a helical wrap is applied whose angle is a function of frame rotation and the mandrel's speed through the frame. It can be seen that a multiplicity of such rotating frames can be added in tandem and the variety of such machines is almost unlimited. Further, low and high angles of wrap can be applied to constantly moving mandrels by using slow and fast rotating mandrels. Non-rotating filaments can also be supplied to provide axial reinforcement, much as in "pultrusion" techniques. As the composite emerges from the wrapping area, the dimple between mandrels is noted and the wrap is wet-cut there. The individual sections are placed in an oven to cure. Mandrels are then extracted from the cured composite by means of a gripping apparatus on the mandrel with the end of the composite pressing against a sphincter plate. The extraction forces put compression on the composite and tension on the mandrel, and this effectively acts to expand the composite, especially if the wrap is helical. This machine is the harbinger of high-production filament winding and one of the first types made in this style. An early application involved the overwrapping of material upon a composite of plywood and balsa strips to construct rails for ladders which were later used by utilities and phone companies. In this case the mandrel was consumed or became part of the product. Later designs used segmented, collapsible mandrels, incorporating inner projections which were bonded to the overwrapped filament-wound RP/C.

Helical Stationary Mandrel. Applications of reinforcements in the helical stationary mandrel are identical to those in the segmented mandrel, but there the similarity stops. In these efficient machines, a highly polished mandrel with a very slight taper is held firmly at the apex of the machine, and a wrap of a suitable release film—usually Mylar, Tedlar, cellophane, or even ABS and/or PVC ribbon is applied with little tension and sealed onto the mandrel. Fibers are then overwrapped. An oven is a necessity with this machine and it can be of the hot-air, radio-frequency energy, gamma-radiation, or other type, as long as the wrapped composite is cured by the time it advances past the end of the tapered mandrel.[31] The taper must be carefully planned because there is some shrinkage as the cure progresses. The draw rolls must not apply too much pressure to the freshly cured tube as it emerges from the oven. In the case of release films such as paper or cellophane, the cut pipe sections are placed in a soak-bath and later flushed with high-velocity hot water. In cases when tubes are used for structural applications rather than for pipes, the paper or cellophane is allowed to remain in the tube. With PVC, ABS, and other release films, the resin usually polyester will react with the materials slightly and bond firmly to them, providing a reasonably smooth bore. New technology for this process includes a film that can be wrapped convolutely around the mandrel just down stream from the mandrel support and then become heat-sealed "on the run" thereby generating a much smoother inner release surface. This method is being seriously considered by the Army MERDC Pipe-line Section at Ft. Belvoir, Va., as a way to quickly construct long-lived and maintenance-free fuel lines under war conditions. The procedure is to prepare slit excavations, generate cure, lay in the pipe, and immediately back fill and tamp. This system provides a fast, efficient cross-country fueling system and also has obvious peacetime uses.

There is great promise in this method because it is a synergistic application of both the RP/C and film materials, and the final product is better and cheaper than the same pipe made from one material. The high production potential is sure to guarantee a price that will extend the use into areas formerly filled by cast iron, ceramic, and concrete pipes.

Helical Wrap, Continuous Mandrel. This process is the same as the continuous-winding method described above, except that a continuous tubular material such as hose, PVC, or ABS pipe or other substrate is fed through the rotating frames. The tube or pipe is left in place, thus providing a thermoplastic liner in a cured thermosetting pressure and temperature-resistant shroud. One notable process now in full production in the United States[32] generates the tube in an extruder, sizes it, cools it, and then runs it through the helical winder without stopping. The resultant product is later cut to convenient sizes. Reasonable bends can be made before the resin cures with the cured composite retaining the induced curves.

An interesting co-development to helical wrapping of RP/C onto straight plastic pipe sections is a method for continuous filament winding over the accompanying plastic pipe fittings—tees, elbows, flanges, and so on. An AMF Versitran® comuterized robot repeats the process ad infinitum after once being led through each step guided or "taught" by manual operation.*

0 to 90° Wrap. These machines are similar to the helical winders, except that more continuous filaments or rovings are applied parallel or almost parallel along the axis than in the helical machines, and the angle of hoop wrap is generally at a high, 75 to 85°. The pipe emerging from the machine looks as though it has been circumferentially wrapped (hoop) in a reciprocating, non-continuous machine. In many cases, the axial-strengthening materials are in the form of orthotropic tapes which are convolutely or helically wound around the mandrel much like Bundyweld® steel pressure tubing.[33] This variation is used for products requiring high axial strength, such as downhole pipe, strain rods, vaulting poles, antennas, and so on. Mandrels are fed to these machines in a manner nearly identical to those described above for the helical-wrap continuous mandrel.

Strickland-B Wrapping Machine. This machine,[34] developed especially to wrap radomes of roughly parabolic shape, consists of essentially a vertical-axis, polar-wrapping device

Source: Courtesy Johns-Manville Pipe Products Division.

combined with a hoop wrapper which applies the circumferential fibers while moving from the large diameter to the small diameter. The ingenious part of this machine is the device which picks out selected axial filaments and cuts them off ahead of the advancing hoop wrap. As the nose portion of the dome is approached, correspondingly more axial fibers per in. are severed and dropped out, and the problem of bunching and overlapping is thereby automatically eliminated. The Brunswick Corp. of Marion, Virginia has machines and owns the patents to this process. (Patent literature describes the process in full detail).

Radomes made by this machine method are characterized by the high degree of uniformity of the weave pattern which, despite the tapered conic or parabolic shape, makes it possible to predict the dielectric constant and loss-tangent values. These radomes also reveal very low bore-sight errors. Because of these advantages, they have captured a major portion of the radome business.

Continuous, On-Site Winding Machines. Most of these machines are built to perform one large job each. Some have been designed as portable factories which are transposed to wherever they are needed in the country to construct tanks and pipes too large for transportation on public highways. There are several designs available:

Composite Type Winding Machine. This machine winds filaments around an inner tank structure which is supported and rotated on an "air bearing," generated by forcing air at low pressure between the concrete tank base and the RP/C tank bottom upon which a collapsible steel mandrel has been erected. The resin and reinforcement are wound upon the surface of the mandrel as it rotates past an elevator that moves vertically, carrying the dispensing head. When the side walls are completed, the mandrel is disassembled, an air-inflated membrane is affixed to the upper edge of the cylinderical section, and the roof is generated as the tank rotates under the dispensing head.

Upon completion of the lay-up, the air is bled from the dome, and the membrane is folded and removed through a manhole. The tank is then positioned in line with the filling and emptying fittings; the air bearing is deflated; and the elevator, pay-off head, rotating mechanism, and blower system are dismantled

and shipped to the next site. The patent describes the process in complete detail.[35]

A variation of this machine permits winding hoop filaments around an inner tank structure made of thin RP/C skins attached to wooden bucks and supported with shoring members. By adding a ply of mat on top of the pre-cured skins and then wrapping prewetted hoop filaments over them, a composite is generated. The pay-off head, roving, and resin supply are mounted on a dolly which is self-powered and rides around the periphery of the tank on a circular steel track. This dolly, or another one, is fitted with rollers which impress the fibers into the resin. Such a system is limited to tanks up to 12 ft high, because, at any greater distance from the ground, the pay-off head becomes unstable and it is difficult to densify the laminate. One variation in pay-off mechanisms includes rapid vertical movement as the dolly travels around the tank. This generates a high-angle helical wind pattern which effectively ties the elements of the wall structure together.

Boot-Strap Type. The modification represented by this machine permits the manufacture of tanks beyond the limit of the procedure and equipment of the similar composite type. The mandrel is designed with jacking provisions which lift the original cured laminate vertically, revealing the mandrel surface below which is then made ready for another lay-up of material.

A separate overwrap of fabric and rovings is added between the original and second layers, thus forming a joint and a stiffening ring. Half-round cardboard tubing is often applied vertically over the joint area and overlayed with more RP/C to provide additional rigidity. Construction of the tank proceeds with as many wraps as the customer dictates.

In especially large tanks, the roof is constructed first and is elevated to the top of the mandrel, where the first lay-up is bonded to it. The jacking system then can push against the roof and balance the lifting action, providing structural rigidity not available with large open-top tanks. The mandrel is usually removed by simply dismantling it and slipping the parts out from under the skirt of the tank laminate. The tank is then lowered into a prepared groove made in the base laminate of the tank and is potted with a suitable resin. Additional layers of tape are applied inside and outside the juncture, and the tank is complete.

Vertical Carrousel Machines. Vertical carrousel machines are merely large cylindrical winding machines designed with features that are purely commercial in nature.[36] They are made as cheaply as possible in all non-critical areas. The turntable or carrousel is usually composed of a large disc made of 1 in. marine plywood. The disc is screwed to a steel channel frame with a series of non-swivel casters affixed to the periphery, which roll on a steel track. The assembly is carefully leveled and grouted to provide a smooth, true surface. The carrousel is rotated by either a pressure roller or a sprocket engaging a chain attached to the periphery of the disc. The pay-off head is set on an elevator (1 unit used a forklift) with a height equal to that of the tank mandrel. The payoff head can be either a tape dispenser, a chopper gun, or both. On some machines, spring-loaded rollers are attached downstream from the material application so that the head applies impregnated fiber glass and densifies as it progresses along the axis of the mandrel. The head is set to apply resin first, chopped fibers and resin next, and continuous rovings last. The roll-down takes place after the application of all fibers. This procedure can be repeated until the proper thickness of material has been applied.

The bottom of the tank has been previously made, cured, and dropped onto the top of the mandrel, where it rests on a movable top plate affixed to a hydraulic ram. The tank bottom is made with downturned flanges which fit the mandrel, passing several inches past the seam between the top plate and the mandrel wall. The mandrel has an automatic ejection system. After the tank bottom is bonded or laminated to the tank wall lay-up, pressure is applied to the ram, and the full lay-up is lifted off the tapered cylindrical section of the mandrel. At this stage standard tackle is applied to completely remove the tank, overturn it, and expedite further manufacturing operations.

Tanks made by this method expand the feasibility of composite construction for the storage of unusual chemicals, for insulative shapes, and for other special applications not fabricated by conventional winding.

Horizontal-Tank Machines. Horizontal-tank machines are similar to vertical carrousel-type machines, except that the pay-off head travels horizontally along the floor next to a rotating mandrel which is supported on 1 end by simple but rugged bearings. The cylindrical mandrel is usually made from rolled steel plate designed to collapse slightly on 1 side, permitting the cured lay-up to drop off in a radial direction onto a wheeled cradle.

In one example of this procedure, three men work at the pay-off head near 1 end of the slowly rotating mandrel. The first man handles a two-component resin gun and sprays resin on the mandrel at the bottom (or end) juncture and cylindrical area. He walks toward the other end of the mandrel at a rate which allows him to spray the resin in a low-pitch spiral pattern with enough overlap at each revolution to eliminate any dry area. The second man in line is one revolution or wrap behind the first. He applies tape, mat, fabric, or roving from a hand-held or dolly-mounted spool of material. The third man rolls the reinforcement which was applied by the first 2 men down into the resin. The trio advances the whole length of the rotating mandrel, applying resin and reinforcement and rolling it down to remove air. They complete as many layers as are required by the design specifications. The second man is very important in this method, because it is his responsibility to either lap or butt the tapes as they are spirally wrapped onto the mandrel.

As in the verical carrousel machine discussed above, 1 end of the structure is made on a separate mold and is affixed to the outboard end of the mandrel. It is then subsequently laminated in with the wall structure. This filament-winding method is about as simple as can be imagined, and yet it is doing a capable and adequate job for commercial applications. A wide variety of materials may be applied efficiently using this simple machine. Consider, if you will, the construction of an insulated tank.

The inner skin is applied as described using the 3-man application. Then foam or balsa wood is applied in tape form (such material is available laminated to a scrim fabric in small 1 or 2 in. squares,[37]) and in tape widths. This scrim is spirally wrapped onto the first skin which has been wet-out (copiously covered with extra resin to be used as an adhesive for the core). The outer skin is then applied over the insulating core in the same manner as the inner skin, and the tank assembly is allowed to cure. A tank of this construction that is 12 ft long and 12 ft in diameter made can be built by a 3-man crew in less than 1 working day. Other materials can be inserted during the rotation of the mandrels, and because the speed is fully controlled, the rotation can be indexed. Longitudinal splines or staves can be applied and later wrapped in. This is truly a versatile method.

Continuous, Large-Pipe Machines. In 1962, a Swedish development comprising a continuous, large-pipe machine was described in a paper[38] given at an SPI conference. An improved version of the machine was described in 1968 by Drostholm Products, of Vedbaek, Denmark and was pronounced ready for international licensing. This offer was accepted by the Glascraft Division of Republic Industries, and the machine is currently being licensed for operation in the United States.[39] Pipe in sizes up to 12 ft in diameter can be continuously fabricated. The fundamental feature of this continuous-production, large-pipe or tank machine is the incorporation of a non-rotating, stationary head. The machine forms the product continuously and the materials are fed to the mandrel from the stationary head as the mandrel rotates slowly. The generated spiral-wrap pipe exudes from the downstream oven at the terminal of the machine in an advancing and rotating manner. The machine can then be programmed to feed all manner of materials onto the mandrel to produce products with many features previously unattainable in continuous production: sandwich construction, tapered lay-up, banded pipes, convex shapes other than cylindrical, unlimited 1-piece construction, different resins applied at will, and longitudinal fibers of all sorts and quantities.

Several additional elements are included in this continuous, large-pipe machine:

Drive Unit. This unit contains a variable speed motor with tight speed control which drives an axle. The axle, which projects rearward about 20 ft, is securely supported on large, spherical roller bearings about 4 ft apart.

Mandrel. This mandrel previously described is capable of smoothly moving the wound laminate in the axial direction. The mechanism amounts to an endless belt of steel strapping, 3⅛ in. wide by about 0.060 in. thick. This belt is wound around a mandrel support that is composed of a series of beams held in a cylindrical form by means of several discs which have slots in them to accomodate the beams.

The center of the mandrel is hollow and it is through this that the continuous belt returns to the starting end after unwrapping from the downstream end of the beam supports. The belt is then guided back around the end of the hollow axle, to the re-wind position. In this manner the steel tape brings about a smooth mandrel surface, rotation, and a simultaneous advance in the axial direction. The mandrel can be quite easily changed from one diameter to another by the addition of spacers, etc. (Figure V-1.3).

The success of this machine depends largely upon the proper operation of the axial-transporting mechanism on the mandrel.

Axial Reinforcement-Payoff Unit. This is a unit which cuts strands of roving and positions them onto a carrier band of mat. After saturation with resin, the mat is wrapped around the mandrel. The cut rovings, from 8 to 24 in. long, are laid in an axial direction, contribute up to 40% of the laminate weight, and effectively provide the required longitudinal strength. This unit is similar to all filament-winding machines, except that it is considerably larger and has the capability of wetting and dispensing up to 144 strands of 60 end roving. With some modification, it can also add fabrics and oriented-fiber tapes, and, if absolutely required, materials other than glass.

Oven. The oven is cylindrical in shape and completely surrounds the rotating mandrel. The heat for curing is provided by radiant heaters which have a total capacity of 80 kw and are mounted on coordinated jack screws so that one lever can set the heaters and reflectors to the proper distance for the various size capabilities of the machine.

Cut-off unit. This unit is mounted on the bed of the machine and is geared to move axially at the same rate of travel as the mandrel mechanism. Three wetted diamond saws are advanced radially onto the exiting pipe, and cut the laminate to the precise dimensions required. There are centering rollers on the carriage to support the cut pipe until the workmen can remove it for the next finishing operation.

The standard machine is capable of producing a 5-ply lay-up, consisting of three layers of hoop windings with 2 layers of axial sandwiched in-between each. It can apply as much as ½ in. of laminate at rates up to 100 ft per hr (6 ft diameter).

The machine output is determined by the capacity of the oven. Present machines are capable of up to 80 ft per hr for small pipes and 20 ft per hr for large units. Material pay-off varies from 660 to 4000 lb per hr for the large pipe or tank sections. Material loss caused by start-up has been observed to be less than 0.5% of the total production.

The method for making pipes with this machine is a combination of axial and circumferential reinforcements arranged in various orders that will satisfy a wide variety of design requirements. The actual winding operation is very similar to that described for the horizontal-tank machine, except that in this case, the system is completely automatic and the uniformity of the pipe or tank section is many times greater. Experience in Sweden and Denmark certainly attests to the superiority and the economy of the method. The product is experiencing ready sales.

In summary, this type of machine with almost complete automation is symbolic of the trend of the future in RP/C.

Elbow and Tee Machines. The specialized elbow and tee machines have been designed and built mostly by the companies involved in producing RP/C fittings for their RP/C pipe. One machine[40] has the capability making elbows 90 degrees or less from 2 to 12 in. in diameter and up to 15 in. in short elbow radii.

The basic operation involves orbiting a toroidal mandrel, capable of being collapsed or washed out, in a vertical plane. During the orbiting, a multiplicity of rovings and tape, held on a rotating table with a vertical axis, are fed onto the mandrel at its sweep center. There they are tightly wrapped, through tension controls, around the oscillating mandrel. By synchronizing the sweep motion and the rotational speed of the pay-off head, a helical or even hoop wind is developed on the toroidal mandrel. An additional feature, for extra reinforcement of the ends of the elbow, is the capability of the pay-off head to oscillate vertically a short distance and helically wrap a band as required. The machine works well with pre-impregnated materials, however wet winding presents somewhat of a clean-up problem.

Tee-making machines follow the design of modified, special-purpose, lathe-type, helical-hoop winding machines. They also provide supports for the mandrel so that cross-fibers can be wound to secure the arms of the tee to the leg. In many cases, the mandrel is a thin molding, over which the filaments are wrapped to provide the required strength and pressure resistance.[41]

Box-Winding Machines. Box-winding machines were designed and built with considerable ingenuity to demonstrate that filament winding was not limited to surfaces of revolution such as tanks and bottles.[42] These machines have demonstrated their feasibility by simultaneously fabricating two 13 qt. dairy cases, back to back, in a fully automatic sequence. To wind a box of 6 sides and 3 axes, the machine had to be able to wind material around the box in 3 axes, actually covering 4 sides in each axial wrap and putting down only 2 cross plies of wrap onto each face of the box in 1 sequence. Accomplishing such a winding operation with no human manipulations involved required a sophisticated mandrel-transfer mechanism and control panel; films are now available that sequentially show the complete process. The reason for including this machine in this chapter is not its commercial importance, (there are much better ways to make a box) but to illustrate that there exists a wide freedom in design of items that can be made from the winding of filamentary materials. Later machines that copied some of the control and transport mechanism of the original box machine are showing great promise in the commercial market.

The most important feature or method of this machine is the control system with its capability to visually "tune" the roving pattern into the proper lay from the operator's console. The spindle of the console rotates at slow speed for good visibility. In most automatic machines there are computers or tape controlled servo-mechanisms, and highly trained and specialized personnel are needed. The box machine, on the contrary, is programmed by the operator. The basic mechanism is a double array of phototransistors which are arranged parallel to the major travel of the pay-heads. A light source synchronized with the payoff head position activates the phototransistors, causing the head to go through the proper fiber laying motions in accordance with the manner in which the operator connects the patch cords in the wiring circuit. The connections for patch cords are arranged to duplicate the stroke of the pay-off head. The operator merely plugs in the phototransistor at dimensional locations duplicating the dimensional locations of any lay pattern changes. Hence control of the lay is accomplished through visual patch-cord insertions—in the same scale as the part being produced. Changes or corrections in pattern are made by merely relocating the patch cord and/or changing the setting of cycle relays.

Some products made with the box-winding machine are RP/C overwrap on blow-molded containers, agriculture field boxes, cages, and filter boxes. These 3 items are possible because the machine can be programmed to wrap strands spaced in an open configuration similar to a grid pattern.

Filament-Placing Machines. In the manufacture of the more advanced composites for the aerospace industry, structural plastics inevitably will be used to a great extent in aircraft and missiles in a few years. Such new materials will have more stiffness than steel and yet will weigh about as much as magnesium. In order to get the most for the money (boron, carbon, and graphite filaments cost $100 to $300 a lb., 1971), the composite must be designed so that the expensive materials are used only where they are required. To do this repeatedly under production conditions, machines must be designed that will accept these materials and apply them to the mold surface in a precise manner. Such machines have been designed, and some are in operation making skins for the wings of high-speed, high-performance airplanes. These machines look like large planning mills with movable tables that pass under a frame which holds the pay-off head. The table and head are controlled by punched computer tapes which compensate for angle, contour reinforcing band width, and other considerations and lay down the reinforcement in a precisely determined pattern. These machines are called *filament placers,* although they are the same as filament winders considering the actual laminate that is made.

There are many other styles of filament placers, and as the public becomes more at ease with the idea that the use of filaments requires winding over a cylinder, then the ability to consider filament-placed and filament-wound products as one and the same will enable the industry to open up many new, interesting, and profitable operations.[43, 44, 45, 46, 47, 48, 49, 50]

Kinetics for Mechanical Chain Drive. It is now appropriate to consider the basic periodicity of fiber placement for all machines versus the mandrel rotation as determined by all of the mechanical elements involved.

The relative carriage and mandrel speeds are adjusted not only to wind the desired helical angle but also to form a repetitive pattern leading to complete coverage of the mandrel in 1 layer. It is desirable to know the number of mandrel revolutions of traversing circuits required to perform this function. Pattern control is the most fundamental feature of any traversing machine. The carriage is driven by an endless link chain to which it is attached. The shape being wound in the following example is a cylindrical pressure vessel with spherical end caps. The machine calculations for the helix angle, winding pattern, and number of revolutions per layer were originally carried out in early work at M. W. Kellogg and Young Development Labs. The mandrel and carriage-drive sprockets are initially geared in a ratio of one revolution of the mandrel to one revolution of the carriage sprocket. Table V-1.2 supplies the nomenclature used in this discussion.

For one mandrel revolution:

Rotational distance $= D$

Transverse distance $= N_s L_k$

$$N_s = \frac{D}{N_s L_k} \qquad (1)$$

For a desired winding angle, the number of teeth in the timing sprocket is:

$$N_s = \frac{D}{L_k \text{ Tan (of winding angle)}} \qquad (2)$$

Roughly, the straight-length part of the carriage chain corresponds to the cylindrical length of the mandrel. Actually a slightly longer chain length is needed, depending upon the angle of wind and the height of the feed discharge above the work. At any rate, the total number of chain links is totaled at:

$$N_c = \frac{2L_c}{L_k} + N_s \qquad (3)$$

To determine the winding pattern, the ratio of sprocket turns per circuit is then reduced to its lowest fractional form. The numerator indicates the number of sprocket turns and the denominator, the number of traversing circuits

Table V-1.2 Nomenclature

D	Diameter, in.
N_s	No. teeth in timing sprocket
L_k	Length per chain link, in.
L_c	Cylinder length, in.
N_c	No. chain links in traverse
C_p	Circuits per pattern
C_i	Circuits per layer
S_n	Reinforcement bandwidth, in.
S_i	Longitudinal component of S_n
T_p	Spindle turns per pattern
S_c	Circumferential component of S_n
Circuit	One complete transit of the traverse carriage, usually measured from the point of filament delivery.
Pattern	The number of circuits related to either spindle or timing sprocket revolutions after which the filament path is repeated, separated only by one bandwidth.

required to complete a pattern. Suppose that $N_c/N_s = 264/48 = 11/2$. In eleven timing sprocket (or spindle) revolutions, two traversing circuits would be completed and the feed delivery point would be exactly at its origin, ready to repeat the pattern. If the traverse chain contained one more link, the ratio 265/48 could not be reduced further. To complete the pattern, 48 traverses or 265 revolutions are necessary. This is tantamount to a random pattern. A small pattern number usually less than 10/1 is desirable. The total number of links or the number of gear teeth in the timing sprocket are juggled to arrive at some reasonable combination without greatly affecting the winding angle or overall traverse distance.

After establishing the pattern, indexing is necessary to move the reinforcements over one bandwidth so that complete coverage results as the pattern continues. The distance the band must be displaced, S_1 or the axial component of the bandwidth, can then be translated to a fractional advance or decrease in the timing sprocket. A new gear ratio can also be calculated which differs slightly from the original 1:1 ratio. This gear ratio can be expressed as:

$$R_c = \frac{N_c C_p \pm \dfrac{S_1}{L_k}}{N_s T_p} = \pm \frac{S_1}{N_s T_p L_k} \quad (4)$$

The gear change will cause a slight change in the winding angle which can then be recal-

culated. For complete coverage of the mandrel with the reinforcing band, the number of traversing circuits is:

$$C_1 = \frac{D}{S_c} \quad (5)$$

The number of mandrel revolutions per layer is:

$$T_1 = \frac{N_c C_1}{N_s} \quad (6)$$

The above analysis has been made for a single layer of relatively thin-walled sections. Calculations based on an average diameter are satisfactory. For heavier sections, the calculations may be altered to handle the thickness of increments. The analysis can be extended to cover more complex situations where the original ratio of spindle rotation to timing sprocket rotation is not 1 to 1.

Processing Variables Affecting Winding:

Winding Tension. Winding tension is a critical parameter in controlling and limiting the void content in a laminate. Optimum winding tension varies from vendor to vendor and ranges from ¼ to 1 lb per end. Other parameters, such as resin viscosity and mandrel material behavior during post-cure cycles, have an effect on the ultimate stress capabilities of the completed structure. Obviously, too high a tension will produce a "rubber-band" effect and considerable prestress, while too low a tension will permit strand wandering and overlapped bands that cause voids and reduced physicals.

Resin Content. A certain amount of outward resin flow occurs as strands are applied, serving to remove trapped air. This is closely related to the winding-tension situation, where excessive strand tension forces the resin outward and causes gross variations in the glass percentage between the inner and outer layers. Using prepreg, as opposed to "wet winding," will do much to insure a consistently proper or designed resin content. Usually prepregs have a high-viscosity impregnate which makes higher mandrel speeds feasible, because there is less resin "throwing" caused by centrifugal force. Prepregged strands also seem to give less slippage at the juncture of the cylinderhead tangency, and the resin content therefore tends to remain within close design limits. Because the

Table V-1.3 Stress Values versus Resin Content

	% Resin			
Stress	*15*	*17.5*	*20*	*25*
Composite stress, KPSI	—	—	195–205	160–185
Hoop stress, KPSI	—	—	450–480	375–435
Longitudinal stress, KPSI	340–365	400	425–445	345–400

resin content is fundamental to the integrity of the structure, much thought should be devoted to insuring its control. Optimum content seems to be 20% by weight.[51] Table V-1.3 presents the stress figures according to the amount of resin used.

Condition of the Reinforcement. It has been well established that glass filaments "at the bushing" demonstrate tensile values approaching 700,000 psi or even higher. However, the acts of winding, tensioning, guiding, or simply re-spooling all contribute to filament damage and the filament tensile values can drop to 300,000 psi. Although filament-winding machines transport the strand from the spool to the mandrel with a minimum of guidance bushings, rollers, wheels, and so on, each handling contact or friction point extracts some tribute and results in lower tensile values. As in the case of other glass manufacture, the glass is only as strong as its surface.

Excess humidity leads to stress corrosion and changes in existing flaws. Stress corrosion is the factor which limits or lowers values in static-fatigue and tensile tests where moisture is present.

From the above, it should be obvious that optimum strengths in the yarn are dependent upon the dryness of the glass and the sophistication of the mechanism to take the strand from the spool to the mandrel with the least amount of contact and handling. Strands of materials other than glass generally require similar treatment.

Winding Angles. High-strength performance is dependent upon proper placement of the strands. Once a design has been selected and a prototype is tested, the winding machine can be expected to repeat the pattern without fail. Bottles or tanks designed through "netting analysis" should have hoop windings placed sequentially between layers of the helical fibers.

The resulting network consists of a system of triangles rather than parallelograms in which the matrix has much less to do in resisting the internal stresses. Such hoop windings help to counteract against expanding and/or contracting forces.

Departures from the original winding angle design will grossly affect the performance of the product. Good practice requires "locked ratios" in the machine and also an inspector who immediately will be aware of any changes in ratios.

Winding Speed. In commercial production, the goal is to wind at as high a speed as possible. In production for the military and aerospace, speed is not a factor, but high specific-strength is. Once a speed has been selected as either high or low, it must be maintained. Changes in speed require another look at the physicals generated in the products. This is true whether the main consideration is the mandrel or the pay-off head as in a polar winder (although speed variation in a polar winder is not as critical as in a helical or hoop winder). Speed anomalies can be costly and time consuming, and often offset any profits that would be gained by a higher production rate.

Cure Schedules. There generally is an optimum cure required in order to obtain the maximum strength for each rate of temperature rise. The rate of cure also depends upon the hardener (in the case of epoxies) and the catalyst (for polyester systems). With epoxies, the initial cure is accompanied by exotherm, low viscosity, and then gelation. The low-viscosity phase usually demands that the product, such as a tank or pipe, be rotated in the oven to control resin flow and prevent incipient drainage until the gel has been established. With polyesters this is no problem because gelation occurs prior to exotherm.[51] To establish optimum physicals, epoxy systems demand post-cures which require that the recommendations of the resin and hardener suppliers be closely followed. Such post-cures are usually made in temperature steps, with the temperature held until thermocouples indicate that the heat has permeated through the full thickness of the part. Obviously, the type of mandrel material and its thermal expansion coefficient must be considered so that undue strain is not imposed upon the curing laminate. Epoxy systems fortunately have considerably less cure shrinkage than polyesters, and most mandrels are designed to be collapsible, thus minimizing the problems of removal. Furthermore, in prepreg winding, there is a stress-relaxation upon initial cure which serves to minimize the mandrel problem.

Faster cure schedules can usually be accomplished in any one of three ways: (1) by making the mandrel hollow, with a provision for ambient oven temperature to impinge upon the mandrel's inner surface; (2) by fitting the mandrel to circulate hot water or oil; or (3) by using electrical "hot rods" which are embedded in or clamped to the mandrel walls.

Curing Agents and Ratios. There is no simple way to describe the wide[52] varieties of curing agents and their behavior. It is sufficient to say that uniform performance is based upon uniform mixing and weighing of these additives and careful attention should be given to the manufacturers' recommendations. Polyester systems are usually cured by peroxides which have different breakdown temperatures and rates of decomposition. These materials must be measured precisely, because the percentages added are directly related to resultant gel times. The hardeners for epoxy systems—to distinguish from catalysts for polyesters—come in a wide variety of types and concentrations. Some systems are on a 50-50 basis, while others can be effective with as low as only 2 or 3 parts by weight per 100 parts of resin. Hardeners in the higher combining quantities such as polyamides are considered to be mostly co-polyymerizers. They impart some of their properties to the basic epoxy and thus perform much like the ingredients in a metal alloy.

Resin Temperature. In optimum epoxy systems, it is necessary to heat the resin to obtain a low viscosity (100-300 cps) for the best winding properties, because the better epoxies have a high viscosity at room temperature. Diluents can be added to reduce the room-temperature viscosity, but always with some sacrifice in physical properties. Most popular epoxy resins attain satisfactory winding viscosities at about 200°F. Obviously, either the hardeners must be stable above this temperature, or the winding must go to completion before the time prior to gelation has elapsed.

Material Variables. The manufacture of filament-wound products depends upon the recognition of variables and the control of those that affect the completed product. Most materials used for filament winding are not made by the manufacturer of the filament-wound product, and he must learn how to control the purchased item. Almost all suppliers of reinforcement and matrix (resin) materials publish specifications governing their products. The fabricator must therefore do 2 things in order to protect himself:

1. Demand certification that materials meet the advertised specifications. In the case of military contracts, such certificates must state that the materials meet certain MIL specifications.
2. Set up minimal test equipment to determine that the supplied material meets the advertised specifications, as follows:
    - Gel-timer test—to check on the response of the resin to a controlled amount of catalyst or hardener.[53]
    - Viscosity test,—to check upon any deviation from specified-viscosity. Gross changes require an immediate follow-up to determine the reason for the change (wrong resin, old resin, and so on).[54]
    - Hardness tests—to check on the cure characteristics and to act as a check on the resin supplied. i.e., same resin should, in the same time increments, always cure to the same hardness (assuming that hardener, and/or catalysts are kept constant).[55]
    - Fabrics and fibers tests—burn-out oven —to check if the required amount of finish, binder, or prepreg resin has been applied, as per the purchase specification, or to make frequent determinations of the glass-resin ratio of the finished filament wound composites.
    - Microscope—to check on the filament

diameter and the weave of the fabric (warp, fill, and so on).

Because of such tests, the suppliers soon become aware that the fabricator is testing incoming shipments of materials, and they tend to be more careful in checking material to be delivered to that customer.

Catalysts and Hardeners. Hardeners for epoxy systems encompass a wide variety of types, viscosities, dangerous and toxic properties, and different shelf lives. Adherence to the manufacturers' suggestions is mandatory if best results are wanted. The catalysts for polyester systems are peroxides. They are unstable compounds and potential explosion hazards and should be handled as such in strict compliance with the manufacturer's suggestions. Most catalysts for filament-winding systems are of a more stable variety, requiring oven cure and temperatures in excess of 200°F before breakdown. For items like tanks and large pipes, room-temperature cure systems are utilized. Substances known as accelerators are used in room-temperature curing polyester systems. They must not come into direct contact with the catalyst or else an explosion, or at least a fire, will occur.[56]

Catalysts must be controlled with care. It is good practice to examine each shipment by making simple checks of the gel time and cured-resin properties. Catalysts must be measured carefully because slight variations in concentration have a great effect on gel time. Also, because catalysts are sensitive to resin and molding temperatures, the ratio should be adjusted continually to reflect temperature changes and keep gel time constant.

Minimum quantities for proper cross-linking must be observed, however. Hardeners for epoxy systems, usually of a co-polymerizing material, impart some of their properties into the epoxy-hardener alloy. While relatively wide variations in the ratio of epoxy to hardener have no effect on gel time, they can have an effect on the cost. Most hardeners cost more than the epoxy resin and the fabricator should use the minimum hardener consistent with the optimum cured properties of the system. As in all chemical reactions, good and uniform mixing is mandatory, and it is distressingly apparent that many failures in otherwise well-engineered products have been traced to improper mixing of hardener or catalyst.

Release Agents for Filament Winding. Regardless of the type, concentration, or other variables of release agents, they must be:

- Non-compatible with the mold surface and material.[57]
- Non-compatible with the resin in all conditions.
- Stable at all levels of exotherm and post-cure temperatures.
- Capable of accepting the thermal expansion of the mold and the cure shrinkage of the material without cracking and wrinkling.
- Strong enough to withstand the application of the gel coat and the first wrap without damage.
- Easy to remove after the cure is complete.
- Non-interfering as far as extraction forces are concerned, such as the removal of tube or pipe mandrels. Some releases have tended to ravel or jam.

Once the proper release has been selected, the control of several variables must be exercised:

- Uniformity of application. Good release application of the sprayed type demands skill and care. Often, poor gel coat surfaces can be traced to spotty, non-uniform, runny, or dirty release films. A poorly applied release can undo the effort and expense of polishing the mandrel.
- Proper drying. A lay-up placed upon a PVA film that is not yet dry will often bond securely to the mandrel with the PVA acting as an adhesive.
- Proper mixing. If a release material is blended in the factory, great care should be exercised to use high-shear mixer blades to insure a thorough wet-out of all the ingredients. Releases that dry with a "grainy" feel are obviously going to perform improperly.

Film or sheet release agents should be clean and free from the static charges that tend to pick up small motes of dirt. The material is usually applied in a spiral manner onto a rotating mandrel in the form of tape from a spool. When applying the agent, be sure that the tape edges have no irregular cuts, tears or raveled edges and reject those that do. Apply the tape with uniform tension, butting it very carefully with the edge of the fresh wrap running as parallel as possible to the edge of the preceeding wrap (a very difficult technique). Most release tapes are lap-wrapped about 50% over the preceeding wrap so that an essentially uniform thickness is generated as well as a good seal to prevent resin from migrating through the lap and onto the mandrel.[58]

Some common tape release materials are Mylar (a polyester film), PVF (poly vinyl fluoride film), PVC (poly vinyl chloride film), PVA (poly vinyl alcohol film or sprayable liquid, cellulose acetate, and polyethylene films.

PVC and PVF films are sometimes treated so that the contacting surface will bond to the resin in the wound part. This action provides a release liner which allows the easy removal of the part from the mandrel and forms a chemical and/or weather resistant liner in the pipe or tank.[59]

Fillers. Fillers are added to the laminating resin to extend the resin and to import some of the filler's properties to the resin. Fillers such as clay, calcium carbonate, or milled glass fibers can lower the cure shrinkage of the resin and also its coefficient of thermal expansion. They can also increase both the chemical resistance and the mechanical properties. Short-fiber asbestos for example is added to increase toughness. Fillers must be carefully specified and their moisture content held in strict adherence to the specifications. Generally, fillers should have no more than ½ % moisture by weight of the resin in the mix.

Fiber Glass Surface Treatment. In the last few years, greatly improved sizings or organic surface treatments have been developed which permit faster wet-out and better glass-resin interface bonds, with resulting improved physicals and less moisture absorption. Moreover, fiber glass manufacturing companies are improving their quality control, and the uniformity of the treatment is better. It makes good sense however, to periodically check the treatment by preparing flexural samples, water-boil tests, and burn-outs for determining the amount applied and its effectiveness.

Summary. Process and material variables must be reduced to the economic minimum. Simple statistical analyses of the relative effect of changes in either processing or materials will reveal those which should be more carefully controlled. A simple quality-control procedure must then be set up and followed. The quality of the final filament-wound product is the result of the quality of all the ingredients.

Successful companies have learned this lesson well.

PRODUCT PROPERTIES

Product Design. Detailed analysis has been made of fiber placement and orientation, winding angles, and netting and has been reported in an excellent manner that precludes repetition here.

Product Types. It is obvious from the foregoing discussion that filament-wound products are primarily oriented toward surfaces of revolution. This encompasses piping, tanks, cylindrical vessels with one or both ends enclosed, pressure bottles, and even chains.[46] A review of the overall requirements shows that filament winding will grow in relation to how well it is able to adapt winding equipment and processes to non-revolutionary shapes such as the aircraft wing and tail section for military and commercial applications. Growth also depends upon how well and rapidly the technique may be incorporated into combinations with other methods of RP/C fabrication.

Physical Properties. Performance data and product strengths for filament-wound struc-

Table V-1.4 Characteristic Properties of Filament-Wound Structures

Property	Specification
Specific gravity	1.7–1.9
Tensile strength, psi	80000–250000
Tensile modulus, psi	Up to 7×10^6
Compressive strength, psi	50000–75000
Flexural strength, psi	100000–200000
Impact strength, ft-lb per in.	50–70 (unnotched)
Moisture pickup, %	0.1–0.5
Burning rate	Slow to self-extinguishing
Heat resistance, continuous °F	Up to 500
Resistance to acids and alkalies	Fair—excellent
Resistance to solvents	Good—excellent
Machining qualities	Poor—fair (should be used as molded)
Resin content, %	15–25

Source: "Handbook of Reinforced Plastics of SPI," Mohr, J. G., Oleesky, S.S., Reinhold, 1964, p. 541.

tures are the highest in RP/C because of the highest possible reinforcement contents. Table V-1.4 lists the physical properties of fiber-glass-reinforced filament-wound structures. Strengths for products utilizing other reinforcements are generally similar, with the exception of the tensile moduli. For advanced composites using carbon, graphite, or boron fibers, these parameters approach 35×10^6 psi.

Concerning methods of testing—the internal burst pressure of a pipe section which is filament wound to duplicate or simulate the product involved represents a rapid and duplicatable method of preparing specimens and obtaining performance data. It also represents the single test that comprehensively gives information about numerous individual parameters in the filament wound structure.

References

1. Kempton, W. H. (To Westinghouse Corp.), U.S. Patent No. 1,400,078 (Dec. 31, 1921). Kemp, R. (To Westinghouse Corp.), U.S. Patent No. 1,393,541 (Oct. 11, 1921).
2. Young, R. E., "History and Potential Filament Lining," Sec. 15-C, Annual SPI, 1958.
3. "Derakane," Resins-Dow Chemical Co., Midland, Mich.
4. "Atlac" Resins, Atlas Chemical Co., Wilmington, Del.
5. Grosh, J. L. Chemical To United Aircraft Corp., U.S. Patent No. 3,470,917 (Oct. 7, 1969).
6. Bozer, K. B., Brown, L. H., and Watson, D. D., "FR Furane Composites, a Unique Combination of Properties," Sec. 2-C, 1971, SPI.
7. Delmonte, J., "Urethane Laminates," Sec. 17-C 25th Annual SPI, 1970.
8. Browning, C. E., and Marshall, J. A., "Graphite Reinforced Composites," Sec. 19-C, 25th Annual SPI, 1970.
9. Bandaruk, B., "Polymers for the Space Age," Sec. 18-D, 22nd Annual SPI, 1967.
10. Drostholm Products, Vedbaek, Denmark.
11. Werts, W. E., Lemons, C. R., and Wiltshire, A. S., "Non-Metallic Pressure Vessel Mfgr. Techniques," SAE (NASE) Meeting, Ambassador Hotel, Los Angeles, Calif., May 9–13, 1961.
12. Tanis, C., "Large Reinforced Plastic Rocket Cases," Sec. 6-F, 17th Annual SPI, 1962.
13. Jube, G., "Winding Wisdom," Sec. 16-B, 20th Annual SPI, 1965.
14. Stockton, W., and Russell, W., "Fabrication Techniques Developed in Filament Winding," SPI, Sec. 1-B, 19th Annual SPI, 1964.
15. Walter, T., and Hinchman, J. R., "Large Segmented FRP Rocket Motor Cases (TU-312 Rocket Motor Case) AFML 69-107 Thiokol Chemical Corp., Brigham City, Utah, June 1969 (Figs. 79–82, p. 207).
16. Asakawa, A., and Peregoy, L. F., "Mfgr. Technology for Large Monolithic FRP Plastic Rocket Motor Case," AFML TR-65-436 Aerojet General Corp., Azusa, Calif., Nov. 1965.

17. Whinery, D. G., and Ady, C., "Mfgr. Methods for Plastics Airframe Structures by Filament Winding," AFML TR-68-378, North American Rockwell Corp., Columbus, Ohio, Mar. 1969.
18. Paraskevopoulos, S. C. A., and Peregoy, C. F., "Research on Potential of Advanced Technology for Housing," pp18-25, Architectural Research Laboratory, University of Michigan, Ann Arbor, Mich., 1968.
19. Boheln, J. C., and Kirkpatrick, H. B., "Use of Polyester Resin in Filament Wound Dwelling Units," Sec. 2-E, 21st Annual SPI, 1966.
20. Hager, J. H., and Ritchey, J. L., "Process Mechanization and Control Yields High Quality, FRP Turbine Blade" Sec. 20-F, 21st Annual SPI, 1966.
21. Baxter, J., Jr., "Filament Winding Future? Good!" Sec. 16-A, 20th Annual SPI, 1965.
22. Baxter, J., Jr., et al (Black Clawson Co.), U.S. Patent No. 3,380,675 (Apr. 30, 1968).
23. List of Machinery Builders of Filament Winding Machines:
   Baje Machine Co., Chicago, Ill.
   Black Clawson Co., Plastics Dept., Hamilton, Ohio.
   Brenner, I. G. Co., Newark, Ohio.
   Capital Equipment Co., Chicago, Ill.
   Center-Line Machinery, Div. Center-Line Industries, Santa Ana, Calif.
   Covema, S. R. L., Milan, Italy.
   Davco Industries, Inc., Framingham, Mass.
   En-Tec, Inc., Salt Lake City, Utah.
   Glas-Craft of California, Sun Valley, Calif.
   Iddon Brothers, Ltd., Brookfield Iron Works, Leyland, Lancashire, England.
   Industrials Plastics Ltd., London, S.E. 11, England.
   Kleinewefers, Joh. Sohne, Krefeld, W. Germany.
   McLean-Anderson, Inc., Milwaukee, Wis.
   Modern Plastic Machinery Corp., A Sub. of Savoy Industries Inc., Clifton, N.J.
   Parkhill, J. E. Ltd., Don Mills, Ontario, Canada.

Roblex Div. Granite State Machine Co., Manchester, N.H.

Ruf Machine Co., Inc., Bronx, N.Y.

Vermont Instrument Co., Inc., Burlington, Vt.

24. Aerojet General, Azusa, Calif.

Bendix Corp., South Bend, Indiana

B. F. Goodrich, Akron, Ohio

United Technology, Div. of United Aircraft, Sunnyvale, Calif. 94088

25. Fiberdynamics, Inc., Gardena, Calif.

Roblex, Inc., Manchester, N.H.

ENTEC, Salt Lake City, Utah 84115

Goldsworthy Engr., Torrance, Calif. 90505

I. G. Brenner, Co., Newark, Ohio

26. De Ganahl, C., et al (To OCF Corp.), U.S. Patent 2,714,414, Aug./2/1955.

27. SPE-RP/C RETEC, Cleveland, Ohio, Sept./28/70, Proceedings, p. 25, A. J. Wiltshire. *See also* "Composites," March, 1971, p. 15, Ainsworth L.

28. Rohr Aircraft, Bendix, South Bend, Ind.

29. Tanis, C., "Large Reinforced Plastic Rocket Cases," Sec. 6-F, 17th Annual SPI, 1962.

30. Asakawa, A., and Peregoy, D. F. (See Ref. 14).

31. Imbert, R., et al, U.S. Patent 2,822,515 (Feb. 11, 1958).

32. "Permastran"® Pipe, Johns Manville Products Corp., N.Y. 10016, Technical Bulletin TR 593A.

33. King, W. J., "Convolutely Wound Tubular Shapes," Sec. 4-B, 20th Annual SPI, 1965.

34. Strickland, E. T., et al. (to Philbrick-Strickland, Inc.), U.S. Patent 2,987,100 (June 6, 1961).

35. Clements, H. R., (to Rohr Corp.), U.S. Patent 3,524,780 (Aug. 18, 1970).

36. Copolymer Corp., Torrance, Calif. 90501.

37. Balsa Ecuador Corp., Northvale, N.J.

38. Wester, A. L., "Production & Marketing of Large Diameter Polyester Tubes in Scandinavia," Sec. 15-D, 17th Annual SPI, 1962.

39. Gilbu, A., "The Drostholm Continuous Filament Winding Process," Sec. 16-D, 26th Annual SPI, 1971.

40. Thaden Engr. Co., High Point, N.C. 27262.

41. Koch Products Co. (formerly Rock Island Fiberglass Prod.), Wichita, Kan.

42. Model BWM-24 Filament Winding Machine-Goldsworthy Engr. Div. (Monsanto Corp., formerly Ferro Corp.), Torrance, Calif. 90505.

43. Posniak, B., "Development of a Directed Fiber FRP Helicopter Motor Blade," Sec. 2-D, 17th Annual SPI, 1962.

44. Abelin, R. E., Wohlberg, E. A., "Glass Fiber Reinforced Plastics in Highly Stressed Aircraft Parts," Sec. 2-A, 17th Annual SPI, 1962.

45. Hutter, M., "Glass Reinforced Plastics as Structural Material for the Aircraft Industry," Sec. 13-A, 16th Annual SPI, 1961.

46. Koch, P. E., Gerhardt, W., "Round Link Chains of Glass Fiber Reinforced Plastics," Sec. 7-B, 13th Annual SPI, 1961.

47. Baxter, J., et al. (to Black-Clawson Co.), U.S. Patent 3,380,675 (Apr. 30, 1968).

48. Browning ,J. E., *Chemical Engr.*, 87–48 (Mar. 22, 1971).

49. Eig, M., and Shibley, A. M., "Commercially Available Filament Winding Machines," Sec. 1-C, 19th Annual SPI, 1964.

50. Vogt, C. W., Haniuk, E. S., Trice, J. M., Jr., AFML-TR-66-274 (1966).

51. Lee, H., and Neville, K., "Handbook of Epoxy Resins," McGraw-Hill, N.Y. (1967).

52. Hardeners for Epoxy Systems-CIBA Tech Service Notes . . . TSN-99 and TSN-94.

53. Randolph Gel Timer, SHYODU Precision Inst. Co., Brooklyn, N.Y.

54. Brookfield Viscometer-Brookfield Engineering Labs., Stoughton, Mass.

Zahn Viscometer, General Electric Co., Schenectady, N.Y.

55. Barcol Hardness Tester, Barber-Coleman Co., Rockford, Ill.

56. Cadet Chemical Co., Burt, N.Y. 10428.

Lucidol Div., Wallace and Tiernan Co., Buffalo, N.Y. 14240.

U.S. Peroxygen, Richmond, Calif. 94814.

57. Oleesky, S., and Mohr, G., "Handbook of Reinforced Plastics of the SPI," Van Nostrand Reinhold, N.Y.C., Table II-1.13, p. 45 (1964).

58. Wester, A. L., "Production and Marketing of Large Diameter Polyester Tubes in Scandinavia," Sec. 16-D, 17th Annual SPI, 1962.

59. Tedlar Film, E. I. DuPont, Co., Wilmington, Del.

Ader, G., "Reinforced Plastics plus Vinyl, a New Method of Plastic Construction," Sec. 16-B, 16th Annual SPI, 1961.

V-2

Pultrusion and Rod Stock

INTRODUCTION

Pultrusion can be defined as a process for producing reinforced-plastic profiles in continuous lengths by pulling the raw materials through combining, shaping, and curing operations. The principal materials used are glass-fiber reinforcements in combination with liquid thermosetting resins such as polyesters and epoxies. A variety of processing techniques are used, especially for the cure or hardening operation.

Some forms of pultrusion date back at least 25 years to the production of fishing-rod blanks. Today, pultrusion is an established manufacturing operation in Japan, Sweden, England, Germany, and France as well as in the United States. A designer can now specify off the shelf end products from $\frac{1}{32}$-in. diameter rods to 8-in. I-beams and other structural elements with long length.

Markets for pultruded products fall principally in the electrical, corrosion-resistant, and sporting-goods fields. The total yearly output is in the 8 to 10 million lb range and is growing rapidly.

MATERIALS

Resin Binders. Polyesters and epoxy resins are the major matrices used and polyesters account for perhaps 90% of the total production. Both may be formulated to suit end-use requirements such as corrosion resistance, electrical characteristics, elevated temperature exposure, flame resistance, and so on.

For continuous die curing, a resin system requires a catalyst-resin balance that produces a gel strong enough to resist coating the die surface. Avoiding a build-up of this kind of resin skin is an important criteria for choice of a particular resin. Benzoyl peroxide is the catalyst most widely used. A reasonable pot life of 8 hr. or more is an obvious production advantage. The avoidance of excessive exotherm is a necessity.

Resin systems at minimum viscosities (under 2000 cps) assist in accomplishing rapid wet-out of the reinforcements and easier air removal.

Reinforcements. Except for decorative flat laminates in which paper is used, glass fiber in various forms is the principal reinforcement for pultruded products.

Glass roving is perhaps the easiest material to process in parallel orientation for pultrusion and/or rod stocks. As the most efficient and strongest form of glass fiber, it is strategically introduced into highly stressed areas of many profiles, such as the edges of angles, channels, and so on. It is also braided, spirally wrapped, and hoop wound for reinforcing round, oval, or rectangular profiles both solid and hollow. Spun roving is sometimes used to give a measure of cross-strength.

Glass-fiber mats are employed to produce shapes in which strength is needed laterally as well as longitudually. Such mats must be strong enough for mechanical handling with binders and insoluble enough in the resin system to resist softening during wet-out and preforming. Three types are generally available: bonded surfacing mat, bonded (insoluble) chopped mat, and a combination of roving mechanically needled to chopped mat.[1] Bonded mats which have partially soluble binders are sometimes sandwiched between roving to prevent mechanical disruption.

Glass cloth is also employed to give cross-strength, and is most often in tape form. Product requirements dictate the specific weaves and fabric weights chosen.

All glass fiber is, of course, surface-treated with the appropriate finish (coupling agents) for the resin system being used.

Additives. Fillers, pigments, flame extinguishers, and other additives are used to produce desired end-product properties, and would be formulated in the usual manner (mixing and blending) for reinforced plastics.

Intermittent cure processes will tolerate somewhat more filler than continuous-operation procedures. High filler loadings (over 20% of resin batch) are apt to cause filtration and separation during continuous pulling through tanks and preforming steps.

Polyester resins cured in closed dies require an internal release agent such as a phosphate ester.[2] This is an important additive which prevents the relatively weak gel from adhering to the die wall rather than to the laminate.

EQUIPMENT AND TOOLING

Although one equipment manufacturer[3] does offer a complete mechanized line utilizing the continuous-pull, radio-frequency cure process, most pultrusion equipment is custom built by each producer. The following discussion covers the component operations in a typical pultrusion line:

Roving Let-off Equipment. Because of the slow linear speeds involved, let-off equipment for dispensing roving is relatively simple and low in cost. Fiber glass roving packages are set up in creels equipped with pigtail guides. For precise guiding into critical die configurations, orifices or plastic tubes are used. Web reinforcements such as mats and cloth or fabric, which are generally in narrow widths, are let off from simple roll stands. End plates improve alignment. At the slow operating speeds, braking or tensioning the web shaft is unnecessary. Tension build-up during impregnation is usually more than adequate for satisfactory operation.

Impregnation Equipment. Most pultrusion lines use simple, narrow tanks 2 to 6 ft long, which are fitted with side guides and wet-out aids such as transverse breaker bars, multiple orifices and squeeze rolls. These devices operate beneath the liquid resin surfaces and serve to expel air and promote fiber wetting which is necessary in order to obtain void-free products with consistently good properties.

Good design of impregnation tanks dictates that all tank components be easy to remove and clean. For example, pins should be used rather than bolts, and wedges rather than screws.

Resin Control Control of the resin-to-glass ratio is, of course, the leading factor in product properties and quality. Fortunately in a moving line, simple orifices of the proper shape serve as good metering devices. They are not, however, independent of variations in resin viscosity, a factor that also should be adjusted and controlled for RP/C pultrusion work. Orifices also impose certain speed limits, because they force excess resin backward and produce hydraulic pressures of sufficient magnitude to disrupt some reinforcements.

Because almost all reinforcement materials have a spring-back or bulk factor, the die itself (or mold or wrap points) is approached with a slight excess of resin in order to compact the fiber and expel any air which may be present.

Preforming of Drawing Dies. All profiles are preformed to the approximate required shape before molding. These may be simple ring orifices for forming rods, sheet-metal guides for forming angles or mandrels to form hollow articles. Preforming is usually done gradually with careful attention given to holding the desired fiber orientation and making full use of mechanical aids such as spiders, tubes, combs, and ring guides.

In one interesting concept, the glass fiber is oriented by preforming on a mandrel, and then collapsing this tubular preform into a shape such such as an I-beam with no exposed fiber ends.[4]

Dies. Die materials for continuous pultrusion and cure are generally steel.[5] Chrome plating reduces the coefficient of friction and concomitant pulling force required, extends die life, and improves resin release. Die lengths vary depending upon production rates, but are usually 18 to 60 in., with multiple side-by-side dies preferred over longer tools.

Heated metal continuous-cure dies require a separate cold junction or cold-entrance section to prevent premature hardening of the resin which, as stated above, is in slight excess at this point. Water cooling a short-cored or jacketed section of the die performs this function well.

Dies used for radio-frequency curing must be made of non-conducting materials which do not themselves heat up when exposed to electrical loading. Although several new candidate materials are being evaluated, Teflon has been the production die material most widely used. These dies are usually machined from solid bar stock, and are strengthened with a glass reinforced silicone resin layer on the outside. The abrasive nature of glass fiber and some types of fillers limit die life to a throughput of between 2000 and 10,000 linear ft. Teflon is also useful in the cooling sections of thermoplastic pultrusion dies. In this application, however, it is used as a thin replacable lining in a steel die.

Dies used in conjunction with film sandwiches (sheet), film wrap (rod), or film-lined troughs can, of course, be much more crude. Corrugated sheet metal and split metal tubes are examples of dies which, in combination with oven heat cure, produce pultrusions at a relatively low equipment cost.

In the liquid-metal cure process, it is interesting to note that the liquid metal is both the die and the heat source.

Curing. Thermosetting resins will, of course, require careful heat control to produce quality materials. For continuous cure in a metal die, the exotherm point must be kept within the die extremities, and temperatures must be controlled to within a narrow range (1 or 2 degrees within the range 220 to 250°F depending upon the resin system). Use of circulating liquid and rather massive dies both tend to remove the heat from peak exotherm, and to make control easier.

After leaving the die, the pultruded product is generally allowed to cool in air or in a wash from simple air jets. The exception is in thermoplastic processes where water-cooled die sections are necessary to harden the laminate.

When the composite has enough mechanical integrity prior to the cure, as with braid or film wrap, oven curing offers a high linear output at a low capital investment.

Curing of thick sections over 1 in., especially with polyester resins, is difficult to do without producing internal cracking and delamination. Radio-frequency cure at 8 to 10 kw, and 70 mhz does permit cure of thick sections without these troubles, because hardening takes place simultaneously throughout the mass with the lowest possible heat input. A typical radio frequency electrode arrangement may be assembled for curing several rods at one time side by side. Epoxy resin composites can be similarly cured at microwave frequencies (2450 mhz).

Pulling Equipment. Pullers for pultrusion processes vary widely depending upon the forces required. Simple belt type pullers are commercially available and are satisfactory for smaller profiles (made for intermittent, film-encased or open-die processes) such as rod stock. Continuous-pull closed die processes require caterpillar-type tractor pullers with more force. Specially shaped contact pads are used to match the profiles being made. This type of puller will sometimes transmit up to 1 million lb of force on a large tubular shape where the surface area is great. One unique commercially available high-force system uses two hydraulically operated reciprocating clamps.

Cut-Off Equipment. Cutoff equipment for pultrusion lines usually consists of standard flying saw systems of a size to suit the profile. Diamond or abrasive saws are most satisfactory, and some form of dust control is highly desirable.

Miscellaneous Equipment. A few additional mechanical operations are often added as part of the process described. These include in-line coating, in-line deflashing, film wrap, braid wrap, centerless grinding and core insertion. Equipment for these steps is highly specialized but is usually straight-forward mechanization.

MOLDING METHODS/PROCESSING DETAILS

A number of pultrusion processes are in commercial use today. These include intermittent (stop and cure) and continuous-pull methods, die-curved, oven-cured, radio-frequence cured[6] and special film[7] or braid-restrained[8] processes.

Although most production lines are horizontal operations, there is some advantage in a vertical arrangement when producing hollow profiles with internal stationary mandrels supported from one end only. Vertical operation eliminates gravity sag of such mandrels but does not eliminate concentricity variations caused by material variations or lateral flow before cure.

Production rates per die are in the order of 1 ft per min. for thin small sections and over 25 ft per min. for film sandwich processes. All pultrusion processes take into account certain common factors dictating practical operation. Resins, for instance, are most often in the liquid phase during impregnation and at the die entrance. Liquid polyesters and epoxies are therefore the two major resins used, although thermoplastics have been applied as emulsions.[9]

For continuous-pull processing, polyesters are particularly sensitive to gel control, and failure to prevent or confine excessive peak exotherm will result in cracks and voids. Conversely, an exotherm that is too low or too late in occurring results in a weak, undercured product which may delaminate in use. One major advantage of a radio-frequency cure is the easier control of exotherm because of simultaneous cure throughout the entire cross section rather than cure by elevated temperature heat conduction from the mold surface inward.

All epoxy resins do not satisfactorily respond to radio frequencies, and require microwave energy to effect polymerization. Because of the lack of a suitable internal mold release agent and generally slower cure rates, epoxy pultrusions are most often film-or braid-wrapped and then oven-cured.

Hollow articles are readily produced with stationary mandrels, removable mandrels, or inserted hollow cores. Cored pultrusions have

been made with wood, foams, metal, and lightweight sandwich structures.

The well-known, continuously laminated, flat- and corrugated sheet operations are both basically classed as pultrusion processes using film-sandwiching techniques (see Section III).

Figure V-2.1 presents a schematic diagram of the basic pultrusion process. Table V-2.1 lists the variations in the basic process with comments on the advantages and disadvantages of each. It should be noted that many of the pultrusion techniques are protected by patents on method and apparatus.[10, 11]

PRODUCT PROPERTIES

The physical properties of pultruded products are dependent upon the type of reinforcement, its directional orientation, and of course the resin system used. Table V-2.2 lists the properties of pultruded products made using three of the most-commonly-used reinforcements: parallel fiber (roving), chopped strand mat plus roving, and 100% chopped mat. All values are based on E glass fiber and a general-purpose polyester resin. A few basic electrical properties are presented together with the mechancal properties.

Applications. Typical volume applications for pultruded products are electrical-pole live hardware, fishing rods, tower guys, boat mooring whips, and corrosion-resistant structural supports. Ladders, public seating, and handrails are growing end-use applications. A recently-developed line of telescoping aerial booms is a good illustration of a massive, well-designed product taking advantage of the combination of excellent electrical and mechanical properties inherent in RP/C. Large quantities of structural beams and girder stock in standard profiles are also shipped out of distributors' ware-

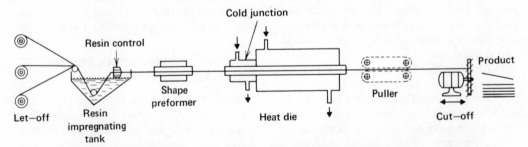

Figure V-2.1 The basic pultrusion process: Intermittent pull, closed heated rigid die, complete cure.

Table V-2.1   Pultrusion Process Variations with Characteristic Advantages and Disadvantages

Process Classification	Advantages	Disadvantages	Comments
Basic process: closed, heated, rigid dies, complete cure.	Excellent surface with continuous resin skin; independent cure allows wide range of resin formulations.	Index line sometimes visible; speed tied to die length.	Cold junction design is critical.
Intermittent pull, heated matched-metal dies, complete cure.	Excellent surface; surface decorations possible (embossing, printed mat, etc.); independent cure allows wide range of resin formulations; mold densifies reinforcement.	Index line sometimes visible; flash line visible and needs removal; profile limited to moldable shapes; tooling costs high.	Requires press equipment; may also be accomplished by pulling into seamless tubing and subjecting to batch oven cure.
Continuous pull, Teflon die, radio-frequency cure.	Surface good but dull; fast linear speeds; produces crack-free thick sections.	Die materials limited; die life limited; resin formulations must take R.F. into account.	For polyesters only; improved dies under development.
Continuous pull, rigid, heated and cooled die, thermoplastic.	Excellent surface, wide property range.	Some resins not available as emulsions or solutions.	Product reformable.
Continuous pull, film-lined trough.	Low die cost; film can add gel coat, embossing, etc.; continuous resin surface (except flash line); high linear speeds.	Requires some machining; profiles limited; pressure limited; film cost high.	Primarily rod and bar stock.
Continuous pull, film spiral-wrapped, continuous or batch cure.	Film can add gel coat, embossing, etc.; continuous resin surface; high linear speeds; tapered products are practical; wide range of resin formulations.	Shapes limited to circular or oval profiles; surface marked with spiral pattern; film cost high.	High modulus products produced; some forming before cure is feasible.
Continuous pull, braid wrap, continuous cure.	Braid adds lateral strength and resistance to splitting; oven cure allows wide range resin formulations.	Shapes limited to circular or oval profiles; surface requires treatment (coating, grinding, etc.).	Good combination of physical properties.
Continuous pull, multiple-ring die, oven cure.	Low-cost equipment;; high linear speeds.	Poor surface requires machining; profiles limited.	Generally high resin, low modulus rod and bar stock.

Table V-2.1 Pultrusion Process Variations with Characteristic Advantages and Disadvantages (Cont.)

Process Classification	Advantages	Disadvantages	Comments
Continuous pull, liquid-melt dies, complete cure.	Fast linear speeds; one bath cures multiple side by side products.	Rough surface; occasional melt occlusions; profiles limited.	Primarily for rod and bar stock.
Continuous laminating.	Film adds gel coat, embossing, etc.; high linear speeds possible.	Thin stock (to 0.060 in.) only; film cost high.	Primarily for flat and corrugated sheet.
Continuous pull, rigid heated metal dies, interrupted cure.	Fast linear speeds; low-cost equipment.	Poor surface; delicate unsupported gel; reinforcements limited (strong cohesive forces required); low resin content difficult to achieve.	Primarily for rod and bar stock.
Continuous pull, rigid heated metal die, complete cure.	Good surface, no machining required, large profile range, including hollow articles; good process reliability.	Resin must be internally lubricated; strong gel required; speed limited.	Polyester resins only.

Table V-2.2 Properties of Fiber Glass Reinforced Pultrusions[a]

Property	Units	Approx. 65% Parallel Fiber	Approx. 45% Parallel plus Mat	Approx. 35% Mat
Tensile strength	psi $\times 10^3$	70–125	30–40	15–25
modulus	psi $\times 10^6$	5–6	1.5–3.0	1.0–1.8
Compressive strength	psi $\times 10^3$	40–70	20–35	20–30
modulus	psi $\times 10^6$	5–6	1.5–3.0	1.0–1.8
Flexural strength	psi $\times 10^3$	70–100	25–40	20–30
modulus	psi $\times 10^6$	5–6	1.5–3.0	1.0–1.8
Shear strength	psi $\times 10^3$	6	6	3
modulus	psi $\times 10^6$	0.5–1	0.25–0.5	0.15–0.25
Impact strength	ft lb per in. notch	18–30	8–18	4–8
Thermal conductivity	Btu per sq ft per hr per °F per in.	2–4	2–3.5	1.5–3
Specific heat	Btu per lb per °F	0.24	0.28	0.30
Thermal coefficient of expansion	In. per in. per °F	3–4	4–5	6–8
Specific gravity	—	1.7–2.0	1.6–1.9	1.5–1.8
Water absorption	24 hr %	0.1–0.6	0.1–0.7	0.1–0.8
Barcol hardness	—	50–55	45–50	40–45
Dielectric strength	Volts/mil	150	175	200
Arc resistance	Sec	120	110	100

[a]Values based on E glass fiber and general-purpose polyester reported in the longitudinal directions. Transverse values are 1/4 to 1/2 of above.

houses for miscellaneous industrial applications much the same as structural shapes of steel and aluminum are handled. The many profiles available include beams, channels, angles, solid bars and rods, round and rectangular tubes, and many special molding-type shapes.

New materials using high-modulus fibers such as graphite, also thermoplastic resins regarded as the new engineering polymers are expected to further expand markets for pultruded products.

Figure V-2.2 illustrates one such new adaption:

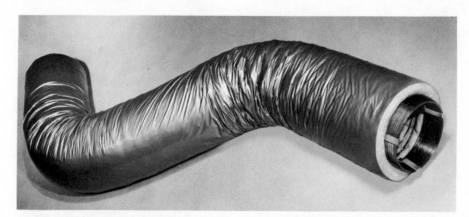

Figure V-2.2 Continuously-pultruded helix made using continuous-filament fiber glass roving impregnated with thermosetting resin. The end application is as the circumferential supporting member in a flexible fiber glass insulating duct. (Photograph courtesy of Johns-Manville Fiber Glass Reinforcements Div., Descriptive Bulletin B-356, Oct. 9, 1970.)

a continuously-pultruded coil with a helical configuration requisite to its end-use. Also recently developed are high-quality musical instrument components (xylophone bars) made from dense, pultruded bar-stock and electronically tuned.[12]

Costs Pultruded products vary widely in cost depending upon raw materials, profile, and linear mass. Rod stock ¼ to ¾ in. in diameter made using a general-purpose polyester resin system is sold for as low as $0.60 per lb. Complicated or extremely small profiles using special reinforcements and premium resins can triple this cost.

References

1. Fiber Glass Industries, Amsterdam, N. Y.
2. E. I. duPont Company, Wilmington, Del.
3. W. B. Goldsworthy, SPI 23rd Reinforced Plastics Technical Conference, 1968.
4. U.S. Patent 3,284,852.
5. Joichiro Segawa and Bungo Nakazawa, SPI 23rd Reinforced Plastics Technical Conference, 1968.
6. U.S. Patent 2,871,911.
7. U.S. Patents 2,571,717 and 2,918,104.
8. *Plastics Technology,* "What's New in Production R.P. Processing," pp. 47–58, Oct. 1964.
9. U.S. Patent 3,470,051.
10. U.S. Patents 3,235,429; 3,244,570; 3,244,784; and 3,374,132.
11. *Plastics World,* p. 42, Mar. 1971, (prepared by editorial staff).
12. Personal communications, J. E. Perko, Glastic Corp., Cleveland, Ohio, May 20, 1971.

V-3

Pre-Impregnation

INTRODUCTION

In earlier sections of this book, frequent reference has been made to procedures for processing the various composite materials at low, intermediate, and high temperatures. Also, discussions have been provided on the various procedures for different pressure levels, from simple contact-molding techniques to the more complex matched-metal tooling used in press molding.

In all these discussions, it has been pointed out that the reinforcements may be combined with the resin matrix in any one of several ways. In spray-up molding, for example, the glass or other fiber is usually chopped at the gun and sprayed out of one nozzle, while the catalyzed resin emerges simultaneously from another nozzle. This procedure does not readily lend itself to close control of the glass/resin ratio. Similarly, wet lay-up required that the reinforcement be impregnated by hand or by machine, with the excess resin usually removed by rollers, squeegees, or similar procedures. In all of these methods, the resin content is held to nominal tolerances purely by virtue of the operator's skill.

Early in the history of the reinforced-plastics industry, innovators conceived the idea of providing reinforcements to which the resin had already been added mechanically. Procedures were developed for processing in which the resin content was controlled within close tolerances. Resin systems were selected and modified so that advancement of the cure could be achieved to the extent that the product could be supplied to the molder in a relatively dry,

tack-free condition. This so-called B-staging made it possible for a molder to purchase his materials from the prepregger in rolls, ready for cutting and immediate use without the usual problems of a sticky mess, resin waste, and uncertain resin/glass ratios concomitant with wet lay-up. These pre-impregnated reinforcements, or "prepregs," have enabled processors to obtain raw materials with predictable properties at reasonable costs and have eliminated much of the waste and guesswork from RP/C procedures.

At the present time, both woven and unwoven reinforcements are available with almost all of the thermosetting resins as impregnants. Prepreg suppliers provide desired resin/glass ratios and will normally guarantee resin content, resin flow, volatiles, and so on within very close tolerances. Normally, for example, resin content can be provided with ± 1 or 2% from nominal.

The advantages of these materials are so numerous and so obvious that listing them would be both impossible and superfluous. Paramount among them, of course, are cleanliness, quality control, and economy of operation.

MATERIALS

The basic materials for the pre-impregnation process are usually reinforcement and resin. Glass fibers in the forms of roving, yarn and various fabrics comprise the bulk of reinforcements used at present, but quartz, graphite boron, and similar, more exotic fibrous materials are making incursions into the market as requirements dictate their need.

The resins, as earlier noted, include polyesters, phenolics, epoxies, silicones, phenylsilanes, polyimides, and several of the more expensive types. Catalysts, promotors, inhibitors, flame retardants, and other additives are combined with the resin according to the specifications of the customer and as the storage, handling and processing requirements dictate. (See also Chapter VII-1).

MOLDING, METHOD/PROCESSING DETAILS

These two materials, reinforcement and prepared resin, are combined in a machine called a "tower" which usually consists of a holding device for the reinforcement roll from which the fibers (woven or random) pass over and under a number of spring-tensioned rollers to maintain uniformity of motion of the material (see Figure V-3.1). The reinforcement moves from these rollers through a resin tank, where complete impregnation is accomplished.

From the resin tank, the impregnated reinforcement is carried over or between a set of "doctor" blades. These blades are merely metal or other rigid bars which scrape the excess resin from the prepreg. It is by adjustment of tension and doctor-blade clearance that most of the control of resin content is achieved.

Next, the still wet material moves through a drying tunnel. Normally, this is a periodic oven

Table V-3.1 Temperature Ranges for Prepregs

Polyester/glass	Room temp. to 450°F
Epoxy/glass	Room temp. to 500°F
Phenolic/glass	Room temp. to 500°F
Phenyl silane/glass	Long-term to 700°F
Polyimide/glass	Long-term to 800°F
Phenolic/silica, graphite, or carbon, phenyl silane/ synthetics or asbestos	Ablative and/or insulative temp. of 2000°F and higher.
Polysulfone/glass	Room temp. to 250°F

which contains a number of short sections, each with a progressively higher temperature. Here again, quality control is achieved by close attention to temperatures (see Table V-3.1). Volatiles are removed and the resin is advanced in stage to the tack-free condition necessary for storage and handling, without over-advancement to the point where cure is achieved.

As the prepreg leaves the tunnel, it is again rolled without air blisters onto cores, usually with a non-stock film, such as polyethylene, interspersed between the layers. Most prepregs are then placed in refrigerated chambers, where they are held prior to shipment. At the molder's plant, they are again held under refrigeration until needed for use. With almost all prepregs shelf life is dependent upon the temperature at which they are stored. Most resins call for holding at 40°F or less for best results and

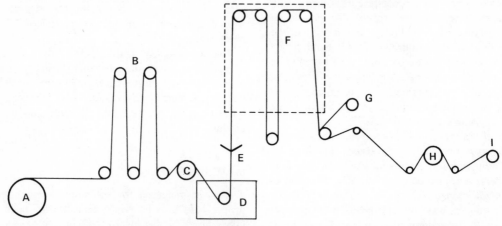

Figure V-3.1 Impregnating Tower. (A) Unimpregnated reinforcement take-off roll; (B) Accumulator rolls; (C) Guide roller; (D) Resin bath; (E) Metering device ("doctor" blade); (F) Single- or multi-stage oven; (G) Polyethylene separator film; (H) Cooling drum; (I) Final prepreg take-off roll. (Courtesy of Cahners Publications.)

longest shelf life. In this regard, it must be pointed out that condensation of moisture on the prepreg must be scrupulously avoided, and refrigerated packages should not be opened for use after removal from cold rooms until the package has had time to reach room temperature.

One particular advantage of the prepreg operation which is sometimes overlooked is the fact that it enables molders to use certain types of resin which would be difficult, if not impossible, to handle in their own plants without a considerable amount of meticulous control not usually attainable. While the situation may change in the future, a good example of such a resin system is the polyimide series. Preimpregnation of reinforcements with these matrices is not really a difficult problem, but the removal of volatiles in pre-determined ratios and the advancement of the resins to B stage are operations requiring controls not usually available to the molding shop operator. By relying on the prepreg supplier for accurate management of variables, the molder can rest assured that his operation may be scheduled on a time-temperature basis which will produce finished parts with predictable and desired physical and mechanical properties.

To attain a more complete understanding of prepregs, it is desirable to consider specific cases. The following discussion, therefore, presents the history of a typical polyester prepreg, including the as-received properties; storage recommendations; stepwise procedure for processing into a typical fabricated item, rolled rod stock in this case, and finally cured and finished (flat) laminated product properties are presented in the discussion of "Product Data" in Table V-3.4. (See Chapter II-3, Vacuum Bag Molding.)

As-Received Properties. For impregnated, semi-unidirectional glass fabrics designed to be used in tubular applications, the resin content is 30 to 34%, the volatile content is 1 to 4% and the resin flow is 0.5 to 5%. Table V-3.2 presents the general as-received properties of polyester prepreg material.

Recommended Storage Conditions:

Bulked Storage in Roll Form. The impregnated cloth is supplied in essentially air-tight polyethylene bags for purposes of retaining the original tested properties. To further insure the retention of these properties, it is recommended that the rolls be stored at approximately 65°F and 50% relative humidity. Under these conditions, the impregnated cloth has a minimum of 6 months of shelf life, starting from the time the material is received in the fabricator's plant. At a storage temperature of 35° to 45°F, the shelf life of the material is in excess of 1 yr. As stated, care should be taken to prevent moisture condensation directly on the material upon removal from cold storage. Storage beyond 6 mo. at 65°F may alter the original rolling or processing characteristics of the preimpregnated material. This alteration is not critical with respect to the quality of the finished laminate but may be sufficient to require slightly higher temperatures at the rolling table, particularly where fly rod tips are being manufactured. The use of material stored for more than 1 yr under any conditions is questionable. The storage referred to starts at the time of receipt of material at the fabricator's plant.

Pattern Storage. The impregnated cloth should be conditioned in roll form prior to cutting for a minimum of 24 hr at approximately 65°F

Table V-3.2 Properties of Polyester Prepreg Material

Property	Specifications for Medium to Heavy Prepreg Stock[a]	Light Prepreg Material[b]
Approx. thickness as delivered, (in.)	0.016	0.011
Approx. thickness of 3-ply laminate molded at 20 psi, (in.)	0.030	0.018
Type of weave	Crowfoot satin	Crowfoot satin
Average weight of glass fabric, oz per sq yd	8.94	5.56
Final impregnated weight, 12 oz per sq yd	12.40	7.80

[a]For heavy rods, hot sticks, pipe, stand-off insulators.
[b]For light rod stock or ducting.

and 50% relative humidity. If it is desirable or necessary to store the cut patterns for more than 2 days, it is recommended that they be wrapped in polyethylene and stored at 65°F and 50% relative humidity. Special care should be taken to re-seal and restore partially used rolls. The use of a slip-sheet of polyethylene is recommended in stacking patterns. The slip-sheet acts as a seal to prevent loss of volatiles, as well as a preventative with respect to sticking of the individual plies.

Processing Procedure:

Mandrel Preparation. The mandrels or other molds are best handled after being cleaned and treated with a release agent. Several release agents have been found to be quite satisfactory—paste floor wax or silicone releases for example.

Following the application of a release agent in most cases is a thin application of catalyzed polyester-resin mix to the mandrel. The purpose of this addition is purely for bonding the impregnated cloth to the mandrel during the tacking operation.

Attaching Pattern to Mandrel. After the patterns have been cut and properly stored, they will be ready for attachment to the treated mandrel. The attachment of the impregnated cloth to the mandrel is done by use of a hot iron. (Wells automatic iron, Style PF 100 watt as manufactured by Wells Mfg. Co., San Francisco, Calif.) After attaching the impregnated cloth to the mandrel, it is desirable to maintain the cloth at a temperature of 75° to 85°F to soften it and ready it for the rolling operation. This suggests a room controlled at about these temperature conditions. To minimize flagging, (ripples) it is not recommended that the impregnated cloth be allowed to remain at these conditions for more than 4 hr. before applying the shrink tape.

Rolling Operation. The impregnated cloth attached to the mandrel should be placed on the rolling table. The table surface must be heated to a temperature somewhere between 80 and 120°F. By the time the mandrel is placed in position on the rolling table and the load is applied (15 to 20 sec), the cloth will have been warmed sufficiently to cause good bending when rolled. In rare cases, particularly those

cases in which the impregnated cloth has aged, higher rolling-table temperatures may be necessary; 3-roll, cigarette machine, and endless-belt methods may alternately be used to roll prepreg onto a mandrel.

Application of Pressure. Cellophane or Tedlar over-wrapping is applied to the mandrel to provide pressure before curing. The dry spiral method is recommended for impregnated material. Pressure may also be applied using matched dies, pressure bags, internal-pressure bags, or centrifugal casting. Any type of cellophane may be used, as long as the same type is used at all times.

The Curing Cycle. The resin formulation employed should give the degree of flow necessary to produce excellent laminates under the pressure developed in the shrinkage-tape enclosure. The temperatures in Table V-3.3 have been found highly satisfactory for producing high quality finishing rod blanks or other molded items of uniform color and properties.

Shrink-tape-wrapped cloth should be placed into the curing oven at 285°F and slowly raised to 300°F. No blowing or out-gassing should be experienced. When it is necessary to raise the temperature of the curing oven over a period of time as opposed to going directly into the hot oven, it is suggested that, once the oven has reached the curing temperature, the period at this temperature be timed. This will offset the period required to bring the oven up to temperature while the impregnated cloth is heating.

A typical cure cycle may be as follows:
1. Enter oven at 285°F.
2. Hold oven temperature at 285°F for 15 min.
3. Raise oven temperature to 300°F.
4. Hold at 300°F for 45 min.
5. Remove cured blanks.

Table V-3.3 Curing-Oven Temperature

Cloth Style	Min. Temp. (°F)	Max. Temp. (°F)	Cure Time at Cure Temp. (min.)
"H" Rod	280	300	45–60
"L" Rod	280	290	45–60

Table V-3.4 Cured and Finished Properties and Test Methods of Flat
Laminates Made from Polyester Prepreg Stock

Property	Specification[a]	Test Methods ASTM	Test Methods Federal 406
Glass by weight, %	64.3	—	7061
Resin by weight, %	35.7	—	7061
Dry tensile strength	102.9×10^3	D638–60T	1011
Dry tensile modulus	—	D638–60T	1011
Dry flexural strength	122.9×10^3	D790–59T	1031
Wet flexural strength	115.0×10^3	D790–59T	1031
Flexural modulus	5.12×10^6	D695–54	1031
Dry compressive strength	101.8×10^3	D695–54	1021
Impact strength, ft lb/per in.	—	D256–56	1071
Moisture absorption for 24 hr, %	0.045	D570–59T	7031
Specific gravity	1.9	—	5011–5012
Dielectric constant (1MC)	5.2	D150–59T	4021
Dissipation factor (1MC)	0.006	D150–59T	4021

[a]F149-5C.

Exceeding the maximum cure temperature may result in non-uniformity of properties and color.

The above findings are based on hot-air temperature. Also, the cure cycle is based on forced draft oven conditions. Some slight alterations in these suggested curing conditions may be necessary to compensate for load mass and the design of the oven.

Product Properties. The product properties of a typical cured polyester prepreg are listed in Table V-3.4.

PRODUCT DATA

There are three major markets for pre-impregnated materials whose requirements at the present time lead others by a relatively wide margin: aerospace, electrical circuit boards, and sporting goods.

In aerospace, the ability of prepregs to control resin content, viscosity, flow, and volatiles makes them ideal for such applications as sandwich construction, especially with honeycomb-core materials. In this application, it is of utmost importance that void-free skins be bonded to the core with a good filleting action at each cell-wall contact line. The void-free condition, so readily provided by prepreg, assures high interlaminar shear strength as well as a base (in wind leading-edge and radome or other nose applications) for rain-erosion-resistant coatings. Again, as in so many other applications, the controllable and predictable resin content makes for high, uniform strength properties and allows the weight-and-balance engineers to add their contribution to vehicle stability.

In the market for electrical circuit boards and copper-clad laminates, strength requirements are dictated by NEMA, MIL, and other controlling specifications. In addition, however, an important factor in maintaining predictable performance from the boards is their electrical properties. These include dielectric constant, loss tangent, and voltage breakdown values. Since the first two are direct functions of the type and concentration of resin in the reinforcement, it is of utmost importance that these variables be held within tight tolerance limitations. Prepregs offer that advantage. Also, the type and amount of resin in the laminate determine the ability of the board to resist breakdown when voltages are applied. If for example, a phenolic resin is not properly cured in the molding process, moisture from the condensation reaction may remain trapped between the layers. Water has a relatively high dielectric constant and would worsen properties if present so that the resulting laminate would probably not be acceptable from the electrical standpoint. More serious, however, would be the fact that the entrapped moisture would vaporize if the laminate temperature were raised above 212°F potentially resulting in a catastrophic failure from exploding steam.

The advantages of prepregs, when properly handled, are apparent.

In sporting goods, many of the smaller items use prepregs exclusively. Thus far, in the boat field, they have not been economically feasible, and spray-up or hand-lay-up procedures still predominate. However, skis, toboggans, fishing gear, and similar applications still use substantial quantities of molded prepregs. Dividing boards, archery equipment, vaulting poles, golf clubs, billiard cues, and tennis rackets all add to the growing market.

Among the specific advantages of prepregs are the following:

- High strength/weight ratio.
- Predictable properties based on control of resin/reinforcement ratio.
- Labor savings in fabrication procedures.
- Reduced materials waste, with resulting cost savings.
- Controllable inventory of raw materials.
- Reproducibility of part weight and dimensional tolerances.
- Wide choice of resin/reinforcement combinations.
- Versatility of design options.

References

1. All data in this discussion is adapted from information supplied by the Coast Manufacturing and Supply Co., Division of Hexcel Corp., Dublin, Calif.

V-4

Centrifugal Casting

INTRODUCTION

In the centrifugal casting method, reinforcement and resin are applied to the inside surface of a rotating tank, cyclinder, pipe, parabola, or even a flat-edged disc serving as a mold. The speed of rotation is controlled in order to introduce the desired centrifugal forces which serve to hold the RP/C material against the mold surface. Spinning the mold causes the reinforcement to compact, and induces the removal of air from the reinforcement and resin prior to cure. Cure follows, and the part is removed from the mold by the easiest means in RP/C—by loosening itself and falling out.

Centrifugal casting is limited essentially to symmetrical cylindrical parts and/or surfaces of revolution. Typical products include small and large diameter piping in lengths up to over 20 ft for use in the chemical and petroleum industries, chemical process and water conditioning cylinders, tanks and ducts, appliance parts, submarine launcher tubes, parabolic elements, tear-drop shaped disposable or drop fuel tanks, and many others.

MATERIALS

Reinforcements. Almost any form of fiber glass reinforcing may be used in centrifugal casting. For pipes, the material can be in the form of chopped fibers, chopped strand mat, woven roving or braided roving, or a combination of these. For tanks, chopped fibers mixed with fabrics in tape or sheet form are commonly used. In molding parabolic forms, a rough reinforcement shape is formed in an earlier operation and placed in the mold. In special cast work, other preforms similar to those used in matched-die moldings are placed in the mold, rotated, and resin is added.

Resins. General purpose isophthalic polyester resins, catalyzed with MEK peroxide and promoted with cobalt naphthenate are in common use. For applications requiring exceptional chemical durability, the resin may be a bis-phenol-A or HET acid type, and the system for catalyzing could be benzoyl peroxide together with a promoter such as dimethyl aniline. For casting parabolas, epoxy resins of low viscosity plus room temperature hardeners or promoters such as diethylene triamine are used.

EQUIPMENT AND TOOLING

Equipment for centrifugal casting is usually designed to match the part being produced. There is little so-called general purpose equipment. Large casting operations for products such as pipe and smaller tanks utilize equipment especially designed for the particular product. The pumps, applicators, rollers and handling tools are all adapted from standard items.

Rotating Means. Usually variable speed drives are provided to permit slow speed during charging of the resin, and high speed during densification and cure.

Heating. As a means of heating, some manufacturers merely permit an open gas flame to

283

impinge on the outside surface of the rotating mold, making certain to keep the styrene fumes separated. Hot air is also used. One manufacturer uses jacketed molds and hot oil or steam as the heating medium. Heat is not always needed for the centrifugal processes, but it is used to speed up the production and achieve quicker cures.

Tooling. For pipes, the molds may be of one-piece construction because the casting will shrink and can be easily removed. The mold surface must be smooth and free from waviness. Split molds are sometimes used when a pipe part has threads molded into the ends. Also, the threaded part of the mold can be removed with the casting and then unscrewed after the casting is out of the mold. The mold is usually rotated by means of cradling drive rollers which are powered in turn by a variable-speed motor. Care must be taken to insure that the mold is dynamically balanced. Because repeated heating and cooling takes place, the mold material must be normalized before final finishing in order to avoid warping at a later time. Steel, cast iron, stainless steel, cast aluminum, as well as RP/C have been found satisfactory as mold materials. For large-scale production, proven investments are chrome plating, Teflon coating, and so on.

For small tanks, molds are usually mounted with an axle and are open on one end. The rim of the open end forms a lip to keep the resin from flying out. This lip is then removed with the cured part. The molds have a common axle size so that molds of different diameters or part sizes can be made alternately on the same machine, although different rotational speeds will be used. These molds are usually made from welded sheet steel. For large tanks, these molds are practically all made from sheet steel and are fitted with outer rails which engage with cradling rollers driven by variable speed motors. Dynamic balance is important here as well as for the smaller tank molds.

For shaped tanks, drop tanks, or tanks with parabolic shaped ends, the mold is set to rotate about a vertical axis. Similar, molds must be dynamically balanced, and can be made from sheet steel, aluminum or the like. One manufacturer has placed strip resistance heaters in a balanced manner around the mold to speed the cure.

For casting parabolas, the tooling usually consists of a metal or RP/C mold in the shape of the part itself. This mold is mounted on a turntable which is controlled to maintain a constant speed. The resin is added to the center of the part, and the rotation of the part causes the resin to flow into the parabolic shape, thereby covering the surface of the part. With constant speed of the correct magnitude, the resin is allowed to cure and the resulting surface is a perfect parabola.

MOLDING METHOD/PROCESSING DETAILS

The general basic steps followed in centrifugal casting of RP/C are itemized in proper sequence below:

- Apply a release (unless the mold is prepared in advance with Teflon mold lining) and gel coat the mold surface if required.
- For both small and large tanks, a tank bottom should be prefabricated with an extension for tie-in to the part to be molded, and installed into the mold, where it will be conjoined with the centrifugally cast resin and reinforcement.
- After bottom placement, add the reinforcement in the form of either mat, fabric sheets, or as mixed reinforcements if required.
- Spin or rotate the mold, referring to the centrifugal casting formulas given below.[1,2]
- Introduce the resin by pouring, jet, spray, oscillation, or tilt-pan systems. The main goal is to get a uniform application of resin as quickly as possible, and densify thoroughly prior to gelation.
- Continue spinning until the resin has gelled, adding heat to the mold to hasten gelation if necessary.
- After cure, remove the part from the mold. With centrifugal casting, part removal is never a problem because shrinkage generates high forces which pull the part from the mold surface (see Figure V-4.1).

Means for Addition of Fiber Reinforcement. For small tanks, the reinforcing mat is loosely rolled and introduced inside the slowly-rotating mold. For piping, the fiber glass reinforcement may be in the form of a braided sleeve which is slid onto a placement mandrel. This can rotate within the rotating mold. As rational speed is increased, the braid expands and transfers from the placement mandrel to the

Figure V-4.1 Photo sequence illustrative of the process for centrifugal casting to produce an RP/C brine drum for a home water-conditioning unit. *Upper left*—The mold is lined with fiber glass chopped strand mat (high solubility binder) that is cut to the exact required width (tank length) as received. The mat also has overlapping feathered edges so that the laps are not visible in the finished drum. The mat overlaps the sanded edge of the molded premix bottom. The top edge of the drum is thickened using cut strips of mat (visible to right of operator). These are placed into a groove at the outer edge of the mandrel just prior to inserting the roll of mat. *Right*—The precatalyzed resin is injected using a jet spray mounted on a traverse (long shaft of air cylinder)which traverses uniformly starting at the outer end of the mandrel as shown, and ending up at the brine drum bottom. *Bottom left*—The brine drum is now cured after rotating in the heated chamber. The operator is removing the rubber resin-retaining ring from the groove in the mandrel. The drum is produced with a resin-rich internal surface. The molded premix bottom is placed into the mold prior to inserting the mat, and is molded in during the centrifugal spinning cycle. *Right*—Here the ejector piston has broken the drum loose from the inner mandrel wall, aided by part shrinkage, and an air-bleed provided assists the operator in slowly sliding the drum out of the mandrel. (Photo courtesy of Structural Fibers, Inc., Chardon, Ohio.)

mold, and the placement mandrel is then withdrawn.

For large tanks, a reciprocating arm supports a chopper gun which applies reinforcement while traversing the length of the rotating mold. Tapes can also be applied from a roll which rotates fast enough to match the peripheral speed of the mold. For tear-drop shaped tanks (airdraft drop tanks), fiber glass mat preforms are placed into the lower end of the vertically rotating mold.

Means for Introducing Resin. For small tanks, the required amount of resin is placed in a trough and poured into the rotating mold area. The trough is tilted and the resin pours and spreads out in a film or sheet onto the reinforcement. The resin is often added first so that the centrifugal force will push the reinforcement into the resin, and make removal of entrapped air more efficient.

For pipes, the resin is either pumped into the mold through a full-length perforated pipe or is introduced from the end of a long pipe moved through the mold at a controlled rate of advancement. For large tanks, either the resin is added at the same time that the fibers are chopped, as in sprayup, or the resin gun is traversed separately, as when preforms, mats, tapes or sheets of reinforcement are used.

Densifying Means. In small tanks or pipes, the rotational speed constitutes the means of densifying the resin and reinforcement. In large tanks, rollers are automatically pressed into the laminate, and traversed the length of the mold. Some hand rolling is also usually required.

For tanks 6 to 15 ft in diameter, a variable-speed drive is usually used, and the mold is rotated just fast enough to keep the dry reinforcement from falling away from the mold. As resin is introduced, the speed is increased to effect air removal and densification.

Centrifugal Casting Formulas. Prior to considering several practical examples or case histories of centrifugal casting operations, it is advisable to review the formulas derived for calculation of the correct "pressure" or rpm required.[2]

The classical formula for centrifugal force is:

$$F = \frac{Wv^2}{gr} \tag{1}$$

where W = weight of the article in lb
v = speed of rotation in ft per sec
gr = acceleration of gravity in ft per sec^2
F = total force on the laminate

For centrifugal casting of RP/C, a designer could not use this formula until it was converted to provide values for rpm of the mold. This may be accomplished by manipulating the formula as follows:

By transposing,

$$v = \sqrt{\frac{Fgr}{W}} \tag{2}$$

and $\quad F = P \times A \tag{3}$
where $\quad P$ = pressure in psi
$\quad A$ = area in sq. in. (πdl)
and $\quad W = A \times t \times q \tag{4}$
where $\quad t$ = thickness in in.
$\quad q$ = density in lb per cu in.

(a 50:50 glass-resin ratio = 0.06 lb per cu in.)

and $\quad \dfrac{F}{W} = \dfrac{\pi dl \times P}{\pi dl \times q \times t} = \dfrac{P}{qt} \tag{5}$

where $\quad d$ = diameter

$\quad l$ = length

Hence, $\quad v = \sqrt{\dfrac{Pgr}{qt}} \tag{6}$

From formula (6) above, it is obvious that the velocity is independent of the weight and length of the article. The speed of the rotation in rpm is equal to v multiplied by 60 divided by the circumference of the article in ft.

or, $\quad \text{rpm} = \dfrac{v \times 60}{\pi d} = \dfrac{\sqrt{\dfrac{Pgr}{qt}} \times 60}{\pi d}$

The value P is very important in this formula, and it should be interpreted as similar to pressure applied via the vacuum-bag process, that is, full vacuum-bag pressure is about 14 psi at sea level, and practical values are about 10 psi. Autoclave pressures are about 50 psi, and so on.

The viscosity of the resin, the solubility of the binder, the finish, and other factors must be considered in selecting the densification desired. Also important is the stiffness of the reinforcing material. Mat products tend to wrap better than fabrics, and they break down easier once the resin contacts them. Fabrics are stronger, but if not inserted properly in the mold, they tend to bridge rather than unwrap properly. The choice of P (pressure) is critical, and those experienced in centrifugal casting claim that, in the final analysis, it is based upon empirical information. However, two specific practical problems using the formulation will be presented here for illustration.

Problem 1: Determine the rotational speed required for centrifugally casting a tub for a washing machine which is 24 in. in diameter and 24 in. deep. The bottom of the tub is not to be considered because it is pre-cast and placed in the end of the rotating mold. Also, consider the laminate to have a 50:50 glass to resin ratio and a thickness of 0.100 in. (or, $q = 0.06$, $t = 0.10$). Because the tank must be leakproof, the value chosen for P should be at least 20 psi, which is approximately equivalent to the pressure induced for a good vacuum-bag layup:

$$\text{rpm} = \frac{\sqrt{\dfrac{Pgr}{0.06t}} \times 60}{\pi d} = \frac{\sqrt{\dfrac{10 \times 32.16 \times 1}{0.06 \times .1}} \times 60}{2\pi}$$

or, rpm $= \dfrac{\sqrt{53,600} \times 60}{6.28} = \dfrac{231.5 \times 9.56}{}$

$\quad = 2212 \text{ rpm}$

Problem 2: Determine the rotational speed required for centrifugally casting a roll tube 8 in. in diameter by 52 in. long and ¼ in. thick. Use a value for P of 5 psi and a laminate density of 0.06 lb. per cu in:

$$rpm = \frac{\sqrt{\dfrac{5 \times 32.16 \times 4/12}{0.06 \times 1/4}} \times 60}{\pi \times 8/12}$$

$$= \sqrt{3400} \times 28.9 = 1700 \text{ rpm}$$

Case Histories. The investigation of the RP/C centrifugal casting process is furthered by studying 4 specific examples of currently-used molding procedures.

Example 1. This concerns production of straight cylindrical pipe in large diameters and long lengths. These casting machines are in excess of 20 ft in length and are also in the shape of long rectangular boxes. These insulated shells contain the mold tube, bearings, and heated oil required in the casting operation.

Oil is continuously pumped from the casting machine through electrical heater coils and is recirculated into the machine. Heat sensors in the machine activate the heating coils as required so that a relatively constant temperature is maintained. The purpose of the hot oil is to supply the heat necessary to polymerize the resin. The oil level is maintained at a point above the mold tube so that constant heat is transmited over the entire circumference of the mold tube.

The mold tube extends the length of the machine and rests on a series of bearings within the chamber which keep it straight while rotating. The mold tube is belt-driven from one end with an electric motor. The desired rpm is maintained throughout the casting process, creating forces in excess of 100 G's.

Warm air blowers are located on top of the machines. Moveable air ducts are lowered over each end of the mold tube during the curing process. Metered air is blown through the mold tube to remove the heat generated by the exothermic reaction of the resin and catalyst. Warm air blowers are designed so that the volume and temperature of air blown through the mold tube may be carefully regulated. The flow of air through the tube is bi-directional and programmed to alternate during the cure cycle so that an almost constant temperature is maintained over the entire length of the pipe. Temperature control is necessary to eliminate exotherm bubbles or creation of a brittle condition in the finished product.

The centrifugally cast pipe in this example is manufactured by first braiding or wrapping the desired type and number of layers of fiber glass reinforcement over a collapsible mandrel. The loaded mandrel is then inserted into the mold tube, collapsed and removed, leaving the fiber glass in place. The fiber glass material forms the reinforced wall thickness of the finished pipe. After removal of the mandrel, donut shaped end-caps are placed on each end of the mold tube. This prevents the resin from overflowing at the ends when it is injected into the mold tube. The mold tube is then rotated at the calculated or specified rpm and the catalyzed resin is injected into the machine through a hose from a resin pot. The precise amount of resin and catalyst are weighed and mixed for each size and grade of pipe produced. The centrifugal force holds the glass in place and forces the resin through the fibers until it reaches the inner surface of the mold tube. The high G forces insure a uniform high-density laminate with complete wetting of the glass fibers. The quantity of resin injected slightly exceeds the amount required to encapsulate the fiber glass, and this results in a pure layer of resin on the inside surface of the finished product. The inner surface is extremely smooth. The outer surface duplicates that of the mold, and the smooth ID of the mold tube allows the finished product to have a well-controlled OD.

Example 2. The product resulting from this operation is a cylindrical tank with tapered ends in which the ends are integrally molded with and into the cylindrical portion. The basic machine is a cylindrical mold mounted on rollers and powered by motor with a belt drive onto a pully that is an integral part of the mold. Mat reinforcement in a tight roll is introduced into the mold after one premolded end-cap of reinforcement with a feathered edge is set in place, and held by an O-ring. The opposite end-cap is then inserted, together with a special pressure fitting and bearing for closing the mold. A perforated probe is inserted through a hole in this fitting. After rotation and consequent centrifugal forces have caused the mat to unroll, a measured amount of resin is pumped through the perforated probe, thoroughly saturating and bonding the mat and both previously cured end pieces together into one integral part. Heat is introduced through a duct into the box or housing surrounding the external portion of mold to speed the cure.

Example 3. This method relates to the production of cylindrical elements using sprayed-up resin and reinforcement. A retractible metal mold cylinder is set in bearings and is driven by a variable-speed motor. Resin and chopped glass are introduced from a head on an arm and dolly casting. The resin is pressure-pumped from a tank to the spray-head through suitable hoses and is sprayed onto the inner mold surface. Glass roving from creels is passed to a cutting head on the end of the extendable arm, where it is cut into short lengths (¾ to 1¼ in.) and deposited together with the resin onto the inner surface of the rotating mold. The arm and dolly travel the full length of the mold on a track. The speed of rotation is sufficient to compact and densify the deposited material. Curing of the resin is hastened by supplying heat onto the outside of the mold from infrared lamps. For better compaction of the glass fiber reinforcement, acetone may be added to the chopped glass layer sprayed into the mold prior to adding the resin. The acetone solublizes the fiber glass sizing and causes the glass to compact and to lay very close to the inner mold surface. The heating lamps may also be utilized to vaporize the acetone prior to adding the resin.

Example 4. This method concerns the manufacture of parabolic-shaped tanks. The mold rotates on a vertical axis, driven by a motor capable of variable speed. Resin is metered and pumped through a vertical boom. A rotating disc serves to blend resin and catalyst through shear action by throwing the material out against the mold wall. Glass fibers from a roving package are chopped between rolls, and are blown down a tube by air blast, and are dispersed in a random manner by another rotating spreader disc. Both the resin and the chopped fiber are simultaneously thrown out against the walls of the rotating mold as the holding boom travels vertically through the full length of the mold. The mold is heated electrically using strip heaters in the space between the mold and the housing or shroud. Ob-

viously the rate of retracting the boom must be slowed as the part diameter increases. Also, the correct speed of rotation must be maintained. Excessive speed will cause non-uniform wall thickness distribution. (see also U.S. patent no. 2,908,039).

PRODUCT DATA

Examples of typical products and current manufacturers are:
Pipe:
Apex Manufacturing Company, Cleveland, Ohio
Youngstown Sheet and Tube Company, Sand Springs, Oklahoma
Tanks and Ducts:
Raven Industries, Sioux Falls, S. Dak.
Apex Manufacturing Company, Cleveland, Ohio
Structural Fibers, Inc., Chardon, Ohio
Parabolic Shapes:
General Electric Company, Philadelphia, Pa.
Electronic Specialties Company, Los Angeles, Calif.
D. S. Kennedy Company, Cohasset, Mass.

Advantages of Centrifugal Casting:
- The centrifugal casting process may be automated and usually is.
- Extremely uniform wall thickness is possible.
- Inner and outer surfaces are smooth. In fact, the inner surface is a first or virgin surface and is glass-smooth.
- Excellent part duplication is possible.
- The only major limitation on size and length of parts to be molded is the cost of the machine to do the rotating.
- Machines can be simple and low-cost, and yet turn out a quality product.
- Centrifugal casting is the only process that can form cured parabolic surfaces merely by rotation.
- There is a low labor factor in production of centrifugally molded parts. Workers possessing only a minimum of skill may be trained to consistently produce high quality parts.
- Two or three resin systems may be used sequentially in the same part one after the other with reinforcement, to give composite properties to the tank or pipe.
- A clear, non-reinforced resin layer may be incorporated onto the inner surface to pro-

vide special chemical durability, elasticity, a non-porous layer, or other properties.

- Various reinforcements or reinforcement combinations may be used in the same part.
- Different sized or shaped parts may be produced on the same rotating equipment.

Disadvantages of Centrifugal Casting.

- Placement of the reinforcement is critical and should be controlled to avoid out-of-roundness and unbalanced parts.

- Densities of the components must be carefully chosen to insure proper arrangement of the ingredients during the centrifugal action.
- The process is limited to surfaces of revolution, usually cylinders, low tapered items, and parabolas.
- High-rate production equipment is costly, and the part design must be carefully worked out to allow for dynamic balance of all the components.

References

1. Amos, H., et al. (to Patushin Aviation), U.S. Patent 2,870,054, Jan. 20, 1959.
2. Amos, H., (to Patushin Aviation), U.S. Patent 2,937,039, Oct. 13, 1959.

Additional information concerning centrifugal casting can be obtained by consulting the following references:

Wiltshire, A. J., (to Apex Div., White Sewing Machine Co.), U.S. Patent 3,012,922 (Dec. 12, 1961).

Wiltshire, A. J., "Centrifugal Molding," SPI 15th Annual RP/C Div. Proceedings, 1960, Sec. 8-D.

Biedbach, W. F., Jr., et al., U.S. Patent 2,915,425 (Dec. 1, 1959).

Voit, J. A., et al., U.S. Patent No. 2,157,580 (1939).

Richardson, H. M., U.S. Patent No. 2,372,983 (1945).

Stearns, H. C., U.S. Patent No. 2,376,831 (1945).

Moyer, H. R., U.S. Patent No. 2,436,726 (1948).

Stephens, J. F., U.S. Patent No. 2,467,999 (1949).

DeGanahl, C., et al., U.S. Patent No. 2,714,414 (1955).

Noland, D. B., et al., U.S. Patent No. 2,718,583 (1955).

Daley, H. S., et al., U.S. Patent No. 2,792,324 (1957).

Ising, G. E., et al., U.S. Patent No. 2,815,534 (1957).

Wentz, E. A., U.S. Patent No. 2,901,190 (1959).

Nerwick, C. M., U.S. Patent No. 2,977,269 (1961).

V-5

Casting, Potting, and Encapsulation

INTRODUCTION

Although cast plastics are often not reinforced and cast polyesters might be considered beyond the scope of this book, the majority of cast products are reinforced with fillers, short fibers, or other materials. Moreover, the major tonnage of cast plastic products today is based on the same polyesters and additives used in fibrous-reinforced products. Cast polyester manufacturing is therefore of intest to the RP/C industry and will be discussed in the following product areas: clay-pipe seals, furniture, decorative buttons, plastic cultured marble, television tube interlayers, and miscellaneous applications.

CLAY-PIPE SEALS

One of the most important recent improvements in the clay pipes is a slip-type sealing joint which has had a major effect on the pipe industry as well as on plastic-resin consumption. With today's high labor costs, manufacturers have found that a plastic seal which permits lengths of pipe to be simply pressed and sealed together greatly reduces the installation cost of their product and increases its attractiveness for use. After witnessing the labor required to seal underground pipes with tamped oakum and cement or asphalt, one can readily appreciate the savings involved in a pipe which literally snaps together and seals at the same time. Figure V-5.1 illustrates this labor-saving concept.

The authors are particularly indebted to Mr. Jacob A. Dvorak, P.P.G. Industries, Inc., for his assistance in preparing this chapter.

Installation costs are not the only savings which the use of plastic seals offer the clay-pipe user. Better performance also promotes savings in areas such as sewage treatment. The sewage treatment plant serving a given area must be large enough to process not only sewage from that community but also the leakage into the piping system. Plastic joint seals make tighter and longer-lasting joints and reduce leakage, thereby permitting a minimum-size plant for a given community. The community thus saves money on pipe installation and also on overall long term sewage treatment.

Plastic pipe seals also provide a counter against the statements made by cast-iron pipe manufacturers regarding the use of clay pipe; namely, that cast-iron pipe sections, being longer, require fewer joints and less labor and entail less leakage than do clay pipes. Plastic-sealed clay pipe is, of course, considerably lower in cost than cast iron pipe.

Materials:

Resin. Three types of cast plastics have been used for pipe seal: polyester, epoxy and polyurethane. Table V-5.1 shows a performance comparison of these materials and explains clearly why today's production primarily utilizes polyester resin. It is low in viscosity and can therefore accept high filler loadings. It is also semi-flexible, thus providing impact strength and resilience. Another advantage is the low cost. The catalysts are the usual room-temperature-curing types, and are chosen for fast gel times over a wide range of plant conditions. Examples are MEK peroxide catalyst and cobalt promoters.

Table V-5.1 Comparison of Resins Used for Clay-Pipe Seals

Unit of Comparison	Polyester	Epoxy	Polyurethane
Cost	Very low	High	Moderate
Resistance	Good acid resistance	Good acid and alkali resistance	Excellent acid and alkali resistance
Process	Versatile and fast	Slow	Rigid controls required
Adhesion	Requires primer	Best adhesion without primer	Requires primer
Product	Uses "O" ring	Uses "O" ring	No "O" ring used

Additives (Fillers). Fillers such as silica flour and clay are added to control both hardness and shrinkage. Silica flour is preferred by most manufacturers.

Red oxide or other pigments are used primarily as a process control to monitor good mixing.

Primers are used to insure adhesion of the cast polyester to the clay pipe. One proven primer* is an air-dry type which is simply applied as a thin coating to the clean dry clay surface and is allowed to harden. It is colored red to assist in visual control of complete coverage of the desired bonding surface.

Formulation. A typical formulation for a pipe-seal casting mix is:

	parts by wt
Polyester resin	100
Silica flour (140-200 mesh)	100
Dimethyl aniline	0.2
Cobalt octoate (12%)	0.2
MEK peroxide (60%)	1.0
Red oxide pigment	1.0

This formulation will have a gel time of 4 to 6 min. depending upon ambient room temperature.

Equipment and Tooling. Automatic mixers and dispensers are used to mix the systems used for the two component (resin and filler) casting plastic clay-pipe ends.

The abrasive effect of the silica flour filler, if utilized, should be taken into account in purchasing the automatic mixer or other handling equipment. Most automatic mixers that are used also dispense measured charges into

*PPG Industries Selectron 5964.

the molds as the pipe sections pass by on simple conveyors.

Following is a typical casting cycle:

	min.
Charge mold	0.1
Time in mold	10.0
De-mold	0.4
Mold preparation	1.5
Total	12.0

Full cure to optimum properties will, of course, require 24 hr or more at ambient temperatures.

Molding Method/Processing Details. Casting is usually carried out under controlled conditons in a clay-pipe manufacturing facility or yard. After casting the plastic sealer, the pipe may be removed to the installation site, the "O" ring applied, and the pipe assembled end-to-end on the job.

The casting process consists of simply mixing and then pouring the mix into the space between the steel molds and the clay pipe bell and spigot ends. Bell casting is done with an inserted mandrel extending into the pipe itself. The spigot ends require split molds in order to form the "O" ring groove and allow for part removal after casting.[1]

The process for casting plastic material into clay-pipe ends involves the following steps:

1. After firing, inspect the clay-pipe sections to specifications and tolerances, particularly to assure freedom from out-of-roundness in the bell and spigot ends.
2. Clean, apply the primer, and allow adequate time for air-drying.
3. Insert the molds: a straight cylindrical mold inside the pipe ID for the bell end, and a split, larger-than-OD mold for the spigot end. (See Figure V-5.1).

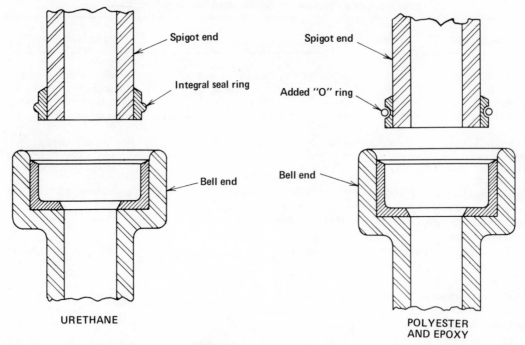

Figure V-5.1 Cast pipe seals.

4. Cast the plastic sealer, and allow time for proper cure.
5. Remove the molds.
6. Inspect.

Product Data:

Testing. Simple long-time immersion in water plus actual in-service performance are the principal tests used to check performance and resistance to delamination and/or cracking. In addition, some producers use a thermal-shock cycle test of 16 hr at $+150°F$ immediately followed by 16 hr at $-20°F$.

Barcol Impressor values of 35 to 45 indicate proper plastic hardness development after aging for 24 hr.

FURNITURE

The rapidly expanding use of cast polyester resins in furniture already consumes 15 million lb annually and is expected to reach 40 million by 1975. Some twenty custom operations and established-line furniture-producing plants take advantage of the fact that polyesters can be formulated with low-cost materials; cast to reproduce explicit detail using low-cost tooling;

hardened at room temperature; and easily finished to look, feel, and work like wood. All these advantages plus low capital and floorspace requirements point to a bright future for cast polyesters in this industry.[2,3,4]

Materials. Polyester resins for furniture casting have several desirable properties:
Low viscosity,
Ability to wet out fillers,
Extendability with monomer without loss of resiliency,
Short de-molding times after room temperature gel and cure, and
Resiliency when cured, especially at lower temperatures.
Table V-5.2 shows typical physical properties of a low-reactivity, resilient polyester resin for furniture casting.

The usual catalysts and promoters are used to effect rapid, room-temperature gel, de-molding, and cure. Table V-5.3 shows two typical systems for moderate and fast gel rates.

Fillers used in casting furniture components range from powders to fibers and are carefully chosen to provide the desired end properties. Even water is employed as a filler,[5] and the use of air as a filler is also on the horizon

Table V-5.2 Typical Properties of a Polyester Resin for Furniture Casting[a]

Property	Specification
Liquid Resin Properties	
Viscosity 77°F, poises	3.5–4.5
Appearance	Clear straw
Solid content, %	70–74
Monomer	Styrene
Specific gravity	1.15
Properties of a 1/8 in. Cured Resin Casting	
Hardness, Shore D	50–55
Tensile elongation, %	47.1
Specific gravity	1.24
Izod impact, ft, lb, per in. notch	4.7
Water absorption, 24 hr, %	0.7–0.8
Volumetric shrinkage, %	7.3

[a]Laminac EXP-279-1, American Cyanamid Co.

Table V-5.3 Typical Polyester[a] **Catalyst-Promotor Concentrations for Furniture-Casting Mixes**

	Moderate Gel-Time System	
% MEK Peroxide	*% Cobalt Naphthenate (6%)*	*Gel Time @ 77°F (min.)*
0.5	0.5	28
1.0	0.3	16
1.5	1.0	11

	Fast Gel-Time System	
% MEK Peroxide	*% Promoter*[b]	*Gel Time @ 77°F (min.)*
0.5	0.3	18
1.0	0.3	10
1.5	0.3	8
1.0	0.5	5
1.0	1.0	3

[a]Laminac EPX-279-1, American Cyanamid Co.
[b]Promoter 411, American Cyanamid Co., Stamford, Connecticut.

(foamed polyesters and urethanes). The most commonly used fillers are pecan-shell flour, calcium carbonate, and expanded volcanic ash (Perlite).® Because fillers have a wide variation of density in the mix, their true cost can be misleading. Table V-5.4 shows the cost of some fillers based on their volumetric relationships. Wood and pecan flour reduce the cost and impart woodlike appearance and feel.

Lightweight fillers, such as volcanic ash and microballoons, lower the cost and density but tend to float and stratify in the liquid mix. A low percentage of a thixotrope, Cabosil for example, may be incorporated to prevent this mixing problem.

Cotton and PVA fibers increase the impact strength and crack resistance, thus bettering the screw-holding power. Inorganic fibers such as asbestos and glass increase molded or cast

*®Trademark of Johns-Manville Corporation.

strengths. Calcium carbonate, talc, clay and other fillers raise the density (sometimes desirable) and reduce the cost.

Water, essentially a no-cost filler, is especially suitable for decorative parts where dimensional tolerances are not too important (0.4 to 3.9% shrinkage). Up to 60% water is emulsified into the resin in a licensed process which produces a material called WEP or water-extended polyester.[6] The 2 to 5 mu water particles form a cellular product which costs as low as $.10 per lb, or $6 to $8 per cu ft. Although still somewhat experimental, this compares favorably with urethane foams

Table V-5.4 Volumetric Cost of Fillers for Polyester Resin in Furniture Casting

Filler	True Density, lb per cu in.	Cost per lb, $$	Cost per cu in., $$
Saran microballoons	0.0013	5.00	0.0065
Phenolic microballoons	0.0134	0.86	0.0115
Glass microballoons	0.0134	0.69	0.0093
Wood flour	0.0538	0.03	0.0016
Pecan flour	0.0534	0.03	0.0016
Expanded perlite®	0.0224	0.575	0.013
Kaolin clay	0.095	0.07	0.007
Calcium carbonate	0.099	0.04	0.004
Glass fiber, 1/8 in. milled	0.082	0.50	0.041

at $8 to $10 and light-weight filled polyesters at $14 to $16 per cu ft.

Table V-5.5 presents several typical furniture-casting formulations, together with cured characteristics, and costs. Most of these produce parts which have densities in the 20 to 55 lb per cu ft wood ranges.

Coloring the furniture-casting formulations is not too extensive, because the choice of fillers usually determines color. Pigments are sometimes added, however, to tone a base color and make final finishing easier.

Equipment and Tooling. Propeller-type mixers are used for all formulations and are especially applicable where fragile fillers such as as Perlite are present.

Automatic mixers and dispensers will, of course, give much faster gel times and shorter casting cycles. However, they cost more ($4000 to $6000) and require more clean-up time. Generally, automatic mixers do not operate efficiently at less than a full shift per day or approximately 5000 lb of material per day.

Molds can be made of a variety of materials including RP/C, metals, and elastomers. The following discussion presents some of the methods in major use at this time for producing casting molds from elastomeric plastics. (Additional information is available in Chapter II-4 as well as from suppliers of the elastomer used.)

Mold Materials:

Silicone Rubber. Although silicone rubber is the most expensive of the commonly used mold materials, it has excellent durability, high tear strength, good mold duplication of the original part, and it requires no mold release for polyester castings. Mold costs can be reduced by pulverizing damaged molds and blending the pulverized material with virgin silicone in equal proportions for casting new molds.

Table V-5.5 Some Typical Furniture Casting Formulations, Characteristics and Costs

Formulation (Parts by Wgt.)	A	B	C	D	E
Polyester resin[a]	100	100	100	100	100
Styrene	5	15	25	40	25
Cobalt naphthenate (6%)	0.5	0.5	1.0	0.5	0.5
Promotor[a]	0.3	0.3	0.5	0.3	0.3
Cabosil M-5 thixotrope	—	—	—	0.75	—
Pecan-shell flour	2.5	55	30	20	65
Calcium carbonate	7.5	—	—	—	—
Glass microballoons	—	—	15	—	—
Perlite® (4 lb Tenn-Sil #50)	—	—	—	8	—
TiO₂ pigment	0.2	2.5	1.5	1.0	3.0
MEK peroxide	1.0	1.0	1.0	1.0	1.0
Mix Properties					
Viscosity	Very Low	Low-Med.	Low-Med.	Low	Medium
Specific gravity	1.18	1.21	0.91	0.95	1.19
Density, approx. lb per gal.	9.9	10	7.6	7.9	9.19
Volumetric shrinkage, %	6–7	4–5	4–5	2–5	6–7
Gel time, 77°F, min.	1–2	2–3	4–6	2–4	2–4
De-mold time, 77°F, min.	5–10	7–10	7–10	7–10	7–10
Approx. cost per lb	$0.23	$0.18	$0.24	$0.19	$0.16
Approx. cost per gal.	$2.27	$1.82	$1.82	$1.50	$1.58
Cured Properties					
Specific gravity	1.26	1.27	0.96	.96–1.0	1.27
Characteristics	Very flexible, nailable	Heavy, nailable	Light in wt. nailable	Light in wt. nailable	Rigid, not nailable

[a]American Cyanamid Company, Laminac EPX-278-1 Resin, #410 Promoter.

When using the silicone rubbers, such as Dow Corning Silastic E or General Electric RTV 630, vacuum degassing at 28 to 29 in. of mercury prior to pouring is necessary to prevent voids and surface blemishes caused by entrapped air. These elastomers cure at room temperature under conditions of high humidity in the surrounding air.

Urethane Elastomer. The material cost of urethane elastomers is approximately one-fifth that of the silicone rubbers. For this reason, these materials are most attractive for mold construction even though they generally are not as durable as silicones and have only moderate tear strength. Mold definition is excellent but still not quite as good as with the silicones. An example is Smooth-on DMC-704.® Saturation of urethanes with aromatic solvents causes hardening and reduced production capability of the molds. Therefore, a bake-out of 8 hr at 300°F is recommended after every 25 castings. SELECTRON 5918® mold release is recommended, because it will have no adverse effects on finishing.

Two or more pours are recommended when using urethane elastomers as mold materials, because the short pot life after catalyzation sometimes results in air entrapment on the mold surface. The first pour is usually painted on to give a very thin, uniform coating, thus enabling any entrapped air to escape. The second pour consists of filling in the remaining void around the master. Alternately, the second rubber pour may be substituted by plaster or any hard castable back-up material for the rubber mold facing. This will provide greater rigidity and longer life for the thin rubber surface shell.

Rubber Latices. These materials are used when there is danger of harming the original part or master. Mold construction with latex is very time-consuming because it has to be painted on in very thin layers, usually 20 to 30 layers per mold. Each layer must be well cured before another can be applied. Usually a flexible strengthening material such as cheesecloth or Dacron tricot is applied after the tenth or fifteenth layer of latex. The molds must be supported with a rigid material such as urethane foam (SELECTROFOAM 65008-6409)® "or RP/C backing."

®Smooth-on Co., Inc., Gillette, N.J.
®PPG Industries, Pittsburgh, Pa.
®PPG Industries, Pittsburgh, Pa.

Latex molds distort and tear easily and generally have poor mold life. The exotherm of polyester castings will cause discoloration and distortion of the mold after a few pours. The discoloration is actually transferred to the casting. A mold release is absolutely necessary. However, despite these disadvantages, latex is the least expensive of the three materials discussed.

Procedures for Mold Construction. Almost all molds may be prepared by fastening the master face upward in a rigid retaining form, or font, allowing ¼ to ½ in. clearance on all sides. Mortite weatherstripping is suitable for fastening the master to the form. A light coating of mold release on the interior sides of the frame and over the master will facilitate release. Prime consideration should be given to protecting the master with the appropriate mold release.

The configuration of the part will dictate which of the following procedures is the most practical and economical:

Procedure A. For masters which are of fairly uniform thicknesses such as plaques, medallions, and so on where there would be no sizeable excess of molding compound in any one area, it is necessary only to pour the compound over the master. In using any of the aforementioned compounds, multiple pours can be made provided that the previous layer has not fully cured.

Procedure B. A shell-type mold should be made for parts which vary considerably in thickness or configuration (see Figure V-5.2).
1. Place ¼ to ½ in. thick layer of pattern wax or clay over the master, covering and eliminating any undercuts or sharp edges.
2. Cover the clay or wax form with plaster of Paris reinforced with cheese-cloth or Dacron tricot. Leave an opening in the plaster for filling and let the plaster cure. A fiber glass-reinforced polyester shell would also be suitable. Such a shell could be prepared with a suitable base and become a permanent part of the mold so that reinforcing with rigid foam would not be necessary.
3. Remove the plaster shell and clean the master thoroughly of all clay or wax.
4. Coat the master and inside of the plaster

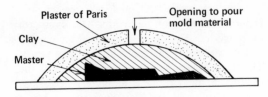

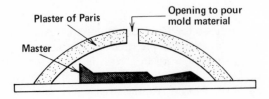

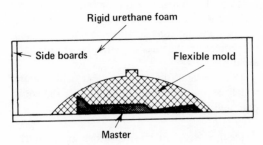

Figure V-5.2   Mold construction—Procedure B.

or RP/C shell with mold release. A silicone-type release agent can be used when molding with urethane elastomers but should not be used with silicone rubbers. Although the silicone rubber will, in most cases, release readily without use of a parting agent, it is advisable to coat the master with an appropriate release agent.

5. Replace the plaster of Paris shell over the master, and fill the void created by removing the pattern wax or clay with the molding compound.
6. After the compound has cured, remove the plaster shell.
7. Coat the mold exterior with mold release and fill the back-up area within the retaining form with **SELECTROFOAM** 65008-6409® for support.
8. After the foam has cured, invert the mold and remove the retaining form and attached master.

Procedure C: Two-Piece Molds (Top and Bottom Halves):
1. Prepare the master as described in the Procedure B for Mold Construction (immediately above).
2. Pour the elastomeric molding compound into one corner of the form until the master is half-covered. Allow the material to "level off," and then resume pouring until the master is covered to at least ½ in. above its highest point.
3. After the material has cured, remove the retaining form and, without removing the master, invert the mold and cut alignment notches in the top surface at the parting line.
4. Coat the face of the mold with a release agent and replace it (top surface) into the retaining form.
5. Pour the molding compound into the master cavity and up to the top of the retaining form.
6. After cure, remove the entire mold from the form, part the two halves, and release the master.
7. Cut the sprue and vent, and the mold is ready for casting.
8. Because of the alignment notches, the two halves match perfectly. Assemble the mold using suitable rigid retaining forms or panels that permit clamping of the two notched, aligned halves together. Pour the casting material.

Procedure D: Butter-On Molds. A thixotropic silicone-rubber molding compound is available for making molds of patterns too large to lay down and pour around, e.g., brick or stone walls, statuary, wall plaques, rock formation, and the like.
1. Thoroughly clean the pattern surface and coat with a release agent.
2. Spread the mold material over the pattern with a trowel or a brush. If the surface is extremely complex, a swirling action, while brushing, will insure a void-free mold.
3. This type of mold can be reinforced by embedding the edges of open-weave fabric strips on the outside of the uncured mold and then brushing a catalyzed polyester resin over the fabric after the mold has cured.

Typically, elastomeric open-split molds for a chair leg might cost $200 each and last for 700 cycles, again depending upon the number of undercuts and detail required.[7,8]

Molding Method/Processing Details. Mixing with automatic mix and dispensing equipment greatly simplifies operator requirements but does not restrict the choice of fillers. In addition, batch mixing of multiple components is necessary before the catalyst can be incorporated using the automatic mixer, and the finished compound dispensed into the mold.

Batch mixing is carried out in the following sequences:

1. Place the additives, such as promoters, pigments, monomer, and so on in the mixer.
2. Add a thixotropic agent—Cabosil, for example.
3. Add the primary solid fillers, such as pecan flour, calcium carbonate, or others.
4. Add the fragile lightweight fillers, such as microballoons, Perlite, and so on, to the mixer at low speed to avoid fracturing.
5. Provide a rest period to allow any foam or trapped air to rise and dissipate.
6. Add the catalyst.

Simple pouring procedures are used for most castings with the following checkpoints observed.

1. Clean the mold.
2. Apply the mold release or barrier if necessary (on urethane elastomer molds).
3. Carefully place structural inserts, cores, and so on.
4. Pour the catalized resin formulation so that air bubbles are not trapped. This usually means a slow pour over the mold surface from lowest point first. Many shapes could require specific pour procedures depending on their geometry. Vibration of the filled mold is sometimes employed to assist air removal.
5. Allow the part to gel (usually 5 to 10 min.) and develop enough strength to de-mold.
6. Remove the part from the mold before exotherm has had time to reach full peak. High temperatures shorten mold life. This time can vary depending upon formulation, part thickness, inserts used, and cycle requirements.
7. Allow the part to cure with enough support to avoid warping, sagging, or other distortions.

Cure is usually accomplished by simple ageing at room temperatures for 24 hr. Cure time however, can be dramatically reduced to seconds if a radio-frequency cure is used.[9,10,11] This method employs a silicone rubber belt conveyor which carries the filled molds between the electrodes of a radio-frequency heating unit.

There are several advantages to this procedure: rapid cure and shorter cycles, controlled cure, longer mold life (silicone molds give up to ten times longer life), savings in manufacturing floor space, and fewer necessary molds.

The formulations for this procedure are limited to applicable materials (water boils out of WEP resins). Mold materials (silicone is acceptable) and insert materials (metal is not acceptable) are also limited.

Rotomolding, an entirely different approach to furniture casting, is a new and particularly attractive process for large, massive parts. It may be used where thick sections would present difficulties because of exotherm dissipation, shrinkage, and often cracking. This process, originally developed for thermoplastics, uses room-setting formulations which eliminate the expensive heating and cooling and save as much as 30% in equipment costs. (Glass reinforcements may be used in rotomolding. See Chapter VI-2.)

Because rotational molds for filled polyesters need little heat conductivity, they can be made of glass-reinforced plastic or an elastomeric detailed surface backed up by RP/C, on a metal structure.

Rotomolded parts, if thin-walled, can be easily filled with lightweight urethane foam which automatically improves the strength of such hollow articles with little increase in cost or weight.

A suggested formulation for rotational molding follows:

	parts by wt.
Polyester resin[a]	400
Styrene	100
12% Cobalt octoate	1.25
Dimethyl aniline	2.50
Asbestos (RG 244)	5 to 10
Glass fibers (milled ⅛ in.)	5
Calcium carbonate (atomite)	255
Pecan-shell flour	195
MEK peroxide	5

[a]Ashland Chemical Co.—EP-034-58124

Finishing cast polyester parts requires no special techniques or materials other than removal of a mold release (if used) with a simple detergent wash. Ordinary wood systems including staining work well. This is one of the advantages of cast polyesters.

Product Data Physical Properties. Physical properties of castings used for furniture vary over a wide range because the end products can be decorative, structural, or both. The many fillers and reinforcements available give the producer an almost infinite variety of formulations.

Table V-5.6 lists a few formulations with some primary fillers showing their effect on physical properties.[12]

Testing. Testing cured castings has largely been a decision for individual manufacturers, and efforts vary widely depending upon end use. The Society of the Plastics Industries Furniture Council has published a list of proposed test methods covering strength, deformation, stiffness, impact, thermal properties, appearance, hardness, abrasion and scratch resistance, defects, water absorption, weight and dimensional conformance, permanence properties, finish quality, and fastening tests. Many of these employ existing ASTM methods. No values are as yet published.

DECORATIVE BUTTONS

Historically, decorative buttons were synonymous with pearl colors and were machined out of marine shells. This expensive and uncertain raw material was long a target for plastics, but it was not until pearl pigments became available that plastic buttons were produced. At first these pigments were made from real fish scales, but today synthetics supply the industry on a reliable basis.

Materials:

Resins. Although acrylic polymers were originally used for this application, the advent of isophthalic polyesters yielded products with the necessary heat resistance, better chemical resistance, and lower cost. In addition, the isophthalic polyesters provide excellent color and optical properties, excellent toughness and machinability, and versatile processability. Typical properties of a polyester resin suitable for buttons are shown in table V-5.7. Up to 10%

Table V-5.7 Properties of a Typical Polyester[a] for Casting Buttons

Liquid at 77°F	
Property	*Specification*
Color, APHA	100
Specific gravity	1.14
Weight per gal., lb	9.5
Viscosity, Brookfield, cps	3750
Stability (in the dark)	Greater than 4 months
Cured Resin	
Cure—Gel at 140°F with 1% benzoyl peroxide plus 1 hr at 170°F and 1 hr at 250°F	
Compressive strength, psi	23,000
Flexural strength, psi	17,000
Flexural modulus, $\times 10^5$ psi	6.6
Tensile strength, psi	7,000
Tensile modulus, $\times 10^5$ psi	5.9
Elongation in tension, %	1.4
Izod impact strength, ft lb per in. notch	0.3
Barcol hardness	42–47
Heat deflection temp. @ 264 psi, °F	230
Water absorption, 24 hr immersion, %	0.2

[a]P.P.G. Industries Selectron 50096.

of a flexible resin is often used to facilitate machining.

Catalysts. Casting processes normally require safe, low-temperature gels in a time short enough for economic cycles. Some useful catalyst and promoter concentrations are shown in Table V-5.8.

Pigments and Colors. Pearlescent effects have been the most popular decorative appearance in this market. They are obtained with pearl pig-

Table V-5.8 Useful Catalyst and Promotor Concentrations for a Typical Casting Polyester[a] at Economical Cycles

Gel Time at 77°F (min.)	Promotor- Cobalt Octoate (12%)	Catalyst- MEK Peroxide (60%)
10	0.03	1.0
20	0.03	0.5

[a]P.P.G. Industries, Inc.; based on Selectron 50096.

Table V-5.6 Formulations and Physical Properties of Some Filled Furniture Castings[12]

Formulations[a] (Parts by Wt)	1	2	3	4	5	6	7	8	9	10	11	12	13
Resins:													
Flexible polyester[b]	40	—	—	—	—	—	—	—	—	—	—	—	75
Resilient polyester[c]	40	76	76	76	72	72	99	72	40	46	65.5	71.7	—
Styrene monomer	20	19	19	19	18	18	—	18	10	11.5	16.4	14.5	—
Fillers:													
Glass fiber 1/8 in.	—	5	5	—	—	—	—	—	—	—	—	—	—
Glass fiber 1/4 in.	—	—	—	5	—	—	—	—	—	—	—	—	—
Asbestos fiber	—	—	—	—	—	—	—	—	—	—	—	—	—
Glass microballoons	—	—	—	—	10	—	—	—	—	—	—	—	—
Phenolic microballoons	—	—	—	—	—	10	—	—	—	—	—	—	—
Saran microballoons	—	—	—	—	—	—	1	—	—	—	—	—	—
Perlite (expanded)	—	—	—	—	—	—	—	10	—	—	—	—	5
Calcium carbonate	—	—	—	—	—	—	—	—	50	—	—	—	—
Pecan flour	—	—	—	—	—	—	—	—	—	42.5	—	—	20
Vermiculite	—	—	—	—	—	—	—	—	—	—	18.1	—	—
Wood flour	—	—	—	—	—	—	—	—	—	—	—	13.8	—
Properties (Initial Mix)													
Mix density:	1.07	1.09	1.09	1.09	0.99	0.864	ᵈ	ᵈ	1.53	1.20	1.14	1.12	ᵈ
Cured castings:													
Cured density	1.16	1.19	1.19	1.20	1.06	0.923	0.949	0.936	1.63	1.25	1.22	1.17	1.06
Linear shrinkage, %	2.7	2.9	2.7	3.0	2.3	2.2	—	ᵈ	2.1	1.4	2.2	1.6	—
Tensile strength, psi × 10³	1.6	2.9	3.1	3.5	2.97	2.3	0.265	1.18	4.8	2.7	2.4	3.2	0.295
Tensile modulus, psi × 10⁵	0.42	1.8	0.85	1.5	0.2	0.1	0.015	1.26	4.4	2.9	2.7	2.0	0.07
Tensile elongation, %	39	1.5	2.5	3.0	3.0	3.0	34	1.3	1.7	1.0	1.5	1.7	17.5
Flexural strength, psi × 10³	—	2.9	3.1	3.7	3.7	2.6	—	1.85	10.7	4.6	3.2	2.8	—
Flexural modulus, psi × 10⁵	—	0.64	0.8	0.97	0.97	0.72	—	0.72	34	2.4	1.5	1.0	—
Izod impact (unnotched)	3.3	0.8	0.4	0.9	0.4	0.3	2.7	0.8	0.7	0.4	0.4	0.6	0.6

[a]Promoters: 0.25% dimethylaniline and 0.25% cobalt octoate on resins. Catalyst: 1.00% MEK peroxide on resin plus styrene.
[b]Aropol Q 6273, Ashland Chemical Co.
[c]Aropol EP 034-58124, Ashland Chemical Co.
[d]Values questionable because of occluded air in mix.

Table V-5.9   Properties of a Typical Pearl-Pigment Paste[a]

Property	Specification
Pearl pigment	Lead carbonate
Pigment, %	60
Brilliance	Excellent
Heat stability	Good
Light stability	Excellent
Acid resistance	Good
Alkali resistance	Good
Sulfide resistance	Poor
Specific gravity	2.61

[a]Ultra-Brilliant Lead Carbonate Pearl Paste #4000-13754, Claremont Polychemical Corp., Roslyn Heights, N. J.

ments composed of minute platelets or flakes. When oriented parallel to one another and to the exposed surface, they produce the typical pearllike appearance.

These pigments are supplied as lead-carbonate pastes, 60 to 80% solids and are added directly to the resin at 1 to 3% by weight. Table V-5.9 lists some typical properties of a commercial pearl paste.

Although some coloring of button stock is carried out by post-soaking in dye solutions, the most brilliant color is obtained by adding pre-dispensed pigment color pastes or concentrates to the resin mix. These color pastes must, of course, be resistant to the oxidizing effect of the catalysts. Also, they should be present only in limited amounts to avoid an overwhelming optical pearl effect.

Fillers. Because of their opacifying effect, fillers detract from the pearllike brilliance and are generally not used.

Release Agents. Mold releases which preserve gloss are preferred because one side of the casting may not be machined. The releases used include liquid systems such as silicones or buffed systems such as carnauba-based waxes.

Molding Method/Processing Details. Some manufacturers in the industry employ compression molding methods; however, casting methods account for 80% of the total production. Polyester resin formulations make up almost all of this output.

A typical casting formulation would contain the following:

	parts by wt
Isophthalic polyester resin	100.00
Promoter—Cobalt octoate (12%)	0.025
Catalyst—MEK peroxide (60%)	0.500
Pearl-pigment paste (80%)	1.500
Color concentrate (depending upon color)	0.100

The mixing sequence starts with the addition of the promoter to the resin, then the catalyst, pearl pigment, and color. To avoid agglomerating the pearl pigment, add a small portion of catalyzed resin to the preweighed pearl paste, and stir slowly until well mixed. Add more resin in small increments until one-fourth of the batch is mixed. The balance of the resin can then be added rapidly. Avoid long periods of intensive shear. Over-processing the pearl pigment breaks down its flake structure and results in a loss of luster.

Simple pour-casting between polished plates is used to produce sheets ⅛ in. thick and up to 10 sq ft in area. Close control of the resin viscosity and temperature are the most important factors. Degassing the resin mix by vacuums of 29 in. or over plus vibration of the molds during and immediately after filling serve to keep entrapped air bubbles to a minimum.

Another process uses centrifugal casting in rotating drums to produce sheets which are gelled in the drum, slit longitudinally, and then layed flat. Cycle times as fast as 30 min. are feasible. Casting in long tubular molds to eliminate waste by cutting transversely is also a possibility.

Both processes use punching equipment to produce button blanks. Punching is performed as the cast sheets reach handleable hardness. The punching must be completed prior to rigid cure. After punching, the blanks are post-cured in ovens and sometimes dyed to color.

Product Data:

Testing. Buttons are, of course tested for the obvious environmental requirements of detergent washing, dry cleaning, ironing, and oven drying. Special tests are sometimes required to evaluate resistance to specific fabric treatments such as melamine sizing and to fulfill certain military requirements.

PLASTIC (CULTURED) MARBLE

Plastic marble is often called cultured marble because it contains up to 85% natural marble in powder form and is "cultured" by the addition of polyester resin, coloring at will, and casting to shape. The resulting material has all the beauty of real marble without the porosity and brittleness. It has been readily accepted for bathroom bowls, vanity and table tops, and many other decorative uses. The market has grown in sales from $3 to $77 million in 6 yr, with a growth forecast at $500 million by 1975. Part of the enthusiasm for this forecast is based on the facts that the two major raw materials, polyester resins and calcium carbonate, are in plentiful supply, and the end products go into our expanding needs for housing. (See Chapter II-5.)

Although some work has been done on press-molded marble and associated processes using sprayed RP/C-backed thermoformed sheet, cast polyesters account for at least 90% of production. The high marble content (as pulverized $CaCO_3$) of cultured marble gives the product both the advantage of weight in service and the disadvantage of weight for shipment over long distances. It is not surprising, therefore, to find today's production and marketing requirements serviced by over 700 strategically located companies.

Materials. Resins for cast marble are formulated for room-temperature cure. They are low in viscosity in order to facilitate good wet-out, fast bubble release, and to accept high filler loadings. They are also low in APHA color when cured so that white and pastel colors may be accurately controlled. As shipped, resins usually contain the promoter, and only a simple addition of the catalyst is required in processing. (Typical properties of a polyester resin for cultured marble production are shown in Table V-5.10.)

Fillers are, of course, the major material by weight. They are primarily calcium-carbonate powders in appropriate degrees of fineness or mesh sizes. Table V-5.11 shows the typical properties for the most popular grades used.

All filler grades which have been selected for assured whiteness are available at a premium cost. Such premium fillers are useful as standards to establish quality control for optimum product uniformity.

Table V-5.10 Typical Properties of a Polyester Resin[a] for Casting Plastic-Marble

Property	Specification
Liquid at 77°F	
Color (disappears on cure)	Blue green
Specific gravity	1.11
Weight per gal., lb	9.23
Viscosity, Brookfield, cps	225
Stability (in the dark)	Over 4 mo.
Cured Resin (no filler contained)	
Barcol hardness	50–55
Specific gravity	1.22
Flexural strength, psi	21,000
Flexural modulus $\times 10^5$ psi	6.0
Tensile strength, psi	9,500
Tensile modulus $\times 10^5$ psi	5.5
Tensile elongation, %	1.8
Compressive strength, psi	21,500
Impact strength, izod ft lb per in. notch	0.25
Water absorption, 24 hr. immersion, %	0.1
Heat distortion, 264 psi, °F	149

[a]Selectron 50069, P.P.G. Industries, Pittsburgh, Pa.

MEK peroxide catalyst is added from 0.5 to 2% by weight of resin to effect a room-temperature gel time of 10 to 45 min., depending upon the ambient temperature and desired speed.

Formulations: A typical formulation for minimum cost is mix A:

	parts by wt
Polyester resin	15.0
XO calcium carbonate[a]	42.0
40-200 calcium carbonate[a]	25.0
Industrial calcium carbonate[a]	17.0
Catalyst MEK peroxide	1.0

Table V-5.11 Calcium Carbonate Polyester Filler for Cast Plastic-Marble

Designation[a]	Particle Size—U.S. Screens
Industrial filler	80% through 200 wet screen
#10 white	99.5% through 325 wet screen
X O	1% on 16, 99% through 40, dry rotap
40–200	1% on 40, 99% #10% through 200, dry rotap

[a]Used at Georgia Marble Co., Tate, Ga.

Another mix providing low viscosity and better flow for easier pouring is mix B:

Polyester resin (promoted)	24.5
40-200 calcium carbonate[a]	37.5
#10 calcium carbonate[a]	37.5
Catalyst (MEK peroxide)	0.5

[a]Georgia Marble Co., Tate, Ga.

Water-extended polyester has had some consideration in formulations for production of lighter-weight grades of cultured marble.

Although some flat sections are easily polished to a high gloss, the use of molds with good surfaces can produce superior high-luster products if they are first coated with a transparent gel coat. A typical formulation is:

	parts by wt
Polyester resin (promoted)	97
Cabosil thixotrope	1.2
MEK peroxide (catalyst)	1

Gel coatings need intensive, thorough and careful mixing to avoid air entrapment. Many companies that mix their binder resins in-house purchase precompounded gel coats from resin suppliers who specialize in this commodity.

The colored striations which are a characteristic of marble are obtained by a separate mix. A typical color-mix formulation is:

	parts by wt
Polyester resin (promoted)	100
Pigment paste	5 to 10
Catalyst (MEK peroxide)	1

Equipment and Tooling. Although any kind of mixer for high-viscosity materials or batches —from the cement to the pony type—can be used, a planetary dough mixer is the preferred tool for the filled-resin formulation.

Gel coats are spray-applied with the usual air-atomizing units.

Molds are of the simple open type and can be made of metal, silicone elastomers, or fiber glass reinforced polyester. RP/C is the preferred production mold material. At least several hundred parts can be cast from one mold set, and also mold surface rework is possible.

Molding Method/Processing Details:

1. Molds are treated with standard polyester release agents and then sprayed with a clear gel coat in the usual manner. Spray operators need to take extra care to visually monitor the 0.015 to 0.030 in. thickness because of the absence of color in this gel coating.

2. The casting resin is first catalyzed and pigmented if a base color other than white is desired.

3. Filler is added which results in a heavy high-viscosity mix, approaching a barely pourable dough.

4. The color mix is added to the stopped mixer, and the desired striations are produced by one or two final turns of the mixer blade. Another method is to pre-pour a filled color mix. Several other proprietary methods also exist for obtaining the required realistic effects of natural marble. These include adding dry color powder to produce striations, incorporation of crushed, finite sized pearl shells, exotic minerals, and other procedures.

5. The mix is then poured into the prepared mold and is allowed to harden at ambient temperatures. Mold vibration is often used to facilitate air removal before resin gelation. The excess mix is screened off to produce a product with a constant thickness.

6. The hardened casting is often buffed and polished after cure and removal from the mold in order to produce the traditionally high gloss of natural marble.

Following is a time breakdown of a typical cycle for molding a lavatory bowl:

	min.
Mold, clean and wax	20
Pour, vibrate, wipe down	5
Cure (average to de-mold)	30
De-mold, de-flash	15
Total	70

Overnight cure is required for development of optimum strength. An operation with a yearly output of 500,000 lb. requires approximately 8,000 sq ft of floor space.

Product Data:

Properties. Typical properties of cured plastic marble are shown in Table V-5.12.

Costs. With calcium-carbonate fillers at $.01 to $.02 per lb and a suitable polyester resin at $.24 per lb, the raw materials now cost $.06 to $.10 per lb as they go into the mold with variations (depending upon color costs). Scrap rates are low and repairs are easy to make.

Table V-5.12 Typical Properties of Cast Plastic Marble[a]

Property	Specification
Barcol hardness	65–70
Specific gravity	2.19
Tensile strength, psi	2,000
Compressive strength, psi	18,000
Impact strength (Gardner Drop Ball), in.	28
Water absorption, 24 hr immersion), %	0.15

[a]Based on Selectron 50069, P.P.G. Industries.

Standards. A committee concerned with cast marble has been formed within the RP/C Institute of the Society of the Plastics Industry, and formal standards will probably evolve in the future.

Surface standards of gel coated products usually conform to those of the United States of America Standards Institute for Gel-Coated Glass-Fiber-Reinforced Polyester-Resin Bathtub Units. [13, 14, 15]

TELEVISION-TUBE INTERLAYERS

Safety interlayers of cast polyester resin are used between the TV tube face and window to provide a cushion which absorbs the impact of a possible tube implosion and prevents flying glass particles from entering the viewing area. Although some mechanical devices are also used, this polyester casting technique is the major safety method employed for tubes over 17 in. in diameter as well as the primary process for tube rebuilding and repair.

Materials. The primary material is a flexible polyester with chemically built-in color stability, low viscosity, and excellent retention of mechanical properties. Some epoxy resin is also used.

The properties of a typical television interlayer resin are given in Table V-5.13. The only required additive is a catalyst such as Lupersol DDM (MEK peroxide). In the process, a pressure-sensitive tape is required such as 3M #Y9082 or the equivalent.[16]

Equipment and Tooling. Tooling for the basic process of fabricating television-tube interlayers involves a method of positively spacing the safety window away from the tube face,

Table V-5.13 Properties of a Typical Polyester Resin[a] for Casting Television-Tube Interlayer

Property	Specification
Liquid at 77°F	
Color	Blue gray
Specific gravity	1.11
Weight per gal., lb	9.3
Viscosity, Brookfield, cps	850
Stability (in the dark)	Greater than 3 mo.
Cured Resin	
Cure: Gelled at 140°F with 1% Lupersol DDM plus 2 hr at 140°F and 48 hr at 77°F	50–55 (Barcol hardness)
Specific gravity	1.17
Hardness, Shore A	50
Tensile strength, psi	40
Elongation in tension, %	30
Adhesion to glass (tension), psi	75

[a]Selectron 5252, P.P.G. Industries Approved by Underwriters Laboratories File E-31058.

and the edge opening, prior to filling the space between tube and facing with catalyzed resin. Three levels of production are in use.

1. Semi-automatic taping machines, conveyorized production lines, and automatic resin-dispensing equipment are used, for high-production rates. With this method, it is possible to process up to 120 tubes per taping machine per hr.

2. For more limited production, a hand-taping machine is available. While not able to fully match the rates of the semi-automatic machine, this device produces a neat, trimmed, taped assembly.

3. A third method, for small quantities or individual tubes, involves using processing techniques done entirely by hand. No special equipment is needed, but hand trimming of the tape on the tube face is usually required after casting.

Molding Method/Processing Details. The process for casting TV tube interlayers consists of the following steps:

1. *Cleaning.* Cleanliness of both the tube and face plate is essential. The two components are first wiped with a damp cloth. Bon Ami cleanser is then applied, and the entire surface

area is thoroughly rubbed. The cleanser is allowed to dry and is then wiped off. The final step in cleaning is wiping the entire surface area with methanol to insure that it is free of oil and water.

2. *Taping*:

■ *Taping by Machine*[17] The tube is placed in the taping machine with the safety window centered on the face plate. Semi-automatic taping machines space the window automatically, but spacing wires[18] must be used with hand-taping machines. If pressure filling is to be done, breather vents[19] (optional) and a valve retainer[20] are attached in the interface with small pieces of pressure-sensitive tape. The machine then wraps the entire tube-safety-window assembly with tape, removing any excess tape from the surface of the window. The assembly is then taken to the heating facility.

■ *Taping by Hand.* The tube is placed face up in some type of holder, and Teflon-coated spacing wires are positioned between the tube face and the safety window. If pressure filling is to be used, the breather vents and a valve retainer are attached as described above. The entire assembly, with spacing wires enclosed, is then wrapped with the pressure-sensitive tape to form a liquid-tight cell. If the optional breathing vents and a valve retainer are used, the spacing wires can be removed after the taping operation. When gravity filling is used, however, the wires are not removed until the cell is being filled and an opening in the taped side wall is provided. Hand taping also requires trimming of any tape which has overlapped onto the viewing surface of the tube. It can be done at this stage or after casting, during the final cleaning of the tube. The excess tape is cut away with a razor blade or a sharp knife. This entire taped assembly is then removed to the heating facility.

3. *Preheating.* The taped tube-safety-window assembly is heated in an oven, using either radiant or convection heating, for a period of at least 30 min. at a temperature sufficient to produce a glass temperature of 150°F.

4. *Resin Preparation.* While the tube assembly is being heated, an amount of polyester resin is weighed into a vessel and heated to 140°F. If pressure filling is used, the resin may be heated in the pressure tank. In all cases, the resin is heated in order to lower the viscosity, facilitate flow within the cell and

also shorten the gel time, so that the interface material sets up quickly. Heating the resin also minimizes chances of damage loss of thickness keying, or loss of optical clarity. After the resin has been heated to 140°F, it is stirred and the catalyst is added and thoroughly dispersed. If a pressure pot is used, it is closed and a pressure of 5 to 10 psi is applied. A small quantity of resin is purged into an accessory container so that no air bubbles will be present when the resin is injected into the cell.

Caution: The catalyst must be added gradually as the heated resin is being stirred in order to avoid the explosive decomposition of the catalyst.

5. *Resin Casting.* Two techniques are used for resin casting: pressure fill and gravity fill.

■ *Pressure filling.* The taped tube assembly is removed from the preheating oven and placed in a rack with the tube axis vertically aligned and the safety window facing down. The tape is then cut away through the aperture of the valve retainer. A cross-shaped slit is cut in the valve facing with a razor blade or a sharp knife. The valve is placed in the retainer and the resin hose nozzle is inserted. The resin is injected into the interface until the space to be cast is almost completely filled (about 2 in. from the edge). The nozzle is then withdrawn. The pressure within the cell and also thermal expansion will cause the resin to flow, filling the gap completely. If breathing vents are not used, air will be trapped near the edges. Tiny holes may be punched in the tape to allow this air to escape. The casting is allowed to cure with the tube in this vertical position.

■ *Gravity Filling.* This method requires no special equipment. When the taped tube assembly is removed from the preheating oven, it is placed in a rack so that the tube axis is horizontal and the safety window is vertical. Some type of funnel should be used so that the resin flows into the tube smoothly without carrying air with it. (An improvised funnel such as a new, empty toothpaste tube or a large laboratory fishtail burner may be used.) The resin is poured into the cell until it is half-filled. The wire spacers are then removed and the holes from which they were withdrawn are sealed with tape. Resin is again poured into the cell until it is nearly full. The funnel is then removed, the glassware is wiped clean, and the opening is sealed with tape. The tube assembly is then placed with the axis vertical and the tube face down. If any air is trapped at the edges of the tube, the tape is punctured slightly to allow it to escape. The tube is kept in this position until the resin has cured (usually 10 to 12 min.).

6. *Cleaning and Trimming.* If a taping machine has been used, no trimming will be necessary; only cleaning of the face to remove fingerprints is needed. If the tube has been processed entirely by hand, excess tape should be trimmed flush with the edge of the safety window. The safety window should then be cleaned with pumice and water or any suitable substitute. Use as little water as possible because too much wetting will adversely affect the adhesion of the tape. Minimum cleaning should be required if the tube has been handled carefully during taping and filling.

7. *Inspection.* Tubes should be ready for inspection of thoroughness of resin cure, optical clarity, voids, and color approximately 60 min. after pouring.

8. *Repair.* If processing techniques are performed carefully as outlined, rejected tubes should be kept to a minimum. If rejects do occur, however, recovery of all glass parts without damage is practically assured. The recovery technique is done as follows:

- Remove the pressure-sensitive tape from the assembly.
- Heat the laminated assembly to 250°F.
- Insert small hardwood wedges between the tube face and the safety window at either end of the tube.
- By an alternate prying and wedging action, separation will occur.
- Peel, scrape, or soak the resin interlayer from the glass parts.
- Clean all glass surfaces to remove traces of resin and adhesive.
- Separation of the tube and protective facing may also be accomplished by resistance-heating a thin nichrome wire to incandescence and passing it by force through the cast resin layer. Heat from the wire burns or degrades the resin so that separation occurs easily.

MISCELLANEOUS APPLICATIONS

A number of miscellaneous cast-resin applications account for good volume in the industry and will be discussed briefly: electronic encap-

sulation and potting; body putty; decorative; massive; and architectural castings.

Electronic Encapsulation and Potting. The need to protect electrical gear, especially electronic components, from environmental exposure is a continuing requirement of the fields in which cast plastics have contributed markedly to successful operation. Aerospace applications as well as land and sea transportation are some of the major areas which use operational and communication equipment.

Filled resinous materials for encapsulation or potting must have the following general characteristics following cure:

- *Low viscosity.* This assists in minimum air entrapment, especially when encapsulating intricate electronic components such as transistors and coils.
- *Low-temperature cure.* Some electronic components cannot tolerate temperatures higher than 180°F.
- *Low exotherm.* Although most electronic components will withstand 250°F for short periods, a low temperature exotherm during resin cure is desirable.
- *Low volatility.* This will prevent objectional gas formation during vacuum de-airation as well as during cure.
- *Low shrinkage.* Shrinkage during cure is objectionable, not only because it can change dimensions and cause cracks but also because it can literally damage delicate components which are sometimes extremely sensitive to pressure.
- *Hardness and resiliency.* In combination with shrinkage, hardness and resiliency have an effect on the potential damage to components. The decision as to material is necessarily a compromise between soft materials which are the most gentle and the more rigid materials which, once in place, protect better.
- *Low thermal expansion.* Ideally, the thermal expansion of the cured casting should match that of the encapsulated components. Although this is obviously not practical, a low thermal movement is desirable and is usually obtained with the use of fillers.
- *Heat conductivity.* Because many electronic assemblies emit heat, the ability of the casting formulation to absorb and transmit this heat is important. Most resins are poor heat conductors and therefore special fillers are

used. One of these is beryllium oxide, which conducts heat but not electricity. A new resin development, aminoquinone polymer (IIT Research Institute, IITRI, Chicago) is claimed to possess four times the thermal conductivity of epoxy resins.

- *Thermal-shock resistance.* Applications in aerospace can require rapid changes over side ranges of temperature. The casting materials should be carefully chosen to resist thermal shock and still meet the above criteria.
- *Environmental resistance.* Although dependent upon the specific application, resistance to aging, chemicals and chemical vapors, and radiation are of obvious importance.

Presentation of the many resins (rigid and flexible), fillers, and reinforcements used in this work would be far beyond the scope of this handbook. Other, more detailed, sources are readily available. However, a general summary and comparison of the properties of the typical encapsulants is presented in Table V-5.14. Specifically, most of the material developments in the past few years have been based on epoxy, silicone, and urethane resins. In the epoxy resin field, basic manufacturers supply resins, hardeners and diluents, and formulating companies sell casting and transfer-molding materials. These transfer-molding materials have been the most important development in epoxy resins for encapsulation. They are one-component formulated systems which can be molded with a pressure as low as 50 psi, a temperature of 250°F, and mold cycles as short as 15 sec.

In the silicone-resin field, some basic manufacturers supply both raw and formulated materials. The silicone rubbers (now peroxide-cured) have found important usage in heat-stable soft encapsulants as well as in coatings to protect the electronic components which must be encapsulated with the harder, more rigid resins.

Formulated urethanes are available as two-part liquid systems and/or frozen one-part systems. Frozen systems are pre-mixed, degassed, plastic packages in discrete shot sizes which are stored frozen until ready for use. They greatly simplify the equipment, process, and quality control problems of the user.

When low mechanical properties are adequate, heat dissipation is not critical, and light-weight desirable foams are used. The density of

foam-in-place urethane can be as low as 2 lb/cu ft. Syntactic foams using plastic microballoons in polyester or epoxy binders are usually 20 lb/cu ft or higher.

The use of polyester casting resins in formulated encapsulating materials has decreased considerably because of inherent hardness, exotherm, and thermal shock resistance. However, these low-cost resins are used internally by many companies having good design know-how, especially for high-volume commercial work.

Certain prerequisites distinguish a polyester as desirable for potting and encapsulation. These include medium to low exotherm, low viscosity (under 10 poise), volume shrinkage of less than 10%, good storage stability, and ability to cure tack-free in air. Also, any polyester resin used for impregnation or encapsulation of electrical components should have low water absorption and good electrical properties. Resilient-type resins are needed to avoid cracking or to lower the exotherm, hence, blends may be made by adding flexible resins or ingredients that impart additional flexibility. Excellent resistance to cracking has been established by blending 10 to 15% polyether resin with a polyester casting resin.

Room-temperature gelation and preliminary cure cycles are called for, with polyester resins, followed by post-cure at elevated temperatures. Cobalt naphthenate MEK peroxide plus benzoyl peroxide or cumene hydroperoxide systems are used.

Fillers include milled glass fibers, calcium carbonate, clays, pulverized silica, mica, powdered metals, asbestos, and talc. The addition of these materials lowers cost, reduces shrinkage and exotherm, and thus minimizes strains which may cause cracking. Casting formulations usually contain 25 to 55% filler, although for small shapes and the lowest water absorption, fillers may be omitted. Gel times can be increased or decreased depending upon the type and amount of filler used. Fillers for electrical-component potting formulations should be carefully selected to avoid materials which absorb moisture. Usually satisfactory for this purpose are electrical talc, silica flour, mica, asbestos, and alumina.

Formulations should be made by incorporating the filler into a small amount of resin with a mechanical mixer; the remainder of clear resin is added after dispersion is complete. It is advisable to prepare the mixture in advance

Table V-5.14 Comparison of Properties of Typical Encapsulating Materials

Material	Linear Shrinkage	Thermal Expansion	Thermal Conductivity	Volume Resistivity	Dielectric Strength
Epoxy:					
Unfilled	very low—medium	low—high	low—medium	good—excellent	very good
Filled (rigid)	very low—low	low	high	very good—excel.	very good—excel.
Filled (flexible)	low—high	low—high	medium	good—very good	very good
Syntactic	very low—low	very low	very low—low	very good	good
Polyurethane:					
Foam	very low	low—high	very low	very good	(not available)
Cast	very low—high	high	very low	good—very good	good—very good
Polyester:					
Filled (rigid)	med.—very high	low—high	medium	good—very good	very good
Filled (flexible)	med.—very high	high	medium	good	good—very good
Silicone:					
Cast (filled)	low	high	very high	excellent	good
RTV rubber	high	very high	medium	very good	very good
Gel	very low	very high	medium	excellent	excellent
Polysulfide rubber:	very high	(——)	(——)	good	good
Alkyd (molded)	low—medium	very low—high	high—very high	very good	good—very good
Vinyl plastisol (filled)	med.—very high	high—very high	low	good—very good	good—excellent

Notes: Linear shrinkage, (in./in.): very low <0.002; low 0.0021—0.004; medium 0.0041—0.010; high 0.0101—0.020; very high >0.0201.
Thermal expansion, (in./in./°C) × 10⁻⁵: very low <2.0; low 2.1—5.0; high 5.1—10; very high >10.1 (figures were referenced against aluminum).
Thermal conductivity, (cal./sec./sq. cm./°C per cm.) × 10⁻⁴: very low <1.5; low 1.6—4.0; medium 4.1—9.0; high 9.1—20; very high >20.1.
Volume resistivity, (ohm-cm.): good 10¹¹—10¹²; very good 10¹³—10¹⁴; excellent 10¹⁵—10¹⁷.
Dielectric strength, (volt/mil): good 225—399; very good 400—500; excellent >500.
Source: Product Engineering, 19 September 1960.

to allow time for air to escape prior to use. If benzoyl peroxide is used, it can be added immediately before the filler, because the catalyzed mix is stable for several days at 70 to 75°F. When room-temperature systems are used, the cobalt naphthenate is usually combined at the same time as the filler, and the MEK peroxide is added just prior to pouring. One polyester formulation which has been particularly effective in potting or encapsulation is presented below:

	part by wt
Polyester resin, resilient	100
Ground silica, 250-mesh	80
Milled glass fibers $\frac{1}{32}$ in.	11
Titanium dioxide powder	0.5
Quaternary amine promoter	0.01

With the addition of 0.005 part hydroquinone inhibitor to counteract any heating effects during blending, the above mixture is ball-milled for 24 hr. After completion of the blending, 0.2 to 0.3 cc of cumene hydroperoxide per hundred grams of mix are added, and the encapsulated item is permitted to gel at room temperature. Final cure is accomplished by heating for 30 min. at 250°F. Typical epoxy, silicone, and urethane casting formulas are readily available in the literature.[21, 22, 23]

Equipment for potting and encapsulation can be as simple as that used for batch mixing, with the addition of a direct pour. Highly reliable electronic-component assemblies, however, can put a premium on void-free castings in spite of the complicated shape of these assemblies. For this work, special vacuum encapsulators[24] can be used which consist of a transparent-sided vacuum chamber into which the mold is placed. A heated thinning and degassing plate between the resin inlet and mold lowers the viscosity, de-gasses the mix, and delivers the formulation into the mold. An oven cure follows.

More elaborate equipment for higher volume would consist of de-gassing units supplying an automatic mixer and dispenser mounted on a conveyor carrying molds which receive a measured charged casting mix and then move through a curing oven.

Where volume can justify the capital and mold expense, special transfer molding machines are used to inject and cure low-pressure molding compounds. One manufacturer has gone a step further and supplies a unit to both inject and cure liquid systems.[25] Molds for these processes use Teflon seals.

The product, including the encapsulated or potted assemblies, is tested for functionality performance.[26] In addition to the required environmental resistance, thermal shock is perhaps the most severe test, with long-time water (electrical performance) resistance a close second.

Although lower cost materials are chosen on an equal basis, electronic encapsulants are generally a minor part of the overall costs of the component or electrical assembly.

Body Putty. Another area in which reinforced plastics/composites are demonstrating their adaptability is that of automobile body and fender repair. Replacing earlier materials, such as heavy low-melting metal with poor adhesion (lead), body-putty formulations offer the advantages of easier application and freedom from rust and corrosion, especially in such vulnerable locations as rocker panels and doors. Two excellent formulations are presented here:

Polyester Body Solders:

	parts by wt
Resin	42
Cobalt naphthenate, 6%	0.4%
Talc (46 μ max.)	30 based on resin
Talc (6 μ max.)	10
Talc (98% through 325 mesh)	10
Talc (90% through 325 mesh)	7
Black iron-oxide pigment	1

The mixing procedure, using a spiral blade or Hobart mixer or equivalent, includes the following steps:

1. Add resin to mixer.
2. Add any additional promoter and color to resin and blend thoroughly (time required—1 min.)
3. Slowly add dry filler to mixer and mix until fillers are thoroughly wetted and dispersed.
4. Mix should be completed to a smooth paste-like consistency ready for packaging.

The finished putty mix should have a room-temperature uncatalyzed shelf life of 20 days at

130°F as a control (with little or no air in test-sample container).

Room-temperature cure may be induced by adding MEK or cyclohexanone peroxides, or benzoyl peroxide if DMA or DEA is used as the promoter. The gel time should be 10 min. and should be followed by rough filing. Smooth sanding and feather-edging is carried out 20 to 30 min. after gelation. Primary and finish painting may follow rapidly, and the whole job can be completed in 2 hr. It is essential to clean the area, remove rust from the parts to be puttied, and use a back-up webbing such as fiber glass cloth.

The polyester resin for putty compounding should be highly flexible and should have a low exotherm, a 3.5 to 4.0 poise viscosity at room temperature, and the maximum possible uncatalyzed stability. It should also provide rapid room-temperature gel and cure.

Fillers selected for body putty have centered almost entirely around talc. The primary reason is because, unlike clay and $CaCO_3$, it permits maximum uncatalyzed storage stability of the mixed putty compound. Also valuable in talcs are the full range of particle-size gradations available, and the low oil absorption for potentially higher filler loading. Favorable thixotropic properties are also provided.

After the putty has been cured in place in its end application, the friability of talc is of prime importance in both initial and final sanding operations. Here again, talc is superior to clay or $CaCO_3$ fillers, which retain a certain amount of stickiness and permit the sandpaper to become rapidly caked with the material removed.

Pigments are included only for the purpose of providing characteristic base colors. Fibrous reinforcing ingredients are seldom used because of higher cost and poor handling and sanding properties. Sisal fibers would, of course, deleteriously contribute to high water absorption and poor weathering properties. Milled glass fibers and short-fibered asbestos could be used, but high-reinforcing materials are not actually necessary because most applications call for good adhesion to the substrate and presentation of a smooth, paintable, fiber-free surface after sanding.

Epoxy Body Solders. As compared with polyesters, epoxy solders have the advantage of superior adhesion, but they are higher in cost and require a longer cure time before sanding is possible. A two-part epoxy solder formulation is presented below:

Part 1

	parts by wt
Epoxy resin, unmodified, high viscosity	100
Glycidyl ether	20
Filler mix, as follows:	
Limestone, C_aCO_3	90)
Milled glass fibers	8) = 100
Color pigment	2)

Part 2

Polyamide resin	25
Mixture of diethylene-triamine and bisphenol A in equal parts	20
Filler mix, as given above	125

When ready to use, parts 1 and 2 should be first combined by mixing equal quantities and then applied to the work area. Allow 1 hr for exortherm to take place before applying external heat. If external heat is used, it should exceed the expected operating temperature of the repaired part. After the mixture has been completely cured, the body patch may be sanded and painted.

Performance. A set of typical properties of polyester putty is presented in Table V-5.15.

Decorative Castings. Decorative castings are generally transparent or highly translucent and are in demand for art objects, hobby activities, and embedments. Opaque decorative castings for plaques, ornamental filagree and simulation of wrought artisan creations are also very popular.

Polyester resins for this work should have the following characteristics: low viscosity, simple formulations (pre-promoted for room-temperature cure), low exortherm, clear and colorless appearance (when cured), tack-free cured surface (without wax), crack and impact resistance. P. P. G. Industries supplies polyester #50111 casting resin which meets these requirements.

Embedding. Small metallic stampings, stones, shells, dried ferns, flowers, insects, and

Table V-5.15 Typical Properties of Polyester Putty

Formulation	
Polyester resin, resilient, promoter-containing, %	40–45
Talc fillers, %	55–60
Color pigments (fused) %	1
Curing properties	
Gel time—1% MEK peroxide, min.	4–7
Sanding time, min.	14–18
Gel time—2% benzoyl peroxide paste (50% concentration), min.	8–11
Sand time, min.	15–17
Physical properties	
Barcol hardness (#935 impressor)	30–70
Elongation, %	0.29–7.5
Lap shear strength, psi	350–700
Stability in the dark of uncatalyzed putty @ 130°F, days	20

Note: The particular properties within the indicated ranges depend upon the base resin from which the putty is formulated.

other objects are quite readily embedded in polyester resin. The uses are varied and include paperweights, jewelry, lamps, penholders, trays, and many other commercial and hobby applications.

The most satisfactory method for embedding objects is the use of multiple pours. The number of pours is determined by the size, shape, and composition of the objects to be embedded. The first pour determines the amount of cover desired for the object. The number of succeeding pours is a matter of judgment as to how the object can best be completely covered and saturated without the entrapment of any air. Lower catalyst concentrations extend the gel time and allow the air to escape by rising through the resin. Higher catalyst concentrations shorten the gel time but must be gently stirred into the resin. Casting imperfections are created when excessive amounts of air originally whipped into the resin have insufficient time to escape during shortened gel time periods. The final pour is the base and may be tinted or dyed for decorative effect with many of the opaque or transparent dyes currently in the hobby shops.

When colors are used, accurate measurement is required for color matching. The best procedure is to color a master batch greater than your immediate need and draw from it for future needs. The colored master batch should not be catalyzed until it is ready for use. Pigments are inert and have no effect, except for carbon black pigment which lengthens the gel time.

Molds. Many of the common household containers such as pie pans or Pyrex dishes make excellent molds. The mold should be wiped with a mold-release agent or liquid wax and the surface rewiped to remove any excess.

Because the mold surface becomes the top of the casting, a careful choice should provide a finished piece that requires no sanding or buffing. Matte or grained-surface molds transfer their finishes, and can be used if such an effect is desired. The ideal mold for jewelry or small items is the ceramic type which can be put into an oven at 150°F for 1 hr to hasten the cure and then plunged into cold water for immediate release of the cast object. Glass containers with tapered sides for easy release may be practical but should not be subjected to thermal shock, such as a cold-water dip following high exotherm due to an accelerated cure rate.

Casting Spheres. Polyester resin was specifically designed as a crack-resistant resin that does not lake or stick to the mold. In casting small spheres such as those used for making grape clusters, the most important items are the catalyst level and mold fill. Although clear glass Christmas tree ornaments can be used, glass spheres of various sizes are sold in many hobby shops for this specific application. These molds cannot be reused because they must be broken to release the cast material.

When it is desirable to embed wires into the spheres, either covered or bare wire may be used, but it should be of a gage light enough to permit the necessary flexibility. The best practice is to underfill the mold to a level below the neck of the glass mold, and then center the wire. A lower level of catalyst extends the working time, and slow pouring helps to avoid air bubbles in the spheres. An egg carton is excellent for holding a dozen molds while pouring and placing wires. After the resin has hardened and cured, the glass molds can be broken away. When the mold is tapped against the palm of the hand and the sphere rattles, the mold is ready to be broken away from the casting.

When jewelry fittings or similar attachments are going to be fastened to the castings, the resin makes the best bond. A thickening agent such as Cab-O-Sil is added to the resin to make a thick paste and then catalyzed with MEK peroxide and this mix can then be spread on the casting. A thin layer of this paste will cure virtually colorless and be unnoticable.

Plaques, Trays, and Statues. There are a variety of molds available for casting plaques, trays and statues. Most suppliers specify the amount of material and catalyst level on the edge of the mold. If no recommendation is made, 0.5% catalyst (MEK peroxide) should be safe. This low level of catalyst lengthens the gel and cure time of the resin in order to prevent heat damage to the plastic molds. The release marks are easily visible and act as a guide for determining when the casting can be removed. Because of atmospheric conditions and the thickness of the casting, it may be advisable to let the casting remain in the mold overnight. If the air-exposed surface is slightly tacky, a light application of furniture polish or liquid wax will make it tack-free. This slight surface tackiness is apt to occur on a particularly warm or humid day.

The exceptional clarity of casting polyester makes for a most attractive "picture on wood." A piece of wood or veneer and a picture of a decorative print (with no printing on the reverse side) is all that is required. The wood can be gouged, shaped, or burned to create any decorative effect desired. A coating of catalyzed resin is brushed over the entire surface and allowed to gel. While the resin is still tacky, the print is placed on the surface and pressed down tightly enough to remove any air underneath. After this layer is fully cured, any number of finish coats may be applied. Dust or dirt picked up on the surface coating can be removed with a fine grade of steel wool. A new finish coat must be applied to completely mask out any scratches made by the steel wool.

General Information. Paper cups may be used for measuring and mixing the resin. Larger amounts of resin can be mixed in throw-away polyethylene gallon containers such as milk cartons. The containers can be reused, because the hardened resins can be flexed out of the container after curing. Leftover resin should not be discarded in a trash or paper container until it has hardened completely and has cooled to room temperature.

Certain other precautions are necessary in handling the polyester resin and the MEK peroxide catalyst. Neither material should be left within reach of small children. It is desirable to work in a well-ventilated area, because the raw material odors and those given off in the curing process may be objectionable to some people. A drop or two of Nilodor may be added to the resin to overcome the odor. Directions for use of the catalyst should be followed explicitly. A pipette with a squeeze bulb or a medicine dropper can be used for measuring the catalyst. In no case should the materials be pipetted by mouth.

Massive Castings. Large, massive castings have always been difficult to produce because of the inherent exotherm, low impact, low elongation, and shrinkage of most polyester resins. Although some fillers and reinforcements can be added to reduce shrinkage, quantities are limited if the required translucent decorative effects are to be retained for heavy art objects, table tops, bowling balls, and so on. Strength and dimensional stability may be designed in by casting around centered bases of wood or other materials. In the case of bowling balls, variation of the ratio of quantities of a porous, filled, inner polyester core and the tough, cast, outer polyester skin (constant diameter) serves to control the ultimate finished weight of the ball.

Polyester resins for large castings should have the following characteristics: low viscosity, simple formulation (pre-promoted for room-temperature cure), very low exotherm, tack-free cured surface (without wax), and very high resistance to cracking and impact.

Table V-5.16 shows typical catalyst concentrations versus gel times and peak exotherm temperatures for various masses up to 5 gal in size.

Table V-5.17 lists the typical properties of a cured resin suitable for massive castings. Note the high tensile elongation and impact strength.

Architectural Castings. An almost unlimited number of decorative architectural castings is made by many small manufacturers, ranging from translucent textured surfaced panels used

Table V-5.16 Catalyst Concentrations versus Gel Time and Peak Exotherm
Temperatures for Various Size Polyester[a] Castings

Mass	% MEK Peroxide	77°F Gel Time (min.)	Total Cure Time (min.)	Peak Exotherm (°F)
100 g.	1.0	13	40	251
1,000 g. (qt. can)	0.5	21	50	282
1,000 g. (qt. can)	1.0	13	43	303
4,000 g. (gal. can)	1.0	13	61	328
21,400 g. (5-gal. can)	1.0	13	65	332

[a]Selectron 50112, P.P.G. Industries.

as interior partitions to stone-faced exterior panels, decorative windows, simulated brick or stone siding, and terrazzo-type floors.

Cast panels often use a core of transparent acrylic sheet on which the polyester is cast. Decorative embedments and striated pigmenta-

Table V-5.17 Typical Properties of a Cured Polyester[a] Resin for Massive-Casting

Property	Specification
Hardness: Barcol Impressor	
(Model #934)	40–45
Rockwell, M Scale	88
Flexural strength, psi	13,700
Flexural modulus, $\times 10^5$	5.0
Tensile strength, psi	8,000
Tensile modulus, $\times 10^5$	4.4
Elongation in tension, %	2.5
Compressive strength, psi	18,400
Izod impact strength,	
ft lb per in., unnotched	1.1
Water absorption, 24 hr	
immersion, %	0.14
Heat distortion, 264 psi, °C	46
Specific gravity	1.2

[a]Selectron 50112, P.P.G. Industries. Specifications are based on a 1/8 in. panel gelled at 140°F with 1% benzoyl peroxide, plus 1 hr at 170°F and 1 hr at 250°F.

tion may be added to the surface texture. These effects are obtained by the judicious use of thixotropic agents or from suitable molds.

Stone-faced panels usually employ the casting resin as a weather-resistant adhesive when crushed stone is bonded to substrates such as asbestos board. Some decorative cast panels have stone or sand embedded into the resin with enough surrounding clear or translucent area to provide a light-transmitting effect.

Terrazzo floors can be cast in place or made into tiles for subsequent placement and bonding. This types take advantage of resilient resin formulations not possible with the inorganic types. Both use high filler loadings similar to formulations for cultured marble and add various colored aggregates for decoration. Some cast-in-place floors can be troweled smooth enough to eliminate the difficult grinding usually necessary with stone terrazzo. Also cast floors comprised of chemically resistant resin plus appropriate filler are superseding ceramic floor tile, etc., because of ease of application, stability, long life, and low maintenance.

Stage armor, trappings, temporary buildings with authentic surface finishing, other props— and so it goes—the list of applications is almost limitless.

References

1. Reichhold Chemical Corp., "By Gum," 1961, Anon.
2. SPE Journal, Vol. 26, Nov., 1970. Entire issue devoted to "Plastics in Furniture."
3. SPE National Technical Conference, "Plastics in Furniture," St. Louis, Mo., Nov. 10, 1970.
4. L. L. Scheiner, "Plastics Processing in Furniture," Plastics Technology, Sept., 1970.
5. Leitheiser, R. H. et al., "Water Extended Polyester Resins," Ashland Chemical Co., SPE Journal, Vol. 25, 1969.
6. Fosnot, H. R., "What You Should Know About Casting Polyesters," Plastics Technology, Jan., 1969.
7. Plastics World, July, 1971, p. 50.
8. Wood, A. S., "Arrividerci, Mediterranean, Hello Plastics," Modern Plastics, Aug., 1970.
9. Jansen, E. W., "Radio-Frequency Curing of Furniture Parts, "Reichhold Chemicals, Inc., Plastics Design and Processing, Dec., 1970.
10. Daniels, D. A., and Baumrucker, J., Modern Plastics, July, 1971, p. 60.

11. W. T. LaRose and Associates, Inc., Troy, N.Y.

12. Petersen, C. R., and Dalluge, M., Ashland Chemical Co., "What Furniture Manufacturers Should Know About Unsaturated Polyesters," *SPE Journal*, Vol. 26, Nov., 1970.

13. Commercial Standards CS-221-59 and CS-222-59, "Polyester Gel Coated Bathtub and Shower Receptors," U.S. Dep't. Commerce.

14. Wood, A. S., "Plastic Marble–A Little Known Market Takes Off," *Modern Plastics,* June, 1970.

15. Anon., "Have You Looked At Your Bathroom Lately?" Reichhold Chemical Corp., "By Gum," 1969.

16. 3M Company, Inc., St. Paul, Minnesota.

17. Taping Machines: Hand Taping Machines—PPG Drawing Set #17-A 13362; Semi-Automatic Taping Machine—PPG Drawing Set #17-A 14869; Available from Recommended Supplier: S. P. Seinkner Co., 910 Sampson St., Newcastle, Pa.

18. Wire Spacers: Teflon or Equivalent Coated Wire, OD = 0.060 to 0.070 in., (EE-MIL-W-16878-B), Military Type EE-20, #20 AWG Stranded 10 × 30, available from The Belden Co., Chicago, Ill., Local Electronic Parts Supplier.

19. Breather Vents: Per PPG Drawing #17-D14-466, Available from Conneaut Rubber and Plastics Co., Commerce St., Conneaut, Ohio.

20. Valve Caps and Retainers: PPG Drawings #17-D14582 and #17-D4586, respectively, Available from Rockwell Manufacturing Corp., 1350 Fifth Ave., East McKeesport, Pa.

21. Lee, H., and Neville, K., "Handbook of Epoxy Resins," McGraw-Hill, New York, 1967.

22. R.T.V. Technical Data Bulletins, Dow Corning Corp., Midland, Mich.

23. Technical Data Sheets, Mobay Chemical Co.

24. Technical Data, Warlock Systems, Inc.

25. Technical Data, Hull Corp., Hatboro, Pa.

26. Plastec Report #29, "Encapsulation of Electronic Parts in Plastics," Plastics Technical Evaluation Center, Picatinny Arsenal, Dover, N.J., See pp. 46, 47, 48, Tables 10, 11, and 12, for requirements of MIL-I-16923, MIL-S-23586, and MIL-P-46076, respectively.

V-6

Fiber Glass-Reinforced Foam Structures

INTRODUCTION

Several methods for incorporating fibrous glass reinforcing into a foam structure have been attempted. Glass-loaded thermostat material (epoxy) has been mixed with partially expanded polystyrene foam beads and the entire mass cured while heat fully expanded the beads (exothermic reaction assisting). In the same system, attempts were also made to adhere the mass to glass cloth on laminate skins on either or both sides of the foam sandwich.

Another procedure involved mixing chopped glass fiber with polyurethane resins prior to the addition of the catalysts and blowing agent with the anticipation of forming a uniform reinforced foam structure. Glass fiber roving has also been chopped into a foam mass during the blowing phase.

These processes are neither 100% technically nor economically successful and it has remained for one of the industry pioneers to innovate a workable process which provides a dimensionally stable, well-dispersed glass-reinforced structural foam with many present and potential applications.[1] The process in its present not fully developed state is described in the following pages.

MATERIALS

Resins. Flexible, semi-rigid, or rigid polyurethane foam resins, either general-purpose or fire-retardant, are suitable for reinforcing.

Reinforcements. "A" glass (soda-lime-silica composition) is used. The form of the glass is a drum wound cross-fiber mat similar to the product prepared in the formation of fiber glass veil and/or surfacing mats. The glass fiber mat is wound in a condensed form and expands upon combination with the foamable resins during blowing to provide a uniform reinforcement throughout the structure.

No treatments or surface-active agents are applied to the glass mat during forming.

EQUIPMENT AND TOOLING

It is not within the scope of this text to include descriptions of the equipment required for preparation of either the resin or the glass fiber mats. Neither is it possible to elaborate on equipment for fabricating the reinforced foam structures, because the methods are either proprietary or in the development stage.

MOLDING METHOD/PROCESSING DETAILS

In fabricating reinforced-foam products, the resin and glass mat are combined by direct impregnation. The glass mat expands together with the foam during its expansion or blowing period.[2]

Heavy slab stock, thin sheets, and laminated or formed (molded) products may be fabricated. By selective placement of the glass reinforcement, a heavy concentration of glass may be kept near one surface or the other, or the glass distribution may be made uniform throughout.

Skins such as corrugated sheet metal, plywood, plaster board, and impregnated building paper may be laminated to one surface or the other by causing the resin in the glass-polyurethane composite to expand while in contact with the surfacing membrane.

PRODUCT DATA

Metal-surfaced building panels for roofing and sidewalls for the construction markets are already in production in Italy.[3] Molded furniture parts, selectivity resilient cushioning, veneers, and types of synthetic wood are either being contemplated or produced experimentally.

Dimensional stability, uniformity of strength, improvement of virgin foam strength, and fire retardancy where required are built-in properties of this unusual composite material. (See also Table VI-1.13, "Properties of Injection-Molded Reinforced Thermoplastic Foam Resins." Chapter VI-1, Injection Molding of Reinforced Thermoplastics.)

References

1. SPI 25th RP/C Proceedings, 1970, Section 8-L, Modigliani, P. See also: Modern Plastics Encyclopedia, 1970–71, p. 272.
   Plastics Technology, Feb., 1970, p. 55.
2. U.S. Patent #3,554,851, Jan. 12, 1971, "Glass Reinforced Foam Structure and Method of Making Same," J. P. Modigliani.
3. Personal Communication, P. Modigliani, Artfiber Corporation, New York, N.Y., Feb. 23, 1971.

SECTION VI
REINFORCED THERMOPLASTICS

VI-1

Injection Molding of Reinforced Thermoplastics[*]

INTRODUCTION

Thermoplastics are resins which do not undergo chemical or heat curing to a state of insolubility or permanent rigidity during their processing in order to form a functional item. Polymers designated as "thermoplastics" can be repeatedly re-heated, forced into a die cavity or through an extrusion die, and cooled to a designed shape.

Reinforced thermoplastics, commonly referred to as RTP (of RP/C), contain the same thermoplastic resins which are used in the standard injection or extrusion processes except that they contain reinforcing fibers to form a composite. The addition of reinforcing fibers brings a significant increase in 1 or more performance characteristics. Reinforcements include materials such as glass, asbestos, talc, and exotic fibers. Glass fibers used to reinforce the thermoplastic resins provide a composite material that otherwise would not be competitive with other structured materials. Special glass fiber-reinforced polymers initially were considered for only limited uses. Today nearly all technically important thermoplastic polymers are reinforced with fibers for injection molding and extruding.

Specifically, the reinforcement of polymers with glass fibers substantially improves mechanical properties, decreases distortion under load, improves dimensional stability, and reduces both % elongation and thermal-expansion properties. Finished parts reinforced with fibers retain their dimensional stability. They are stiffer and are used in applications requiring a wider temperature range than the same article without such reinforcement.

The cornerstone of the reinforced thermoplastics industry was laid in the early 1950's by Rexford Bradt, founder of Fiberfil, Inc. In 1952 this embryonic company marketed a styrene molding compound containing 30% glass[1] which was first used in molding a non-metallic, non-detectable housing for a military land mine. The first commercially available reinforced thermoplastic contained glass fibers ⅜ to 1 in. long. Long glass fibers in RTP were used for applications requiring high performance. Research on RTP has been active since 1960, but it achieved an important additional impetus with the advent of the screw injection-molding machine to aid in fiber dispersion. Thermoplastics reinforced with short glass fibers were developed to assist molders with plunger machines to obtain improved fiber dispersion. A general loss of 10% in strengths under those for long fibers occurs in most short-fiber thermoplastics. Izod impact strengths become decreased by as much as 50% in the short-fiber compounds.[2]

The characteristics of resins used in RTP can be improved by 3 separate methods: (1) Existing resins may be changed by chemical modification or may be alloyed with other compatible resins. (2) Research is providing new molecular structures in polymer form which process unique properties. (3) Currently available materials can be enhanced by reinforcing them with filamented or particulate materials of suitable configurations.[3] The resulting prod-

[*]This chapter prepared by S. B. Spencer, Johns-Manville Fiber Glass Reinforcements Division, Waterville, Ohio.

uct mix of thermoplastic resins, reinforcements, and fillers has expanded the selection of materials with a wide range of properties and prices.

MATERIALS

The base thermoplastic polymer is purveyed in a variety of physical forms and chemical compositions, ranging from powdered reactor flake, as in the case of polypropylene, to extruded pellets. After processing to combine the reinforcement or fillers with the thermoplastic resin, the final compounded material may be a pellet or diced cube conveniently handled by injection-molding or extrusion equipment. In other technology, resin and reinforcement may be handled separately and combined right at the press feed hopper. Physical properties can be varied by the percentages of fiber or filler, fiber length and dispersion, and the compounding processes to form the finished RTP part. Compounds based on a constant rate of reinforcement which can be let-down or "diluted" with unreinforced polymer are also available to obtain the glass percentage desired in the finished item. As an example, 1 portion of a polystyrene concentrate mixed with 3 parts of unreinforced styrene resin will provide a 20% fiber glass-reinforced polystyrene.

Reinforced thermoplastics are accepted engineering materials bridging the difference between unreinforced thermoplastic resins and die-cast metals.

The following thermoplastic composites are available: Acetal resins—homopolymer, copolymer; nylon resins—type 6, type 6/6, type 6/10; polystyrene, SAN (styrene-acrylonitrile), ABS (acrylonitrile-butadiene-styrene), polypropylene, polyethylene, polysulfone, polyphenylene oxide, polycarbonate, urethane, thermoplastic polyester, chlorinated polyether, PVC (polyvinyl chloride).

The unreinforced properties of each individual polymer are compared with the properties of the reinforced or filled polymer version in Tables VI-1.4 through VI-1.13.

Reinforcements and Fillers for Thermoplastics.
The major reinforcements and fillers used in thermoplastics are glass fibers, asbestos, talc, and calcium carbonate. Used to a lesser extent are materials such as glass beads, calcined clays, and wallastonite (calcium metasilicate). The exotic fibers such as graphite and carbon should also be included (see Chapter VII-1). Glass is the primary reinforcing fiber, and it upgrades the thermoplastic-composite properties to a level competitive with structural metals. Asbestos and talc are often considered reinforcing fillers and attract attention from a favorable cost-to-stiffness ratio.[4] The description and properties of fiber glass reinforcing chopped strands and roving are listed respectively in Tables VI-1.1 and VI-1.2.

There are three major classifications of fillers for thermoplastics: reinforcements, functional fillers, and low-cost fillers or extenders. These components can be applied to any non-organic portion of the plastic composite.

The extenders contribute nothing to the composite except low cost. Functional fillers have added utility in addition to the low cost. The features of a functional filler consist of the electrical insulation properties of calcined clay, the processing advantages of glass beads, or the improved chemical resistance of stearate-coated calcium carbonate.

A list of fillers commonly used in thermoplastic compounds is presented in Table VI-1.3.

Filled and Reinforced Polyolefins.
Filled and reinforced polyolefins (polyethylene and polypropylene) became commercially important in the early 1960s.[5] Prior to that time, the early development of polyethylene resins indicated increased stiffness and less mold shrinkage with fillers. However, impact and tensile strengths were lowered. It was found that glass fiber upgraded the mechanical properties without adversely affecting impact strengths.

The addition of fillers and reinforcements to polypropylene produced a greater increase in stiffness and hardness than it did in polyethylene. However, sustained thermal stability was the major deficiency with the initial compounds. Anthophylite asbestos-polypropylene composites were combined to achieve good, long-term heat stability. The next step involved adding glass fiber reinforcement which produced a polypropylene composite with the highest rigidity, Izod impact strength, and lowest mold shrinkage of any other filled or reinforced polypropylene on an equivalent loading basis.

Still further improved polypropylene composite properties were obtained by Hercules, Inc.[6] with a modified polypropylene matrix resin. The modified matrix polymer was able

Table VI-1.1   Fiber Glass Chopped Strand for Reinforced Thermoplastics

Description	High density chopped strand for ease of mixing and feeding.
Use	Reinforcement for engineering thermoplastics for normal and high temperature applications.
Types available	⅛ and ¼ in. standard lengths.
Typical properties	Glass type—E glass. Moisture content—.08% maximum. Loss on ignition—range—1.0 to 2.2%. Fiber bulk density—114 max. for ¼ in. fibers.
Requirements for maximum physical properties in finished molded parts	The high density factor will aid dispersion in resin whether in powder or pellet form. The high film strength of the glass will protect against broken filaments and "balling up" in the mixing and injection molding process. The chemical coupling system should provide maximum bond between the glass surface and the thermoplastic resin type. The controlled bulk density should provide accurate glass feeding for close control of glass to resin ratio. High temperature resistance will reduce the possibility of discoloration of molded parts and also corrosion of the mold.

Source: Johns-Manville Fiber Glass Reinforcements Division, Waterville, Ohio.

to couple chemically with a glass surface containing a silane coupling agent. The resulting material provided mechanical strengths 50% higher than general-purpose polypropylene homopolymer which was not chemically modified for glass reinforcement. Other attempts with de-polymerization or activation catalysts applied either onto the glass reinforcement or blended with the RTP compound have not been commercially successful. Free-radical coupling

Table VI-1.2   Fiber Glass Roving for Reinforced Thermoplastics

Properties	Nominal yd per lb—227.
	Glass type—E glass.
	Strand integrity—high as possible. Roving integrity—minimum ribbon (max. = 4) for maximum dissemination if chopped, or maximum pick-up if hot-dipped.
	Stiffness rating—6.3 nominal.
	Static—hold to minimum—less than 1000 v.
	Ignition loss—1.95% nominal.
	Strand quality—should have finish with maximum abrasion resistance for best resistance to fuzzing at guide eyes, etc., during processing.
Performance	Chopability—should be excellent and chop cleanly with minimum filamentation thereby permitting maximum resin flow.
	Compatibility—must be especially compatible with the type resin to be reinforced.
	Wet-out—Should show rapid wet-out with minimum filamentation and minimum fiber degradation.

Source: Johns-Manville Fiber Glass Reinforcements Division.

Table VI-1.3 Fillers Used in Thermoplastics

Type	Particle Type	Particle Size, (mu)	Oil Absorption	Sp. Gr.	Bulking Value	Cost, per lb	Remarks
Calcium carbonate							
Dry-ground	Crystalline	5–44	5–6	2.71	0.044	1.0	Extremely low in cost.
Wet-ground	Crystalline	0.5–10	15	2.71	0.044	2.0	Used widely in floor tiles; fine grinds used in molded and extruded products.
Precipitated	Crystalline	0.06–8	14–63	2.65	0.045	2½–9.0	Expensive types used to supplement TiO_2.
Stearic-coated	Crystalline	—	—	—	—	3–4	Widely used in rigid pipe.
Clay	Plates	0.2–4.0	44	2.58	0.046	1.5	Used in floor tile, but absorbs plasticizer.
Calcined clay	Plates	0.5–6.0	45	2.63	0.046	2.5	Provides insulation.
Western talc	Platy	0.5–16	38	2.79	0.043	2.5–3.5	In PVC, imparts hot strength in calendering.
New York talc	—	—	35–41	2.78	0.043	2.0	In PP, provides stiffness and good heat aging.
Chrysotile asbestos	Fiber	(100% through 10 mesh)	—	2.56	0.047	3–5.5	In PVC, used in floor tile.
Anthophyllite asbestos	Short Fiber	—	—	3.0	—	9.5	In PP, used almost exclusively in molded parts. Heat stability good.
Diatomaceous Earth	Diatom	(96% through 325 mesh)	81	2.0	0.06	2.0	Improves compression, reduces shrinkage; also used to solidify PVC plasticizers.
Colloidal silica, pyrogenic	Spherical		150	2.0	0.057	66.0	Thixotropic agent.
Quartz	Angular	(99% through 200 mesh)	19	2.65	0.045	1.0	
Wollastonite	Fibers	0.5–30	26	2.5	0.041	2.5	Not used much in TP, but does impart good impact strength.
Barytes	Granular	0.5–30	7	4.46	0.027	4.5	Used to increase specific gravity of foam.

Source: Adapted from *Plastics Technology*, Vol. 15, No. 12, Nov. 1969, pp. 38 to 43.

mechanisms generally involve chain scission of the polyolefin polymer. Consequent reduction in the polymer molecular weight and degree of polymerization offsets the coupling benefits.

The initial properties of the thermoplastic polymer provide base for controlling the properties of the finished composite. In addition, the concentration of glass fiber, the tensile and flexural strengths of the fibers, and the sizing agent on the glass influence the composite properties.

The decisive or controlling composite parameters are the glass fiber length in the end product, the distribution uniformity of the glass fibers in the melt, and a uniform wetting by the plastic melt of the fiber coated with a compatible sizing agent.

Reinforcement Theory. The improved strength properties of glass-reinforced thermoplastics are obtained by transferring applied stresses from the resin to the supporting fibers. The length of the fiber is particularly important in determining the degree to which the stress is transferred. This critical fiber length value should be determined in order to establish optimum reinforcement properties for a given matrix resin.[7, 8]

The relationship of the critical fiber length to other parameters is expressed as

$$L_C = \frac{D \times T_f}{2S}$$

where

L_C = critical length of fiber
D = fiber diameter
T_f = tensile strength of fiber
S = strength of adhesive bond between fiber and resin, usually assured to be equal to the shear strength of the resin

Theoretical considerations involving critical fiber length may be substantiated and verified by studying the stress-optical behavior of polymers reinforced with fibers of varying lengths. The photoelastic patterns visible at fiber ends interfere and cancel out if the fiber length is too short. An optimum L/D ratio can be readily determined by observation.

In addition, the weight of the reinforcement (actually the volume fraction) in the resin has a primary influence on the level or magnitude of mechanical strength.

The ultimate tensile strength of a glass nylon composite can be approximated as reported by Lees:

$$\sigma_{uc} \cong V_f \left(1 - \frac{\ell_c}{2\ell}\right) \sigma_f + V_m \sigma_m$$

where

σ_{uc} = ultimate tensile strength of the composite
V_f = volume fraction of glass fibers
ℓ_c = critical fiber length
ℓ = length of fiber
σ = tensile strength of fiber
V_m = volume fraction of resin
σ_m = tensile strength of resin at strain corresponding to composite failure

Note: $\sigma = \sigma$ (sigma)

In a reinforced thermoplastic composite, therefore, a desirable balance of good processing characteristics and enhanced physical properties is obtained by the optimum combination of fiber lengths. The fiber lengths should exceed the critical length but should not be sufficiently long to interfere with the processing capability of the compound. When the reinforcing fiber lengths approach the critical length, the tensile strength decreases and approaches 50% of the ultimate tensile value.[9]

EQUIPMENT AND TOOLING FOR RTP

Injection Presses and Molding Machines. Injection molding of thermoplastics is an established field of long-standing success and repute. It is beyond the scope of this treatise to attempt to improve upon the excellent literature and background of technology on injection molding. It is sufficient to reiterate here that the molding of fiber glass and other reinforced thermoplastics has been successfully adapted to all sizes and types of injection equipment from 125 ton—6 oz to 650 ton—80 oz and even larger if and when required.[10]

The most critical factors in molding RTP are the resin temperature and the manner in which this temperature controls the homogeneity (or lack of it) of the glass-resin phases. Excessively high molding temperatures or extra work done on the molding compound cause resin and glass to separate in the barrel and molds, resulting in defects, and thus a correspondingly high molded-part reject rate.

Itemized below are several elements that

Table VI-1.4 Properties of Reinforced Acetals

| Property | ASTM | Homopolymer | | | Copolymer | |
		Standard (unreinforced)	20% Glass-Reinforced	22% TFE-Reinforced	Standard (unreinforced)	25% Glass-Reinforced
Physical Properties						
Specific gravity	D792	1.425	1.56	1.54	1.410	1.61
Ther. cond., Btu/hr/sq ft/°F/ft	—	0.13	—	—	0.16	—
Coef. of ther. exp., 10^{-5} per °F	D696	4.5	2.0–4.5	4.5	4.7	2.2–4.7
Specific heat, Btu/lb/°F	—	0.35	—	—	0.35	—
Water absorption (24 hr), %	D570	0.25	0.25	0.20	0.22	0.29
Flammability, ipm	D635	1.1	0.8	0.8	1.1	1.0
Mechanical Properties						
Tensile strength, 1000 psi	D638					
Ultimate	—	10.0	8.5	6.9	8.8	18.5
Yield	—	10.0	—	—	8.8	18.5
Elongation, %	D638					
Ultimate	—	25	7	12	60–75	3
Yield	—	12	—	—	12	3
Mod. of elast. in ten, 10^5 psi	D638	5.2	—	—	4.1	12.5
Flex. strength, 1000 psi	D790	14.1	—	—	13	28
Mod. of elast. in flex., 10^5 psi	D790	4.1	8.8	4.0	3.75	11
Impact str. (Izod, notched), ft-lb/in.	D638	1.4	0.8	0.7	1.2	1.6
Compr. str. (1%), 1000 psi	D695	5.2	5.2	4.5	4.5	—
Electrical Properties						
Volume resistivity, ohm-cm.	D257	1×10^{15}	5×10^{14}	—	1×10^{14}	1.2×10^{14}
Dielectric str. (short-time), vpm	D149	500	500	—	>400	580
Dielectric constant	D150					
60 cycles	—	3.7	4.0	—	3.7	3.9
10^6 cycles	—	3.7	4.0	—	3.7	3.9
Dissipation factor	D150					
60 cycles	—	0.0048	0.0047	—	0.001	0.003
10^6 cycles	—	0.0048	0.0036	—	0.006	0.006

	ASTM					
Heat Resistance						
Max. rec. service temp., °F	—	195	195	195	220	220
Deflection temp., °F	D648					
66 psi	—	338	345	329	316	331
264 psi	—	255	3.5	212	230	325
Fabricating Properties						
Injection molding						
Pressure, 1000 psi	—	15–25	15–25	15–25	15–20	15–20
Temperature, °F	—	380–420	380–420	380–410	360–440	375–440
Shrinkage, in./in.	—	0.025	0.01–0.025	0.025	0.020	0.018–0.004
Extrusion temperature, °F	—	370–400	370–400	370–400	360–440	360–440
USES		Appliance parts, gears, bushings, aerosol bottles, auto, plumbing, textile, consumer uses.	Same as homopolymer. Where high stiffness and dimensional stability are required.	Same as homopolymer. Where low friction and high resistance to wear are required.	Appliance parts gears, bushings, aerosol bottles, various automotive, plumbing, textile machinery and consumer products.	

Source: Adapted from "1971 Materials Selector," *Materials Engineering, 72,* No. 6, Nov. 1971, pp. 201–227.

Table VI-1.5   Properties of Reinforced Thermoplastic ABS

Properties	ASTM	Medium Impact Type
Physical Properties		
Specific gravity	D792	1.05
Thermal conductivity, Btu./hr/sq ft/°F/ft	C177	0.08–0.18
Coef. of thermal exp. 10^{-6} per °F	D696	3.2–4.8
Spec. ht. Btu./lb/°F	—	—
Water absorption (24 hr), %	D570	0.2–0.4
Flammability, ipm	D635	1.0–1.6
Heat deflection temp. (264 psi), °F	D648	185–223
Mechanical Properties		
Modulus of elasticity in tension, 10^5 psi	D638	3.3–4.0
Tensile strength 1000 psi	D638	6.3–8.0
Elongation (2 in./min) %	D638	5–20
Hardness (Rockwell)	D785	R108–115
Impact strength (1300) ft-16/in. notch	D256	2.0–4.0
Impact strength (−40°F) ft-16/in.	D256	0.8–1.0
Modulus in elasticity in flex. 10^{-5} psi	D790	3.5–4.0
Flexural strength 1000 psi	D790	9.9–11.8
Compressive strength 1000 psi	D695	10.5–11.0
Electrical Properties		
Vol. res., ohm-cm.	D257	2.7×10^6
Dielectric strength (short-term) V./mil	D149	385
Dielectric constant	D150	2.8–3.2
10^6 cycles	D150	2.75–3.0

have been found to represent good practice and procedure in adapting press and mold components to the processing of RTP.[11] Further details may be obtained from the excellent references available or from the equipment manufacturers and compound suppliers.[12, 13]

Whether a powdered, pelletized, or other form of resin and reinforcement is used, any normal means for introducing the material into the press and for heating the material are applicable to RTP. Optimum processing temperatures for the various reinforced thermoplastics are indicated in Tables VI-1.4 through VI-1.13.

Screw L/D ratios of 20:1 are acceptable and are most favorable for the gradual, controlled heat-up of the glass and resin. Chrome-plated 4140 material is preferred for screw fabrication, as are Xaloy barrel liners, although nitrided barrel liners are widely used with RTP.

A "bull-nose" cone on the front of the screw is preferred over either the sliding ring or ball valve because of its lower tendency or action to separate glass and resin.

The use of little or no (100 psi maximum) back-off pressure after the shot is also effective in keeping resin heat controlled at the desirable low state.

Molds of H-13 or the equivalent steel material hardened to a Rockwell value of 50 to 55 are recommended for long-run mold performance.

In order to promote high rates of flow and to prevent glass degradation and separation from the resin, sprues and runners should be of maximum practical size and not be skimpy dimension-wise. Runner diameters of ⅜ in. are recommended for RTP.

Tunnel gates of a circular configuration with a 0.060 in. diameter are the most acceptable. Rectangular and trapezoidal cross-section gates are also used with RTP.

MOLDING METHODS/PROCESSING DETAILS

The methods of pelletizing to produce the raw RTP material to feed to injection ma-

Table VI-1.6   Properties of Reinforced Thermoplastic Polycarbonate

Properties	ASTM	Standard (unreinforced)	Reinforced (40% Glass Fiber)
Physical Properties			
Specific gravity	D792	1.20	1.51
Thermal conductivity Btu./hr/sq ft/°F/in.	—	0.11	0.13
Coef. of thermal exp., 10^{-5} per °F	D696	3.75	1.0–1.1
Water absorption (24 hr), %	D570	0.15	0.08
Mechanical Properties			
Tensile strength, 1000 psi	D638	9.5	18
Yield strength, 1000 psi	D638	8.5	—
Tensile modulus, 10^{-5} psi	D638	3.45	17
Impact strength, (Izod notched) ft-lb/in.	D256	16	—
Flexural strength, 1000 psi	D790	13.5	27
Modulus of elasticity in flex., 10^{-5} psi	D790	3.4	12
Compressive strength, 1000 psi	D690	12.5	18.5
Electrical Properties			
Volume resistivity, ohm-cm.	D257	2.1×10^{16}	1.4×10^{15}
Dielectric strength, (S.T.)	D149	400	475
Dielectric constant	D150	—	—
60 cycles	—	3.17	3.80
10^6 cycles	—	2.96	3.58
Dissipation factor	D150		
60 cycles	—	0.0009	0.006
10^6 cycles	—	0.010	0.007
Fabricating Properties			
Injection molding			
Press, 1000 psi	—	15–20	15–20
Temperature, °F	—	525–625	575–650
Mold shrinkage, in./in.	—	0.005–0.007	0.002
Extrusion temperature, °F	—	475–580	—
Uses		Electrical parts, housings structural parts, electronic components, safety helmets, streetlight globes, portable tool housings.	Military parts, module cases, pump impellers, weapons components, aircraft parts, automotive parts, portable tool housings.

chines are of prime interest. Discussion of these methods precedes information on actual molding practices.

There are 2 main methods for making glass-fiber-reinforced thermoplastic pellets: the coating process and the compounding process, including several variations.

The Coating Process. In the coating process, 6 to 8 strands of continuous glass-fiber roving are die-coated with a thermoplastic melt. The procedure is similar to a wire-coating operation. The coated strand is then cooled and cut into pellets.

The length of the glass fiber corresponds approximately to the length of the cut pellets. The distribution of the glass fibers in the thermoplastic is not homogeneous and many fibers in the roving may not be wetted by the plastic melt. This lack of homogeneity has disadvantages for some applications, such as complex or thin-walled parts. Fiber clumps may cause

Table VI-1.7   Properties of Reinforced Nylons—Type 6

Properties	ASTM	General Purpose (unreinforced)	Reinforced (30% Glass Fiber)
Physical Properties			
Specific gravity	D792	1.14	1.37
Thermal conductivity, Btu./hr/sq ft/°F/in.	—	1.2	1.2–1.7
Coef. thermal exp., 10^{-5} per °F	D696	4.8	1.2
Water absorption (24 hr) %	D570	1.7–1.8	1.3
Flammability, ipm	D635	Self-extinguishing	Slow burn
Mechanical Properties			
Tensile strength (2 in./min.)	D638		
Ultimate		9.5–12.5	21–23
Yield		8.5–12.5	—
Elongation (2 in./min.), %	D638	30–220	2–4
Modulus of elasticity in tension, 10^{-5} psi	D638	—	10–12
Flexural strength, 1000 psi	D790	Unbreakable	26–34
Modulus of elasticity in flex., 10^5 psi	D790	1.4–3.7	10–12
Impact strength (Izod notched)			
ft-lb/in.	D256	0.8–1:2	3–2.3
Compressive strength (1%) 1000 psi	D695	9.7	19–20
Electrical Properties			
Vol. resistivity, ohm-cm.	D257	4.5×10^{13}	$0.28–1.5 \times 10^{15}$
Dielectric strength (S.T. V./mil)	D149	385	400–450
Dielectric constant	D150		
60 cycles		4.0–5.3	4.6–5.6
10^6 cycles		3.6–3.8	3.9–5.4
Dissipation factor	D150		
60 cycles		0.06–0.014	0.022–0.008
10^6 cycles		0.03–0.04	0.019–0.015
Heat Resistance			
Max. recommended service temp. °F	—	250–300	250–300
Deflection temp., °F	D648		
66 psi		360	425–428
264 psi		155–160	420–419
Fabricating Properties			
Injection molding			
Pressure, 1000 psi	—	10–20	10–20
Temp., °F	—	440–550	500–570
Shrinkage, in./in.	—	0.12–0.20	0.003
Extrusion temp.	—	450–600	500–600
Uses		Bearings, gears, bushings, coil forms, brush backs, rod, tubing tape.	General-purpose type, 6 parts requiring greater stiffness and dimensional stability.

Table VI-1.8   Properties of Reinforced Nylon—Type 6/6

Properties	ASTM	General-Purpose (unreinforced)	Reinforced* (Glass Fiber)	Reinforced† (Glass Fiber and Molybdenum Disulfide)
Physical Properties				
Specific gravity	D792	1.13	1.37, 1.47	1.37–1.41
Thermal conductivity, Btu./ hr/sq ft/°F/in.	—	1.7	1.5, 3.3	—
Coef. of thermal exp. 10^{-5} per °F	D696	4.5	2.1, 1.4	1.75
Specific heat, Btu./ 16/°F	—	0.3	—	—
Water absorption (24 hr), %	D570	1.5	0.9, 0.8	0.5–0.7
Flammability, ipm	D635	Self extinguishing	Slow burn	Slow burn
Mechanical Properties				
Tensile strength, 1000 psi	D638			
Ultimate		11.8	25, 30	19–22
Yield		11.8	—	—
Elongation, %	D638			
Ultimate		60	1.8, 2.2	3
Yield		5	—	—
Modulus of elasticity in tension, 10^{-5} psi	D638	4.75	14, 20	—
Flexural strength, 1000 psi	D790	unbreakable	26, 35	26–28
Modulus of elasticity in flexure, 10^{-5} psi	D790	410	10, 18	11–13
Impact strength (Izod notched) ft-lb/in.	D256	1.0	2.5, 3.4	—
Compressive strength (1%), 1000 psi	D695	4.9	20, 24	—
Electrical Properties				
Volume resistivity, ohm-cm.	D257	10^{14}	5.5×10^{15}	—
Dielectric strength (S.T.), v./mil	D149	385	400, 480	300–400
Dielectric const.	D150			
60 cycles		4.0	4.0, 4.4	—
10^6 cycles		3.6	3.5, 4.1	—
Dissipation factor	D150			
60 cycles		0.014	0.018, 0.009	—
10^6 cycles		0.04	0.017, 0.018	—
Heat Resistance				
Max. recommended service Temp., °F	—	250–300	250–300	250–300
Deflection temp.	D648			
66 psi F		470	507	—
264 psi F		220	495	—

Table VI-1.8   Properties of Reinforced Nylon—Type 6/6 (cont.)

Properties	ASTM	General-Purpose (unreinforced)	Reinforced* (Glass Fiber)	Reinforced[†] (Glass Fiber and Molybdenum Disulfide
Fabricating Properties				
Injection molding				
Pressure, 1000 psi	—	10–20	10–20	—
Temp., °F	—	520–650	520–650	—
Shrinkage, in./in.	—	0.015	0.004, 0.003	—
Uses		Bearings, gears, bushings, coil forms, brush backs, rod, tubing, tape.	Mechanical parts where lubrication is undesirable or difficult.	

*First value 30% glass fiber; second for 40% glass fiber.
[†]30% glass fiber.

poor molded surface characteristics. The completeness of wetting of the glass fiber surfaces by the polymer may be insufficient for desired, optimum or ultimate long-term performance. Generally, however, homogeneous fiber distribution is improved in the injection-molding operation.

The Compounding Process. In the compounding process, the thermoplastic polymer is mixed with the glass fibers in either single- or twin-screw extrusion machines extruded and cut off during cooling to form the pellets. This process may utilize either cut lengths of glass-fiber strands, or the continuous-roving form of glass fiber reinforcement. In both cases, a more homogeneous pelletized compound can be produced than in the coating process. The compounding of the fiber and the thermoplastic polymer causes the bundles of glass fibers to separate and become wetted by the plastic melt. Another advantage of the compounding process includes the removal of volatile matter which might ultimately impair the complete wetting of the glass surface by the polymer. Screw-compounding equipment usually has degassing vents to remove volatile components from the fiber-reinforced compound. In the compounding process, the fiber length is shorter than in the coating process. Proper design of the compounding euipment more specifically the screw geometry, and suitable operating conditions make possible an average glass-fiber length of between 0.020 and 0.12 in.

Glass-fiber compounds are pelletized in the conventional way either by strand cutting or by hot-melt cutting processes. Twin-screw compounding equipment has several advantages for processing glass fiber composite materials. The twin-screw consists of co-rotating shafts which preferably inter-mesh and clean each other along a special curve, Dead corners are avoided and uniform working of the material is possible.

Compounding Materials. Glass fibers may be either metered separately into the feed throat or pre-blended with the powdered or granular polymer. These components are gravity-fed into the feed section by any one of several suitable metering devices. The polymer is plasticized in the plasticizing zone while the glass fibers are being blended. During the plasticizing or softening of the thermoplastic material, the glass fiber bundles become reduced in size. In this plasticizing zone, the screws and barrels may be subjected to severe abrasion and wear caused by the mechanical action of the unmixed ingredients. The wear is not necessarily contributed by the glass fibers. An excellent discussion of causes of wear and its prevention in injection machine cylinders and screws is available in the literature (see ref. 14).

As an alternate to mixing the fibers in the plasticizing zone, glass rovings or chopped strand may be added downstream through a de-gassing port into the already molten polymer (Werner-Pfleiderer system). The continuous unchopped glass rovings may be added without any special metering unit. The glass is pulled

Table VI-1.9   Properties of Reinforced Nylon—Type 6/10

Properties	ASTM	General-Purpose Dry, as Molded, (unreinforced)	Reinforced (30% (Glass Fiber)
Physical Properties			
Specific gravity	D792	1.07	1.30
Thermal conductivity, Btu./hr/sq ft/°F/in.	—	1.5	3.5
Coef. thermal exp., 10^{-5} per °F	D696	5	2.5
Specific heat, Btu./lb/°F	—	0.3	—
Water absorption (24 hr), %	D570	0.4	0.2
Flammability, ipm	D635	Self-extinguishing	Slow burn
Mechanical Properties			
Tensile strength, 1000 psi	D638		
Ultimate		8.5	19
Yield		8.5	—
Elongation, %	D638		
Ultimate		85	1.9
Yield		5	—
Modulus of elasticity in tension, 10^{-5} psi	D638	2.8	10–12
Impact strength (Izod notched) ft-lb/in.	D256	0.6	3.4
Compressive strength (1%) 1000 psi	D695	3.0	18
Electrical Properties			
Volume resistivity, ohm-cm.	D257	10^{15}	—
Dielectric strength (S.T.) v./mi.	D149	470	—
Dielectric constant	D150		
60 cycles		3.9	—
10^6 cycles		3.5	—
Dissipation factor	D150		
60 cycles		0.04	—
10^6 cycles		—	—
Heat Resistance			
Max. recommended service temp., °F	—	225–300	250–300
Deflection temp., °F	D648		
66 psi		300	430
264 psi		135	420
Fabricating Properties			
Injection molding			
Pressure, 1000 psi	—	10–20	10–20
Temp., °F	—	450–600	500–620
Shrinkage in./in.	—	0.015	0.0035–0.0045
Uses		Jacketing for wire and cable, special molded parts.	

into the molten polymer mass by the rotating screw and is thoroughly broken into short lengths and dispersed before the mass is extruded for pelletizing. A special kneading section or element with zero pitch in the screw flights is located just downstream from the fiber glass roving entry point, and is responsible for breaking up the glass into finite lengths and mixing it with the polymer. Varying the geometry of this kneading element results in production of

Table VI-1.10 Properties of Reinforced Polypropylene and Polyphenylene Sulfide

Properties	ASTM	Polypropylene			Polyphenylene Sulfide	
		General-Purpose (unreinforced)	Asbestos-Filled	Glass Reinforced	Standard (unreinforced)	40% Glass Reinforced
Physical Properties						
Specific gravity	D792	0.900–0.910	1.11–1.36	1.04–1.22	1.34	1.64
Thermal cond. Btu./hr/sq ft/°F/in.	—	1.21–1.36	—	—	2.0	—
Coef. of thermal exp., 10^{-5} per °F	D696	3.8–5.8	2–3	1.6–2.4	3.0	—
Specific heat, Btu./lb/°F	—	0.45	—	—	0.26	—
Water absorption (24 hr), %	D570	<0.01–0.03	0.02–0.04	0.02–0.05	—	—
Flammability, ipm	D635	0.7–1	1	1	Non-burn	Non-burn
Mechanical Properties						
Tensile strength, 1000 psi	D638, C					
Maximum		4.8–5.5				
Yield		4.8–5.2	3.3–8.2	6–10	11	21
Elongation, %						
Break		30–>200	3–20	2–4	3	3–9
Yield		9–15	5	—	—	—
Modulus of elasticity in tension, 10^{-5} psi	D638, B	1.6–2.2	—	812	4.8	11.2
Flexural yield strength, 1000 psi	D790, B	6–7	7.5–9	8–11	20	37
Modulus of elasticity in flex, 10^{-5} psi	D790, B	1.7–2.5	3.4–6.5	4–8.2	6.0	22.0
Impact strength (Izod notched), ft-lb/in.	D256	0.4–2.2	0.5–1.5	0.5–2	—	—
Compr. yield strength, 1000 psi	D695	5.5–6.5	7.0	6.5–7	—	—
Electrical Properties						
Vol. resistance, ohm-cm.	D257	$>10^{17}$	1.5×10^{15}	1.7×10^{16}	—	—
Dielectric strength (short time), vpm	D149	650(125 mil)	450	317–475	595	490
Dielectric constant	D150					
60 cycles		2.20–2.28	2.75	2.3–2.5	—	—
10^6 cycles		2.23–2.24	2.6–3.17	2–2.5	3.22	3.88

	ASTM					
Dissipation factor	D150					
60 cycles		0.0005–0.0007	0.007	0.002	—	—
10^6 cycles		0.0002–0.0003	0.002	0.003	0.0007	0.0041
Heat Resistance						
Max. recommended service temp., °F	D648	230	250	250	500	500
Deflection temp., °F						
66 psi		205–230	270–290	275–310	—	—
264 psi		135–140	170–220	250–300	278	425
Fabricating Properties						
Injection molding						
Pressure, 1000 psi		10–20	10–20	15–20	10–15	10–15
Temp., °F		400–550	400–550	450–575	600–700	600–700
Shrinkage, in./in.		0.010–0.025	0.003–0.008	0.001–0.008	0.008	0.004
Extrusion temp.		380–430	—	430–575	600–700	600–700
Uses		Hospital ware, housewares, appliances, radio and TV housings, film, fibers.	Housings, automobile fan shrouds, covers.	Housings, shrouds, cases, panels and mechanical parts.	Corrosion resistant pump components, valves and pipe.	Pump vanes, valve parts, gaskets, fuel cells and auto parts req chem res at higher temps.

Table VI-1.11   Properties of Reinforced Polyphenylene Oxide and Polysulfone

Properties	ASTM	Modified Phenylene Oxide		Polysulfone	
		Standard (unreinforced)	Glass Fiber Reinforced*	Standard (unreinforced)	Glass Fiber Reinforced†
Physical Properties					
Specific gravity	D792	1.06	1.21, 1.27	1.24	1.41, 1.55
Thermal conductivity, Btu./hr/sq ft/°F/in.	C177	1.5	1.15, 1.1	1.8	1.6, 1.2
Coef. of thermal exp., 10^{-5} per °F	D696	3.3	2.0, 1.4	3.1	—
Specific heat, Btu./lb/°F	—	—	—	0.24	—
Water absorption (24 hr), %	D570	0.07	0.06	0.22	0.22, 0.18
Flammability, ipm	D635	Self-extinguishing	Self-extinguishing	Self-extinguishing	Self-extinguishing
Mechanical Properties					
Tensile strength, 1000 psi	D638				
Ultimate		—	—	—	—
Yield		9.6	14.5, 17.0	10.2	17, 19
Elongation, %	—				
Ultimate		60	4–6	50–100	2, 1.6
Yield		—	—	5.6	—
Modulus of elasticity in tension, 10^{-5} psi	D638	3.55	9.25, 13.3	3.6	10.9, 14.9
Flexural strength, 1000 psi	D790	13.5	20.5, 22	15.4	25, 28
Modulus of elasticity in flex, 10^{-5} psi	D790	3.6	7.4, 10.4	3.9	12, 15.5
Impact strength (Izod notched), ft-lb/in.	D638	5.0	2.3	1.3	1.8, 2.0
Compressive strength, 1000 psi	D695	16.4	17.6, 17.9	13.9	—
Electrical Properties					
Volume resistivity, ohm-cm.	D257	10^{17}	10^{17}	5×10^{16}	10^{17}
Dielectric strength (short-time), v/mil	D149	550 (⅛ in.)	1020 (¹⁄₃₂ in.)	425	480
Dielectric constant	D150				
60 cycles		2.65	2.93	3.06	3.55
10^6 cycles		2.64	2.92	3.03	3.41
Dissipation factor	D150				
60 cycles		0.0004	0.0009	0.0008	0.0019
10^6 cycles		0.0009	0.0015	0.0034	0.0049

Heat Resistance

	ASTM				
Max. recommended service temp., °F		212	340	—	350
Deflection temp., °F	D648				
66 psi		279	293, 317	358	389
264 psi		265	282, 310	345	365

Fabricating Properties

	ASTM				
Bulk factor	D1895	—	—	1.8	—
Injection molding					
Pressure, 1000 psi		12–18	15–20	15–25	15–25
Temp., °F		450–600	525–600	625–750	625–750
Shrinkage, in./in.		0.005–0.007	0.001–0.002	0.007	0.003
Extrusion temp.		450–600	—	600–750	—
Uses		Housings, cabinets, consoles, covers, tape cartridge platforms, coil assemblies, electric connectors, bus bar insulators, switch housings, terminal blocks, tuner bars, light fixtures, impellers, valve bodies and discs.		Coil bobbins, switches, terminal blocks, battery cases, connectors, circuit carriers, sockets, tube bases, range hardware, coffee maker parts, sight glasses, auto parts, lamp bezels, aircraft ducts, housing and side wall panels, meter housings and components projector transparencies.	

*Where two values are given, first applies to 20% glass fiber and second to 30%, otherwise same value applies to both. †Where two values are given, first applies to 30% glass fiber and second to 40%, otherwise same value applies to both.

Table VI-1.12 Properties of Reinforced Polystyrenes

Properties	ASTM	Polystyrene		Styrene Acrylonitrile	
		General-Purpose (unreinforced)	Glass Fiber (30%) Reinforced	(SAN) (unreinforced)	Glass Fiber (30%) Reinforced SAN
Physical Properties					
Specific gravity	D792	1.04	1.29	1.04–1.07	1.35
Thermal conductivity, Btu./hr/sq ft/°F/ft	—	0.058–0.090	0.117	—	—
Coef. of thermal exp., 10^{-5} per °F	D696	3.3–4.8	1.8	3.6–3.7	1.6
Specific heat, Btu./lb/°F	—	0.30–0.35	0.256	0.33	—
Water absorption (24 hr), %	D570	0.03–0.2	0.07	0.20–0.35	0.15
Flammability, ipm	D635	1.0–1.5	—	0.8	—
Mechanical Properties					
Tensile strength, 1000 psi	D638				
Ultimate		5.0–10	14	9.5–12.0	18
Yield		5.0–10	14	—	18
Elongation, %	D638				
Ultimate		1.0–2.3	1.1	0.5–3.7	1.4
Yield		1.0–2.3	1.1	—	1.4
Modulus of elasticity in tension, 10^{-5} psi	D638	4.6–5.0	12.1	4.0–5.0	17.5
Flexural strength, 1000 psi	D790	10–15	17	—	22
Modulus elasticity in flex., 10^{-5} psi	D790	4–5	12	—	14.5
Impact strength (Izod notched), ft-lb/in.	D638	0.2–0.4	2.5	0.30–0.45	3.0
Compr. strength, 1000 psi	D695	11.5–16.0	19	—	2.3
Electrical Properties					
Volume resistivity, ohm-cm.	D257	$>10^{16}$	3.6×10^{16}	$>10^{16}$	4.4×10^{16}
Dielectric strength (short-time), v./mil	D149	>500	396	400–500	515
Dielectric constant	D150				
60 cycles		2.45–2.65	3.1	2.6–3.4	3.5
10^6 cycles		2.45–2.65	3.0	2.6–3.02	3.4
Dissipation factor	D150				
60 cycles		0.0001–0.0003	0.005	>0.006	0.005
10^6 cycles		0.0001–0.0005	0.002	0.007–0.010	0.009

	Method				
Heat Resistance					
Max. recommended service temp., °F	—	160–205	190–200	175–190	—
Deflection temp., °F	D648				
66 psi		—	230	—	230
264 psi		220 max	220	210–220	220
Fabricating Properties					
Injection molding					
Pressure, 1000 psi	—	10–24	0.5–1.2 (line)	10–24	0.5–1.2 (line)
Temp., °F	—	325–650	450–625	375–550	430–550
Shrinkage, in./in.	—	0.002–0.008	0.001–0.003	0.003–0.007	0.0005–0.002
Extrusion temp. (Vicat soft)	—	194–224	—	—	
Uses		Thin parts, long flow parts, toys, appliances, containers, film, mono-filaments and housewares.	Auto dash-board skeletons, camera housings and frames, tape reels, fan blades.	Kitchenware, tumblers, broom bristles, ice buckets, closures, film, containers, lenses, battery cases.	Camera housings and frames, auto bezels, electrical components, handles, auto panels.

Table VI-1.13   Properties of Injection-Molded Reinforced Thermoplastic Polyester Resins

Property	ASTM	Not Reinforced	Reinforced
Specific gravity	D792	1.31	152; 1.58
Water absorption, %	D570		
24 hr		0.08	0.07
Equilibrium		0.3	0.3
Shrinkage, in./in.		0.017–0.023	0.002–0.004
Tensile strength, psi	D638	8,000	18,000
Elongation, %	D638	200–300	4–5
Flexural strength, psi	D790	12,000	28,000
Flexural modulus, psi	D790	340,000	1,300,000
Compressive strength, psi	D695	12,500	18,000
Izod impact strength, ft lb/in.	D256		
Notched		1.0	1.8–2.2
Unnotched		No break	15–20
Coefficient of friction	D1894		
Against metal		0.13	0.14
Against self		0.17	0.16
Heat deflection temp., °F	D648		
66 psi		310	420
264 psi		150	416
Flammability	UL94	SB	SB; SE-0
	D635	SB	SB; non-burning
Deformation under load, %	D621	0.17	0.12
Arc resistance, sec	D495	130	150; 70
Volume resistivity, ohm-cm	D257	10^{16}	10^{16}
Dielectric strength, v/mil	D149	590	750
Dielectric constant	D150	3.3	3.8
Dissipation factor	D150	0.002	0.002
Fabricating Properties:			
Injection molding pressure, psi	—	7,000	7,000
Stock temp., °F	—	460–520	460–520

Source: Modern Plastics, vol. 49, no. 4, April, 1972, pp 72–74, Plastics Department, General Electric Co., Pittsfield, Mass.

longer or shorter fiber lengths in the pellets as desired.

A balance of fiber length and homogeneity of compound is required. Reduced fiber length also occurs as a result of additional mixing and shearing of the glass-reinforced polymer. Reducing the amount of work on the material provides longer fiber lengths but also a less homogeneous material.[15] After pelletizing by either strand or hot-melt cutting and cooling the compound is ready to be injection-molded to form a finished article.

Dry-Blend Compounds. A second approach to molding fiber glass reinforced thermoplastic is to use pre-compounding techniques right at the injection machine or press. This is done directly without first compounding fiber and resin into moldable pellets. With a dry blend, the resins and fibers are mixed together by the plasticizing action of the screw injection-molding machine. One possible technique is to proportion the resin and chopped strand directly into the hopper of the injection-molding machine. The action of the reciprocating screw performs the only mixing and fiber-dispersion functions. Other techniques provide for pre-blending the glass and resin before adding them to the hopper on the injection-molding machine.[16] A third adaptation in common usage comprises pelletizing an extremely glass-rich (35 to 50%) mixture of polymer and reinforcing fibers and mixing these with pellets of clear polymer at the injection press. By utilizing this technique, the ratio of reinforced to clear pellets may be readily varied without elaborate cleanout and recharging. Also, preblending at the hopper is greatly simplified because of the physical similarity between the reinforced and non-reinforced pellets.

Re-addition of Re-ground RTP Stock. In practice for either pelletizing or direct molding of RTP, it has been found satisfactory to re-incorporate re-ground RTP material into new stock in the proportion of 20% re-ground to 80% new material. The use of 100% re-ground material would result in serious loss of molded physical properties. Blending as much as 35% re-ground material into new RTP stock is possible. The economic advantages are quite obvious.

Reinforced Foam Polymers. Fiber glass re-inforced-thermoplastic foam resins (Table VI-1.14) were developed in answer to the continuing demands for increased rigidity (higher modulus of elasticity) in molded RTP parts. They are also characterized by significantly lower specific gravity. The higher part rigidity is accomplished by establishing a higher density part surface skin with a lower density core and a thicker molded cross section. The fiber glass reinforcement becomes randomly aligned

parallel to the cell walls during molding, providing uniform strength in all directions. Properties such as deflection temperature, deformation under load, and moduli are substantially improved over those of non-reinforced foam moldings.

The following brief description abstracts the method by which preparation and molding of fiber glass-reinforced thermoplastic-foam is accomplished.[17, 18, 19, 20, 21, 22, 23]

Preparation of Material. Base-resin molding pellets are prepared by plasticating, introducing a blower agent of the carbamide (urea) or equivalent type under pressure, and pelletizing. The glass reinforcement may be either included in the pellets or added at the molding operation.

Molding Operation. The molding operation is carried out by re-plasticating the material in the press chamber; maintaining chamber pressure with a shut-off nozzle ($\frac{3}{16}$ in. minimum diameter); and providing a short, metered shot with extremely rapid injection. During injection the blowing agent decomposes because of the suddenly released pressure and produces the foamed plastic. Smaller voids form beneath the surface skin and larger voids are surrounded by the glass reinforcement in

Table VI-1.14 Properties of Injection-Molded Reinforced Thermoplastic Foam Resins

Property[a]	Foam Polystyrene 20% Glass	Fire-Retardant Vinyl-Foam Resin 20% Glass	Foam Polypropylene 20% Glass	Foam Nylon 15% Glass
Tensile strength, psi	5,000	5,000	3,000	10,000
Flexural modulus, psi $\times 10^{-5}$	7.5	8.0	4.0	6.5
Flexural strength, psi	8,500	9,000	6,000	16,000
Izod impact strength[b] ft lb/in.	0.6	0.6	0.6	1.0
Unnotched impact strength, ft lb/in.	1.5	1.7	3.5	3.2
Deflection temp. under load @ 264 psi, °F	190	210	162	390
Density[c] g/cc	0.84	0.95	0.73	0.87
Mold shrinkage, in./in.	0.002–0.003	0.002–0.003	0.006	0.006–0.007

[a]Values measured from ASTM specimen 1/4 in. thick.
[b]Not valid test for foam as notch cuts into cellular structure.
[c]Density is measured on test samples at a 30% foam level.

the core or body of the part. No nozzle extensions are necessary and it is advisable to prevent drooling.

Mold Components. Because only 100 to 500 psi of molding pressure (instead of the standard of 500 to 15,000 psi) are developed in molding reinforced-thermoplastic foam resins, low-cost mold materials such as aluminum, kirksite, or aluminum-filled epoxy may be used. Molded parts are obviously lower in cost than non-foamed injection molded reinforced composites.

High-Pressure Foam Molding. High-pressure foam molding requires a full-size shot into a closed injection mold at normal injection pressure. The foaming is caused by permitting the expanding material to escape back into a runner system or the equivalent. Good surface reproduction is possible, although tooling metal and mold costs are as high as those for conventional injection molding.

Actual Injection Molding Conditions. Generally speaking actual injection molding conditions or press operating requirements are particular to each individual reinforced polymer type, and as explained previously, are stated together with properties for each resin system in Tables VI-1.4 to VI-1.13. Molding requirements for reinforced thermoplastic injection moldable foam systems are presented in conjunction with Table VI-1.14.

PRODUCT DATA

Product Properties. The properties of finished molded RTP products are presented in Tables VI-1.4 to VI-1.14, together with the comparative properties of the unreinforced resin.

Design Data. It is difficult to generalize about part design for the entire gamut of RTP engineering resins because of their different inherent properties and processing requirements. However, part design rules for any specific unmodified resin become the starting point for design in the glass- (or other) reinforced version of that resin.[24]

The most significant factor to note is that glass RTP parts are stronger than their non-reinforced counterparts and consequently, a part can be designed using less material to achieve the same function. The basic precept

for planning, therefore, is not to over-design. It is always easier to add ribs and stiffeners or to increase part thickness by removing material from the mold than it is to add material to the mold to reduce part size or dimensions. Over-designing results in molding problems and loss of speed. Working with reliable resin and glass suppliers, mold designers, and toolmakers provides an excellent starting point even if they are partially inexperienced in RTP.

The different mold and part design parameters are discussed below, together with the best-known current practices.

Molded Part Wall Thickness. In designing walls the general rule to follow is that part walls should be as uniform in thickness as possible in order to prevent part distortion and promote rapid and uniform thermal solidification of the material (see Figure VI-1.1).

Depending upon the resin, the wall thickness can be 10 to 50% less than would be necessary when using unreinforced material. To avoid heavy wall sections, use ribbing and other strengthening devices which help achieve faster molding cycles and better part strength.

The optimum wall thickness for nylon material is approximately ⅛ in. Polycarbonate and styrene-based materials are about the same thickness. Polyolefins, however, require walls about 50% heavier.

Contrary to the prevailing opinion, thin wall sections are not necessarily more difficult to mold with glass RTP. There is actually a better flow into thin sections, because glass-reinforced materials are generally molded at somewhat higher temperatures and pressures. The minimum wall thickness with nylon materials can range from a 0.018 to 0.020 in. For polycarbonate and styrene-based materials, minimum thickness of 0.025 in. is possible.

Dimensional Tolerances. The tolerances that can be achieved with glass RTP approach those possible with metal parts. The mold shrinkage of reinforced material is $\frac{1}{10}$ to $\frac{1}{2}$ that of the unreinforced resin. Various small threads and gears have been designed and molded to tolerances of 0.00025 to 0.0005 in. (Commercial Class III gears).

Fillets and Radii. As with non-reinforced materials and metals, sharp corners should be avoided in all parts made of RTP. Elimination

Design this:

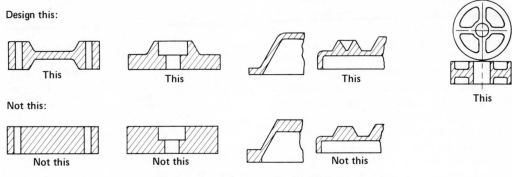

Not this:

Figure VI-1.1   Wall thickness.

of sharp corners helps to promote the resin flow during molding and also to reduce the possibility of notch-sensitive failure of the finished part. Reinforced materials possess greater rigidity and are more susceptible to notch-type impact failures than are the non-reinforced types.

A radius that is 25 to 75% of the adjacent wall thickness should be added on all inside corners, with the minimum radius set at 0.005 in. In general, the designer will find that he will be successful in designing parts for glass RTP if he handles them as though he were designing for hardened tool steel. (See Figure VI-1.2.)

Ribs. For unreinforced plastic it is always recommended that ribs be located at the juncture of two walls or that decorative beads or surface effects be added opposite a rib to hide the effects of sinkage. Because the reinforcement tends to reduce shrinkage on cooling, this is less of a problem with reinforced thermoplastics than with nonreinforced.

It is recommended that the part be designed without ribs at the start, because in many cases the greater inherent strength in the reinforced material will make ribs unnecessary. The part and mold should instead be designed with pro-

vision for later addition should ribs be found necessary.

The standard recommendation for unreinforced material is that ribs should be no higher than three times the wall thickness and less than one-half the thickness of the adjacent wall. These guidelines are considered safe for reinforced material also. With reinforced materials, a rib design should be originated with a height only 1 to 1½ times the wall thickness. A fillet should be added at the point at which the rib joins the wall, and a rib should have a 0.5 to 3° draft for easy ejection from the mold.

Bosses and Studs. It is generally recommended that bosses be no higher than a dimension twice their diameter. With unreinforced materials, studs should always be placed at the convergence of an angle in order to minimize the effects of sinking. This problem assumes less significance with glass RTP because, with 50% less shrinkage, the problem of sink marks is ameliorated.

Long studs should usually be designed with adequate taper in order to facilitate removal from the mold. Deep holes and core pins should be polished to remove all undercuts caused by lathe turning. This should be emphasized for the glass RTP materials, because they do possess greater rigidity and do not respond with the resiliency of their unreinforced counterparts.

Adequate venting of the mold on studs, bosses, or any extended section is also important for reinforced material. Because glass RTP material is molded at faster rates and higher temperatures, adequate venting is necessary to prevent the entrapment of hot gases in the mold.

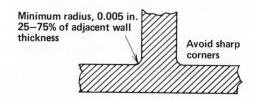

Figure VI-1.2   Fillets and radii.

Screw holes in studs or bosses should be designed with less interference because of the greater hardness of the RTP material. It is recommended that the original part design be made with 0.007 in. interference, as opposed to 0.030 in. for unreinforced material. The interference should be built up in increments of 0.004 or 0.005 in. at a time until the correct interference is achieved.

Holes should be designed with a chamfer at the opening to lead the screw in and with a radius at the bottom of the core. The wall thickness around the hole should be 75 to 100% of the diameter of the screw to be used. The usable depth of the core is 1 to 1½ times the diameter of the screw. (See Figure VI-1.3.)

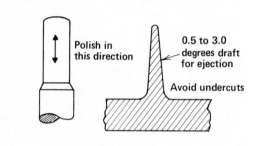

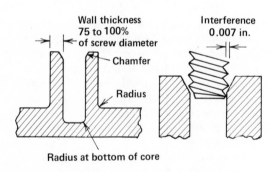

Figure VI-1.3 Ribs, bosses and studs.

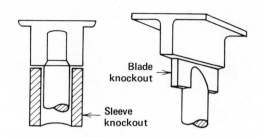

Figure VI-1.4 Draft.

Draft. It has been emphasized that adequate taper must be designed into the part for easier ejection from the mold, because of the comparatively high modulus of elasticity and small amount of shrinkage obtained with reinforced resins. This should not be considered a hard and fast rule. Many glass and other RTP parts have been designed and molded with little or no draft in the sidewalls and other segments. Achieving the no-draft condition requires that a sufficient number of positive knock-out pins be strategically placed. If possible, sleeve knock-outs should be used on long core pins, and blade knock-outs should be used for long, deep walls or ribs. These knock-out facilities also permit venting of trapped air and gases. (See Figure VI-1.4.)

Surface Finish. Grain or texture etched or embossed into the mold surface for transfer to the molded parts should not exceed 0.002 to 0.0025 in. in depth. This parameter is related to the ejection of the part from the mold. The greater the amount of texture designed into the part, the larger should be the amount of draft provided for ejection of the part from the mold. A reliable factor to follow is to allow 1° of draft per in. for each 0.002 in. of texture depth.

References

1. Anon, *Business Week,* 71, Jan. 3, 1970. Automotive Industries, 70, July 15, 1965.
2. Lachowecki, W., *Plastics Design & Processing,* 23, June 1969.
3. Murphy, T. P., SPE Retec Reprint, p. 199, July 1956.
4. Hunt, R. E., *Plastics Technology,* 38, Nov. 1969.
5. *SPE Journal,* Aug. 1968, Vol. 24, p. 24, Jones, R. F.
6. Cessna, L. C., et al., "Chemically Coupled Glass Reinforced Polypropylene," *SPE Journal,* Vol. 25, Oct. 1969, pp. 35–39.
7. Williams, J. C. L., et al., "Short Glass Fiber Reinforced Nylon," Proceedings of 23rd Annual Tech. Conference, SPI Reinforced Plastics/Composites Division, Washington, D.C., 1968.
8. Lees, J. K., "A Study of the Tensile Strength of Short Fiber Reinforced Plastics," *Polymer Engineering & Science,* Vol. 8, No. 3, July 1968.

9. Filbert, W. C., Jr., *SPE Journal*, Vol. 25, "Reinforced 66 Nylon-Molding Variables vs Fiber Length vs Physical Properties," Jan. 1969, p. 65.

10. Personal communication, Swearingen, John, Aero Molded Plastics Co., Napoleon, Ohio, May 1, 1971.

11. Personal communication, Kohl, Paul D., Consultant, E. I. duPont Co., Jan. 29, 1969.

12. Plastics Mold Engineering, Sors, L., Pergamon Press, 1967, p. 79, et seq.

13. *Modern Plastics,* June 1967, p. 140, Paci, R. *See also* notes prepared by Simons, N. D., Stokes Equipment Div., Pennsalt Chemicals, Philadelphia, Pa.

14. "How to Prevent Machine Wear," *Plastics Technology,* Vol. 18, No. 4, Apr. 1972, Lachowecki, W.

15. U.S. Patent 3,304,282, BASF, priority 1964.

16. Schlich, W. R., et al., "Critical Parameters for Direct Injection Molding of Glass Fiber Thermoplastic Powder Blends," SPE Antec, May 1967.

17. "Injection Molding of Thermoplastic Rigid Foams," Frailey, N. E., Shell Chemical Co.

18. Technical Bulletin No. 71.9871, Fiberfil Division, Dart Industries.

19. *Plastics World,* Nov. 1971, p. 44.

20. "Serving Up A New Plastic Sandwich," *SPE Journal,* Vol. 27, Sept. 1971, p. 38, Sandiford, D. J. H., and Oxley, D. F.

21. "Glass Reinforced Foam for Injection Molding," *Plastics Design and Processing,* Jan. 1972, p. 12, Wilson, M. G.

22. "Guide to Working with Reinforced Thermoplastic Foam," *SPE Journal,* Vol. 27, June 1971, p. 35, Wilson, M. G.

23. "The Action is Swinging to Structural Foams," *Plastics Technology,* Vol. 18, No. 4, Apr. 1972, p. 37, Weir, C. L.

24. *Plastics World,* May, 1971, p. 61.

VI-2

Rotational Molding of Fiber Glass Reinforced Thermoplastics

INTRODUCTION

In many modern day industries, some ingenious methods have been devised for producing hollow parts, especially those which are both hollow and undercut.[1] In the glass industry, open-end glass bottles are blown from a hot melt in a closed mold, the mold pulling away after rigidity is reached. In ceramics, hollow and undercut parts may be slip-cast. In plastics, some undercut parts may be produced by resorting to complicated hand lay-up, compression, or blow-mold design. Also now in plastics, both hollow and undercut parts may be fabricated by rotational casting, and what's more, at acceptably high, profit-worthy production rates. Rotational molding came into prominence in the early 1950's and was later adapted to RP/C use.

In rotational molding to produce RP/C parts, powdered resin and chopped fiber glass reinforcement are placed in a single-cast, closable mold or set of gang molds, and these molds are rotated and tumbled, first in heating then in cooling chambers or areas. This action causes the plastic and glass fiber to flow together with the plastic softening and encapsulating the reinforcement. A uniform-walled layer builds up against the inner mold wall to form the desired RP/C plastic part or shape. In this interesting method, small, irregularly shaped parts may be molded, as well as larger items, the size of which is limited only by the capability or handling power of the equipment to support and rotate the molds.

At the beginning of the development of rotational molding of RP/C, a central Ohio man desired to produce life-like scale model statuary by roto-molding using only low-density polyethylene. After much distressing failure of these parts due to sagging or cold-flow at room temperature, the technical head man incorporated a finite percentage of chopped glass fibers, and his thermal-structural stability problems were at an end.

Presently at a neophyte stage, glass fiber reinforced rotationally molded thermoplastics are far from being a full-blown commercial entity. However, great promise exists for well-engineered, high-performance products made by this method. Some very interesting technical information has been generated, and is presented in this chapter together with brief descriptions of the process and accompanying variables.

MATERIALS

Resinous materials for fiber glass reinforced roto-molded (or roto-cast) thermoplastics are chiefly polyethylene, polypropylene, and polyvinyl chloride resins in the dry or plastisol state. Some butyrate, styrene, and nylon are also being used. Other ingredients are fiber glass reinforcement and some pigments. No complicated formulating at the molding site is necessary.

Resins. Powdered (35 mesh) low-density polyethylene with a medium-to-high melt index has been found to give the optimum cost/performance combination in roto-casting of RP/C. It is possible to use cross-linked polyethylene for superior properties, especially for tempera-

Table VI-2.1 Typical Properties of a Low-density Polyethylene Resin
for Rotational Molding[a]

Property[b]	Value	ASTM Test
Density, g per cc	0.955	D1505–68
Melt index, g per 10 min.	6.5	D1238–65T
Flow index (C.I.L.), 190°C, 1500 psi, g per 10 min.[c]	7.4	—
Environmental stress cracking resistance, F_{50}, hr	5 to 15	D1693–66
Tensile strength @ yield, 2 in. per min., psi	3900	D638–68 Type IV specimen
Ultimate elongation, 2 in. per min., %	greater than 650	D638–68 Type IV specimen
Brittleness temp., °F	less than −180	D746–64T
Flexural modulus, psi	195,000	D790–66
Mesh size	35	—

[a]Source—Phillips Petroleum Company, Bartlesville, Oklahoma.
[b]Per ASTM D1928, Procedure C.
[c]Gas plastometer—0.01925 in. orifice diameter and 0.176 in. land.

ture resistance. Dry PVC, or formulated vinyl plastisol resins, similar to those used for slush molding, are also employed in rotational molding, and can be reinforced. The services of a formulator should be used when considering vinyl resins. Other pulverized thermoplastics such as polypropylene, butyrate, nylon, polystyrene, polycarbonate, and others have been evaluated and successfully used on a commercial scale. Also, pulverized thermosets have been found to impart greater rigidity and impact strength to roto-molded RP/C parts. Typical resin properties are presented in Table VI-2.1.

Fiber Glass Reinforcement. Fiber glass chopped strands up to ¼ in. in length have been successfully roto-cast. Lengths of ⅛ in. and 1/32 in. are also used. Shorter lengths favor faster, more uniform assimilation with fewer fibers protruding from the inside surface after molding. Strands of fairly high integrity are preferred for prevention of disintegration and fuzzing. Generally, a smaller fiber bundle results in presenting maximum surface area for coverage by the resin, and hence, the greatest reinforcing potential. Nodule-free, processed or milled glass fibers have considerable virtue as a completely assimilated reinforcement for roto-cast parts. While such fibers do not contribute substantially to higher finished-part mechanical strengths, they greatly increase thermal and dimensional stability. The usual glass loadings are in the range of 2 to 10% and may go as high as 20%.

Precombining of Materials. Materials for reinforced roto-casting may be either separately charged to the mold, premixed in a dry blend in a predetermined fixed percentage prior to charging the molds, or compounded by heating the resin, incorporating the glass fiber (as in a Banbury mill, for example), and then repulverizing the resin-glass mix and weighing out mold charges. Of the three possibilities, the dry-blending of materials has provided the most consistent finished product properties. However, separate charging is most economical of time and materials and is probably the most widely used method.

Release Agents Stearates constitute the major on-the-job release agent. In some plastics for roto-casting, both for polyethylene powders and plastisols, a release agent is added by the resin supplier. Preliminary mold conditioning for release, however, constitutes baking on a silicone mold release resin on all molds in the molding shop at least once per week.

EQUIPMENT AND TOOLING

Rotational Equipment. Four types of rotational casting equipment exist: the rotary automatic, the autobatch, the shuttle type, and the jacketed mold type.[1]

Rotary Automatic Equipment. For successful roto-casting, the molds must be positioned in a heated oven and slowly rotated simultaneously in two directions—rotation and tum-

Figure VI-2.1 Rotary automatic machine for rotational casting. (Photo courtesy of NcNeil-Akron-McNeil Corp.)

ble—at 90°. By this means, all surfaces of the mold are sequentially rotated to the bottom point and, because of the action of gravity, are contacted by the glass-resin mix. Gears inside a series of shafts and power transmission trains provide the rotation, while the plastic material becomes fluid, coalesces with the fiber glass reinforcement, and builds up a uniform coating on the mold walls.

Next, the machine indexes and exposes the molds to a cooling medium in another enclosed chamber (cool air plus water spray or mist). At the third stage, the molds are loaded and unloaded. Figure VI-2.1 shows typical automatic-indexing roto-casting equipment. Table VI-2.2 presents a compilation of engineering data descriptive of several sizes of automatic rotational molding equipment.

Auto-Batch Equipment. The auto-batch method is less costly than the rotary automatic because it is manually operated. The charged mold is rolled into a multi-station oven, rotated and heated, and then manually transferred to a non-confined cooling station.

Shuttle-Type Equipment. In shuttle-type equipment, the molds are supported from the sides and not cantilevered as in the continuous automatic machinery. The molds shuttle from sation to station along a continuous track, sequentially being loaded, heated, cooled, and taken out.

Jacketed-Mold Equipment. With jacketed-mold equipment, heating and cooling are accomplished by circulating thermally suitable liquids through a jacketed chamber fabricated into the mold wall. The mold remains in situ, rotating and tumbling on permanent stanchions.

Molds. Cast aluminum with a nominal wall thickness of ½ in. are most widely used for combined lightness of weight and high rate of thermal conductivity. However, several other metals such as cast iron, stainless steel, and electroplated or chromium plated machined stock may be used.

The main requirements for a set of molds are: (1) they should be as uniform as possible in wall thickness in order to provide a condition of uniform heat flow to the plastic-glass charges in the mold during the fusion cycle, (2) they should have designed-in means for rapid and dependable clamping or closure, (3) they should be provided with an outer

Table VI-2.2 Engineering Data for Typical Automatic Rotational Molding Equipment (All 3-Arm Units Unless Otherwise Indicated)

	Machine Sizes				
Maximum weight supported by arm in lb./sperical diam. mold swing in in.	500/48	800/64	1500/88	1700/110	5000/204
Maximum mold area, in. (diam. × ht.)	32 × 34	50 × 46	72 × 66	92 × 81	180 × 150
Outer shaft speed in rpm	7 to 35	0 to 32	0 to 32	0 to 32	2.6 to 15.6
Inner shaft speed in rpm	6 to 15	0 to 12	0 to 12	0 to 12	1.7 to 10.2
Maximum heat release of oven Btu/hr	600,000	1,000,000	2,000,000	4,000,000	8,000,000
Maximum internal oven temp., °F	900	900	900	900	900
Mold cooling method	Air/water	Air/water	Air/water	Air/water	Air/water
Cooling water consumption, gal/min	0.8	1.0	1.6	2.5	5.0
Space requirements, ft, (l × w × h)	20 × 14 × 10 (two arm)	19 × 22 × 12	30 × 25 × 12	39 × 30 × 14	52 × 75 × 20
Est. total cost of operation, $/hr	0.68	1.24	1.93	3.36	6.25
Approx. basic machine cost, $	20,000	31,000	75,000	88,000	120,000

Source: Technical data bulletins, McNeil Akron McNeil Corp., Akron, Ohio.

structure to prevent contact or damage, and (4) the mold material should have highest possible thermal conductivity to accommodate rapid heat-up and cool-down cycles.

The interior surfaces of the molds should be non-porous and should be castable or machinable for a smooth finish in order to adequately accept mold release and prevent part sticking. The interior of the molds may be embossed for decorative effects, or tooled to extremely complicated designs. Other surfaces or configurations may be incorporated to produce many different types of molded articles.

Vents in the form of Teflon tubes are frequently used and are provided for in the mold design. The vents normally extend far enough in between the mold halves at the mold juncture or parting line to extend inside the inner wall of the part. The purpose of the vents is to alleviate any pressure build-up and permit the escape of any volatiles generated. Other baffles of a natural release material (Teflon, for example) may be used as desired to prevent the plastic material from being roto-cast onto any surface, such as an open end, top, or side of a part to be molded.

MOLDING METHOD/PROCESSING DETAILS

The steps in the roto-casting method comprise charging, rotating in a heated chamber or oven, chilling while rotating under air blast or water fog or mist, unloading, recharging and recycling, part finishing and shipping.

Materials Handling. Plastic and reinforcement may be received either in bulk or in smaller separate containers and stationed near the loading platform. If pre-blending is required, precautions against separation during handling should be taken. Mold charging may be carried out either after weighing or by simple volume measurement of the ingredients. Weighing is preferred for uniformity, especially if the glass reinforcement and resin are added separately. Plastisols (polymers blended into plasticizers) may be treated similarly to the manner in which the powders are handled except that they are charged in the liquid state.

Charging. A pre-production mold release should be baked onto the mold surface. It should be renewed on a weekly basis. Zinc stearate may also be added (1 to 1½%) during charging for release during individual molding cycles.

The amount of the charge (plastic plus glass reinforcement) is naturally dictated by the part thickness required. Both experience and customer specifications eliminate guesswork, and determine the best economics and maintenance of quality.

All manner of clamping devices are utilized for mold closure.

Heating. After loading the charge, and clamping the molds, the mold arm is indexed to the heating chamber or oven. With controls provided, the heating time cycle may be varied. With three-stage equipment, however, it is best to allow for equal time at all stations. Heating is usually carried out at the shortest time and fastest cycle commensurate with good operation and part quality. Typical production heating cycles are presented in Table VI-2.3.

If a slower or intermediate heating cycle is utilized, the lower production rate is compensated for by reduced labor and more acceptable part quality. At the start of a campaign, shift, or day's production, molds should be heated to 150°F prior to start-up. Rotation and tumbling during the heating cycle may be programmed. Generally a 1:1 ratio of rotation to tumbling is desirable. Coordination of temperature and rotation should be carefully planned to prevent non-compatibility of or protrusion of the chopped glass fiber lengths from the inside surface of the part.

Cooling. In the most versatile equipment, air cooling and water fog spray may be programmed as desired. The real criterion is the volume or weight of the part. It is necessary to cool the molds to set and sinter the roto-cast material and to shorten the takeout cycle.

Table VI-2.3 Typical Production Heating Cycles for Rotational Molding of Thermoplastic RP/C

Oven Temperature, °F	Cycle Time, Min.	Probable Temp. Inside Mold, °F
550	10	400 or less
600	8	450 to 500
750	5 to 6	580

Note: In newer equipment, the temperatures in an air-heated oven reach 900°F within 8 sec after closing the door.

The rotation to tumbling ratio is generally 1:2 in the cooling chamber. The polyethylene part temperature should be approximately 120 to 150°F upon completion of the cooling cycle. Plastisol formulations are slightly hotter out of the mold.

Part Removal. After cooling and unclamping the mold, part removal is favored by natural shrinkage away from the mold surfaces. Sticking is encountered when the mold-release coating wears, as at the end of a week's production. Sometimes sticking is associated with cross-linked, high temperature resins. Many symmetrical parts are molded double and are separated at the mold-join to form two identical parts. Blown high-pressure air, mild hammering, shaking, and so on are all employed to hasten mold break-away. A minimum of cleanup is required. Extra or standby molds for any specific part should always be on hand to prevent time delays in the event that a mold is damaged in use, or if other difficulties occur such as dirtiness, sticking or parts broken in the mold, or surface damage.

Process Variables. Several technical parameters for roto-cast thermoplastics RP/C including melt index, viscosity, resin density, and fiber length and their relationship to the process are discussed below:

Melt Index, Viscosity, and Density of the Resin. In order to gain the best benefit from the reinforcement, it is desirable to completely encapsulate the fiber glass during the roto-molding cycle. For a given glass length, resins with a higher melt index (6 to 15) and lower density (polyethylene) generally possess lower viscosity, and exhibit higher flow. They therefore have greater glass-encapsulating ability.

Resins with a lower melt index and higher density do not behave in this manner. The results in molding are protrusion of fibers from the inner mold surface, naturally detracting from the power of the glass to reinforce the part because of non-uniform or inhibited glass distribution.

In the case of plastisols, the lower initial formulated viscosity would favor early glass wet-out and thorough, uniform encapsulation.

Glass Loading and Variation in Fiber Length. When glass fiber is used, the roto-casting mold-

Table VI-2.4 Appearance of Inner Surface of Roto-cast RP/C with Variation in Fiber length and Loading

Chopped Fiber Length, in.	Glass Loading		
	5%	10%	15%
1/32 (milled)[a]	Smooth in inner molded surface	Smooth	Smooth
1/8	Smooth	Smooth	Surface rippled
1/4	Smooth	Surface rippled	Extreme fiber protrusion

Source: Sowa, M. W., SPE Journal, 26 (July 1970), 31.
[a]Shorter-length free-flowing milled fibers may be incorporated at percentages higher then 15%.

ing cycle is 1 or 2 min. longer than when plastic is used alone. With a high-melt-index/low-density resin, glass lengths of $\frac{1}{32}$, $\frac{1}{8}$, and $\frac{1}{4}$ in. will yield the results listed in Table VI-2.4 for variations in the level of glass loading.

In intricately-shaped molds, complete fill-out, such as the depths of a narrow groove, tends to be inhibited by longer fiber lengths. It is practical to inject a small added charge of resin by itself toward the end of the heating cycle so that the inner surface is sealed, and no glass fibers protrude.

Quality Control. Rotationally cast parts that are properly molded should have a slight pebbly texture in the plastic on the inner surface. A shiny inner molded surface indicates excessive heating, and excessively high warpage, shrinkage and part embrittlement may result. Also, parts chilled too rapidly tend to become brittle. The best assurance for a high rate of part selection and acceptance is a capable, well trained staff and crew. Rotational molding is not a pure science. Hour-to-hour and day-to-day variables exist in the equipment and its operation. Because of the long cycles, it is necessary to keep personnel to a minimum. Therefore, all operators on the job must be able to perform many important functions.

Plant Lay-Out. The plant layout for a roto-molding operation depends upon the part and hence machine size and scale of the operation contemplated. Various equipment companies can be consulted for assistance in plant design and lay-out.

PRODUCT DATA

Product Properties. As is true for other branches of RP/C, the properties of thermoplastics are generally improved by higher glass levels and longer fiber lengths. Higher impact strengths would greatly expand the use of roto-cast RP/C parts, but unfortunately, the addition of glass fiber reinforcement has not yet contributed any substantial improvement in impact strength values. The use of higher modulus thermoset resins in roto-casting promises to upgrade somewhat the physical properties of the molded parts. Table VI-2.5 shows reinforced roto-cast part properties compared to those of the unreinforced resin.

Product Applications. The chief benefits of incorporating glass fiber into thermoplastics are the improvement of properties and better engineered performance. Figure VI-2.2 illustrates a grouping of products manufactured by one company in which glass levels of $\frac{1}{4}$ in. fiber of 3 to 6% were used. Figure VI-2.3 presents a group of products in which glass reinforcing levels up to 25% were utilized.[3]

For standards and codes, the selection of glass reinforced roto-cast thermoplastic parts is still based on individual properties and specific property requirements. As the development of this phase of the industry proceeds, greater knowledge will become available concerning rotomolded RP/C part properties and performance, and ease of molding, this method will take its rightfully deserved place among the other "greats" in reinforced plastics.[4]

Figure VI-2.2   Below-grade service junction enclosure, statuary, and machine shroud for paint equipment all in glass reinforced rotationally cast thermoplastic resin. (Photo courtesy of Sherwood Plastics, Inc.)

Table VI-2.5   Comparative Properties of Unreinforced and Reinforced Rotationally Molded Plastic Materials

Property	Low-Density Polyethylene	Glass-Reinforced LDPE (10% 1/8 in. chopped strand)	High-Density Polyethylene
Density (specific gravity)	0.919	0.990	0.951
Tear resistance	555	660	500
Tensile strength	1,540	1,800	3,150
Ultimate elongation	195	20	41
Yield strength	1,260	1,700	3,030
Flexural strength	1,340	2,300	4,520
Flexural modulus	34,400	77,000	145,000
Secant modulus	28,000	53,800	98,300
Izod notched impact, ft/lb	No break	6.6	65
Heat-distortion temp. °F	89	103	155–182
Cost (cents per lb)	21–26	22–27	25–30
Approximate cycle times for typical parts molded at the same temperatures (min.)	7	8	10
Reject rate	Low	Low	High because of warpage and lack of fill-out

Source: Plastic World, June 1970, p 72–73 (editorial staff).

Figure VI-2.3 Fiber glass reinforced rotationally cast containers and tanks for use in the chemical processing and food production industries. (Photo courtesy of Resco Division, Interplastic Corp.)

References

1. *Plastics Design and Processing,* July and Aug. 1969, Wittnam, C. A., and Nickerson, J. A.
2. "How to Plan a Rotational Molding Facility," *Plastics Technology,* Jan. 1972, p. 19, Ramazzotti, D. J.

Also, personal communication, E. B. Blue, E. B. Blue Co., Norwalk, Conn., Apr. 20, 1972.
3. Personal Communication, B. W. Yutrzenka, Resco Division, Interplastic Corp., Minneapolis, Minn., Oct. 20, 1970.
4. *Modern Plastics,* Oct. 1970, p. 86.

VI-3

Cold-Forming of Reinforced Thermoplastics

INTRODUCTION

The process of cold forming should be considered in contradistinction to the well-known process of cold molding and its variations.[1] In cold molding, powdered materials including an adhesive, a polymerizable resin, fillers, and a solvent are first mixed and homogenized, then pressed into shape in cold pressure dies, and finally baked in an oven to cure to a hard infusible mass. The cold molding process for plastics involving latent cure corresponds generally to the dry pressing of wall and floor tile or to the blending of asbestos and cement in the ceramic industry.

In cold forming, a prepared reinforced thermoplastic sheet material is cut to shape, heated to softening temperature, and quickly placed into a cold die set in a compression or stamping press. The press is cycled shut, held for a few seconds to congeal the thermoplastic material, and then opened for part removal and re-loading. By the use of metal stamping equipment, extremely rapid molding cycles are possible.

Cold forming of RP/C actually evolved as a result of the initial incorporation of reinforcements into thermoplastic molding materials. It is naturally related to the injection molding of thermoplastics where hot, fluid material fills and conforms to the shape of colder dies. In cold forming, however, there is less movement of the material, and longer reinforcing fibers may therefore be incorporated to produce structurally stronger parts.

MATERIALS

The materials used in cold forming consist of thermoplastic resins such as phenoxy (formerly used), polyvinyl chloride, styrene-acrylonitrile, high-density polyethylene,[2] and others.[3] Requirements for processing dictate that the glass reinforcing materials used possess high integrity and compatibility with the resin systems with which they are to be used. Fiber lengths are 2 in. and over in length and are discontinuous. The glass content of the prepared sheet may be as high as 40%. The moldable sheet may be supplied in 4 ft x 8 ft blanks and may be cut to moldable size. Even continuous-roll sheets have been contemplated.

EQUIPMENT AND TOOLING

Two categories of equipment are necessary in cold forming: that required for original sheet formation, and more important to this discussion, that required for heating and fabricating the reinforced thermoplastic sheet into molded articles.

Sheet Formation and Preparation. Sheet formation of cold forming RTP sheet is generally a proprietary process. The methods used include: (1) alternating, and ultimately fusing and compressing layers of thermoplastic sheet or film and glass fiber mat, preformed glass, or chopped roving, (2) mixing resin powder and chopped glass fiber, fusing and rolling into sheet form, and (3) impregnating glass reinforcement with hot-melt resins, solutions or latices and emulsions of the desired polymer, and drying to form a sheet.[4,5]

Preparation or Blanking. The cold, moldable RTP sheet may be prepared for molding by cutting or otherwise properly shaping into blanks of suitable size for feeding to the press. Methods or equipment include metal shears, hand-machine cutting, or stamping die in a punch press (preferred for small or irregular shapes).

Equipment for Molded-Part Fabrication.

Ovens. Heating facilities must be capable of heating the RTP charge to 400, 450, or 500°F in as short a time as possible. Radiant electric infrared-producing elements such as sheathed coils, silicon-carbide radiant resistance elements, or microwave and radio-frequency heating units may be used.

It is necessary for the heat source to decay rapidly when turned off (within 2 sec) to prevent overheating. A gridded wire frame on rollers is required to support the charge through the heating chamber. The performance of applicable types of heating units is shown in Table VI-3.1.

Vacuum-Transfer Equipment. Cold, unheated RTP sheet may be transferred either manually of by using mechanical aids such as stamping, vibrating, clamping, sliding down incline, pushing, or transferring by vacuum, all of which are employed in sheet-metal transfer. When heated RTP sheet must be transferred from the heating frame to the cold press for forming, a low-conductivity low-heat-absorbing fixture on the end of a tube is preferred. Molded

Table VI-3.1 Performance of Heating Units for Cold Forming Sheet

Type of Lamps in Oven	Come-up Time to 450°F, sec
Super-high-density tungsten-quartz, 50 kw per sq ft	10
High-density tungsten-quartz, 25 kw per sq ft	26
Low-density tungsten-quartz, 4.5 kw per sq ft	58
Low-density nichrome-quartz	200

phenolic (sp ht = 0.38 to 0.42 cal per g) is one such suitable material. A vacuum cup end of 1 sq in. in diameter will support 1.0 to 1.3 lb of material at a vacuum of 12 to 13 lb.

Molds. Because it is desirable to produce parts requiring no trimming, dies which telescope at least ¼ in. with a 0.001 to 0.002 in. clearance on all sides are used. Some conforming or free-press dies without telescoping sides are used, but the parts produced require trimming. Another type of mold utilized is the kirksite or tool-steel type. Aluminum molds are applicable for prototype production, but are not capable of holding close-tolerance closure. Epoxy or metal-epoxy prototype stamping dies are also used.

Production dies must be cored for cooling. Three possible means of thermal treatment exist: (1) maintain the ambient temperature with the mold used as a heat-sink (no cooling water, (2) flow sufficient water through the dies to maintain a temperature of 70 to 80°F (room temperature) during the production cycling, (3) circulate chilled 50°F water through the dies. This latter procedure probably favors the most rapid and uniformly reliable production conditions.

Press Equipment. Conventional metal-stamping presses are preferred for the production of cold formed RTP items. The press must be slightly altered to permit dwell for approximately 6 to 15 sec. at the bottom of the stroke. This assures complete rigidizing of the material in the mold. If the process cycle is carried out properly, as in the following description, excessive clutch and brake wear are not problematical:

- To hold pressure on the charge long enough to solidify the RTP material, the press must stop at "bottom dead center (bdc)" for the 6 to 15 sec. dwell time.
- To accomplish this, the material acts as a shock absorber as it is contacted by the plunger, and then formed, compressed, and increased in area to fill out the mold cavity.
- As the press reaches bdc, the clutch must disengage the flywheel to remove the source of power.
- The brake then engages, holding the press ram stationary in the bdc position.
- The RTP material shrinks 30% in the thick-

ness direction upon cooling, thereby relinquishing the forming pressure and permitting the press ram to re-start readily and without excessive clutch wear on the flywheel face.
- It is possible to stop the press when bottoming out within a ¼ in. lateral range each cycle. This represents a total variation in part thickness of only 0.0011 in.

With this method, cycles up to six parts per min. are possible and pressures up to 1000 psi are developed. The presses used may vary in size from 75 to 2000 tons. A ram travel of 600 in. per min. is desirable.[6]

MOLDING METHOD/PROCESSING DETAILS

The basic procedure for pressing or stamping structural items using fiber glass reinforced RTP sheet comprises the following steps:
1. Cut cold sheet to required size or sizes.
2. Pre-heat.
3. Transfer to stamping die with an absolute minimum of time out of the pre-heating oven prior to pressing.
4. Operate the press through the required single molding cycle.
5. Remove part from mold.
6. De-bur, finish, inspect and pack.
7. Repeat the cycle.

Several adaptations and variations exist in the method of handling the moldable sheet and in the type of mold used.

Short-Flow Molding. The blank is cut to a peripheral size ¼ to ⅜ in. less all around than the inner mold extremities. During stamping, the plastic material fills out the complete mold and moves the reinforcement with it without separation of the resin and glass. With the telescoping edges of the mold, trimless and flash-free parts are produced. The best results for in-mold laminating of overlays and cover coatings are obtained using short-flow stamping. Blanks may be stamped or pierced in the mold, or holes may be stamped or punched into the cold sheet prior to pre-heating and molding.

Drape Forming. A heated single-thickness sheet of material for cold forming may be draped over a die designed with a spring-loaded clamping ring which holds the sheet in position as a restraint while the plunger or plug descends. This ring should be made using a non-heat-absorbing material such as wood or cured phenolic resin. For shallow draws, the restraining ring may not be necessary, but it is advantageous for preventing wrinkling. A disadvantage to drape forming is that parts produced have a copious flash and must be trimmed to size after molding or stamping. However, this method permits dies originally built and intended for metal stamping to be used for RTP cold forming.

Matched-Die Molding. Optimum molding conditions for an RTP sheet of moldable fiber glass reinforced PVC fabricated to 0.065 in. thick in a matched-die compression press were found to be 350°F and 750 psi.[7] The material flows properly to fill out the mold and provides an excellent surface, but the molding cycle is undesirably longer than possible in the stamping or cold forming presses. The sheet of reinforced PVC was also found moldable by cold forming methods in a stamping press at low pressure (70 to 140 psi) when the starting or heating temperature was increased to 375°F and the pressing time was extended by 5 or 6 sec.

Center-Flow Forming. In center-flow forming, one or more 1 in. wide extra strips of blank moldable sheet are used. Cutting the charge in half and overlapping are also possibilities. The extra strips are placed across the center of the main charge. Proper cold forming of parts with varying or differential thickness is favored in this method. The center strips result in less air entrapment. They also adjust or minimize distant flow into bosses or thicker sections and lower the resin viscosity, thereby more thoroughly wetting the glass fibers. Moreover, the strips improve the part surface finish by shear-type traverse across the die surfaces. The extra strips should preferably be heated separately and then be combined at the press.

When deeper and larger bosses must be designed into the mold, they may be more readily filled by increasing the pre-heat temperature, by adding a shaped slug of RTP sheet to the blank during pre-heating, or by adding a cold slug into the die at stamping.

Overlays. Coatings or desirable decorative or protective media may be permanently adhered to or surfaced onto the RTP material during cold forming by treating the material as an overlay and permitting it to be stamped directly with the heated charge. In most cases, the heated material will adhere to the overlay without separate bonding.

Embossed or decorative vinyl trim (some fabric-backed), some metal fabrics, foams, vinyl sheeting plus foam, polytetrafluorethylene high-weathering film, and other surfacings or membranes may be used. Surfaces of the mold may also be embossed to contain decorative effects which then become transferred to the finished molding. Many cold formed parts, especially those for automotive applications, may be painted.

Sandwich Construction of RTP Skins with a Thermosetting Epoxy Core. One process variation described, strictly developmental at this stage, is geared for improved properties and high speed production in RTP cold forming. Skins of 0.010 in. PVC backed with 0.003 in aluminum foil surrounding an 0.080 in. fiber glass reinforced epoxy core form the moldable sheet. An adhesive was used to bond the aluminum to the PVC. One particular sample box shape was made by this process and thoroughly evaluated. The sheet material was cut and stamped in a cold press. The shaped blanks were post-heated from 290 to 380°F for 12 to 30 min. in an oven to cure the epoxy resin and set the vinyl films. Springback and non-conformity were problematical after cold stamping. Resistance to elastic deformation made retention of the stamped shape difficult. Properties and part density, however, were enhanced by post-curing in an oven under 20 or more psi.

Miscellaneous Cold Forming Methods. The following cold forming methods have been used to fabricate end-use articles using only RTP sheet.

Hand Lay-up:
- Two workable methods are hand lay-up and bag-molding of a glass mat staturated with thermoplastic resin in a solvent solution. Wet-out is followed by drying and vacuum-bag autoclave molding at 500 psi under steam heat for 10 min. Exterior automotive body components have been produced in this manner.

- Flat-Sheet Pressing. Much of the original RTP moldable sheet intended for re-heating and stamping could be cured as-received for flat sheet structural-type products. Control of the resin and thorough wet-out of the glass fiber could result in highly-acceptable flat or corrugated translucent industrial or decorative panels.
- Bending of Pressed Sheet. The method of bending pressed sheet is similar to metal forming or sheet-metal working in which heat is applied for pliability and then removed.
- Welding of Pressed Sheets. Welds can be made by using hot gas flame or high frequency heating techniques. Straight butt welds are difficult, and overlaps with heating or adhesive bonding should be used.
- Roller-Laminating of Pipe. Piping may be fabricated by heating RTP blank strips as they are rolled onto a mandrel with pressure simultaneously applied. The edges of the strips are lapped and not butted.
- Pressing Pultrusion or Channel Parts. Pressing pultrusion or channel parts using RTP sheet for cold forming is accomplished by compression press.

PRODUCT DATA

Advantages. The advantages of cold forming RTP are many:
- Many parts may be formed or stamped per unit time, i.e., from 180 to 360 parts per hour depending upon size. The forming or stamping cycle may be as low as 6 sec. for a well-designed operation.
- Parts may be molded with differential cross-sectional thicknesses or with ribs and bosses in order to provide extra strength for load-bearing parts or areas.
- Cold formed RTP parts are formed in one operation that may combine or supersede fabrication of as many as 4 separate metal stamping parts.
- Cold formed RTP parts weigh only half as much as steel stampings, from 0.65 to 0.70 lb. per sq. ft.
- Large-sized sheets may be lap-jointed to fabricate excessively large moldings and parts.
- Fasteners can be insert-molded, a method that is much less costly than the post-mold-

Table VI-3.2 Physical Properties of Cold Formed RTP Parts Based on Various Plastics

Property	Material and Specification			
	SAN[a]	Polypropylene[c]	PVC[b]	Reinforced PVC–Epoxy–PVC Sandwich, Cold Formed
Flexural strength, psi	32,400	26,900	24,200	13,000
Flexural modulus, psi	1.3×10^6	1.03×10^6	0.88×10^6	1.76×10^6
Wet flexural strength, psi	26,900	24,600	—	—
Wet flexural modulus	1.3×10^6	1.0×10^6	—	—
Flexural wet-strength retention, %	83.0	92.0	—	—
Tensile strength, psi	18,400	16,500	16,100	7,000
Tensile modulus, psi	1.1×10^6	0.86×10^6	0.84×10^6	1.3×10^6
Impact strength, ft lb per in. notch	15.7	11.2	9.2	1.3
Specific gravity	1.38	1.27	1.52	1.93
Heat distortion temp., 264 psi, °F	255	327	201	233
Maximum short-term use temp., °F	200	300	—	—
Coefficient thermal expansion, in. per in. per °F $\times 10^{-5}$	1.3	1.5	1.9	4.55
% glass reinforcement	38	42	28	11%-1/4 in. glass fibers in epoxy core

[a]G.R.T.L. Co.
[b]Data from T. Ishimatsu et al., SPI 25th RP/C Proceedings, (1970), Sec. 6-B.
[c]Data from H. J. Oswald et al., SPI 23rd RP/C Proceedings, (1968), Sec. 6-C.

ing attachment of fasteners and other items to metal stampings.

- The operation of cold forming RTP in limiting dies is essentially scrap-free. In a roll-fed stamping operation, the scrap could be recovered and re-processed.
- For purposes of finishing, parts may be provided a surface treatment by embossing the dies, by molding in decorative cover or overlay fabric, or by painting without surface reworking and without priming (except for polyethylene).
- Inexpensive die materials such as kirksite and cast aluminum are satisfactory and are used to advantage because of their high thermal conductivity.
- Parts may be pierced, punched, trimmed, sheared, blanked, drilled, and so on, after forming or even prior to forming in some cases. Parts may also be joined by adhesive bonding or by welding.

Disadvantages. There are also several disadvantages associated with cold forming RTP sheet:

- The basic material is costly: $0.60 to $0.70 per lb in moldable form prior to processing.
- The original sheet material is difficult to heat and to support while heating. No clamps or double layers may be situated to intercept the heat source, which would cause nonuniform heating of the sheet.
- The material is sticky when hot, difficult to handle, and subject to differential cooling while being transferred from oven to stamping press.
- Complete automation would be difficult.
- Undesirable glass fiber orientation results with any increase in the lateral flow distance to which the material may be subjected in molding. The charge must fill the mold as much as possible and be accurately placed prior to press closing. Resin-glass separation actually occurs if the material is caused to flow too far.
- Shrinkage sink marks occur on molded surfaces opposite bosses and ribs.
- Trim results from certain of the cold form-

ing molding methods and the stock is difficult to reclaim and re-process.

- A variation in fiber length in the original sheet material is required for optimum strength and minimization of resin-glass separation.
- Flow difficulties may be encountered in molding parts with thinner cross sections (0.060 in.), and lowering the glass content to 20% to compensate may be required.[8]
- The molded surfaces are comparable to those which result in some phases of the injection molding of RTP. The molded surfaces are not comparable in quality to those possible using premix BMC or matched-die molded SMC with low-profile thermostat (polyester) resins.

Cold Formed RTP Products. The types of cold formed RTP products produced to date include hidden car and truck parts such as door panels, trim pads, lamp housings and fender liners; chemical gear such as utility and storage tank covers and piping; molded structural shapes such as ells, U's, and so on.

PHYSICAL PROPERTIES

The nominal molded part physical properties for a cross section of four separate types of cold formed RTP structures are summarized in Table VI-3.2.

PRODUCT COSTS

Material cost as received is currently in the $0.60 to $0.75 range. Long range expectations point to a reduction bottoming at $0.50 per lb. This material price is high and results in a high product cost. The cost and slightly longer molding time (compared to sheet metal) are counterbalanced, however, by savings accumulated by eliminating the multiple-stage stamping or drawing operation required for metal parts. Costs are also higher than those for injection molded RTP parts for comparable moldings, and also for premix BMC and matched die SMC parts.

References

1. SPI Plastics Engineering Handbook, New York, Van Nostrand Reinhold, 3rd Edition, 1960, p. 39.

2. Society of Automotive Engineers Congress, Paper #690009, Detroit, Mich., Jan. 3, 1969, Hofer, P. H. et al.

3. SPI 20th RP/C Proceedings, 1965, Sec. 6-G, Wicker, G. L.

4. SPI 22nd RP/C Proceedings, 1967, Sec. 8-A, Gluck, M. L.

5. SPE RP/C Retec Proceedings, Cleveland, O., Oct. 4, 1971, Norwalk, S.

6. SPI 26th RP/C Proceedings, 1971, Sec. 16-E, p. 15, Keown, J. A.

7. SPI 23rd RP/C Proceedings, 1968, Sec. 2-B, Hani, H., and Hiraga, H.

8. SPI 27th RP/C Proceedings, 1972, Sec. 14-F, Saczawa, J. S., and Slayton, J. L.

SECTION VII
UNIQUE COMPOSITE MATERIALS

VII-1

Unique RP/C Materials for Aircraft, Aerospace, and Other Environments

INTRODUCTION

A host of new materials is being generated today which continuously extends the upper environmental performance of RP/C materials supplied to meet the requirements for aerospace, hydrospace, and rigorous commercial applications. These include new or improved resins and fibers, chiefly for resistance to high temperatures, but also including protection against unusual radiation, particle bombardment, extreme pressure and numerous other conditions.

Materials, methods of fabrication, and product performances are discussed in this chapter. Equipment and tooling are generally similar to those required for the more conventional RP/C processes and are adequately described in other chapters. The processes and fabricating techniques, fully discussed herein, are ingenious adaptations as new as the advanced materials themselves.

MATERIALS

This study will treat only those materials (fibers and resins) which are in the vanguard of development, particularly from the standpoint of heat resistance, but also intended to improve laminate elastic moduli. Although costs for these materials are high, economy is secondary to performance and reliability in such fields as aerospace. For this reason, the RP/C fabrication processes used often seem involved and cumbersome. Sintering, plasma sprays, long-cycle molding, multiple-coated prepregs, and other anomalies are not unusual.

High-Temperature Resins. In the race for temperature resistance, resin development has been spurred by the advent of fibers such as graphite and boron. With the exception of ablative plastics which are chosen for their ability to absorb and block energy as they char and decompose, there has been a significant gap in temperature resistance between fiber and matrix. Recent developments promise to close this gap. Some are portrayed in Figure VII-1.1. which shows how the various plastics perform on the basis of temperature and time. The eight zones in the figure include the following materials:*

Zone 1

Acrylic
Cellulose acetate (CA)
Cellulose acetate-butyrate (CAB)
Cellulose acetate propionate (CAP)
Cellulose nitrate (CN)
Cellulose propionate
Polyallomer
Polyethylene, low-density (LDPE)
Polystyrene (PS)
Polyvinyl acetate (PVAC)
Polyvinyl alcohol (PVAL)
Polyvinyl butyral (PVB)
Polyvinyl chloride (PVC)
Styrene-Acrylonitrile (SAN)
Styrene-Butadiene (SBR)
Urea formaldehyde

*D. V. Rosato, *Plastics World* (Mar. 1968), 30, Vol. 26.

Zone 2

Acetal
Acrylonitrile-butadiene-styrene (ABS)
Chlorinated polyether
Ethyl Cellulose (EC)
Ethylene Vinyl acetate copolymer (EVA)
Furan
Ionomer
Phenoxy
Polyamides[1]
Polycarbonate (PC)[1]
Polyethylene, high-density (HDPE)
Polyethylene, cross-linked
Polyethylene terephthalate (PETP)
Polypropylene (PP)[2]
Polyvinylidene chloride
Urethane
Aromatic polyamines
Poly-para-xylylene
Polyaryl ether

Zone 3

Polymonochlorotrifluoroethylene (CTFE)
Vinylidene fluoride

Zone 4

Alkyd
Fluorinated ethylene propylene (FEP)
Melamine-formaldehyde
Phenol-furfural
Polyphenylene oxide (PPO)[3]
Polysulfone
X-917 aromatic condensation polymer[4]

Zone 5

Acrylic (Thermoset)
Diallyl phthalate (DAP)
Epoxy
Phenol-Formaldehyde
Polyester[3]
Polytetrafluorethylene (TFE)
Polybutadiene[3]
Polybutadiene glycol
Polyphenylenesulfide
Polymethylenediphenyl oxide

Zone 6

Parylene
Polysulfone
Polybenzimidazole (PBI)[3,11]
Polyphenylene
Silicone
Polybenzothiazole[5]
Polyaryl sulfone[6]

Zone 7

Polyamide-imide[7]
Polyimide[3,8,9,10,11]
S.A.P.[4]
Polyphenyliquinoxalines[3]
Para-oxybenzoyl polymer[4]
Polyquinoxaline[3,5]
Polyimidazoquinazoline

Zone 8

Plastics now being developed using intrinsically rigid linear macromolecules' principle rather than usual crystallization and cross-linking principles, plus ladder polymers, and possible replacement of carbon by metallic elements in polymer network.[3]

It is noteworthy that as the temperature resistance of the resin increases, fabrication methods become more exotic, specialized, and difficult. The capability of the plastic to be formed with conventional equipment, however, is a prime advantage. A few molding materials can be formed at temperatures significantly below the temperature of their maximum operating resistance. These are referred to as upgraded thermosets. Some of the newer-glass reinforced polyimide resins are good examples. Their success is a result of polymerization by addition rather than by condensation reaction.

Aluminum Phosphate Matrix for High-Temperature Laminates

An inorganic, ceramic matrix may be prepared by reacting orthophosphoric acid with aluminum hydroxide.[12] The resulting paste-like mixture may be used to form a prepreg with glass fiber reinforcement. It is applied by spatula, doctorblade, or other suitable means. This embryo composite is passed through a heated oven zone at 250°F to induce B-staging, after which it possesses a shelf life of over 4 wks at 45°F. B-staged properties include tackiness, high flow, and drapability. Final laminate structures are obtained at 300°F and 100 psi pressure for 1 hr., followed by a post-cure to 600°F at the increasing rate of 100°F per hr.

Reinforcements:

Glass Fiber. Although there is justifiable interest property and performance-wise in the newly-applied fibrous reinforcements with exotic properties in aerospace RP/C, almost all of the molding methods described later are

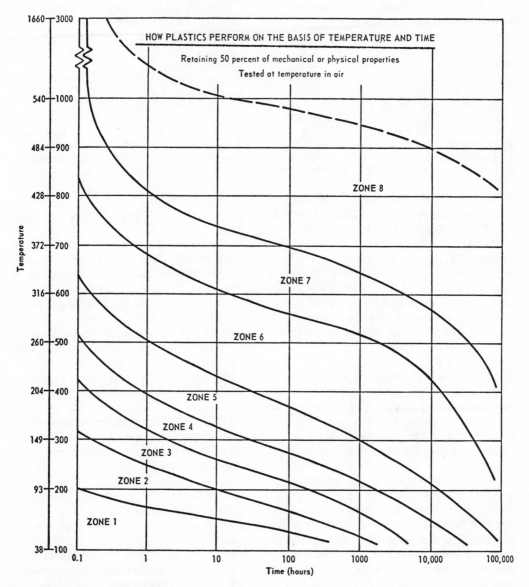

Figure VII-1.1 How plastics perform on the basis of temperature and time. (From D. V. Rosato, *Plastics World* (Mar. 1968), 30.)

basically related to the use of glass fiber. Forms of glass fiber used are rovings, preimpregnated rovings, cloth of all weaves, tapes and many others. Likewise, many new fiberizable glass compositions have been developed: S-glass, D-glass, L-glass, M-glass, and others aimed at approaching 850,000 psi virgin tensile strength, and over 15 million psi tensile modulus. Latest interest and effort is placed upon utilizing the inherent stiffness of glass for greater strain tolerance in laminates for aerospace by forming

composites with glass fibers that approach 4 mils in diameter. Collimated and directed fiber or filament placement methods are used, and laminate performance approaching or superior to those made with the high-cost boron fiber is claimed.[13,14]

Carbon Fiber. Carbon fiber is a fibrous reinforcement for aircraft and aerospace composites which has been converted from a carbon-rich organic precursory fiber such as

Table VII-1.1 Physical Properties of Carbon Fiber

Property	Specification
Specific gravity	1.13
Filament diameter, in.	0.00038
Tensile strength, psi	120,000
Tensile modulus of elasticity, psi	6,000,000
Surface area, sq m per g	130
Crystallinity	Amorphous
Fiber porosity, %	66

Table VII-1.2 Physical Properties of Graphite Fiber

Property	Specification
Specific gravity	1.32
Filament diameter, in.	0.00030
Tensile strength, psi	90,000
Tensile modulus of elasticity, psi	6,000,000
Surface area, sq m per g	less than 4
Porosity, %	66
Crystallinity	Cubic, as microcrystallites
Cost (1970, per lb)	$250

Source: Celanese Corp., New York, N.Y. Technical Bulletin #CAC1A, June 1973.

rayon.[15,16,17,18] It results when rayon continuous filament, cloth or tow is reduction-heated to 700°F and then flash-heated to 2500°F. A 45 to 55% weight loss occurs. Table VII-1.1 presents the physical properties. Carbon fiber is available in cloth of all weave styles plus felts, yarns, and whiskers. Carbon fiber oxidizes readily. Graphite forms, however, show greater promise. Some newer forms of carbon fiber are available with a thermal stability up to 4500°F and good chemical durability.

Graphite Fiber. Graphite fiber is derived from carbon fiber as the result of extra or post-thermal treatment which converts the fiber to the crystalline phase. It results when bulk carburized cloth, tow or other materials are passed through the graphitization stage by reheating to 2000°F for a period of 2 wk, or by flash heating to 5000°F. Direct graphitization has been accomplished on yarns of polybenzimidazole, polyacrylonitrile, aromatic polyamide, and polyoxidiazole.[19] Table VII-1.2 lists the physical properties of graphite fiber. Graphite fiber is available in cloth in many weave styles, felts from random and needled precursors, and also rovings, yarns and whiskers.

Both carbon and graphite fibers oxidize slowly at service temperatures. To reduce or eliminate this characteristic, finely powdered silicon dioxide may be incorporated either by dispersing in the original rayon filament to be heat-treated or by coating the finished carbon or graphite fiber surface. Either SiC[20,2] or silica treating improves the fiber tensile strength, reduces porosity and abradability, and greatly increases resistance to oxidation. Similar treatments with other substances such as nylon[22], boron, hafnium, tantalum and zirconium chlorides, carbides and nitrides are carried out by methods including chemical

vapor plating. These treatments greatly increase the thermal end-point limit of graphite fiber and show potential for strength increases to 350,000 psi with a tensile modulus of elasticity[23,24] of 30,000 psi. However, the interlaminar shear strengths possible with graphite as a reinforcement are low and the accompanying fiber-resin bonding is poor. Hence, graphite fibers and composites are still considered materials of research.[25,26]

Fibers from Silicon Dioxide. Another type of fiber being used comprises filamentary materials for high-performance composite reinforcements consisting primarily of the quartz or high-silica forms of silicon dioxide, SiO_2. Fibers referred to as high-silica are produced by leaching existing glass fiber forms in hot $HCl + H_2SO_4$ to yield a material which is a minimum of 95% silica. The quartz fiber form is 99.5% pure silica and results from drawing fibers directly from a melt of glassy quartz in a process simulating the production of glass fibers. Table VII-1.3 presents the physical properties of silicon-dioxide fibers. The fibers are available in cords and yarn, thread, braided sleeving and tape, woven fabrics, non-woven mats and blankets, and knit fabrics. The quartz (drawn) fibers are slightly more flexible and stronger and are therefore processable into a greater variety of reinforcing forms.

Silica fibers are high in price because of processing limitations and difficulties in maintaining a high quality of fiber from commercial melts at the melting point of silica. The resultant composites have a variety of uses but are limited by the thermal end-point of the silica.

Table VII-1.3 Physical Properties of Silicon-Dioxide Fibers

Property	Specification
Specific gravity	2.2
Filament diameters, in.	0.0004–0.0006
Tensile strength, psi	130,000 (slightly lower at elevated temperatures)
Tensile modulus of elasticity, psi	10,000,000
Melting point, °F	3300
Crystallinity	Both forms amorphous

Boron Fiber. Boron reinforcing fiber consists of a continuous filament of tungsten used as a substrate over which a coating of elemental boron is deposited. The resultant continuous filament is desirable as a reinforcing fiber because of its light weight, high strength, and high modulus of elasticity.[18, 27, 28] Boron fiber is produced by chemical vapor plating over the tungsten filament. Decomposition of boron trichloride supplies the boron, which builds up at the rate of 0.006 in. per min. at 2000°F. Table VII-1.4 lists the physical properties of the fiber. It is produced in basic continuous filaments only.

Hot nitric-acid etching of finished boron filaments is effective in narrowing the spread of tensile-strength values and in upgrading the mean tensile values.[29, 30]

Organic Fibers. The newest entry into the family of specialty fibers finding significant application in advanced composites for aerospace is an organic fiber with properties that make it of interest as a high tensile strength, high modulus material, (source: "Applications of PRD-49 Fibers," Moore, J. W., SPI 27th RP/C Proceedings, 1972, Sec. 17-F, E. I. du Pont Co.) Probably formed from a polymer solution pres-sure-extruded into a chemical coagulating bath in the standard manner for synthetic textile fibers, these are optimized at 0.00045 in. filament diameter (1.4 denier), possess a basic material density of 1.45 g. per cc., and an elongation of 2% (linear stress-strain relationship). The fiber is supplied in strands or yarns containing up to several hundred filaments. Cloth is the preferred form for laminating, but rovings for filament winding are also supplied. The sub-type of this organic fiber recommended for RP/C (other types are used in cables and special fabrics) exhibits 400,000 psi virgin tensile strength with retention of 90% after weaving, thus providing comparatively excellent processability. The tensile modulus of elasticity is 19 million psi. However, an unfavorably low compressive yield strength is characteristic.

The basic fiber has excellent resistance to chemicals and solvents, has a 1.5% moisture regain, is electrically non-conductive, and is flammable but will not melt.

As regards behavior under elevated temperature conditions, there is a loss of fiber tensile strength of 25% at 400°F, and tensile modulus and elongation are respectively lowered and increased by a factor of 0.18. Recommended

Table VII-1.4 Physical Properties of Boron Fiber

Property	Specification
Specific gravity	1.73
Filament diameter, in.	0.004 (diameter of host tungsten filament = 0.0005 in.)
Tensile strength, psi	450,000; range: 200,000–600,000
Tensile modulus of elasticity, psi	60,000,000
Tensile strength at 1200°F	Almost 0 strength
Tensile modulus at 900°F, psi	33,000,000
Crystallinity	Amorphous to fine crystallites at higher coating temperatures
Cost (1973) per lb	$390

maximum temperature for longterm exposure is 465°F (30% tensile loss in 500 hr).

Extremely strong, rigid, and lightweight composites are achievable using this new organic fiber in both directed fiber layup and filament wound structures. Combinations with other reinforcements have been proposed and evaluated as a means to make this type of fiber more applicable. Selling price (1973) is $50 per lb.

Whiskers. Whiskers are a general class of extremely small-diameter, inorganic materials grown as single crystals which possess tensile strengths up to 3.5 million psi and moduli up to 350 million psi. They are produced from a variety of materials, each with a specific use as reinforcement for advanced aerospace composites. Whiskers are grown by any one of four main processes: high stress, chemical reduction of metallic halides, electrolytic deposition, or vapor deposition. Single-crystal formation may be occasioned by either basal or tip growth.[31, 32, 33] Aluminum-oxide (sapphire) single crystals may be grown by passing H_2 gas plus water over molten Al_2O_3 at 2600°F. Whiskers are grown from aluminum oxide, magnesium oxide, aluminum nitride, beryllium oxide, boron carbide, carbon and graphite, silicon carbide (a and b phases), and silicon nitride. Table VII-1.5 presents the physical properties of whiskers. They are available in needles, blades, clusters, wool, fabricated mats and paper, and combinations thereof. The "whiskerizing" process has been employed to deposit silicon-carbide whiskers on quartz, boron, silicon carbide, or tungsten filament substrates at 2500° to 3000°F.[34]

Larger output processes for carbon whiskers should lead to lower cost, higher strength whisker reinforcements.[35] Whiskers in general wet well in many resin systems without binders. Uniformity of distribution and elimination of clustering, balling, and parallelism constitute the main problems in fabricating whisker composites. Corrective measures have been suggested and undertaken, however.[36] Whiskers have also been evaluated as a superior reinforcement to improve metal structures.

Ceramic Refractory Fibers. Inorganic metal oxides provide fibrous materials which answer some of the requirements of the aircraft and aerospace industries in their continuing search for lighter, stronger, more chemically durable, heat-resistant, reinforced structural composites. Ceramic refractory fibers are produced principally by steam-blasting a melted ceramic composition. The accumulated fiber may be processed into a continuous staple fiber.[37] Some varieties are produced by extrusion[38] or by drawing from a melt.[39]

Alumina, alumina-silica, or alumina-silica-chromia fibers are the most widely used types, and form mostly insulation blankets. Fibers are generally amorphous but may develop a crystalline structure after exposure to 3270°F for several hours.

Ceramic refractory fibers are available in felts, batting,[40] continuous filaments (Al_2O_3 with ZrO_2) plus 20% organic fiber to overcome brashness, and as a carrier fiber for weaving. These materials are used to reinforce other ceramic and metal matrix composites.[37, 41]

Beryllium-Wire Reinforcement. Beryllium constitutes another type of continuous filament with the promise of providing a light weight, strong reinforcement material for the production of aerospace composites with these characteristics.[42] Beryllium wire for reinforcement is produced by drawing through successive dies starting with an ingot of ⅜ in. diameter which either may be cast metal or may be formed by powder metallurgy. The die temperatures are set at 750°F., and all beryllium wire is nickel-clad after passing through dies 0.300 in. diameter and below down to the ultimate 0.005 in.

Table VII-1.5 Physical Properties of Whiskers

Property	Specification
Specific gravity range	1.8 (BeO lowest)—3.9 (Al_2O_3 highest)
Filament diameter range, microns	0.2 (Al_2O_3)—30 (Al_2O_3 + AlN mixed)
Tensile strength range, psi	500,000–3,500,000 (MgO whiskers highest in 2.0 micron diameter range)
Tensile modulus of elasticity, psi	45,000,000–350,000,000 (Al_2O_3 highest)
Melting points, °F	3450 to 6500 (MgO highest)

Table VII-1.6 Physical Properties of Beryllium-Wire Reinforcement

Property	Specification
Specific gravity	1.84
Filament diameter	0.005"
Tensile strength	160,000 psi
Tensile modulus of elasticity	38,000,000 psi

Table VII-1.6 lists the physical properties of this reinforcement.

Beryllium wire is made in continuous-strand form. It shows great potential as a high-modulus reinforcement, but no elaborate use in aircraft or aerospace composites has yet been consummated.

Hard-Drawn Steel Wire for Reinforcing. Hard-drawn steel wire has previously been used as continuous-filament reinforcement. It has also been evaluated as a sheet-type reinforcement, i.e., parallel steel wires adhered to a cloth carrier film.[43, 44] In fabrication, steel wires are drawn and laid in a parallel configuration onto a carrier sheet such as paper or glass cloth using polyester or epoxy resin to impregnate. This composite is subsequently B-staged. Table VII-1.7 presents the physical properties of steel wires.

In developing this material, it was intended to provide the capability for designing RP/C structures with a virtually permanent life expectancy.

MOLDING METHODS/PROCESSING DETAILS

There are processes for fabricating composites for aircraft and aerospace which are as unique and inventive as the materials used and as the resultant products. Needless to say, each of the highly developed or well-known methods for fabricating thermosets described so far are applicable. The difference is in the adaptation of special methods or "tricks" to accommodate high-performance reinforcements, to orient for directional design requirements or to fulfill other extraordinary requirements. The following discussion sets forth these unusual methods with the expectation that the reader will be in a better position to apply them for improving products or finding solutions to complicated molding problems.

Hand-Lay-up Adaptations and Variations. The careful placement of pre-cut reinforcement segments in specific patterns provides the desired directional strength.

If pre-impregnated material is used, the material lay-up onto the set contour is assisted by the judicious use of a heat-gun to soften the material and by a roller to force it into place over or around the desired contour and tack it into place before cure.

If wet lay-up is used, the fabric or reinforcement segments are dipped into the resin bath, worked with a squeegee, and pulled into place on the part.

Sheet-Skin Lay-up.[45] In a manner simulating filament winding, fiber glass roving is impregnated through a dip-bath with a restricting or limiting die and wound onto a force-rotated cylindrical form or mandrel. Winding may be either parallel or traversed.

The wet mass is slit transversely, peeled off, and deposited onto a nonporous membrane or film, so that it does not disturb the fiber orientation or bundles.

Strips or patches are cut and applied as wet-lay-up material over a prepared mold or form.

This method provides predetermined dimensional and directional control of the glass filaments in unsymmetrical parts. It also provides excellent drape characteristics by permitting the sheet skin lay-up to be cross-plied or conformed to almost any irregular shape.

Table VII-1.7 Physical Properties of Hard Drawn Steel Wire

Property	Specification
Optimum wire diameter, in.	0.010
Wire tensile strength, psf	400,000 nominal 800,000 possible
Width of composite wire sheet, in.	To 40
Wire content, %	60, 80, 90 possible
Thickness of bonding cloth or sheet, in.	0.002–0.004

Chopped or Macerated Cloth. Either dry or impregnated cloth may be formulated into a molding compound to provide exceptionally high and uniform reinforcement. Reinforcement contents up to 70 or 75% are possible. The resin fraction may be formulated to provide thermal resistance, chemical durability, fire retardancy, high filler content, and so on.

Lap Winding of Prepared Reinforcement. There are essentially four methods of fabricating a high-density structure (heat shields):[46]

1. *Flat lap winding.* This method employs pre-impregnated tape. A heated roll is used to impress the tape against a rotating mandrel as it is turned onto the form. Tension is applied to the tape by a friction brake, and pressure is applied to the heated roll to provide compaction. "Puckers," which weaken the structure, occur if initial pressure has not been consistently applied.

2. *Fabrication or machining from a laminated plate.* A laminate may be formed with the reinforcement oriented at right angles to the ultimate stressing or eroding force. It is then machined to remove the unnecessary material. Although this process was one of the first used, it is obviously inefficient and wasteful.

3. *Edge or oblique winding.* Full-width tape is either edge-wound (perpendicular to the main axis of the part) or oblique wound (plane of the tape is at an angle of approximately 45° to the main axis of the part. The tape may be bias-cut for better conformity or tapered or wedge-shaped in pre-impregnating for adaptation to curvatures.

4. *Convolute lay-up or winding.* For use on a surface of revolution or laid up longitudinally, convolute lay-up involves applying tape or other pre-impregnated reinforcement to alternately intersect the inner and outer surfaces of the structure. Bias tape may also be used for rotational wraps. The structures may be laid up flat or in either the male or female tools. However, it is recommended that subsequent cure be made in a female or cavity tool with pressure exerted onto the inside diameter to accomplish sequential and maximum debulking.[45]

Interface Machining:

- Interface machining is a method for combining two dissimilar components to satisfy the requirements for mechanical interaction depending upon accurate location of the interface.
- In accomplishing the sequence, one material is wrapped onto a mandrel and is subjected to partial cure.
- The first layer is machined to the required contour.
- The second component material is overwrapped, and the two materials are cured together.

Pre-Forming:

- As applied to aircraft and aerospace structures, pre-forming involves combining duplex moldings in which dissimilar materials are pre-molded, handled separately through machining operations, and mated by curing together under pressure.
- One example is graphite and silica nose-cone components, which are fabricated separately in molds at low pressure prior to combining.
- Pre-forming requires precise knowledge and control of the separate material properties. It also demands close adjustment of the process variables such as temperature, pressure, and cycle time so that the preforms are firm enough to be handled, yet will retain an accurately located interface as mated parts at the time of the final cure. The individual components cannot be too far advanced in cure because a strong chemical interlock is required at the interface following the final cure.

Adaptations of Filament Winding:

- Filament winding is frequently used in conjunction with other RP/C fabrication processes in aircraft and aerospace for the purpose of adding its specific benefits to complex assemblies.
- The chief benefit is strength with minimum weight added.
- Filament winding may be used to lock in several other components or types of lay-up.
- The filaments are usually wet-wound, but pre-impregnated fibers may also be used.
- Filament-winding techniques have been successfully used to fabricate a boron B-staged epoxy prepreg tape subsequently used for experimental re-entry vehicles. Boron filaments were 0.004 in. in diameter over tungsten filament, and the prepreg tape was ⅛ in. wide. It was necessary to prevent torquing by applying the tape over a heated

tensioning and delivery roll at right angles to the surface of the mandrel during mandrel wrapping for fabrication of the structures.[47] [48]

Pre-Tensioning Nylon Overwrap:
- Pre-tensioning nylon overwrap satisfies the requirements for pressure application structures to be fabricated by tape wrapping.
- Following the original tape wrap, nylon or other suitable strand material is pre-tensioned by being passed over a friction-type wheel or brake applied over the original winding in a parallel or near-parallel wind. The desired compressive force on the part to be subsequently cured is developed by controlling the number of overwrap plies and the yarn tension.
- Heating for the cure is accomplished in a circulating-air oven.
- The oven temperature causes the nylon overwrap to relax. It is possible to control oven temperature to make the yarn relaxation temperature accurate and predictable without deleterious effect on cure of the structure.
- The pressure can be varied according to the number of plies laid on and the strand pretensioning.
- To establish maximum densification and minimize entrainment of volatiles, membrane bags may be sealed over the part following overwrap, and a vacuum may be drawn during cure.

Encapsulated Short-Fiber Molding Compounds. In order to produce a molding compound utilizing unique fibers for high-performance composites, the following process was devised.[49]
- Boron fibers 0.0013 in. in diameter and ⅛ in. long (tungsten core) are slurried in a water bath with mild agitation. The bath should be in a heater-jacketed kettle.
- Emulsified epoxy resin containing a hardener is added to the water.
- Wetting agents, couplers, interstitial fillers, or other ingredients to control such factors as Zeta-potential are added prior to, curing, or following the resin addition to control agglomeration, fiber-resin wetting, and so forth.
- Agitation of the mix is continued and the temperature of the bath is raised to a point

at which B-staging of the resin occurs. It is then cooled.
- The water is filtered off and the compound is dried. By the nature of the process, almost all the resin is deposited onto the fibers. The natural tendency of the fibers is to collect forming small balls around a resin droplet. By use of the proper wetting agents and other materials, however, fiber-resin bundles may be induced to grow so that the fibers are laid parallel and the desired collimated structure results.
- The resultant compound may be molded using extrusion, transfer, or compression methods.
- Fiber lengths may be varied up to ½ in. Also, various types of fibers may be mixed in the original slurry to provide variations in molded end-product properties.

Use of Interlaminar Graphite Fabric as Heat Source for Curing RP/C Composites. Plain-weave graphite cloth, 0.025 in. thick, made from 0.00033 in. continuous graphite filaments, with a cloth weight of 7.6 oz per sq yd, and specific electrical resistance of 1.3 ohms per lineal in. was laminated into 12, 30, and 54 ply phenolic-glass and polybenzimadazole-glass (prepregs) laminates. It was used as an electrical resistance curing means as follows:[50]
- The graphite cloth was used as the middle ply for the laminate. Two layers were used in the 54 ply laminate.
- One inch of the graphite cloth was peripherally exposed, and brass shims with electrodes were attached to supply power.
- The lay-up was vacuum-bagged with PVA film after the application of an overall release cloth and peripheral bleeder strip.
- Power was sequentially increased from 1 to 3.8 watts per sq in. of laminate over a period of 45 min. to provide an ultimate curing temperature of 277°F (range—264-290°F).
- Ultimate strength values compared favorably with those of identical oven-cure laminates.
- The graphite fabric withstands temperatures of 680°F in air and over 5000°F unexposed. Also, graphite fabrics possess a negative coefficient of electrical resistance with the temperature, so that current surges during warm-up are prevented.

Mixed Reinforcements. The addition of 3% continuous carbon fiber in one direction of a fiber glass woven-roving material was found to increase the ultimate tensile modulus from 2.6 $\times$ 10^6 to 4.0 $\times$ 10^6 psi at less than 0.5% strain.[51] At 0.5% strain, complete failure of the carbon fiber occurred, but it was not catastrophic and the load was transferred to the fiber glass. The stress-strain relationship continued then as in a 100% glass laminate.

Carbon-fiber laminates are virtually unaffected by water. However, carbon acts as a conductor and permits the formation of cells of metallic material. With sea water as the electrolyte, therefore, severe corrosion occurs in these laminates.

Techniques for Fabrication of Fiber-Metal Composites. The question of whether or not a discussion of fiber-metal composites belongs in a handbook on reinforced plastics may be answered by reiterating the fact that the fiber-metal structures have been developed hand-in-hand with the advanced fiber-plastic composites and, in the main, enhance or complement the use of the plastic counterparts.

The following methods of metal fabrication have been pressed into service for the purpose of fabricating composites using metallic or nonmetallic fiber, flake, whiskers, or coatings:[44, 52,53,54,55,56,57] powder metallurgy techniques, pneumatic impaction, plasma-spray deposition, vapor disposition, electroforming, vacuum infiltration by molten metal, casting, crystal growth from a melt, co-extrusion, hot or cold rolling, and diffusion bonding.[53]

PRODUCT DATA

The various extreme environments to which aerospace and unique RP/C structures and components are subjected in service are treated in the following discussion, along with representative examples for each area of application.[58, 59, 60, 61, 62]

Mechanical—Structural Requirements. A great lag naturally exists between research and end-product application. This is pointed up critically in aerospace and high-performance RP/C parts. Where people's lives and safety are involved and where millions of dollars go into a single aircraft, absolutely reliable design data must be developed and the performance of fabricated parts proven by exhaustive testing prior to setting up anything like assembly-line production.

Glass-epoxy and boron-epoxy are the only RP/C materials accepted for main structural elements, with boron-epoxy being used in the primary (main load bearing) structures because of its acceptably higher modulus of elasticity and superior fiber-resin bonding for high interlaminar shear.

Specialty, high-performance RP/C structures are useful in aircraft design where the requirement is always to reduce the overall payload by using lighter weight, stronger, stiffer materials of construction. In high-velocity space travel, the impact of meteorites is an ever-present hazard. Multilayer sandwich configurations and foamed-plastic interlayers between RP/C skins offer the best protection against penetration and general damage from high-velocity impact.[11] The advantages of self-sealing and the facility of repair are also of interest. In addition, the resistance of RP/C vehicles to damage in travel over rough terrain makes the material desirable for critical non-air use.

Almost all aerospace and unique composites are fabricated using pre-impregnated fiber-resin stock in sandwich construction. Exhaustive testing of composite properties has been carried out by the military and is reported in MIL-HDBK-17A* Summaries of the mechanical properties of the various typical reinforcement-matrix combinations tested are reprinted in Tables VII-1.8 to VII-1.13.[63] An excellent discussion of the design and analysis of composites and of the stress-strain curves backing up the data summaries is presented in the original MIL-HDBK-17A handbook.

Testing of fabricated prototype parts follows the development of the original test data. When the minimum number of tests has been completed for a material, its properties are assigned values on a B-basis, i.e., the value above which 90% of the population of values is expected to fall within a confidence limit of 95%. Stresses developed in fabricated parts must therefore be related to the ultimate strengths of the individual materials.

It is of interest to itemize briefly the structural end-product applications for which many of the RP/C materials have been adapted.

Semi-and Non-Structural Applications (Secondary Structures) Radomes. Radomes constitute the oldest and one of the original critical

*(Sept. 1970).

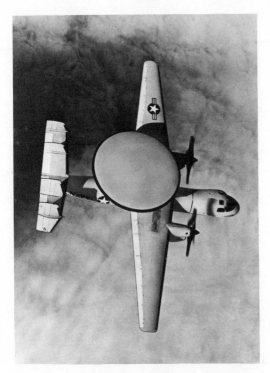

Figure VII-1.2 Rotating radome ("rotodome") of RP/C mounted atop Grumman Navy E-2A "Hawkeye" search plane. The radome is 24 feet in diameter and 30 inches thick. The four vertical tails in the empennage of this aircraft were manufactured of fiber glass RP/C to achieve greater radar transparency as well as structural integrity. (Photo courtesy of U.S. Naval Air Command and Grumman Aircraft Engineering Corp.)

uses for fiber glass: polyester or epoxy sandwich constructions (see Figure VII-1.2).

Helicopter Blades and Wing Tips. Components fabricated using sandwich construction are now in full use as the leading edge or face of blades in craft such as the Navy "Dash" and CH-47A "Chinook" helicopters, and also on wing-tip sections for several aircraft. The construction is fiber glass-epoxy.

Missile Fusilage and Pylon Tanks. External pylon tanks for the A-70 high performance aircraft and the missile fusilages for wing-mounted pylon missles are fabricated using boron-epoxy.

Vertical Tail Fins. Fiber glass-epoxy sandwich construction has been used to fabricate the tail pod of the Navy-P-3 submarine search plane. The vertical tail-fins of the Grumman E-2A Navy search plane are of similar construction, as is the rotodome illustrated in Figure VII-1.2. Boron-epoxy construction is used in fabricating the vertical rudders of the McDonnel Air-Force F-4 service aircraft.

Load-Bearing (Primary) Structural Aircraft Applications: Empennage Structures. The entire tail assembly of the Grumman Navy F-14 and the McDonnel Air Force F-15 are fabricated using collimated boron-epoxy prepreg tape. The full-depth sandwich thickness of the F-15 is 8 in. Boron-epoxy is also approved for the horizontal stabilizer of the General-Dynamics Air Force F-111 but has not superseded the originally used and more economical aluminum.

Leading-Edge Slabs. More than 50 leading-edge slabs for the huge C-5A Air Force transport plane have been fabricated using boron-epoxy.

Present prognostications are that the use of boron-epoxy composites for aircraft will grow. Complete primary wing structures and also unitized RP/C aircraft fusilages are under development. Other structural elements for aircraft and aerospace include small space-craft housings, small high-gain antennas, cryogenic tankage, missile-support systems, and gas-turbine housings.[53]

Temperature. The temperature requirements of RP/C in aircraft and aerospace are critical and may be summarized as follows:

Plane Take-Off. Plane take-off may induce temperature changes of $+100°F$ to $-80°F$ in a very few minutes.[64] Air speed also develops temperature because of the friction involved. It is estimated that the supersonic transport plane now in the planning stages will produce friction temperatures above the boiling point of water during high-speed travel in the atmosphere.

Planetary Entry or Re-entry. In planetary entry or re-entry, the temperature may be immediately increased to $+10,000°F$ and last several seconds until slow-down. Ablative structures have been developed to withstand this condition. During burning, a resin char layer develops which both insulates and structurally maintains its integrity for periods well over the

Table VII-1.8 Summary of Mechanical Properties of a Typical Fiber Glass Epoxy Composite

Fabrication	Layup: Parallel Balanced	Vacuum: 14 psi	Pressure: 35-50 psi	Bleedout: Pinched Edge	Cure: 1 hr/250-265F	Postcure: None	Plies: 8

Physical properties	Weight Percent Resin: 34.4	Avg. Specific Gravity: 1.89	Avg. Percent Voids: 0.5	Avg. Thickness: 0.086 inches

Test methods	Tension: ASTM-D638 Type 1	Compression: MIL-HDBK-17	Shear: Rail	Flexure: ASTM-D790	Bearing: ASTM-D953	Interlaminar Shear: Short Beam

Temperature		-65F				75F				160F			
Condition		Dry		Wet		Dry		Wet		Dry		Wet	
		Avg	SD	Avg	SD	Avg	SD	Avg	SD	Avg	SD	Avg	SD
Tension													
ultimate stress, ksi	0°	115.1	6.70	75.8	2.83	93.6	3.33	71.4	1.63	57.3	3.93	41.8	2.07
	90°	11.23	0.64	8.54	0.29	10.76	0.27	7.14	0.34	6.64	0.34	4.21	0.91
ultimate strain, %	0°	2.44	0.25	1.66	0.14	2.02	0.10	1.52	0.10	1.41	0.11	0.96	0.10
	90°	2.07	0.29	1.66	0.11	1.31	0.17	1.23	0.13	2.19	0.16	1.09	0.24
proportional limit, ksi	0°	79.0		67.3		86.9		58.0		50.2		37.6	
	90°	7.5		4.2		4.8		4.1		4.2		3.9	
initial modulus, 10⁶ psi	0°	4.85		4.56		4.66		4.72		4.2		4.61	
	90°	1.84		1.18		1.14		0.88		0.36		0.40	
secondary modulus, 10⁶ psi	0°												
	90°												
Compression													
ultimate stress, ksi	0°	82.6	4.77	57.1	3.60	50.4	4.17	34.4	3.07	17.0	1.60	9.01	0.83
	90°	37.8	1.60	25.6	1.20	25.2	0.83	17.9	0.47	7.05	0.96	3.30	0.34
ultimate strain, %	0°	1.88	0.11	1.27	0.11	1.29	0.14	0.91	0.11	0.62	0.09	0.51	0.11
	90°	2.42	0.20	2.60	0.19	3.06	0.17	2.61	0.12	1.87	0.30	1.22	0.12
proportional limit, ksi	0°	71.7		52.0		41.9		25.2		13.5			
	90°	20.9		8.6		7.9		6.1		5.2			
initial modulus, 10⁶ psi	0°	4.49		4.54		4.17		3.95		3.30		1.85	
	90°	1.97		1.47		1.43		0.94		0.46		0.31	

		-65F Dry			75F Dry			160F Dry		
		Avg	Max	Min	Avg	Max	Min	Avg	Max	Min
Shear										
ultimate stress, ksi	0°-90° ±45°	13.2			6.7	0.2		2.1		
Flexure										
ultimate stress, ksi	0°	188.7	194.0	185.5	108.1	115.9	99.0	29.3	30.9	26.6
proportional limit, ksi	0°	144.4	152.6	129.7	90.2	101.4	89.2	12.7	17.2	9.0
initial modulus, 10^6 psi	0°	5.39	5.51	5.27	5.07	5.29	4.98	2.64	3.14	2.22
Bearing										
ultimate stress, ksi	0°	64.6	71.0	59.1	37.2	38.1	35.9	12.7	14.7	11.5
stress, at 4% elong., ksi	0°	29.2	44.0	21.0	15.5	18.1	8.2	7.9	9.6	5.7
Interlaminar Shear										
ultimate stress, ksi	0°	11.00	11.22	10.66	7.92	8.26	7.70	3.37	3.60	2.93

Table VII-1.9 Summary of Mechanical Properties of Typical Fiber Glass Phenolic Composite

Fabrication	Layup: Balanced	Vacuum:	Pressure:	Bleedout: Vertical	Cure:	Postcure:	Plies: 8

Physical properties	Weight Percent Resin: 25.3–32.3 · Avg. Specific Gravity: 1.72–1.85 · Avg. Percent Voids: · Avg. Thickness 0.071–0.095 inch
Test Methods	Tension: ASTM-D638 TYPE 1 · Compression: MIL-HDBK-17 · Shear: Rail · Flexure: ASTM-D790 · Bearing: ASTM-D953 · Interlaminar Shear: Short Beam

Temperature		-65F				75F				160F				450F	
Condition		Dry		Wet		Dry		Wet		Dry		Wet		Dry	
		Avg	SD	Avg	SD	Avg	SD	Avg	SD	Avg	SD	Avg	SD	Avg	SD
Tension															
ultimate stress, ksi	0°	48.1	2.4	49.8	3.3	38.9	1.5	37.2	1.8	35.3	1.4	30.6	3.0	21.6	1.6
	90°	37.9	1.8	40.0	2.7	31.5	1.5	32.1	1.4	27.9	1.7	26.2	2.2	21.6	1.7
ultimate strain, %	0°	1.76	0.07	1.76	0.13	1.33	0.14	1.34	0.13	1.19	0.10	1.15	0.14	0.69	0.05
	90°	1.63	0.08	1.65	0.13	1.26	0.15	1.32	0.07	1.11	0.07	1.11	0.14	0.78	0.06
proportional limit, ksi	0°	13.6	0.9	18.1	1.2	13.5	0.6	17.0	1.0	13.9	1.0	14.9	0.70	9.7	1.1
	90°	9.9	0.4	12.5	0.9	9.2	0.8	12.8	0.7	10.3	0.8	11.6	0.70	8.6	0.5
initial modulus, 10^6 psi	0°	3.40	0.21	3.35	0.20	3.94	0.69	3.14	0.26	3.74	0.41	3.01	0.19	3.57	0.24
	90°	3.08	0.29	3.04	0.22	3.54	0.41	2.81	0.24	3.33	0.37	2.78	0.21	3.18	0.30
secondary modulus, 10^6 psi	0°														
	90°														
Compression															
ultimate stress, ksi	0°	66.7	6.2	65.9	5.0	59.7	4.7	54.5	7.1	50.6	2.3	49.2	4.2		
	90°	57.7	5.8	56.2	5.8	49.0	4.6	48.7	4.0	43.0	4.3	42.9	3.7		
ultimate strain, %	0°	1.85	0.09	1.69	0.18	1.58	0.14	1.49	0.12	1.45	0.06	1.40	0.12		
	90°	1.70	0.21	1.63	0.13	1.40	0.09	1.43	0.07	1.37	0.12	1.31	0.15		
proportional limit, ksi	0°	45.8	3.8	38.5	7.9	39.0	2.4	41.2	4.6	39.9	2.4	35.0	1.7		
	90°	35.2	3.8	34.4	5.0	32.6	4.4	35.5	3.0	32.4	3.1	31.1	3.3		
initial modulus, 10^6 psi	0°	3.90	0.19	4.17	0.29	3.95	0.28	3.89	0.26	3.68	0.21	3.67	0.12		
	90°	3.69	0.25	3.68	0.17	3.70	0.20	3.57	0.20	3.30	0.23	3.45	0.21		

		-65F Dry			75F Dry			160F Dry		
		Avg	Max	Min	Avg	Max	Min	Avg	Max	Min
Shear ultimate stress, ksi	0° - 90° ± 45°	13.8		12.3	0.97		11.4			
Flexure										
ultimate stress, ksi	0°	68.2	72.8	65.2	58.4	64.0	52.1	52.7	56.3	47.4
proportional limit, ksi	0°	59.3	66.1	54.6	48.9	56.8	42.5	42.4	46.2	38.8
initial modulus, 10^6 psi	0°	2.97	3.04	2.88	2.89	2.99	2.78	2.97	3.06	2.82
Bearing										
ultimate stress, ksi	0°	65.7	73.2	57.0	58.9	64.0	46.8	49.5	55.8	44.5
stress, at 4% elong., ksi	0°	25.1	26.0	23.7	24.5	24.9	23.8	21.6	22.6	20.7
Interlaminar Shear										
ultimate stress, ksi	0°	4.83	5.10	4.29	4.64	4.92	3.94	4.62	4.88	4.08

Table VII-1.10 Summary of Mechanical Properties of Typical Fiber Glass Silicone Composite

Fabrication		
Layup: Balanced Parallel	Vacuum:	Pressure: 60 psi
Bleedout: Vertical & Edge	Cure: 350F/2 hrs 180F/40 min	Postcure: 400F/2 hrs 480F/16 hrs 600F/10 min
Plies: 8		

Physical Properties		
Weight Percent Resin: 31.9	Avg. Specific Gravity: 1.83	Avg. Percent Voids:
Avg. Thickness:		

Test Methods		
Tension: MIL-HDBK-17	Compression:	Shear:
Flexure: ASTM-D790	Bearing: ASTM-D953	Interlaminar Shear: Short Beam

Temperature		-65F				75F				160F				600F	
Condition		Dry		Wet		Dry		Wet		Dry		Wet		Dry	
		Avg	SD	Avg	SD	Avg	SD	Avg	SD	Avg	SD	Avg	SD	Avg	SD
Tension															
ultimate stress, ksi	0°	36.8	2.0	38.1	1.6	29.2	1.5	26.9	1.0	24.2	0.8	23.6	1.1	27.7	0.5
	90°	34.6	2.3	35.9	1.4	27.6	1.2	24.9	0.8	23.5	0.9	22.0	0.7	19.4	
ultimate strain, %	0°	1.49	0.10	1.53	0.07	1.12	0.06	1.12	0.07	1.03	0.05	1.03	0.05	1.33	0.04
	90°	1.55	0.09	1.62	0.09	1.21	0.08	1.17	0.06	1.13	0.08	1.08	0.07	1.35	
proportional limit, ksi	0°	24.5		27.9		20.8		20.5		17.9		17.5		15.1	
	90°	22.3		24.0		19.1		18.1		14.7		15.7		11.1	
initial modulus, 10^6 psi	0°	2.56	0.07	2.60	0.12	2.70	0.17	2.47	0.07	2.44	0.11	2.38	0.13	2.45	0.05
	90°	2.30	0.07	2.31	0.04	2.39	0.12	2.21	0.09	2.18	0.09	2.17	0.11	2.26	
secondary modulus, 10^6 psi	0°														
	90°														
Compression															
ultimate stress, ksi	0°														
	90°														
ultimate strain, %	0°														
	90°														
proportional limit, ksi	0°														
	90°														
initial modulus, 10^6 psi	0°														
	90°														

		-65F Dry			75F Dry			160F Dry		
		Avg	Max	Min	Avg	Max	Min	Avg	Max	Min
Shear										
ultimate stress, ksi	0° - 90°									
	± 45°									
Flexure										
ultimate stress, ksi	0°	38.5	43.2	34.8	31.5	32.3	30.3	26.3	28.7	25.0
proportional limit, ksi	0°	25.1	27.1	23.2	24.1	25.1	23.1	18.8	21.5	17.4
initial modulus, 10^6 psi	0°	2.46	2.56	2.22	2.26	2.37	2.11	2.24	2.36	2.15
Bearing										
ultimate stress, ksi	0°	23.2	24.0	21.8	22.4	23.9	20.7	16.6	19.0	15.1
stress at 4% elong., ksi	0°	16.8	17.8	16.0	17.9	19.5	14.1	14.8	15.8	13.6
Interlaminar Shear										
ultimate stress, ksi	0°	2.66	2.75	2.57	2.27	2.33	2.16	1.58	1.68	1.50

Table VII-1.11   Summary of Mechanical Properties of Typical Styrene-Alkyd Polyester—Fiber Glass Laminates

Fabrication

Layup: Parallel	Vacuum:	Pressure: 14 psi	Bleedout:	Cure: Press 100 min 220–250F	Postcure: None	Plies:

Physical Properties

Weight Percent Resin: 36.8–38.1	Avg. Specific Gravity: 1.76–1.77	Avg. Percent Voids:	Avg. Thickness: 0.244–0.265

Test Methods

Tension:	Compression:	Shear:	Flexure:	Bearing:	Interlaminar Shear:

Temperature		−65F				75F						160F			
Condition		Dry		Wet		Dry			Wet			Dry		Wet	
		Avg	SD	Avg	SD	Range	Avg	SD	Range	Avg	SD	Avg	SD	Avg	SD
Tension															
ultimate stress, ksi	0°					43.2	49.1		36.9	40.9					
	90°					38.9	45.4		31.0	38.8					
ultimate strain, %	0°														
	90°														
proportional limit, ksi	0°					7.0	8.2		5.0	6.7					
	90°					6.7	8.2		6.5	7.0					
initial modulus, 10^6 psi	0°					2.8	2.95		2.7	2.76					
	90°					2.8	2.80		2.4	2.77					
secondary modulus, 10^6 psi	0°					2.2	2.63		2.3	2.66					
	90°					2.1	2.42		1.9	2.56					
Compression															
ultimate stress, ksi	0°					34.5	44.5		22.0	23.9					
	90°					36.4	39.4		22.3	22.5					
ultimate strain, %	0°														
	90°														
proportional limit, ksi	0°					22.5	26.6		15.6	18.6					
	90°					22.8	23.6		13.3	17.2					
initial modulus, 10^6 psi	0°					3.1	3.3		2.8	2.9					
	90°					2.8	3.2		2.6	2.9					

		-65F Dry			75F Dry			160F Dry		
		Avg	Max	Min	Avg	Max	Min	Avg	Max	Min
Shear										
ultimate stress, ksi	0° - 90°									
	± 45°									
Flexure										
ultimate stress, ksi	0°					56.5	55.3			
proportional limit, ksi	0°					39.4	34.2			
initial modulus, 10^6 psi	0°					2.81	2.58			
Bearing										
ultimate stress, ksi	0°				49.7					
stress at 4% elong., ksi	0°				40.3					
Interlaminar Shear										
ultimate stress, ksi	0°				4.78					

Table VII-1.12 Summary of Mechanical Properties of Typical Fiber Glass Modified DAP Polyester Composite

Fabrication	Layup: Parallel	Vacuum:	Pressure: 50 psi	Bleedout:	Cure: 80 min at 257–295F	Postcure:	Plies: 12

Physical Properties	Weight Percent Resin: 34.5–35.2	Avg. Specific Gravity: 1.99–2.00	Avg. Percent Voids:	Avg. Thickness:

Test Methods	Tension: FED-STD 406	Compression: FED-STD 406	Shear:	Flexure: FED-STD 406	Bearing:	Interlaminar Shear: FED-STD 406

Temperature		-65F				75F				160F			
Condition		Dry		Wet		Dry		Wet		Dry		Wet	
		Avg	SD	Avg	SD	Avg	SD	Avg	SD	Avg	SD	Avg	SD
Tension													
ultimate stress, ksi	0°					64.4	2.7	66.0	1.1	56.3	2.5		
	90°					52.4	2.9	51.9	2.6	46.9	2.8		
ultimate strain, %	0°					2.5		2.5		2.2			
	90°					2.6		2.3		2.0			
proportional limit, ksi	0°												
	90°												
initial modulus, 10^6 psi	0°					3.38		3.40		3.11			
	90°					3.15		3.02		3.12			
secondary modulus, 10^6 psi	0°												
	90°												
Compression													
ultimate stress, ksi	0°					50.7	3.5	52.0	1.8	43.4	2.2		
	90°					42.3	2.6	43.8	1.7	35.3	1.9		
ultimate strain, %	0°												
	90°												
proportional limit, ksi	0°												
	90°												
initial modulus, 10^6 psi	0°					3.57		3.56		3.36			
	90°					3.34		3.34		3.01			

		-65F Dry			75F Dry			160F Dry		
		Avg	Max	Min	Avg	Max	Min	Avg	Max	Min
Shear ultimate stress, ksi	0° - 90° ±45°									
Flexure										
ultimate stress, ksi	0°				76.6			64.5		
proportional limit, ksi	0°									
initial modulus, 10^6 psi	0°				3.25			2.91		
Bearing										
ultimate stress, ksi	0°									
stress at 4% elong., ksi	0°									
Interlaminar Shear										
ultimate stress, ksi	0°				2.0			2.0		

Table VII-1.13   Summary of Mechanical Properties of Typical Boron-Epoxy (100%—0° Direction) (Tentative)

Fabrication	Layup: Parallel	Vacuum:	Pressure: 85 ± 5 psi	Bleedout: Vertical	Cure: 150F - 2" Vac 350F/1 hr	Postcure: 380F/4 hrs	Plies: 6
Physical Properties	Weight Percent Resin.		Avg. Specific Gravity:	Avg. Percent Voids:	Avg. Percent Resin.		Avg. Thickness: 0.005 in/ply
Test Methods	Tension: Tab-ended	Compression: Sandwich Beam	Shear:	Flexure: 4 Point Loading	Bearing:	Interlaminar Shear: Short Beam	

Temperature	-67F		75F				260F				375F			
Condition	Dry		Wet		Dry		Wet		Dry		Wet		Dry	
	Avg	SD	Avg	SD	Avg	SD	Avg	SD	Avg	SD	Avg	SD	Avg	SD
Tension														
ultimate stress, ksi 0°	183.8	12.7			187.5	11.7			185.9	8.0			162.4	3.6
90°														
ultimate strain, μ in/in 0°	5960				6260				6375				5690	
90°														
proportional limit, ksi 0°	3640				4870				8080				11400	
90°														
initial modulus, 10^6 psi 0°	30.6				30.0				29.2				29.3	
90°														
secondary modulus, 10^6 psi 0°														
90°														
Compression														
ultimate stress, ksi 0°	454.2				443.5				272.3				85.8	
90°														
ultimate strain, μ in/in 0°	12280				12240				7960				2450	
90°														
proportional limit, ksi 0°														
90°														
initial modulus, 10^6 psi 0°	36.6				35.5				34.3				33.3	
90°														

		-67F Dry			75F Dry			375F Dry		
		Avg	Max	Min	Avg	Max	Min	Avg	Max	Min
Shear ultimate stress, ksi	0° - 90° ± 45°									
Flexure										
ultimate stress, ksi	0°	271.3	282.0	260.0	256.6	271.1	249.0	Horizontal shear failure at 153.7		
proportional limit, ksi	0°									
initial modulus, 10⁶ psi	0°	30.3	30.8	29.8	29.5	30.2	29.0	16.9	17.2	16.5
Bearing										
ultimate stress, ksi	0°									
stress at 4% elong., ksi	0°									
Interlaminar Shear ultimate stress, ksi	0°	6.13	6.79	5.35	6.29	7.02	5.70	1.97	2.02	1.87

short exposure time required. Silica fiber-phenolic laminates and others fulfill these requirements.[65]

Cryogenic Temperatures. In housing cryogenic fuels at temperatures below $-400°F$, RP/C materials have been found to perform satisfactorily, exhibiting mechanical properties at the lower temperatures which are superior to those at room temperature.[66]

Radiation and Outer Space. The problems of shielding both man and delicate instruments in atmospheric or space travel are critical. Ultraviolet and actinic radiation, Van Allen Belt particles, and solar flares all cause polymer degradation through molecular weight change or cross-linking. Fortunately, all radiations except solar flares are effective in penetrating only a thin layer of the RP/C component. Surface coatings or fillers in the polymeric matrix such as TiO_2 have been found to cut down the deleterious effects of radiation. Loss of volatile plasticers in the plastic-matrix materials may result from exposure to the high vacuum of outer space. Polymers also degrade under radiation at an advanced rate when the system is in vacuum.

Chemical Environment. RP/C components for military use are subjected to water exposure at all temperatures,[67] including tropical or arctic exposure or land and sea exposure.

Compatibility of the RP/C with fuels and other chemicals without degradation is also a requirement. As an assurance, plastic protective boxes and other gear are used to house and transport non-plastic materials. Weathering also creates problems—usually embrittlement—especially as the result of rain erosion during exposure to high radiation. Elastomeric coatings over RP/C surfaces have been found to reduce damage from rain erosion and weathering.

Pressure. Hydrospace technology has dictated product requirements that only RP/C can fulfill. Applications for undersea duty take the form of housings, tips, antenna, hatch covers, fairings, and large fairwaters. Experimentation is continuing for adapting man-carrying RP/C structures[68] to the undersea environment. RP/C material originally showed great potential for undersea hulls because of higher strength-weight advantages than any other existing material.

The major structural problems yet to be overcome are weakness of compression joints, static fatigue, low interlaminar shear of filament-wound structures, low bending and impact strengths, high sensitivity to surface abrasion, inability to identify critical defects by non-destructive testing, and tendency for water absorption. It is still necessary to attain a high correlation between actual performance and results obtained from non-destructive testing. Several interesting design concepts have been worked out, including spherical and other doubly curved shells. Because the problems with RP/C pressure hulls are no more formidable than those associated with other high-strength, low-density materials, the prognosis is good if new research, money, and more sophisticated analytical tools are brought to bear on their solution. Pressure is also a requirement in cryogenic work where fuel tanks at $-400°F$ must withstand an internal-bursting pressure of several thousand psi to maintain the desired physical state of the fuel material.

Non-Military Applications. Presently, high cost is the major deterrent to the use of high-performance or unique RP/C in commercial applications. As manufacturing methods improve, and costs diminish, many interesting civilian applications that are "waiting in the wings" will emerge. Structural applications will no doubt be the most sizable. Thermal insulation and high electrical resistivity to 750,000 v. are also some of the proposed uses for the materials.

Another application which takes extreme advantage of the high-modulus/low-weight factor of RP/C is a long-fiber carbon-epoxy slotted disc. This piece of equipment functions in nuclear research as a neutron-beam chopper. The RP/C component performs its function without disintegrating by rotating at much higher speeds and lower centripetal forces than available metals.[69] Whisker reinforcements show potential as composites for jet-engine compressor blades and also for metal reinforcement.

References

1. Kunstoffe, Vol. 60, Dec. 1970, p. 924, Roos, G.

2. Ibid, p. 931, Joisten, S.

3. Product Engineering, Nov. 3, 1969, p. 84, see also C & EN, Mar. 31, 1969, p. 43, and Chemical Week, Mar. 4, 1970, p. 63.

4. Chemical Week, Apr. 7, 1970, p. 74; see also Modern Plastics, May, 1970, p. 26.

5. Narmco R&D Div., Whittaker Corp. Contract NOW-660144-c Final Reort, July, 1966, Hergenrother, P. M., and Levine, H. L.

6. Plastics Design and Processing, Dec. 1969, p. 29.

7. Amoco Chemicals Corp. Technical Bulletin D0169, and Bulletin HT-9a.

8. E. I. duPont Company Technical Bulletins A-62389 and A-62113; see also Dixon Corporation Technical Bulletin HCG 7310 10M 2/67.

9. SPI 25th RP/C Proceedings, 1970, Sec. 19-C, Browning, C. E., and Marshall, J. A.

10. SPE ANTEC Proceedings, 22nd, Mar. 7–10, 1966.

11. Seventh AIAA & ASME Conference, Apr. 18, 1966, AF 33(6570-8047, Vicars, E.C.

12. SPI 20th RP/C Proceedings, 1965, Sec. 11-G, Reinhart, T. J., Jr., and Chase, V. A.

13. U.S. Navy Bureau Ships Report #92213 on Project SR-007-0304, "Engineering Composites Utilizing Large Diameter Fibers," Keogh, J. C., Jr., Narmo Corp., order no. AD-487-659.

14. Foreign Technical Division of U.S. Air Force Systems Command Report No. FTD-MT-24-42-69 on effect of large diameter reinforcing fibers by Osnach, N. A., Podov, V., et al., U.S.S.R., order No. 696521.

15. Morganite Research and Development, Ltd., technical publication: " 'Modmor' High Modulus Carbon Fibres."

16. Journal of Material Science, 6 (1971), p. 60, Cooper, G. A. and Mayer, R. M.

17. Mechanical Engineering, Feb. 1971, p. 21, McDonald, J. E.

18. Kunstoffe, Vol. 60 (Sept. 1970), p. 623, Grunsteidl, W.

19. Technical Report AFML-TR-70-53, "Direct Graphitization Polymer Yarns," Ezekiel, H. M., June, 1970.

20. SPI 23rd RP/C Proceedings, 1968, 16-B, Simon, J. S., Prosen, S. P.

21. Modern Plastics, Sept. 1948, p. 227, Simon, R. A., and Prosen, S. P.; see also Chemical Engineering Progress, Oct. 1969, p. 46, Economy, J.

22. SPI 26th RP/C Proceedings, 1971, Sec. 14-E, Tock, R. W.

23. Technical Report AFML-TR-65-160, "Filamentous Carbon and Graphite," Schmidt, D. C. and Hawkins, H. F., Aug. 1965; see also SPE retec Preprint, July 14, 1965, Seattle, Wash.

24. SPI 25th RP/C Proceedings, 1970, Sec. 12-E, Speyer, F. B.; see also Product Engineering, June 2, 1969, p. 49.

25. Panel Discussion, "Advanced Composites for Aerospace," SPI 25th RP/C Proceedings, 1970, Sec. 15.

26. Plastics Design and Processing, Nov. 1970, p. 22.

27. AIAA Journal, Vol. 5, Feb. 1967, pp. 289–295, Schwartz, R. T., and Schwartz, H. S.

28. SAMPE 10th Annual symposium, 1966, p. C-47, AF (615)-3279, Saffire, V. N. and Shenker, L. H.

29. SPI 21st RP/C Proceedings, 1966, Sec. 8-C, Joffe, E. H.

30. SPI 23rd RP/C Proceedings, 1968, 16-C, Gutfreund, K., Kutscha, D., Broutman, L. J.

31. "Handbook of Advanced Composites," Lubin, et al., New York, Van Nostrand Reinhold, 1970, pp. 201, 285.

32. SPE, 1965 Retec Preprint, Seattle, Wash., July 14, 1965, Milewski, J. V.

33. SPI 20th RP/C Preprint, 1965, Sec. 6-C, Milewski, J. V., and Shyne, J. J.

34. SPI 24th RP/C Preprint, 1969, Sec. 18-D, Milewski, J. V. and Shyne, J. J.

35. Chemical and Engineering News, Aug. 26, 1968, p. 35; see also American Rocket Society Journal, Apr. 1962, p. 593, W. H. Sutton.

36. Composites, Vol. 1, No. 1, Sept. 1969, p. 25, Parrat, N.J.

37. Ceramic Industry, Nov. 1966, p. 53.

38. SPI 22nd RP/C Proceedings, 1967, Sec. 7-C, Bortz, S. A., and Li, P. C.

39. Chemical and Engineering News, Oct. 2, 1967, p. 28.

40. Technical Bulletins in 571-A and 738-A, Johns-Manville Corp.

41. Ceramic Industry, Feb. 1969, p. 51.

42. SPI 22nd RP/C Proceedings, 1967, Sec. 7-E, Schwartz, H. S. and Mahieu, W.; see also Industrial and Engineering Chemistry, Vol. 56, #4, Apr. 1964, p. 9.

43. SPI 21st RP/C Proceedings, 1966, Sec. 8-G, Jaray, F., et al.

44. Kunstoffe, Vol. 60 (Dec. 1970), p. 937, Heinrich, W., Nixdorf, J., and Wehr, W.

45. SPE RP/C Retec, Oct. 8, 1964, p. 41, Whittaker, R. A. and Gillespie, P. M.

46. SPE Journal, Sept. 1962, p. 1172. Lunn, R. H. and DeRoach, T. R.

47. SPI 22nd RP/C Proceedings, 1967, Sec. 20-E, Juneau, R. W., et al.

48. Chemical and Engineering News, Feb. 2, 1970, p. 7.

49. SPI 23rd RP/C Proceedings, 1968, Sec. 17-E, Anderson, H. M. and Morris, D. C.

50. SPI 21st RP/C Proceedings, 1966, Sec. 15-D, Bandaruk, B.

51. "Engineering Materials and Design," Sept. 1970, p. 1109. *see also* R. Dukes, PHD (British Admiralty Materials Laboratory).
Modern Plastics, Aug. 1968, p. 153. DeBrunner, R. E.
CPI 20th RP/C Proceedings, 1965, Sec. 7-C, Toth, L. W., RP/C in 20°K Range.
SPI 23rd RP/C Proceedings, 1968, Sec. 17-B, Petker, I., Sakakura, R. R., Segimoto, M.
SPI 23rd RP/C Proceedings, 1968, 16-A, Herrick, J. W.

52. *Ceramic Industry,* Feb. 1969, p. 51, Davis, W. L.

53. *Modern Plastics,* Sept. 1970, p. 136, Forest, J. D.; *see also: SPE Journal,* Aug. 1961, p. 743, Vondracek, C. H. and Sampson, R. N.

54. *Journal American Rocket Society,* Apr. 1962, p. 593, Sutton, W. H.

55. *Ceramic Industry,* Nov. 1968, p. 61, Davis, W. L.

56. *Journal Composite Materials,* Vol. 2, No. 1, (Jan. 1968), p. 32, Thornton, H. R.

57. Research and Development, Aug. 1970, p. 62, Mullin, J. V.

58. Picatinny Arsenal, "Plastics in Government," (N67-18436-08-18), July 1966, Postelnek, W.

59. SAE Meeting, Los Angeles, Cal., Oct. 3, 1966, paper 66075, Singletary, J. B. and Rittenhouse, J. B.

60. "Composite Materials Testing and Design"—ASTM Publication STP 460—Papers from Committee D-30 Symposium on High Modulus Fibers and Their Composites, New Orleans, La., Feb. 11–13, 1969.

61. "Concepts of Fiber-Resin Composites," McCullough, R. L., Marcel Dekker, Inc., 1971.

62. Personal communication, Stander, Maxwell; U.S. Naval Air Command, Washington, D.C.

63. 1971 Edition of MIL-HDBK-17, "Plastics for Aerospace Vehicles; Part 1, "Reinforced Plastics," Tables reprinted by permission.

The format used in preparing these tables represents an agreement between government and aerospace industry representatives as to the most useful and necessary information. The working group working on the handbook MIL-HDBK-17 has granted permission to use this data in the hope that this format for recording data will be adopted by all those engaged in evaluating reinforced-plastic composites and laminates.

64. SPE RP/C Retec, Oct. 8, 1964, p. 35, Gruntfest, I. J. and Shenker, L. H.; *see also:* SPI 25th RP/C Proceedings, 1970, Sec. 15.

65. SPI 21st RP/C Proceedings, 1966, Sec. 4-A.

66. Advances in Cryogenic Engineering, Vol. II, Rice Institute, Houston, Tex., Aug. 23, 1965; *see also:* ASTM 70th Annual Meeting Proceedings, June 25, 1967, paper #16, Toth, L. W., et al.

67. Trans. Inst. Chem. Engineers, Vol. 47, 1969, P. T-188, Bott, T. R. and Barker, A. J.

68. *SPE Journal,* Vol. 24, Dec., 1968, p. 56, Kiernan, T. J., and Krenzke, M. A.

69. *Chemical and Engineering News,* Aug. 26, 1968, p. 35.

SECTION VIII

POTENTIAL AND FUTURE GROWTH OF RP/C

VIII-1

Potential and Future Growth of RP/C

Any reader who has made his way through this substantial volume to the final chapter is clearly devoted to reinforced plastics/composites, has absorbed a massive body of fact on materials and processes, and can himself envision countless new applications for this family of structural plastics. We shall contribute a quick tour through the major markets looking at products and developments that seem significant, a relatively informed look at the road immediately ahead, and a more visionary glimpse over the horizon.

Introducing a new material is a kind of obstacle race. The contestant's strengths and advantages provide the motive power. The material's weaknesses, obscurity, and the customer's resistance to change are the obstacles. We reach conclusions on its potential by appraising the strengths and success in surmounting the obstacles. Reinforced plastics moved from the military into civilian markets with an impressive list of advantages: a good balance of physical properties, excellent strength-to-weight ratio, corrosion resistance, good electrical properties, low tooling cost, and many more. These were enough to attract a growing band of technically oriented enthusiasts who could envision fiber glass reinforced plastics capturing great chunks of such markets as transportation, construction, marine industry, appliances, electrical and corrosion-resistant products, aerospace parts, and so on.

General obstacles to the growth of the new family of materials included relatively high costs of raw materials, slow molding cycles,

This chapter was prepared by William H. Gottlieb, Morrison-Gottlieb, Inc., Public Relations, New York.

need for hand labor in an era of growing automation, and costly finishing operations to achieve the mirror-like finish required for many products. There was also the inevitable lack of long-term test data.

Each major market presented its own blend of obstacles. The automotive and appliance industries limited applications to low-volume parts until mass-production economy and high-quality finish could be achieved. In the vast construction market, the thousands of building codes barred most uses of reinforced plastics. Architects and builders thought in terms of heavy tonnage of lower cost traditional materials. Organized labor took a dim view of materials that did not utilize their craft skills—products molded in a factory that shortened on-site construction time.

However, a multitude of compelling factors have favored the accelerating acceptance of reinforced plastics/composites. On the one hand, the industry itself has improved materials and processing methods and actually reduced costs, enabling its products to surmount the old obstacles and compete effectively. The simple passing of time has helped; the industry can point to one quarter of a century of experience and the test and engineering data amassed in that time.

On the other hand, the developing requirements of society have been lowering or demolishing artifical barriers to needed materials. Population growth of itself creates increased demand, and in this age of "consumerism" the buyer is more sophisticated and demanding. He wants less maintenance, more durability. Operating efficiency and economy may loom larger than initial price. He is science oriented and inclined to accept innovation.

The current preoccupation with ecology indicates better utilization of natural resources and protection of the environment against noise and pollution. Our success in extending human life puts the burden of supportive production on a diminishing percentage of the population, and electro-mechanical energy becomes not only more essential but more economical than increasingly expensive human energy. All these factors—and others—enhance the value and build the markets for RP/C products.

It is this writer's opinion that most poundage predictions for 1975, '76 and '80 (including some in earlier chapters of this book) tend to be conservative. They deal realistically with market penetration and growth in today's terms without sufficient allowance for the cosmic forces that are changing the very nature of our world and society.

TRANSPORTATION

Ours is a mobile population and is getting more so every year. The production of "people-movers" is a major industrial activity, and their control is a major safety and environmental problem. In terms of market classification, we include in "transportation" automobiles, trucks, trains, and freight containers. We think of "marine" and "aerospace" as separate classifications, although in the future the lines between the three segments are bound to blur. Already, the intermodal container travels by land, sea, or air. Carriers can be expected to follow.

Looking first at today's vehicles, we see a proliferation of reinforced-plastic applications. In 1970 models of American-built passenger cars, there were approximately 110 RP/C parts. In 1971 models, there were 200 RP/C parts, a significant increase in penetration in this high-volume market. Included were interior and exterior body parts—tail-lamp and head-lamp housings, fender extensions, hoods, valance panels, trim, instrument panels, consoles, grilles, fender liners, and so on. The 1971 cars also have such working parts as fans, air-cleaner trays, windshield washer gears, and other items. And, of course, there's the whole Corvette body.

It is important to note that a great many of these are mass-production parts, manufactured in volume up to 400,000 units. The ability of RP/C to compete with sheet steel and zinc castings at this level can be attributed to a number of developments that improved the product, sped production, reduced costs, or put together a combination of the three. Low-shrink polyesters improved dimensional stability and gave a smooth out-of-the-mold surface that could go straight to the paint line with little or no preparation. Sheet-molding compounds too were well adapted to mechanized production, mostly in compression presses and some in injection-molding machines. Reinforced thermoplastics brought a whole new family of resins and the speed and automation of injection molding. Parts for 1971 were large—an 8 lb. instrument panel, for example. The new cold-molding system offers reinforced-thermoplastic sheet that is preheated, then formed by cold-stamping dies in the same presses used to stamp sheet steel.

RP/C parts are rust-proof, corrosion-resistant, and lighter than the metals they replace. The manufacturer saves on molds; he can get large, intricate parts in one piece and thus save on assembly. As designers and engineers learn to fashion cars and parts to make the best use of plastics, the advantages will multiply. Cost accountants, too, need to figure every factor to assess the full value of RP/C.

Weight saving is important today in cost and vehicle performance. It will become more important as the pollution and safety campaigns progress and as the industry moves to lower horsepower, to electric cars, and to radar-guided vehicles. With improving production and increasing pressure from a new generation of RP/C-conscious consumers for more durable, low-maintenance cars, we can look for more complete fiber glass RP/C bodies in the decade ahead.

It is reasonable to expect further improvements in polyesters, fibrous glass, and coupling agents. We see increasing success in the injection molding of reinforced high-strength thermoplastics. Low-pressure injection molding of homogeneous skin-core structures of glass-reinforced thermoplastics looks promising. Boron and carbon fibers are still very expensive, but their use for high-strength, high-modulus applications is an established fact in aerospace. Their use in ground transportation, possibly in combination with glass fiber, is a safe medium-range prediction.

Eventually it is inevitable that physical, market, and economic factors will combine to force the mass production of RP/C bodies. Growing investment by car manufacturers in RP/C production facilities should smooth the road and speed the day.

Trucks. An improved strength/weight ratio is money in the bank to the trucker. Any time cargo density (plus tare weight) uses up the legal load limit without filling the available cube, reduction in vehicle weight translates into additional revenue for freight capacity. Reinforced plastics provide many plus factors including the required toughness for a variety of truck applications with savings in weight compared with competitive materials.

Fiber glass-polyester just about owns the market for liner panels in refrigerated trucks and trailers, adding low maintenance, cleanability and thermal insulation to the basic strength-weight advantages.

Truck manufacturers are turning increasingly to RP/C for fender-hood units. Besides the weight saving, they find reduced tooling costs, improved appearance, more resistance to damage, lower maintenance cost, and easier access to engine compartment. The great success of the fender-hood is causing the truck manufacturer to re-assess the all-RP/C cab. The industry has learned a great deal since the White 5000 (which in retrospect turns out to have been a very good RP/C job), and we are certain to see many more RP/C cabs on the highway.

Containers. When we talk about containers and truck-trailers, we are discussing the same thing: the intermodal freight box that travels the highway, the seaway, and the rails. RP/C liner panels are well established, and RP/C skins with a plywood core are moving toward a majority position as the basic structural component. There is a weight saving compared to steel or aluminum, and long seamless panels make for manufacturing economy, long service, and easy maintenance. Freedom from rust and corrosion is important, too.

The effort to cut needless pounds continues. One company imports a strong, lightweight plywood for panel cores. Others use a rigid foam core between fiber glass-reinforced polyester skins. The panel with an RP/C foam core has been around for years, used logically for air transport containers and for refrigerated truck bodies. Interest in this construction is growing among shippers and carriers as the volume of air freight increases and the shipment volume of fresh and frozen foods continues to climb.

Trains. Reinforced plastics are out front in modern mass transportation. The BART trains for the new San Francisco Bay Area Rapid Transit system will follow the example of New York-New Jersey PATH trains in the major use of weight-saving, low-maintenance RP/C. The lead car in each BART train will have a 750-lb. cab of RP/C and urethane foam. Interior walls and ceilings in all cars will be of RP/C.

The compact RP/C lavatory makes sense for train or bus. RP/C seating withstands wear and tear, including vandalism.

The last hope of the inter-city passenger train in competition with air travel is the high-speed modern unit capable of cutting down the time differential from downtown to downtown. AMTRAK is bound to take a close look at reinforced plastics.

Transportation in the 1980s—Road and Rail.[1] The following section on Transportation is a visionary prediction of the shape of things a decade or two ahead. It is taken from a much longer unpublished manuscript by Stewart Byrne of Owens-Corning Fiberglas Corp. Ensuing major paragraphs in quotations are also abstracted from Mr. Byrne's excellent work.

"Since movement through the air requires less energy than surface transportation and the speed permits fuller utilization of vehicle and crew, all except local or heavy transportation will of necessity become to some degree airborne. Air-cushioned boats, induction-motivated 'flying' trains and aircraft freight will replace all but the heaviest rail and road transportation.

"The induction-powered 'flying' train will accelerate to 200 mph in 10 seconds, cruise at 300 mph pulled through induction rings by magnetic power. Lightweight design and quiet operation will permit straight tracks to be built between cities, either over super highways or across industrial and commercial sections. They

[1]Stewart Byrne. Personal Communications, Owens-Corning Fiberglass, Toledo, Ohio .

will cross rivers and valleys via a suspension system supported by boron-reinforced plastic cables. A 3-rail system will assure a completely smooth ride.

"Such advances will offer many opportunities for fiber glass-reinforced plastics. Lightweight and unitized construction are essential for the trains. The electrical induction system requires the body to be of a non-conducting material with conductive rings. RP/C parts would include the top shell, bottom shell, horizontal lift-up doors running the length of the car, floor and seating unit, ceiling liner including air conditioning and fold-down luggage racks. Stations and right-of-way would use RP/C for structural beams, pilons to support the induction rings, platforms. Again, non-conduction is a primary property along with strength and freedom of design. The induction rings would consist of RP/C bands filament wound for maximum circumstantial strength. The steel conduction band would fit inside the ring. . . .

"Automobiles will only be driven from the garage to the road where central computer control will take over and guide them to the destination dialed by the driver. . . .

"Commuter cars with inflatable fronts and backs will compact for train transportation and garage storage."

AIRCRAFT AND AEROSPACE

Production figures of the aerospace industry entering the decade of the 1970s made one think of a stunt pilot in a power dive. However, the airplane is our basic vehicle for long-distance passenger travel and is increasingly important in hauling many classes of freight. With more reinforced plastic used in each major new plane, there is no way to go but up—at least in the civilian market.

The important factors are the increasing number of applications for RP/C in the airplane, the progression from interior trim to structural parts, and the ability of the industry to come up with new and improved materials to meet more rigorous performance requirements. Constantly probing new ground, RP/C has had to triumph over a lack of standardization and relevant design experience.

On the Boeing 737, RP/C sandwich panels fastened to metal ribs provided a rudder with savings of 70% compared with sheet-metal

assembly costs. The DC-10 trijet uses more RP/C for its size than any commercial plane yet built—not just paneling throughout the passenger and cargo areas, but wing-to-fuselage fairings and filament-wound tanks.

In the military field, Lockheed's gigantic C-5A gives us a notion of what the air-cargo carriers of the future may look like with the extensive use of RP/C laminates and sandwich panels, both glass and boron composites, in fairings and in the primary structure. High-strength, high-modulus, boron-reinforced epoxy saves 550 lb. in the F-111 and cuts weight and cost in the F-14 horizontal stabilizer. Composites with boron or carbon reinforcement show up in jet turbine blades and in helicopter rotor blades. Boron-epoxy prepreg tape gives double the stiffness and 30% less weight than aluminum in a Bell helicopter driveshaft.

Thinking small in size but big in applications and volume, the Windecker 4-seat, all-plastic airplane boasts a complete glass-epoxy airframe; FAA certification; and fast, economical production techniques.

Air Transportation in the '80s. "Aircraft coming into service in the late 1980s will be largely reinforced plastics with special coatings to dissipate heat. Verticraft taxis will become common for travel between close cities. Lightweight air taxis will be molded for fiber glass-reinforced plastic foam and engines will power RP/C rotors. Total weight of the vehicle will be less than the passengers. An inflatable wing will be provided for emergency landing. Computer-controlled laser highways in the sky will automatically control traffic."

MARINE

It is common knowledge that fiber glass-reinforced polyester (the boatman knows it as "fiber glass") has gobbled up the lion's share of the pleasure boat market. RP/C overwhelmed conventional wood and ambitious aluminum in power boats from 14 to 20 ft long and in sailboats of any size. The take-over of larger cruisers is well advanced and is accelerating. Virtual domination by RP/C seems inevitable.

In the small boat field, notably the 12-ft fishing boats, pressmolded RP/C is squeezing aluminum hard. Some competition today comes from vacuum-formed ABS skins with foam cores and tomorrow will come from low-pressure injection-molded structural foam. RP/C is no stranger to these plastics cousins. Some boat manufacturers spray glass and polyester on vacuum-formed thermoplastic skins. Structural foam hulls may well be reinforced with fibrous glass.

The same qualities that enabled reinforced plastics to take over the pleasure boat market—strength, durability, formability, low maintenance—are moving glass polyester into contention for commercial work boats and fishing vessels. Here, additional values take on importance: weight saving that translates into fuel saving and speed, hull construction that leaves more cargo space, and fewer overhauls and repairs that permit more profitable work days.

At least three companies in the United States are building stock shrimp trawlers over 70 ft—one a 72-footer, one a 73-footer, one an 80 footer. In the fishing grounds off Peru, a fleet of 93-ft purse seiners grows larger as it demonstrates the commercial advantages of fiber glass. The first advantage was an initial cost of $145,000; less than that of a steel boat of the same size.

How big can we build a boat with RP/C? Many years ago, a prominent naval architect reported on a study that found no physical limitation on size. The Navy has studied non-magnetic RP/C for minesweepers up to 200 ft in length. The Royal Navy is a step ahead with actual construction started on a 153-ft RP/C minehunter, a ship that will use 130 tn. of polyester resin and glass fiber for hull, decks, bridge, deckhouse, and mast.

Obviously the big ship requires more sophisticated engineering. Navy studies and experience with ever-larger vessels is providing valuable information on hull structure and secondary bonding. They may also develop new shipbuilding techniques applicable to commercial vessels.

Marine in the 1980s. "RP/C ships of many types and sizes for passengers and mail will make increasing utilization of air cushion planing to get up off the water and reduce energy requirements. Heavy freighters and tankers will still be in steel while fishing, service and rescue craft will be RP/C. Amphibious inflatacraft with fiber glass hulls will take off and land on land or water, expand to sizeable houseboats.

"Submarines will be reinforced plastic and a whole new family of underwater equipment will be required for undersea farming. Such farming will be in the shallows of coastal waters, some of these shallows artificially created by landfill operations (to bring soil level within reach of the sunshine). Other farms will be constructed of foam ballasted by wastes and covered with soil that can be floated around like enormous ships to take advantage of rich nutrients available in specific water areas such as the Gulf Stream. The harvesting vehicles may not be too different from present-day farm equipment but will all have to be of RP/C construction to withstand sea water corrosion."

Construction. Probably no market offers greater potential for reinforced plastics than does construction, and certainly no field interposes more obstacles to acceptance of the materials. We have mentioned the innumerable local building codes, the entrenched position of conventional materials, the opposition of craft unions, and the conservatism of the consumer. Add to this the fact that building is traditionally a fragmented industry with perhaps as many as 50,000 builders in the United States.

But the nation's needs are enormous. The Housing Act of 1968 set a 10-yr-goal of 26 million new housing units, and we will have to double the present building volume to attain it. Clearly it will not be accomplished with conventional techniques and materials.

There is pressure for change and progress is being made. New building codes specify performance rather than materials. Quite a few companies, including some major corporations, have developed factory-built homes. Labor unions, at least at the national level, are showing a willingness to accept factory wages far below the lofty on-site rates. The Department of Housing and Urban Development is encouraging industrialized building systems with Operation Breakthrough. It should be noted that much of the advantage comes from assembly under factory conditions, and most of the systems in the HUD program use the same old materials. But plastics are well suited to factory production and assembly, and the three reinforced-plastics entries could be significant winners.

All companies steered clear of curves and concentrated on straight sandwich panels easily joined to make the well-accepted rectangular box. Materials Systems Corp. bonded RP/C skins to an insulation-filled, deeply corrugated RP/C sheet to make panels 8 ft high and up to 52½ ft long. TRW Inc. puts a honeycomb core betweeen structural panels of RP/C bonded to gypsum wallboard and finally a gel-coated exterior RP/C skin. One-piece panels form the floor and walls of the 8 ft × 22 ft × 32 ft module. Christiana Western Structures uses gel coated RP/C for interior and exterior surfaces over plywood skins and a wood-stud core.

Outside the HUD program, at least two more companies are offering factory-built systems based on reinforced plastics. Great Western Consolidated Inc. forms gel-coated RP/C sheets 16 by 90 ft, cuts them to size, and laminates them to a polystyrene foam core. Panel/Comb Industries bonds RP/C skins to a phenolic-impregnated honeycomb core.

To give just one example of the time and cost factors, Great Western ships the 26 panels required for a 960 sq ft three-bedroom house on one truck at a price of $9500 FOB factory and says it can be erected on a prepared site by 5 men in 8 hr.

The news is not that we can make RP/C panels for buildings—the industry has nearly a quarter century of data on panels—but that conditions finally are shaping up for substantial acceptance by government, builders, and consumers.

In line with the effort to satisfy present tastes, the reinforced-plastic house does not have to look like plastic. Some of the factory units described can look like wood or brick or stone. Many of the modules serve not only for single-family homes but for town houses and low-rise apartments. In RP/C panels for high-rise structures, we may be catching up with our British cousins. An 11-story office building has been erected in California using molded fiber glass-reinforced polyester panels 8½ ft high × 11 ft wide for the entire exterior skin.

While waiting for structural RP/C housing shells to build volume, the industry has been doing very well with housing components. Molded tubs, showers, surrounds, and complete bath units first took over the major share of the growing mobile-home market and then moved into larger homes and apartments, hotels, and motels. It is estimated that in 1973, one of every three or four new bathrooms will use RP/C fixtures.

You will also see more and more RP/C as fascia on public and commercial buildings—though it may look like carved wood or cut stone—and in church steeples, skylights, gutters, and downspouts.

Some day we will see the big switch in the basic housing structure when modern architects and changing tastes bring acceptance of imaginatively shaped modules. Then efficient mass production of RP/C units may actually meet world housing needs. To quote Habitat architect Moshe Safdie, "The more you make with less materials and labor, the more you can give to people and therefore the better environment you can have. Unless you think of architecture in terms of process, any attempt to improve the environment is academic." Safdie built in pre-cast concrete, but his words could be the rationale for a plastics building boom.

City of the Future. "The city of the future might be called Pod City. To preserve the land for agriculture to feed a growing population, we need to get the living space way up high (utilizing exotic filament reinforced plastic cables as tension members to withstand wind load). This will allow the people to live within park areas that will have splendid views and as the sun traverses, all areas will get sunlight and be able to grow grass or harvestable crops.

"The columns on which these buildings are supported will be continuously-molded RP/C tubes or joined Dixie-cup-type sections. Transparent molded plastic shells will give the high-rise buildings weather protection and a controlled climate."

CORROSION-RESISTANT PRODUCTS

In corrosion-resistant applications, reinforced plastics have been too good to be true. In pipe, storage tanks, ducts, hoods, scrubbers, and fans, engineers had to verify case histories before they could believe the claimed performance and cost advantages. A young industry was developing both the materials and the fabrication techniques best suited to this demanding market. Improved materials meant improved strength, dimensional stability, heat and fire resistance. Improved processing meant better quality control and faster production at lower cost. The result was a product line competitively priced, with great weight saving, lower shipping and installation cost and, most important, longer service life.

So compelling are the advantages that this could become the fastest growing market for reinforced plastics. Tailor-made RP/C draws on a variety of glass and asbestos fibers; polyesters, epoxies, vinyls, and other resins; a variety of constructions and production methods from hand lay-up and spray-up to filament winding and rotational molding. Some use extruded PVC wound with glass filaments and polyester for pipe applications. Others mold pipe rotationally from a combination of glass roving, polyester, and sand mortar (RMP). The end result of this tailoring is a product suited to its service environment whether acid or alkali, built to handle specified heat and pressure, perfect for a waste line, and capable of winning FDA approval for food handling.

Pipe may be 1 in. in diameter in a standard 20 ft length, or it may be a huge 144 in. diameter and 60 ft long. Ducts can be fabricated in virtually any size: exhaust stacks may be 12 ft. in diameter and rise hundreds of feet. There are more than 30,000 RP/C tanks of 10,000 gal capacity in service, and others up to 70,000 gal capacity.

Corrosion costs American industry millions of dollars a year, and economics therefore favors reinforced plastics. In chemicals, papermaking, and other corrosion-prone industries, the competition is stainless steel, rubber-lined steel, or other expensive materials. RP/C either matches the performance for less money or matches the price for better performance; often RP/C has both cost and use advantages.

The advantage and potential for RP/C expand impressively when all cost factors are considered: initial cost, shipping cost, installation cost, maintenance, and cost of downtime. Long sections of RP/C pipe, for example, can be handled easily, reducing the number of joints. One 60 ft length of 48 in. RP/C sewer pipe weighed about 8,400 lb compared with 12,000 lb for just a 6 ft length of concrete pipe. RP/C cost more to buy, but the installed cost was lower. Glass-reinforced vinyl epoxy pipe for oil field service is actually cheaper than bare steel and should take over the bulk of this market.

RP/C tanks compete with steel for underground gasoline storage at the local service station. The average life of a steel tank is 10 to 12 yr, and in corrosive soil, leakers show up as early as 1½ yr. RP/C tanks in corrosive soil in Colorado show no deterioration at all after 7 yr. They are installed confidently awash with salt water in Florida.

Heat resistance is still a limiting factor in some applications in process industries. Proper choice of materials and structure, however, will meet a great many demands for service with hot liquids. The RP/C structure maintains integrity at substantially higher gas temperatures. One manufacturer came up with pipe utilizing RP/C inner and outer skins and urethane foam between them.

The demand for corrosion-resistant products must continue to grow, spurred by the interest in pollution control. The economics of the situation favor RP/C more each year. Steel prices keep moving up. Labor costs keep rising. Meanwhile, we see more high-speed processes for the machine production of RP/C products, and the economies increase the competitive advantage of RP/C in the corrosion market.

APPLIANCES

The same factors that opened the door to the mass production of RP/C parts in the automotive industry have accelerated progress in appliances. Low-profile polyesters provide the smooth surface without necessity for costly finishing. Sheet-molding compounds and bulk-molding compounds are suitable for mechanized production methods. Reinforced thermoplastics permit fast, automated injection molding.

There have been a lot of applications in the last year or two: housings, bases, frames, tanks, and working parts such as gears, tubes, fans, valves, and hinges. We see them on washers, vacuum cleaners, air conditioners, floor polishers, garbage disposal units, and the like.

But the current volume is just a hint of what lies immediately ahead. We are about to see air conditioners of reinforced plastic inside and out, except only the motor and compressor. The large molded sections will save on tooling and assembly and will provide better thermal efficiency.

It took a long time to get this far and the industry still has a lot of selling to do, but major appliance manufacturers have accepted reinforced plastics and in many cases have confirmed this acceptance with investment in molding equipment.

CONSUMER PRODUCTS

With more leisure hours, man will spend more time at home or in recreational pursuits. Either way, he seems destined to encounter more RP/C products.

Modern furniture design seems to have been gravitating to fiber glass ever since the Eames chair. A lot of modern is being shown and sold now, everything from simple tables, molded chairs, lamps, and shells for cushioned sofas, to exotic sculptured pieces from France. Those who prefer traditional decor find tables and case goods in RP/C with the appearance of wood—complete with grain and carving.

The riding mower and the golf cart probably both have RP/C housings. The fishing rod, snow and water skis, the exerciser, surf board, pool slide, and possibly the golf club shaft are of RP/C too. At the playground, the children (if they are lucky) will find great, colorful, imaginative, safe play equipment of reinforced plastic. Here RP/C is so superior to the old wood and metal that it just has to take over.

As with housing, the growing population and greater number of family units will build the market potential.

ELECTRICAL INDUSTRY

Reinforced plastics are standard and necessary materials for the electrical industry, but they have realized only part of a much greater potential. The combination of electrical insulation, physical strength, light weight, moldability, and dimensional stability makes increasing use of RP/C inevitable.

RP/C housings, air shields, and end bells for turbines, generators, and motors save weight and ease maintenance. Switchgear, transformers, and printed circuits make varied use of the electrical and physical properties of the materials. Out in the field, the man-buckets and booms are an eye-catching application of RP/C strength and insulation.

The transmission line appears to be a major field for the expanded use of reinforced plastics in such applications as pole top pins, guy strain rods and insulators, line spacers, conductor supports, crossbars, even entire poles. Higher voltages, more demanding performance requirements, and the push either for underground lines or for slender, attractive poles insure a growing role for reinforced plastics/composites. One Florida utility ordered a thousand RP/C poles because steel, aluminum, and wood could not stand the corrosive, salt-laden atmosphere.

The future would seem to lie in both the prosaic and the exotic. RP/C has obvious advantages for non-conducting switch boxes and covers and is beginning to make inroads in the traditional metal market. A less common and more demanding use is the pressure shell for the first stage of a new 200-billion volt proton accelerator. This filament-wound glass/epoxy part measuring 46 in. in diameter and 78 in. long has exceptional arc resistance plus the strength to support heavy electrodes and contain gas pressure. Closer to everyday needs is the reverse-osmosis filter for conversion of brackish and sea water to fresh water. The filter incorporates porous RP/C tubes that must stand a pressure of 5000 psi.

CONCLUSION

This review of current and prospective applications obviously was not all-inclusive or it would have been a book in itself. We think it indicates, though, the proliferation of new products and the likelihood of more. The SPI Reinforced Plastics/Composites Institute characterized 1971 as "the application explosion," and this process of development and expansion has accelerated. With a more favorable economic climate, the new RP/C products will be translated into unprecedented production volume.

Index